Fundamentals and Transport Properties of Nanofluids

Soft Matter Series

Series editors:
Ian W. Hamley, *University of Reading, UK*
Howard A. Stone, *Princeton University, USA*
Karen J. Edler, *University of Bath, UK*

Titles in this series:
1: Functional Molecular Gels
2: Hydrogels in Cell-based Therapies
3: Particle-stabilized Emulsions and Colloids: Formation and Applications
4: Fluid–Structure Interactions in Low-Reynolds-Number Flows
5: Non-wettable Surfaces: Theory, Preparation and Applications
6: Wormlike Micelles: Advances in Systems, Characterisation and Applications
7: Electrospinning: From Basic Research to Commercialization
8: Polymer-modified Liquid Crystals
9: Polymer Colloids: Formation, Characterization and Applications
10: Bijels: Bicontinuous Particle-stabilized Emulsions
11: Peptide-based Biomaterials
12: Droplet Microfluidics
13: Soft Matter for Biomedical Applications
14: Drying of Complex Fluid Drops: Fundamentals and Applications
15: Soft Matter in Plants: From Biophysics to Biomimetics
16: Fundamentals and Transport Properties of Nanofluids

How to obtain future titles on publication:
A standing order plan is available for this series. A standing order will bring delivery of each new volume immediately on publication.

For further information please contact:
Book Sales Department, Royal Society of Chemistry, Thomas Graham House, Science Park, Milton Road, Cambridge, CB4 0WF, UK
Telephone: +44 (0)1223 420066, Fax: +44 (0)1223 420247
Email: booksales@rsc.org
Visit our website at www.rsc.org/books

Fundamentals and Transport Properties of Nanofluids

Edited by

S. M. Sohel Murshed
University of Lisbon, Lisbon, Portugal
Email: smurshed@tecnico.ulisboa.pt

ROYAL SOCIETY OF CHEMISTRY

Soft Matter Series No. 16

Print ISBN: 978-1-83916-419-4
PDF ISBN: 978-1-83916-645-7
EPUB ISBN: 978-1-83916-646-4
Print ISSN: 2048-7681
Electronic ISSN: 2048-769X

A catalogue record for this book is available from the British Library

The Royal Society of Chemistry is a charity, registered in England and Wales, Number 207890, and a company incorporated in England by Royal Charter (Registered No. RC000524), registered office: Burlington House, Piccadilly, London W1J 0BA, UK, Telephone: +44 (0) 20 7437 8656.

For further information see our web site at www.rsc.org

Printed in the United Kingdom by CPI Group (UK) Ltd, Croydon, CR0 4YY, UK

Preface

Nanofluids is a promising multidisciplinary field that encompasses fields like nanoscience and technology, thermofluid science and engineering, and colloidal science, as well as soft matter physics and chemistry. This emerging field has constantly been evolving over the past two decades due to the atypical thermophysical properties and transport features of these materials, as well as their manifold potential applications in numerous fields such as thermal management, energy, transportation, MEMs and biomedicine.

The book addresses a broad range of fundamental and applied research and development on nanofluids starting from their synthesis, stability, rheological and thermophysical properties and performance characterization to advanced applications. It also focuses on challenges and impact of this new class of fluids from socio-economic and environmental perspectives.

The book consists of fifteen chapters structured under four main sections, which are subsequently summarized based on each section.

Section I *Fundamentals of Nanofluids* comprises the first three chapters that cover the most pressing issues of nanofluids, which are the synthesis of nanofluids and their stability.

Section II *Properties of Nanofluids* consists of the next six chapters (from chapters 4 to 9), which deals with various key properties, such as the rheological behavior and viscosity, thermal conductivity, heat capacity and dielectric and magnetic properties of nanofluids.

Section III *Thermal Features of Nanofluids* focuses on single-phase and two-phase transport features, which include the experimental and numerical development of convective heat transfer performance in channel and heat exchangers and pool boiling heat transfer in the subsequent three chapters (from chapters 10 to 12).

Soft Matter Series No. 16
Fundamentals and Transport Properties of Nanofluids
Edited by S. M. Sohel Murshed

Published by the Royal Society of Chemistry, www.rsc.org

Section IV *Applications and Impact of Nanofluids* presents selective applications in thermal management and solar energy as well as the socio-economic and environmental impact of nanofluids in the last three chapters.

This book is intended to be a valuable reference source for graduate students, researchers, engineers and scientists from different fields. However, it is not claimed that this book contains "all a reader needs to know" about nanofluids.

I wish to thank all the authors for their high-quality contributions and the reviewers for their reviewing and feedback which have made possible the successful completion of this book. I also thank the staff of the Royal Society of Chemistry for their cooperation and support from the beginning to the production of this book. Finally, I am indebted to my wife and daughter for their inspiration, support, and sacrifice.

S. M. Sohel Murshed
Instituto Superior Tecnico
University of Lisbon, Portugal

Contents

Soft Matter Series No. 16
Fundamentals and Transport Properties of Nanofluids
Edited by S. M. Sohel Murshed

Published by the Royal Society of Chemistry, www.rsc.org

Section I: Fundamentals of Nanofluids

CHAPTER 1

Synthesis of Nanoparticles and Nanofluids

A. ANGAYARKANNI[a] AND J. PHILIP*[b]

[a] Department of Physics and Nanotechnology, SRMIST, Kattankulathur – 603203, India; [b] SMART Materials Section, Metallurgy and Materials Group, Indira Gandhi Centre for Atomic Research, Kalpakkam – 603102, India
*Email: philip@igcar.gov.in

1.1 Introduction

The term "nano" originates from the Greek word "nanos", which means "dwarf". When the size of a material is on the nanoscale (1–100 nanometers), in at least one of its dimensions, it is referred to as a nanomaterial.[1] Examples of materials with one, two and three dimensions on the nanoscale are layers, nanowires and particles, respectively.[2] The unique properties of nanomaterials have been exploited for a wide range of applications. The need for efficient cooling materials has led to the emergence of a new field of research on nanofluids.[3–34] Several types of nanofluids have been investigated due to their thermal properties, which include suspensions of spherical,[32,35] rod-shaped, nanowire, nanotube,[36,37] and magnetic nanoparticles.[3,9,38] The heat transport in nanofluids has been studied by many researchers to understand the mechanisms involved.[27,39] Moreover, there have been a number of reviews on this topic.[40,41] Conventional heat transfer fluids have very low thermal conductivity compared to solids such as silver, copper, aluminum, and diamond. Thermal energy is sufficient to keep nanoparticles suspended under Brownian motion in base fluids,[42] which is

Soft Matter Series No. 16
Fundamentals and Transport Properties of Nanofluids
Edited by S. M. Sohel Murshed

Published by the Royal Society of Chemistry, www.rsc.org

one of the attractions of using nanofluids for heat transfer applications. The properties that make nanofluids a promising coolant are good thermal conductivity and heat transfer, good stability and good critical heat flux.[43] Studies have shown that relatively small amounts of nanoparticles can enhance the thermal conductivity of base fluids significantly. Different types of nanofluids have been used in a broad range of engineering applications, such as in automobiles, as coolants,[44] in brake fluids,[45] in domestic refrigerators,[46] in solar devices,[47,48] cosmetics, drug delivery,[49] defect sensors,[50] optical filters,[51] hyperthermia,[52] and sealants.[53] Lately, magnetic nanofluids containing superparamagnetic nanoparticles have been shown to exhibit interesting properties that have been exploited in heat transfer,[54,55] sensors[56–60] and magnetic hyperthermia,[61,62] and dye removal,[63,64] among other applications.

1.2 Preparation of Nanoparticles

Nanoparticles can be prepared either *via* top-down or bottom-up approaches.[65] The top-down approach includes methods such as ball milling, etching, laser ablation and electro explosion. The bottom-up approach includes methods such as precipitation techniques, solvothermal processes, hydrothermal processes, and microwave-assisted synthesis. Schematic representations of the synthesis of nanoparticles *via* top-down and bottom-up approaches are shown in Figure 1.1.

1.2.1 Top-down Approach

In the top-down approach, micron-sized particles are reduced to nano-sized particles using different techniques. The reduction of micron-sized colloidal particles *via* mechanical milling in the presence of a liquid carrier and dispersing agent is referred to as the size reduction method. The influencing parameters for this method are plasticity, hardness, composition of the chemical reagents, presence of micro-cracks and defects in a crystalline structure. Particles synthesized *via* this methods are more stable because metal nanoparticles are directly synthesized from their respective metals. Some of the popular techniques used for the preparation of nanoparticles *via* the top-down approach are discussed below.

1.2.1.1 Mechanical Milling

Mechanical milling is the most widely used method for generating materials on the nanoscale. It is used to generate blends and also to manufacture nanocomposites. This method involves the reduction of micron-sized colloidal particles *via* mechanical milling in the presence of a liquid carrier and dispersing agent. The most commonly used method in mechanical milling is the ball milling method, which involves the grinding of a powdered material into extremely fine nanoparticles of 2–20 nm in size. The ball milling

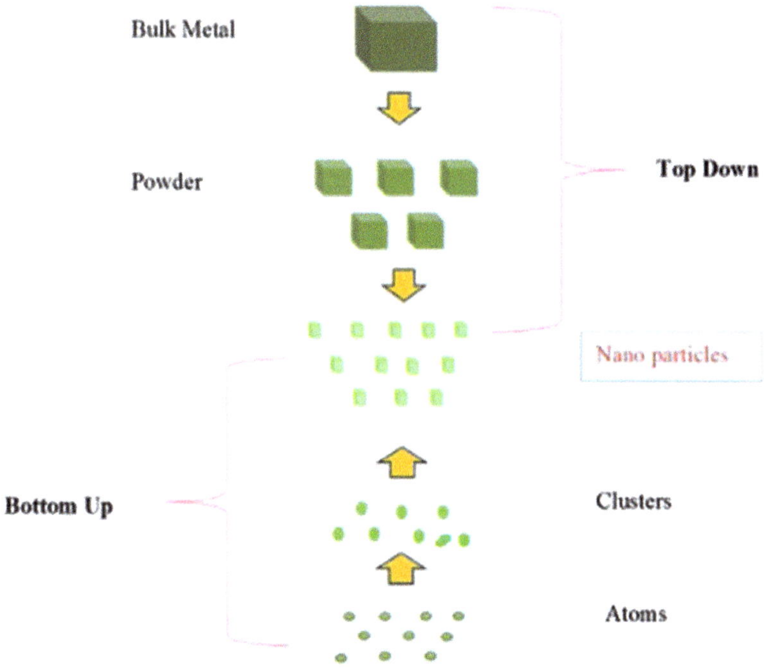

Figure 1.1 Schematic representations of the synthesis of nanoparticles *via* top-down and bottom-up approaches.

chamber consists of a stainless-steel container and many small balls of iron, silicon carbides or tungsten carbides. The balls are allowed to rotate inside the mill along with the material to obtain fine nanoparticles.[66] Carvalho *et al.*[67] synthesized magnetite nanoparticles *via* high energy ball milling using iron powder at a ball-to-powder mass ratio of 20 : 1. Salah *et al.*[68] prepared zinc oxide nanoparticles *via* a ball milling method from its microcrystalline powder in a mixing ratio of 15 : 1 by weight percentage of steel balls and ZnO powder. Alaa El Din Mahmoud *et al.*[69] demonstrated the sustainable synthesis of graphite oxide using a dry ball milling technique employing graphite flakes. A schematic representation of the principles of mechanical (ball) milling is shown in Figure 1.2.

1.2.1.2 *Etching*

Etching is the process of removing a layer from the surface of a material *via* a chemical reaction in the presence of a strong acid, base or suitable gas. Li *et al.*[70] synthesized concave polyhedral platinum nanoparticles using this method, wherein HCl was used for etching the surface of the platinum nanoparticles. Cheng *et al.*[71] synthesized a Fe–Si binary oxide composite using an organic template, where the Si units were then etched using NaOH to leave iron oxide nanoparticles to adsorb arsenic from water.

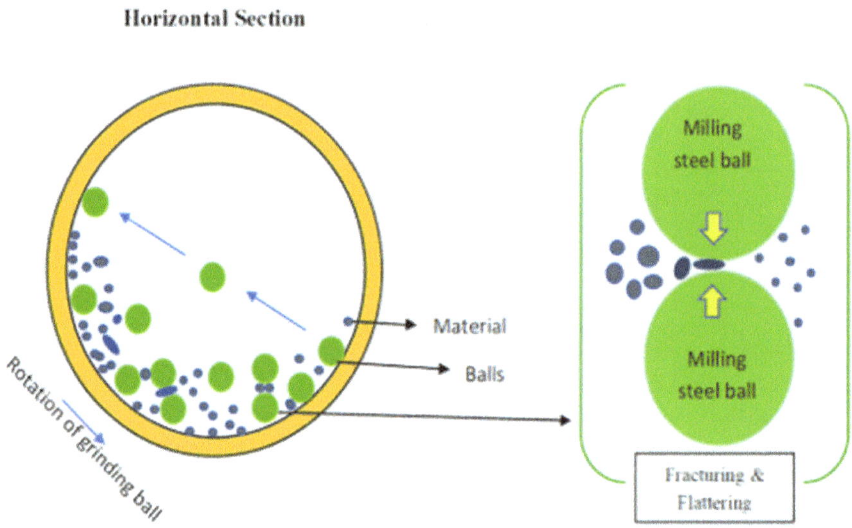

Figure 1.2 Schematic representation of the principles of mechanical ball milling.

1.2.1.3 *Laser Ablation*

Laser ablation involves the irradiation of a target material with a laser beam, which leads to its vaporization and the formation of nanoparticles. The rate of deposition is dependent on the penetration of the incident laser pulse on the surface of the target material within the penetration depth. Pulsed laser ablation of a liquid is an interesting way to manufacture a colloidal solution of uniform-sized nanoparticles without the further addition of surfactants or ligands. Duque *et al.*[72] used this method to prepare colloidal metal oxide nanoparticles in a liquid environment. The method is also used as a replacement for traditional chemical reduction methods for the synthesis of noble metal nanoparticles. This technique does not require any additional chemicals for the stability of the nanoparticles, hence, the synthesis of stable noble metal nanoparticles using pulsed laser ablation is referred to as a green technique.[73] Ismail *et al.*[74] prepared iron oxide nanoparticles *via* pulsed laser ablation using iron in dimethylformamide and sodium dodecyl sulphate solutions as a target.

1.2.1.4 *Electro Explosion*

Electro explosion involves the synthesis of nanoparticles *via* the electrical explosion of metal wire. In this method, nanoparticles are synthesized *via* the explosion of a metal wire when a high-density current pulse is passed through it. This method has many advantages over all other methods due to its high energy efficiency, variation in characteristics according to the process parameters, small variation in the size of the synthesized nanoparticles, stability of the nanoparticles under normal conditions, and high probability to obtain metals, alloys, and oxides and nitrides of metal nanoparticles. Sen *et al.*[75] synthesized Cu, Ag, Fe and Al nanoparticles from their conducting parent materials using an electro explosion of a wire method. Berasategi *et al.*[76] synthesized magnetic nanoparticles *via* the electric explosion of a wire.

1.2.2 Bottom-up Approach

The bottom-up approach includes methods such as electrodeposition, vapor condensation, thermal decomposition, laser pyrolysis, co-precipitation, mechanochemical, sol–gel, solvothermal, hydrothermal, bacterial, and reverse micelle techniques. Some of the popular techniques used for the preparation of nanoparticles *via* a bottom-up approach are discussed below.

1.2.2.1 Precipitation Techniques

Precipitation reactions consist of a nucleation step, followed by particle growth.[77] The simultaneous occurrence of the nucleation and growth processes dictates the particle size and morphology. Due to difficulties in isolating nucleation and growth processes for independent study, the fundamental mechanism of precipitation reactions is still not understood fully. When multiple elements are involved in precipitation (co-precipitation), the process becomes all the more complex. The synthesis of ferrite nanoparticles was carried out by precipitating iron salts in alkaline medium. The advantage of co-precipitation process is that it does not produce or use any toxic intermediates or solvents, and does not require precursor complexes and high temperatures. It has become a popular technique due to its ability to be scaled up, its reproducibility and its eco-friendly reaction conditions.[78]

1.2.2.2 Solvothermal

Solvothermal synthesis is usually performed in autoclaves or reactors by maintaining a pressure of >2000 psi at a temperature of ~200 °C. There are two types of solvothermal processes conducted for the synthesis of ferrite nanoparticles; the solvolysis of metal salt solution and oxidation or neutralization of a mixed metal hydroxide.[79] A typical experimental set up of the solvothermal method is shown in Figure 1.3. In this method, metal oxide nanoparticles are prepared by decomposing organometallic compounds such as metal acetylacetonates in high boiling organic solvents containing surfactants. Fe_3O_4 nanoparticles are prepared by decomposing salts such as [Fe(acac)$_3$] in the presence of 1,2-hexadecanediol, oleylamine and oleic acid in phenol ether.[80,81] Factors such as the ratios of the starting reagents, such

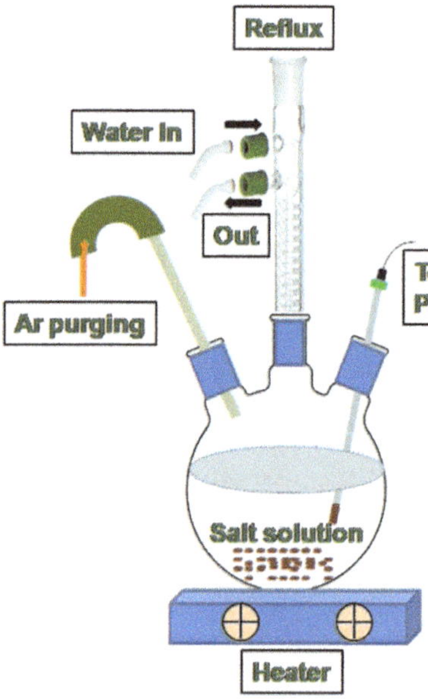

Figure 1.3 Schematic representation of solvothermal synthesis.

as organometallic compounds, surfactant and solvent, are the decisive parameters that control the size and morphology of magnetic nanoparticles. Also, the nature of particles formed in this method depends on the reaction conditions, such as temperature and time.[82]

In this method, the particle size is controlled through a nucleation and growth process. At high reaction temperature, the nucleation might be faster than the growth, while over a long reaction time, the growth process is favored. This method is suitable only for small scale but not for industrial applications due to the high-temperature requirements. Fe_3O_4 nanoparticles prepared in this way are usually dispersed in organic solvent. Recently, it was demonstrated that water soluble Fe_3O_4 nanoparticles (4–60 nm) can be prepared by decomposing $FeCl_3 \cdot 6H_2O$ at 245 °C in the presence of 2-pyrrolidone or α,ω-dicarboxyl-terminated poly(ethylene glycol) as coordinating agents under reflux.[83,84]

1.2.2.3 Flow Injection Method

This method was developed by Alvarez *et al.*[85] and involves the continuous or segmented mixing of precursors under a laminar flow regime in a capillary reactor. The experimental set up includes a multi-channel peristaltic pump, Teflon tubing lines, a T-shaped injector and an injection valve. The advantages of this method are high reproducibility due to plug flow and laminar conditions, a high mixing homogeneity and the precise external control of the process.

1.2.2.4 Decomposition of Metal Carbonyls

This technique is based on the formation of ferromagnetic nanoparticles *via* the thermal decomposition of metal carbonyls in various liquid organic media, gases or under vacuum. At higher temperatures, metal carbonyls decompose and yield metal particles in of the micron-sized range. However, it has been found that in the presence of polymers, these carbonyls decompose into nanosized particles and that it is feasible to change the particle size by selecting the molecular mass and composition (polar and non-polar groups) of the polymer.

1.2.2.5 Reverse Micelle Method

In this method, reverse micelles formed from nanometer-sized droplets of an aqueous phase are stabilized by surfactants in an organic phase and used for the synthesis of nanoparticles.[86] In this technique, surfactant-stabilized water in-oil-emulsions containing the aqueous solution of the precursor are prepared.[87] By mixing two identical water-in-oil micro-emulsions containing the desired reactants, the microdroplets continuously collide, coalesce and finally form a precipitate in the micelles. The precipitate can then be extracted using solvents such as acetone and ethanol by centrifuging the mixture. In this sense, a microemulsion can be used as a nanoreactor for the formation of nanoparticles. Although this method has been used for the synthesis of a wide variety of nanoparticles, the shape and size usually vary over a wide range. The working window is very narrow and only a small amount of material can be prepared using this technique. Moreover, the methods also requires the use of a large amount of solvent.

1.2.2.6 Electrodeposition Technique

This technique was developed on the basis of known practical experience on the subject of preparing metals by electrolysis. The metals are deposited on the cathode as a dense continuous layer, distinctly delimiting single crystals and finely dispersed powders.[88] Using this method, ferromagnetic metals and their alloy-based ferrofluids can be synthesized.[89] The influencing factors of this technique are the cathode surface area, current density, composition of the electrolyte, presence of surfactants and/or additives (such as thiourea, urea, tartaric acid, *etc.*), temperature and agitation of the electrolyte. Electrolysis is performed in a two-layered tank using a needle-shaped cathode sealed in a glass tube, which is lowered and raised periodically. In the two-layered tank, the top layer contains an organic medium, which also contains an appropriate surfactant, and the bottom layer is an aqueous electrolyte solution. After the deposition of nanoparticles at the moving cathode, the metallic nanoparticles are collected in the top non-polar solvent layer containing the surfactant.

1.2.2.7 Sol–Gel Process

Generally, a sol–gel process refers to the hydrolysis and condensation of alkoxide-based precursors. This process contains distinct steps, *i.e.*, (a) the formation of stable solutions of alkoxide or solvated metal precursor, (b) gelation resulting from the formation of an oxide- or alcohol-bridged network *via* a polycondensation or polyesterification reaction, (c) ageing of the gel until the gel transforms into a solid mass, accompanied by concentration of the gel network and expulsion of solvent from the gel pores, (d) drying of the gel, (e) dehydration and (f) decomposition of the gel at high temperature.[90,91] Long *et al.*[92] prepared iron oxide nanoparticles using an iron oxide aerogel obtained from $FeCl_3$ in anhydrous ethanol and epichlorohydrin at a 1 : 10 ratio as a precursor. The dried iron oxide aerogel was decomposed *via* heat treatment at ~300 °C for 20 h. Nanosized iron oxide (ε-Fe_2O_3) with a giant coercivity field (2 T) was prepared by combining micro-emulsion and sol–gel processes.[93]

1.2.2.8 Mechanochemical Reactions

In mechanochemical synthesis, chemical reactions occur at the interfaces of the nanometer-sized grains that are continuously re-generated during milling.[94] Mechanochemical processes that make use of chemical reactions activated by high-energy ball milling have been successfully used for preparing ferrite nanoparticles. Ding *et al.* synthesized $BaFe_{12}O_{19}$ *via* the mechanical milling of a $BaCl_2$, $FeCl_3$ and NaOH powder mixture.[95] Later, other ferrites were prepared *via* the same process using a stoichiometric mixture of an MO (M = Fe, Ni, Zn) and α-Fe_2O_3 as the precursor material.[96–98] Druska *et al.*[96] prepared inverse spinel $ZnFe_2O_4$, instead of normal spinel $ZnFe_2O_4$ *via* the high-energy ball milling of a stoichiometric mixture of ZnO and α-Fe_2O_3.

1.2.2.9 Hydrothermal Synthesis

Hydrothermal synthesis exploits the solubility of almost all inorganic substances in water at elevated temperatures and pressures and results in the subsequent crystallization of the dissolved material from liquid.[90,91] At elevated temperatures, water exists exhibits a high vapor pressure and reactant properties such as solubility and change in reactivity. These changes provide more parameters to tune nucleated particles with different sizes and shapes, which is not possible during low-temperature reactions. The influencing parameters of the process are reaction temperature, pressure, reaction time and salt composition. Recently, Wang *et al.*[99] prepared single-crystalline magnetite nanowires by applying an external magnetic field during the ageing process. It was found that with increasing field, the wire concentration also increases. Zhang *et al.*[100] reported that $NiFe_2O_4$ crystallization in the presence of polyethylene glycol (PEG) produces nanorods.

1.2.2.10 Solvothermal Synthesis

The solvothermal synthesis process is similar to the hydrothermal method but uses solvents other than water. Sun *et al.*[80] reported an organic phase process for synthesizing monodisperse Fe_3O_4 nanoparticles *via* the reaction of $Fe(acac)_3$ and a long-chain alcohol. This reaction was extended to synthesize MFe_2O_4 nanoparticles (M = Co, Ni, Mn, *etc.*) with a tunable size of 3–20 nm in diameter by adding a different metal acetylacetonate precursor to a mixture of $Fe(acac)_3$, 1,2-hexadecanediol, oleic acid and oleylamine.[81] Barker *et al.*[101] studied the ripening process in a solvothermal technique and showed that the size of the nucleated nanoparticles could be varied by changing the reaction temperature, ageing time and metal precursors. Recently, the effects of organic solvents on iron oxide nanoparticles in a solvothermal method were studied.[102]

1.2.2.11 Bacterial Synthesis

In 1975, Blackmore observed that certain bacteria found in marine marsh mud tended to rapidly navigate along a specific direction due to iron rich crystals (~100 nm) forming chains in the major axis.[103] Later, it was found that an unclassified magnetotactic spirillum, when cultured in a solution containing ferric salts, precipitated uniform single-crystal spinel magnetite.[104]

1.3 Preparation of Nanofluids

Nanofluids can be prepared either *via* single (one-step) or two-step processes. The details of these two methods are presented herein.

1.3.1 Two-step Method

In a two-step method, the nanoparticles are first produced as dry powders by either a physical or chemical process, as discussed in Section 2. The prepared nanoparticles are then dispersed into a fluid with the aid of mechanical or ultrasonic agitation, high-shear mixing, homogenization, or ball milling. Nanoparticles are prepared by transition metal salt reduction,[105] ligand reduction and displacement from organometallics,[106] microwave synthesis[107] and coprecipitation.[108]

Nanoparticles have a tendency to aggregate due to their high surface area and surface activity. To overcome this, nanoparticles are often stabilized with surface active species.[109] Bonnemann *et al.*[110] developed a method for the production of palladium (Pd) particles of 2.2 nm in size *via* a chemical reduction pathway. Copper nanofluids can be prepared by dispersing Cu nanoparticles in both water and transformer oil *via* sonication in the presence of stabilizers.[111] Murshed *et al.*[23] prepared a water-based titanium dioxide (TiO_2) nanofluid *via* ultrasonic dispersion. Kim *et al.*[112] prepared copper oxide (CuO) nanofluids by directly dispersing commercially-obtained CuO nanoparticles in ethylene glycol (EG) by sonication, without stabilizers. Multi-walled carbon nanotubes (MWCNTs) are usually prepared using a two-step method.[37,113,114] The CNTs are first synthesized *via* a pyrolysis method and then suspended in a base fluid with or without the use of a surfactant.[37,113,115,116] Lee *et al.*[35] prepared metal oxide nanoparticles (Al_2O_3, CuO) using a sol–gel method and then dispersed the obtained nanoparticles in water using a sonicator. Suresh *et al.*[117] prepared an alumina–copper hybrid powder *via* a thermochemical synthesis method followed by dispersion of the nanopowder in deionized water using sodium lauryl sulphate (SLS) as a dispersant employing an ultrasonic vibrator. A water-based silver (Ag) nanofluid was prepared *via* a two-step technique.[118] Hwang *et al.*[119] prepared carbon black in water and Ag in silicon oil nanofluids *via* a two-step method. Zinc oxide (ZnO) nanofluids have been prepared *via* a two-step technique using a magnetic stirrer.[120] Choi *et al.*[121] used zirconium dioxide (ZrO_2) bead milling in a vertical, super-fine grinding mill, for the preparation of transformer-oil-based aluminum oxide (Al_2O_3) and aluminum nitride (AlN) nanofluids using *n*-hexane as a dispersant. Water-based TiO_2 nanofluids have been prepared using a two-step method.[122] A schematic representation of a two-step method involving microwave-assisted synthesis and a direct mixing technique is shown in Figure 1.4.

1.3.2 One-step Method

In a one-step method, nanoparticle preparation and dispersion in a base fluid are conducted simultaneously.[123] A one-step process can thus be used to prepare uniformly-dispersed nanoparticles stably in a base fluid.[124] The methods used to prepare nanofluids *via* a one-step process include thermal decomposition,[125] physical vapor condensation,[32] chemical reduction[126–128] and polyol synthesis.[129]

1.3.2.1 Direct Evaporation Techniques

Akoh *et al.*[130] reported a single-step direct evaporation method for the preparation of nanofluids. Direct evaporation and condensation are used to produce stable nanofluids.[131] Choi[132] developed a 'direct-evaporation' technique involving a cylinder containing a fluid that is rotated and a source material that is vaporized in the middle of the cylinder. The vapor condenses after it comes into contact with the cooled liquid, as shown in Figure 1.5. However, the drawbacks of this technique are the requirement of low-vapor-pressure liquids and a low yield.

1.3.2.2 Chemical Reduction

Metallic nanoparticles can be prepared in various solvents.[133] Cuprous oxide nanofluids have been prepared *via* the chemical reduction of copper acetate

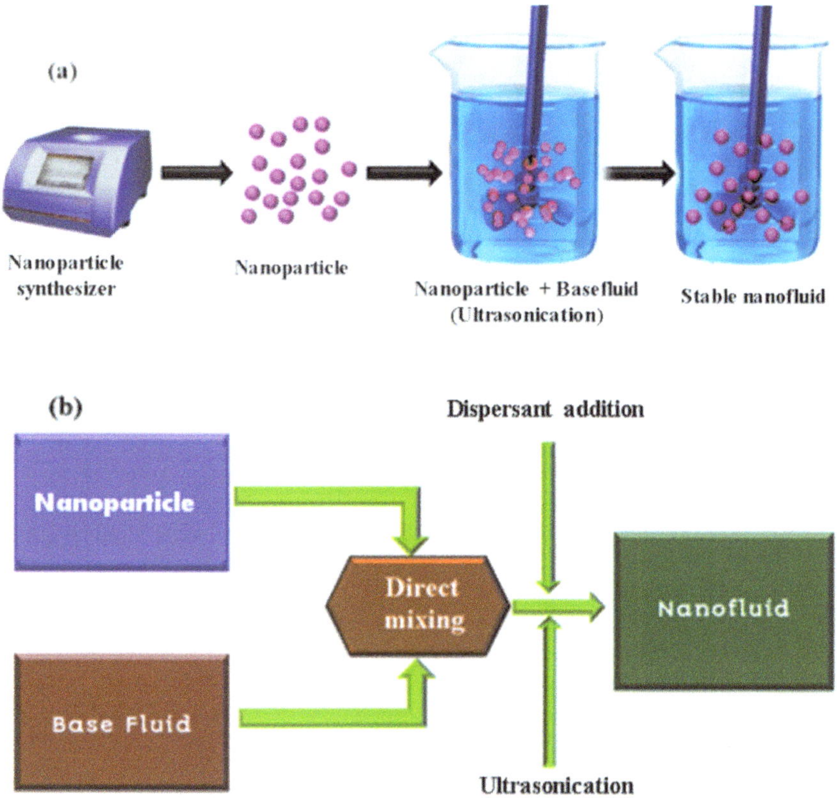

Figure 1.4 Two-step preparation process of nanofluids. (a) Nanoparticles prepared by a microwave synthesizer and particles are dispersed in a base fluid using a mechanical stirrer. (b) Direct mixing of nanoparticles followed by the addition of a dispersant and ultrasonication.

by glucose.[127] Liu *et al.*[134] prepared a water-based copper nanofluid *via* a chemical reduction method, where copper acetate was used as a precursor and hydrazine (N_2H_4) as a reducing agent. Garg *et al.*[6] synthesized copper nanoparticles *via* a chemical reduction method using water as the solvent. In this method, copper acetate was used as a precursor and sodium hydroxide as the reactant. Kumar *et al.*[128] prepared stable EG-based copper nanofluids by reducing copper sulfate pentahydrate with sodium hypophosphite as a reducing agent by means of conventional heating. Copper nanofluids were prepared *via* a routine one-step reduction method in which copper nitrate was reduced by glucose in the presence of sodium lauryl sulfate.[135] Tsai *et al.*[136] prepared aqueous gold nanofluids with different nanoparticle sizes *via* the reduction of aqueous hydrogen tetrachloroaurate ($HAuCl_4$) with trisodium citrate and tannic acid. Xun *et al.*[137] prepared kerosene-based silver nanofluids *via* an extraction reduction method. Said *et al.*[138] prepared

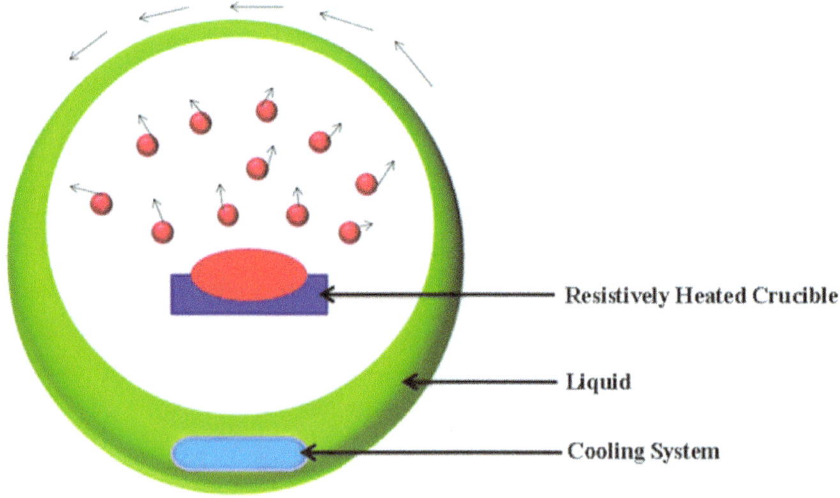

Figure 1.5 Schematic of a one-step nanofluid production system, which simultaneously produces and disperses nanoparticles in a low-vapor-pressure liquid.[147]

water-based carbon nanofibers (CNFs), functionalized carbon nanofibers (F-CNFs), reduced graphene oxide (rGO), and rGO coated over F-CNFs (F-CNFs/rGO) nanofluids using a modified Hummers' method and chemical reduction methods, as well as a hydrothermal technique. Estelle *et al.*[139] prepared reduced rGOs by chemically reducing graphene oxide using various concentrations of sodium borohydride. Santos *et al.*[140] reported the one-pot synthesis of an aqueous nanofluid with improved thermal conductivity and high colloidal stability. An aqueous nanofluid based on thioglycolic acid (TGA)-coated copper sulfide nanoparticles ($NPCu_{(2-x)}S$) was synthesized *via* a chemical reduction method.

1.3.2.3 Submerged Arc Nanoparticle Synthesis System (SANSS)

A stable CuO nanofluid was prepared using a SANSS.[141] The vaporized metal in this process undergoes nucleation, growth and condensation, resulting in nanoparticles being dispersed in deionized water. Using this approach, nanofluids containing CuO particles of 49.1 ± 38.9 nm in size were obtained.[141] EG-based nanofluids containing silver nanoparticles of 12.5 nm in size were synthesized using SANSS.[142] A schematic representation of the SANSS is shown in Figure 1.6.

1.3.2.4 Laser Ablation

Laser ablation is another single-step technique that can be used to produce and disperse nanoparticles directly in base fluids. Various nanofluids have

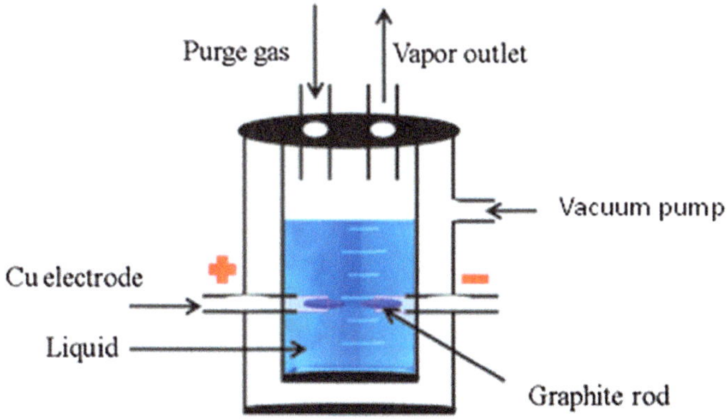

Figure 1.6 Schematic representation of the Submerged Arc Nanoparticle Synthesis System (SANSS).

been prepared *via* laser ablation methods[143–145] by ablating solid metals such as Cu, Ag, and Au submerged in a base fluid.[144] Kim *et al.*[143] prepared bare Au-water nanofluids *via* pulsed laser ablation in liquids. Lee *et al.*[144] prepared a water-based Cu nanofluid *via* a pulsed laser ablation in liquid (PLAL) method using Cu pellets. Phuoc *et al.*[145] used a multi-beam laser ablation technique for the synthesis of water-based silver nanofluids. One-step laser ablation techniques have also been used to synthesize a variety of nanofluids with different nanoparticles including such as Cu[146] and Ag.[147] Tran *et al.*[147] prepared water-based Ag and Al nanofluids with a particle size of 9–21 nm using a laser ablation technique, wherein a Nd:YAG laser was used to ablate Ag and Al in deionized water. A tin solar nanofluid was prepared *via* PLAL, wherein bulk tin immersed in EG was ablated using a femtosecond laser. It was reported that the stability and properties of the solar nanofluid was enhanced when compared to nanofluids prepared *via* conventional synthesis methods. Lee *et al.*[144] prepared a CuO nanofluid *via* PLAL using a single-pulsed laser beam ($k = 532$ nm) as follows: (1) a copper pellet was placed in a beaker filled with deionized water (DIW); (2) a Nd:YAG laser was used to produce a CuO/DIW nanofluid over 8 h. Al_2O_3 nanoparticles were synthesized using laser ablation (at different energies of 1, 3, and 5 J) using an Al target in DIW.[148] Mbambo *et al.*[149] synthesized an silver nanoparticles decorated ethylene glycol based nanofluid using a hybrid chemical–physical laser based approach, where PLAL was conducted in parallel with well-established standard vacuum pulsed laser deposition (PLD) technology to synthesize nanostructured and nanophase materials in the form of thin film coatings (LLSI/PLAL). The advantage of this approach is the combination of two major effects; namely, the standard ablation process due to the laser–matter interaction and the acoustic process caused by the explosion of the native gas bubbles during the local overheating at the target–liquid interface. Kazakevich *et al.*[150] discussed in detail the experimental setup for laser ablation in a liquid environment. A solid target placed under a thin layer of liquid is exposed to laser radiation. Irradiation of the metal surface results in fast removal of the material and the ejected NP remains in the liquid that surrounds the target, resulting in the formation of a colloidal solution. Due to accumulation of nanoparticles in the surrounding liquid, their prolonged interaction with laser radiation is possible. Riahi *et al.*[151] prepared water-based Al_2O_3 nanofluids using a nanosecond Nd:YAG pulsed laser operating at 1064 nm. Woo *et al.*[152] prepared water-based graphite oxide (GO) nanofluids with enhanced thermal conductivity using PLAL and studied the effect of laser frequencies on the variation in the size and morphology of the GO nanoparticles. It was reported that the laser frequency significantly affects the size and morphology of the GO nanoparticles during laser ablation, leading to a profound variation in the thermophysical properties of the GO nanofluids. Maximum thermal conductivity enhancement of 82% was achieved for the nanofluids prepared at a laser frequency of 10 Hz. A comparative study on the role of the nature of the laser on nanoparticle preparation was reported by Rafique *et al.*[153] where water-based Ag nanoparticles were prepared *via* a laser ablation method using a continuous wave (CW) diode laser and a pulsed Nd:YAG laser. Vegetable oil based Ag nanofluids were prepared *via* laser ablation.[154] The target Ag was immersed in a beaker containing 55 mL of vegetable oil. The laser pulse energy/area gives the fluence that is used to prepare the nanostructures in various vegetable oils. A schematic representation of the laser ablation process is shown in Figure 1.7.

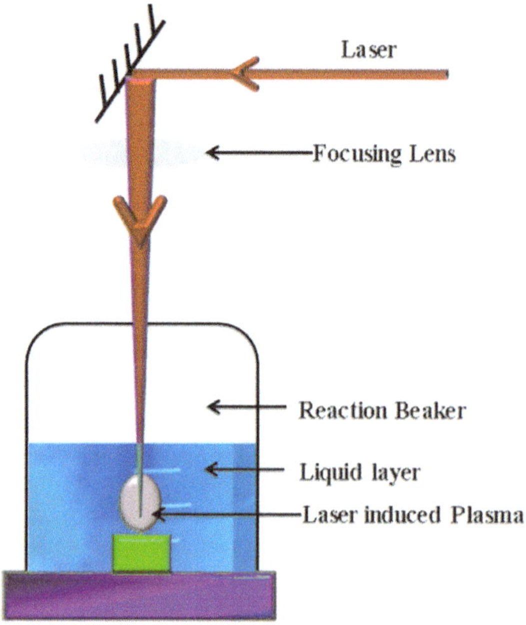

Figure 1.7 Schematic illustration of the experimental apparatus used in laser ablation.

1.3.2.5 Microwave Irradiation

Another one-step approach adopted for nanofluid synthesis is microwave irradiation.[155] Copper nanofluids have been prepared by reducing $CuSO_4 \cdot 5H_2O$ with $NaH_2PO_2 \cdot H_2O$ in EG under microwave irradiation.[156] Mineral oil (MO) based silver nanofluids have also been prepared using this method.[157] Ethanol-based silver nanofluids were prepared using a microwave-assisted one-step method.[158] Polyvinylpyrrolidone (PVP) was used as a stabilizer and reducing agent for colloidal silver in solution.[158] Seifikar *et al.*[159] synthesized a carbon quantum dots (CQDs) nanofluid *via* the microwave heating of PEG 200 (PEG 200), wherein PEG acts as a carbon source, base fluid and dispersant simultaneously. Nikam *et al.*[160] prepared CuO nanoparticles using a microwave-assisted continuous-flow synthesis technique.

1.3.2.6 Polyol Process

In the polyol process,[161,162] a metal precursor is dissolved in a liquid polyol (usually EG) to reduce the metallic precursor, followed by nucleation and growth.[161] Other one-step techniques used for nanofluid preparation are the step physical vapor condensation method[32] and plasma discharging technique.[163] Zeroual *et al.*[164] synthesized EG-based silver nanofluids *via* a polyol process. The synthesis process is based on the reduction of $AgNO_3$ in the presence of the aqueous emulsion of latex copolymer along with EG. Oliveria *et al.*[165] prepared stable EG-based diamond–silver hybrid nanofluids *via* a polyol process. The stability of the nanofluid can be attributed to the ions in the dispersion that contribute toward electrostatic stabilization, preventing the agglomeration of the nanoparticles and their sedimentation.

1.3.2.7 Phase-transfer Method

In the phase-transfer method, the reactant migrates from one phase into another phase where the reaction occurs. This method has the advantage of overcoming the problems associated with the insolubility of precursor materials in base fluids during the preparation of nanofluids.[166] Yang *et al.*[167] developed a general protocol to transfer metal ions from an aqueous solution to an organic medium. Combining phase-transfer technology with a wet chemistry method for the synthesis of nanoparticles is an effective way to prepare nanofluids.[166] A simple process based on amine chemistry for the phase-transfer of platinum nanoparticles from an aqueous to an organic solution was reported by Kumar *et al.*[168] By vigorous shaking of a biphasic mixture of platinum nanoparticles in an aqueous medium and octadecylamine (ODA) in hexane, an aqueous platinum nanoparticle complex was prepared with the ODA molecules present in the organic phase and the prepared particles are hydrophobic in nature.[168] In a similar way, gold and silver nanoparticles are transferred from the aqueous to organic phase using alkylamine as a surfactant.[169] Graphene oxide nanosheets (GONs) were successfully transferred from water to *n*-octane after modification by oleylamine.[170] Apart from surfactants and polymers,[171] surface treatment[172] has also been used to produce stable nanofluids. However, the addition of additives can modify the surface properties of the particles, which may have a synergistic or antagonistic effect on thermal properties.

1.3.2.8 Wet Mechanochemical Techniques

Mechanical actuation that involves the top-down approach for synthesizing nanoparticles is a simple, cost-effective and environmentally benign alternative for synthesizing nanofluids. This method includes crushing, followed by milling and grinding. The enormous developments in the technique of high-energy ball milling (HEBM) have made it possible for the production of ultrafine nanoparticles of different materials.[173] Inkyo *et al.*[174] developed a bead mill process for the preparation of well-dispersed suspensions of TiO_2 (titania) nanoparticles (5%) in methyl methacrylate (MMA). TiO_2–PMMA nanocomposites were synthesized *via* subsequent polymerization of the TiO_2–MMA suspension. Harjanto *et al.*[175] prepared TiO_2–water nanofluids *via* a one-step process in which titania nanoparticles were ball-milled together with distilled water for 15 h. Nine *et al.*[176] prepared water-based copper and copper oxide nanofluids *via* ball milling in an aqueous medium. To prepare stable nanofluids, the ball milling was performed using different ball sizes (1 mm balls for 30 min and 3 mm balls for 1 h). Almasy *et al.*[177] prepared a ferrofluid (10 and 15 nm) using a vibrating ball mill. It was reported that a ferrofluid prepared *via* a wet-milled technique showed higher saturation magnetization than that of a ferrofluid suspension prepared *via* a co-precipitation synthesis.

1.3.2.9 Sol–Gel Method (Hydrolysis)

Sol–gel process is a technique widely used for the synthesis of nanofluids. There are two stages in this process, (i) hydrolysis: in this process the functional binders of the precursors are substituted with hydroxyl groups. (ii) Condensation reaction : a continuous network is formed as the hydroxyl groups of the monomers become connected.[178] For the preparation of water-based silica nanofluids using a sol–gel process, tetraethyl orthosilicate (TEOS) was taken as the precursor, ethanol as a solvent, DIW for the hydrolysis reaction, and ammonium hydroxide as a base catalyst.[179] Samed *et al.*[180] prepared stable water-based TiO_2 nanofluids using a sol–gel technique. Jing *et al.*[181] prepared water-based silica nanofluids containing different particle sizes (5, 10, 25, and 50 nm) by varying the TEOS concentration.

1.3.2.10 Emulsion Polymerization

Emulsion polymerization is a polymerization process in which emulsification of hydrophobic polymers is achieved using an emulsifier in aqueous solution, then free radicals are generated using either water or oil soluble initiators. Goud *et al.*[182] prepared nanoencapsulated phase change materials (nEPCMs) *via* mini-emulsion polymerization. Petroleum jelly (45 g) and paraffin wax (5 g) were heated in a beaker. The composite was then mixed with AIBN, oleylamine and styrene using an ultrasonicator for 1 h at 55 °C. Triton X-100 and DIW were then added to this mixture and mixed using an ultrasonic probe for 30 min. The solution was kept under an inert atmosphere (nitrogen purging) for the polymerization process. After nitrogen purging, the solution was kept at constant temperature at 75 °C and stirred continuously using a mechanical stirrer at 400 rpm for 15 h to obtain nEPCM. Finally, a PCM nanofluid was prepared by dispersing the nEPCM in deionized water using an ultrasonic shaker. Han *et al.*[183] prepared a suspension of indium nanoparticles in polyalphaolefin (PAO) using a one-step, nanoemulsification method. An indium pellet was added to PAO oil and then the mixture was heated 20 °C above the indium melting temperature (157 °C). A PAO aminoester dispersant was injected into the reaction vessel, which also acted as a stabilizer, and the mixture was allowed to stirrer for 2 h to form an emulsion with micron-sized droplets. Then, the emulsion was exposed to high-intensity ultrasound radiation for more than 2 h until a stable nanofluid was formed. Using an emulsification and polymerization route Kim *et al.*[184] prepared biphasic nano colloids *via* the free-radical polymerization of a mixture of an acrylate copolymer precursor consisting of methyl acrylate, methyl methacrylate, and vinyl acetate which was dissolved in poly(dimethyl siloxane) (PDMS). The increasing molecular weight leads to the incompatibility of the acrylate copolymer in the PDMS and drives the phase separation on the nanometer scale and at the end of the polymerization process, with one phase partially or entirely surrounding the other phase, and the resulting structure referred to as a biphasic structure or an acorn structure due to its morphology. Joseph *et al.*[185] synthesized a polystyrene encapsulated PCM *via* a mini-emulsion polymerization. In this method, *n*-octadecane was used as PCM, styrene as the monomer for encapsulation of the PCM, α-α' azoisobutyronitrile solution as the free radical initiator for polymerization, TritonTM X-100 as the surfactant, and oleylamine as the short-term stabilizer to protect the polymer against temperature and oxygen during the synthesis. Jitheesh *et al.*[186] prepared a composite PCM by homogeneously mixing paraffin wax and petroleum jelly to encapsulate polystyrene shells *via* a mini emulsion polymerization process, Bhanvase *et al.*[187] synthesized water-based (polyaniline) (PANI) nanofiber nanofluids *via* an ultrasound-assisted emulsion polymerization method. Yao *et al.*[188] prepared a novel functionalized microgel-based hybrid nanofluid, composed of a composite core of poly(2-dimethylamino ethyl methacrylate) microgel and silica (PDMAEMA/SiO_2) and a shell of polyetheramine, fabricated *via* the combination of inversed emulsion polymerization, biomimetic silicification and grafting approach.

1.3.2.11 Green Synthesis

Many research studies have proved that the preparation of nanofluids by green methods is more effective than traditional methods. Various physical and chemical approaches for synthesizing nanofluids impact the environment badly due to the toxic nature of chemicals. The plant-based synthesis of nanofluids is a quick and easy procedure and does not require high temperature. Green synthesis approaches can be scaled up easily in the preparation of nanoparticles and nanofluids. Green synthesis methods have attracted much attention in recent years as they include the usage of vitamins, enzymes, amino acids and plant extracts. Sarafaz *et al.*[189] prepared different volume fraction (0.1%, 0.5% and 1%) silver nanofluids (ethylene-glycol/water (50 : 50 by volume) as a base fluid) using a plant extract method from green tea leaves and silver nitrate. Kumar *et al.*[190] prepared metallic and non-metallic nanofluids from green synthesized silver and selenium nano-particles (NPs) using the Azadirachta indica plant (Neem). It has been reported that at room temperature, a low concentration of Neem leaf extract is sufficient to biosynthesize silver NPs in the lab from silver nitrate solution and selenium NPs from sodium selenite. Jameel *et al.*[191] prepared platinum nanoparticles *via* both sonochemical and conventional methods (one-pot synthesis methods), using a Prosopis farcta fruit (PFF) extract as a reducing agent and stabilizer. Rufus *et al.*[192] synthesized hematite nanoparticles using an aqueous leaf extract of Anacardium occidentale as a reducing and capping agent. It has been reported that synthesized nanoparticles enhance the thermal conductivity of conventional base fluids; water and EG at room temperature. Gautham *et al.*[193] prepared iron (Fe) nanoparticles using extracts of Terminalia bellirica (TB), Moringa oleifera fruit (MOF) and Moringa oleifera leaves. Among these, TB extract based NPs showed superior antioxidant activity when compared to the other biosynthesized Fe samples. Ranjbarzadeh *et al.*[194] prepared silica nanoparticles using a rice plant source and prepared silica nanofluids. The prepared nanofluids exhibited well-defined nanostructure and long-term stability. Okonkwo *et al.*[195] prepared silica (SiO_2) and titanium (TiO_2) nanoparticles using olive leaf extract (OLE) and barley husk (BK), respectively, and prepared water/BH–SiO_2 and water/OLE-TiO_2 nanofluids. Omiddezyani *et al.*[196] prepared a cobalt ferrite/reduced graphene oxide ($CoFe_2O_4$/rGO) nanocomposite using gallic acid (GA) as a green agent at room temperature. It was shown that π–π interactions between the graphene and the aromatic ring of GA enhanced the stability in a suspension. Sone *et al.*[197] synthesized CuO nanoparticles using *Callistemon viminalis* flower extracts at room temperature as a chelating agent and prepared water-based CuO nanofluids. Govindasamy *et al.*[198] prepared magnesium oxide nanoparticles (MONPs) using Pisonia alba leaf extract. This study reported that the MONPs showed good antioxidant and antifungal properties. Luna-Sanchez *et al.*[199] prepared silver nanoparticles using Jalapeno chili extract. The reduction of the silver nanoparticles was achieved by the flavones and flavanols present in the extract of Capsicum annuum var annuum. Sadri *et al.*[200] prepared graphene nanoplatelet nanofluids *via* a one-pot free radical grafting method using gallic acid. It was reported that the nanofluids showed good stability even after 63 days of preparation. Sadri *et al.*[201] prepared multi-walled carbon nanotubes (MWCNTs) using a free-radical grafting reaction and dispersed them in water to form an aqueous suspension of MWCNTs. It was reported that the thermophysical properties and stability of the MWCNTs was better in aqueous suspension. Shirazi *et al.*[202] synthesized nitrogen-doped activated carbon/graphene using high nitrogen levels employing empty fruit punch pulp and formulated it into a nanofluid for the analysis of its thermoelectric properties. Pauzi *et al.*[203] prepared ZnO nanofluids using gum Arabic as a stabilizer. It was reported that the stability of the ZnO nanofluids was extended to 6 months due to the effect of gum Arabic. Hosseini *et al.*[204] prepared highly-dispersed, covalently-functionalized MWCNTs using clove extract *via* a one-pot free-radical grafting method. Then, the prepared nanotubes were dispersed in water to form nanofluids. Kulkarni *et al.*[205] synthesized silver nanoparticles using neem leaf extracts. Mostafa Yusefi *et al.*[206] prepared Fe_3O_4 nanoparticles using *Garcinia mangostana* fruit peel as a stabilizing and capping agent *via* a co-precipitation method. Zainona and Azmi prepared hybrid nanofluids by dispersing TiO_2–SiO_2 nanoparticles in water and a bio-glycol mixture. The bio-glycol based nanofluids prepared in a ratio of 40 : 60 showed good stability and enhanced thermal properties.[207] Sarafraz *et al.*[208] prepared silver nanoparticles using green tea leaf extract and the prepared nanoparticles were dispersed in coconut oil.

1.4 Enhancement of Nanofluid Stability

The long-term stability of nanofluids is an essential requirement for their applications. Unlike molecular fluids, the production of well-stabilized nanofluids is a difficult task. Particle–particle and particle–fluid interactions account for the stability of nanofluids. Owing to the high surface area-to-volume ratio of fine nanoparticles, the dispersed particles (even with stabilizing entities) have a tendency to aggregate because of van der Waals attraction forces that dominate the other repulsion forces, namely; double layer electrostatic repulsions, hydration forces and steric forces.[209,210] For the production of stable nanofluids, one has to produce fairly monodisperse nanoparticles and functionalize them with a suitable stabilizing moiety to prevent interparticle attractions. Even with the best available production route, it is difficult to produce nanoparticles with a polydispersity of $<5\%$.[211] Chemical methods (surfactant addition, pH adjustment, and surface modification) and physical methods (ultrasonic agitation, homogenization, and ball milling) have been applied to sustain the long-term stability of nanofluids.

1.4.1 Chemical Treatment Methods

From an application point of view, the stability of nanofluids is a major concern. The stability of nanofluids depends on the nature of the base fluid and nanoparticles. By the addition of surfactants or modification of the particle surfaces, the stability of nanofluids can be improved.

1.4.1.1 Surfactant Addition

Nanofluids are thermodynamically unstable systems. However, they can be made kinetically stable with the aid of suitable surface-active species. Surfactants are widely used in nanofluids research to disperse the nanoparticles evenly in the base fluid. The type of surfactant can be selected based on the choice of the base fluid. Surfactants can be anionic (negatively charged), cationic (positively charged), non-ionic (neutral), or amphoteric (both negatively and positively charged) based on the charge on their head group. Commonly used surfactants in nanofluids research are sodium dodecyl sulfate (SDS),[212] oleic acid,[213–218] cetyl trimethyl ammonium bromide (CTAB),[219] and polyvinyl pyrrolidone (PVP).[219] Almanassra *et al.*[220] studied the effect of three different types of surfactants; gum arabic (GA), PVP, and SDS on the stability and thermo-physical properties of carbon nanotube (CNT)/water nanofluids. Choi *et al.*[221] studied the effect of various surfactants such as sodium dodecylbenzenesulfonate (SDBS), CTAB, SDS, and Triton X-100 (TX-100) on the suspension stability and solar thermal absorption characteristics of water-based nanofluids containing MWCNTs that can be used as working fluids for volumetric solar thermal receivers. Vekas *et al.*[222] prepared nano-sized particles (typically 3–15 nm particles of magnetite, maghemite or cobalt ferrite) in a magnetic nanofluid stabilized with various chain length surfactants. Das *et al.*[223] studied the effect of different surfactants (SDS, CTAB, oleic acid and acetic acid) on the stability of water-based TiO_2 nanofluids.

1.4.1.2 Surface Functionalization

In this method, the surfaces of the nanoparticles or nanotubes are functionalized before dispersing them into the base fluids. Functionalization introduces electrochemically active sites on the nanoparticle/nanotube surfaces, thus improving their dispersion in the base fluid.[224] Nanoparticles are more stable when the pH of the solution is far from the isoelectric point, at which the particles have surface charge and the zeta potential values are zero.

1.4.2 Physical Treatment Methods

In this method, high energy is applied to nanofluids through ultrasonication, homogenization or ball milling to break the nanoparticle clusters and form a well-dispersed colloidal suspension.

1.4.2.1 Ultrasonication

In this method, the liquid sample is irradiated with ultrasonic (>20 kHz) waves. Sound waves propagate in the liquid media resulting in alternating high-pressure (compression) and low-pressure (rarefaction) cycles. During rarefaction, high-intensity sonic waves create small vacuum bubbles or voids in the liquid, which then collapse violently (cavitation) during compression, creating very high local temperatures. Ultrasonication (bath or probe) is a commonly used method by which to physically disperse nanoparticle clusters. Probe sonication is the most widely employed stabilization method among the physical treatment methods. Sound energy at an ultrasonication level of 20 kHz and above is applied for a predetermined period of time to disperse nanoparticles into a base fluid and to break clusters of nanoparticles.[225] Mahbubul *et al.*[226] studied the influence of ultrasonication time on the stability of a water-based Al_2O_3 nanofluid. Garg *et al.*[227] investigated the effect of ultrasonication time (20 min, 40 min, 60 min, and 80 min) on the viscosity and heat transfer performance of a MWCNT nanofluid. Kwak and Kim *et al.*[228] studied the effect of ultrasonication time for different time periods from 1 h to 30 h on EG-based CuO nanofluids. Lee *et al.*[229] stabilized water-based Al_2O_3 nanofluids by applying ultrasonic vibration (5 h, 20 h and 30 h) at sound frequencies of 30–40 kHz. The influence of varying the sonication time on the stability of CuO–water nanofluids was studied and reported by Nemade *et al.*[230]

1.4.2.2 Ball Milling

This is a process where micron-sized materials are reduced to ultrafine powders. In a ball mill, the size reduction is achieved by impact and attrition, where a hollow cylindrical shell rotates about its axis either horizontally or at a slight angle and is partially filled with balls made of steel or stainless steel.[231] The collision between the tiny rigid balls in the cylinder generates localized high pressure. Farbod *et al.*[232] synthesized engine oil-based CuO nanofluids, wherein oil, nanoparticles and grinding balls were put into a container and milled for 3 h by a planetary mill.

1.5 Techniques for Characterizing Nanofluids

It is necessary and important to understand the physicochemical properties of prepared nanoparticles and nanofluids. For this purpose, there are a wide range of characterization techniques, such as X-ray diffraction (XRD), phase contrast microscopy, and transmission electron microscopy (TEM), which are used to gain information on nanoparticle size, shape, composition and porosity. In this section, the basic characterization techniques are discussed in detail.

1.5.1 XRD

This is one of the most extensively used techniques for the characterization of nanoparticles of any size. Since the wavelengths of X-rays are similar to the size of atoms, they are useful for determining crystal structures, the nature of phases, lattice parameters and crystalline grain size, achieved using XRD.[233] The principles of crystallography are based on Bragg's law. The Scherrer equation is used to determine crystallite size. For spherical particles the equation for crystallite size can be written as:

$$d = \frac{0.89\lambda}{\beta \mathrm{Cos}\theta_{\mathrm{max}}} \tag{1.1}$$

where d is the crystallite size of the particles, 0.89 is a proportionality constant for spherical particles; β is the full width at half maxima, θ is the angle of diffraction and λ is the wavelength of the X-ray source (for CuKα, $\lambda = 1.5416$ Å).

1.5.2 Dynamic Light Scattering (DLS)

DLS, also referred to as quasi-elastic light scattering (QELS), is a non-invasive technique for measuring the hydrodynamic size and size distribution of molecules and particles typically in the submicron range.[234] Typical applications of DLS are the characterization of particles or molecules that have been dispersed in a liquid. The Brownian motion of the particles or molecules in suspension causes laser light to be scattered at different intensities. Analysis of these intensity fluctuations yields the velocity of the Brownian motion and the particle size can be calculated using the Stokes–Einstein equation. In light scattering experiments, a monochromatic light source, usually a laser, passes through a polarizer and onto the sample. The scattered light then goes through a second polarizer where it is collected by a photomultiplier tube (PMT). The instantaneous scattered field can be regarded as the superposition of waves scattered from the individual scattering centers. The scattered electric field at a given scattering wave vector q is expressed as:

$$E(q,t)\alpha \sum_{j=1}^{N} \exp\left(i\vec{q}\cdot\vec{r}_j(t)\right) \tag{1.2}$$

where $\vec{r}(t)$ is the center-of-mass position of the jth scatterer. The scattered field given by eqn (1.2) therefore fluctuates in response to the motions of the scatterers. If the fluctuations are faster than 10^{-6} s, a filter method is used to detect and analyze the timescale of these fluctuations. Optical mixing or beating methods are usually used for processes that occur on timescales slower than about 10^{-6} s. In optical mixing methods, no "filter" is inserted between the scattering medium and the detector (PMT). The scattered light impinges directly on the PMT cathode. In the homodyne (self-beat) method only the scattered light impinges on the photocathode, while in the heterodyne method a local oscillator (a small portion of the unscattered laser beam) is mixed with the scattered light on the cathode surface. Since PMT is a square-law detector, its instantaneous current output is proportional to the square of the incident electric field $I(t) \propto \|E(t)\|^2$. The PMT output is fed to a digital correlator, which calculates the intensity–intensity auto correlation function (also known as homodyne correlation function), which is defined as:

$$\langle I(q,0)I(q,t)\rangle = B\langle |E(q,0)|^2 |E(q,t)|^2\rangle \tag{1.3}$$

where B is the proportionality constant. The normalized intensity auto-correlation function, $g^{(2)}(q,t)$, is defined as:

$$g^{(2)}(q,t) = \frac{\langle I(q,0)I(q,t)\rangle}{\langle I(q)\rangle^2} \tag{1.4}$$

and the normalized electric field auto-correlation function (also known as the heterodyne correlation function) is defined as:

$$g^{(1)}(q,t) = \frac{\langle E(q,0)E(q,t)\rangle}{\langle |E(q)|\rangle^2} \tag{1.5}$$

with the assumption that the scattering volume can be divided into a large number of statistically independent sub-regions and the scattered electric field is a Gaussian random variable, $g^{(2)}(q,t)$, related to $g^{(1)}(q,t)$ by:

$$g^{(2)}(q,t) = 1 + \beta\left|g^{(1)}(q,t)\right|^2 \tag{1.6}$$

The above equation is known as the Siegert relation and the constant β depends on the coherence area set by the optics of the instrument and is obtained by fitting. The heterodyne correlation function, $g^{(1)}(q,t)$, which is related to the position of the particles, can be redefined as the intermediate scattering function (dynamic structural factor), $F(q,t)$, using eqn (1.7):

$$F(q,t) \propto \langle E(q,0)E(q,t)\rangle \tag{1.7}$$

$$F(q,t) = \frac{1}{N}\sum_{I=1}^{N}\sum_{J=1}^{N}\langle \exp\left(iq.\left[\vec{r}_i(0) - \vec{r}_j(t)\right]\right) \tag{1.8}$$

hence,

$$g^{(1)}(q,t) = \frac{F(q,t)}{S(q)} \tag{1.9}$$

where, $S(q) = F(q,0)$. For an non-interacting suspension, $S(q) = 1$, by measuring $g^{(1)}(q,t)$ (either using a homodyne or heterodyne method), one can determine the self-diffusion coefficient, D_0, of the colloidal particles.

1.5.3 TEM

TEM is a technique that uses an electron beam to image a nanoparticle, providing much higher resolution than is possible using light-based imaging techniques such as optical microscopy.[235] TEM also provides information on nanoparticle size, grain size, size distribution, and morphology. TEM exploits three different electron beam–specimen interactions, *i.e.*, un-scattered electrons (transmitted beam), elastically scattered electrons (diffracted beam) and inelastically scattered electrons. The intensity of the transmitted electrons is inversely proportional to the specimen thickness. Areas of the specimen that are thicker will have fewer transmitted electrons and so will appear darker; conversely the thinner areas will appear lighter due to more transmission. The scattered part of the incident beam is transmitted through the remaining portions of the specimen. All incident electrons have the same energy and enter the specimen normal to its surface. All incidents that are scattered by the same atomic spacing will be scattered by the same angle. These scattered electrons can be collimated using magnetic lenses to form a pattern of spots, with each spot corresponding to a specific atomic spacing (a plane). This pattern can then yield information about the orientation, atomic arrangements and phases present in the specimen. Also, incident electrons can interact with the specimen inelastically and lose their energy during the interaction. These electrons are then transmitted through the rest of the specimen.[236]

Typically, a TEM consists of three stages of lensing[235] comprising condenser, objective, and projector lenses. The condenser lenses are responsible for primary beam formation, whilst the objective lenses focus the beam that comes through the sample itself (in scanning TEM (STEM) mode, there are also objective lenses above the sample to make the incident electron beam convergent). The projector lenses are used to expand the beam onto a phosphor screen or other imaging device, such as a film. The magnification of the TEM is due to the ratio of the distances between the specimen and the image plane of the objective lens. Additional quad or hexapole lenses allow for the correction of asymmetrical beam distortions, known as astigmatism. It should be noted that TEM optical configurations differ significantly with implementation, with manufacturers using custom lens configurations, such as in spherical aberration corrected instruments or TEMs that utilize energy filtering to correct electron chromatic aberration.

1.5.4 Zeta Potential Measurements

Zeta potential measurements provide information about the surface charge of nanoparticles dispersed in solution. Hence, zeta potentials play a significant role in stabilizing particle suspensions (short- and long-term stability). When a charged particle is introduced into a polar medium, the depletion of co-ions (ions with the same charge as the particle surface) from the surface and the adsorption of counter ions (ions with the opposite charge to the particle surface) occurs.[58] The development of a net charge at the particle surface through ionization or by dissociation of surface groups affects the distribution of ions in the surrounding interfacial region, resulting in an increased concentration of counter ions (ions of opposite charge to that of the particles) close to the surface. Whatever the charging mechanism, the final surface charges of the co-ions are balanced by an equal but oppositely-charged region of counterions. Some of the counterions are bound, usually transiently to the surface within the so called *Stern layer* or *Helmholtz layer*, while other ions form an atmosphere of ions in rapid thermal motion close to the surface, known as the diffuse electrical double layer, as shown in Figure 1.8a. The difference between the bound and free ions in the diffuse layer is analogous to the difference between a water molecule in a sea and in the atmosphere.[237] Within the diffuse layer there is a notional boundary inside which the ions and particles form a stable entity. When a particle moves (*e.g.*, due to gravity), ions within the boundary move with it, but any ions beyond the boundary do not travel with the particle. This boundary is called the surface of hydrodynamic shear or slipping plane. The potential that exists at this boundary is known as the zeta potential, as shown in Figure 1.8b.

The magnitude of the zeta potential gives an indication of the potential stability of the colloidal system. If all the particles in a suspension have a large negative or positive zeta potential then they will tend to repel each other and there will be no tendency to flocculate. However, if the particles have low zeta potential values then there is no force to prevent the particles coming together and flocculating. The general dividing line between stable and unstable suspensions is +30 mV or −30 mV. Particles with zeta potentials above ±30 mV are normally considered stable[238] (Figure 1.8c). However, if the particles have a large density that is different from the dispersant, they will eventually sediment.[238] The important factors affecting the zeta potential are pH, conductivity and concentration of formulations.

Zeta potential measurements can be conducted using an electrophoresis approach. When an electric field is applied across an electrolyte, charged particles suspended in the electrolyte are attracted towards the electrode of opposite charge, as shown in Figure 1.8d, where the viscous forces acting on the particles tend to oppose this movement. When equilibrium is reached between these two opposing forces, the particles move with a constant velocity. The velocity of the particle is dependent on the strength of the electric field or voltage gradient, the dielectric constant of the medium, the viscosity of the medium and the zeta potential. The velocity of a particle in an electric field is referred to as its electrophoretic mobility, which is related to the zeta potential by the Henry equation:

$$U_{\mathrm{E}} = \frac{2\varepsilon\zeta f(ka)}{3\eta} \tag{1.10}$$

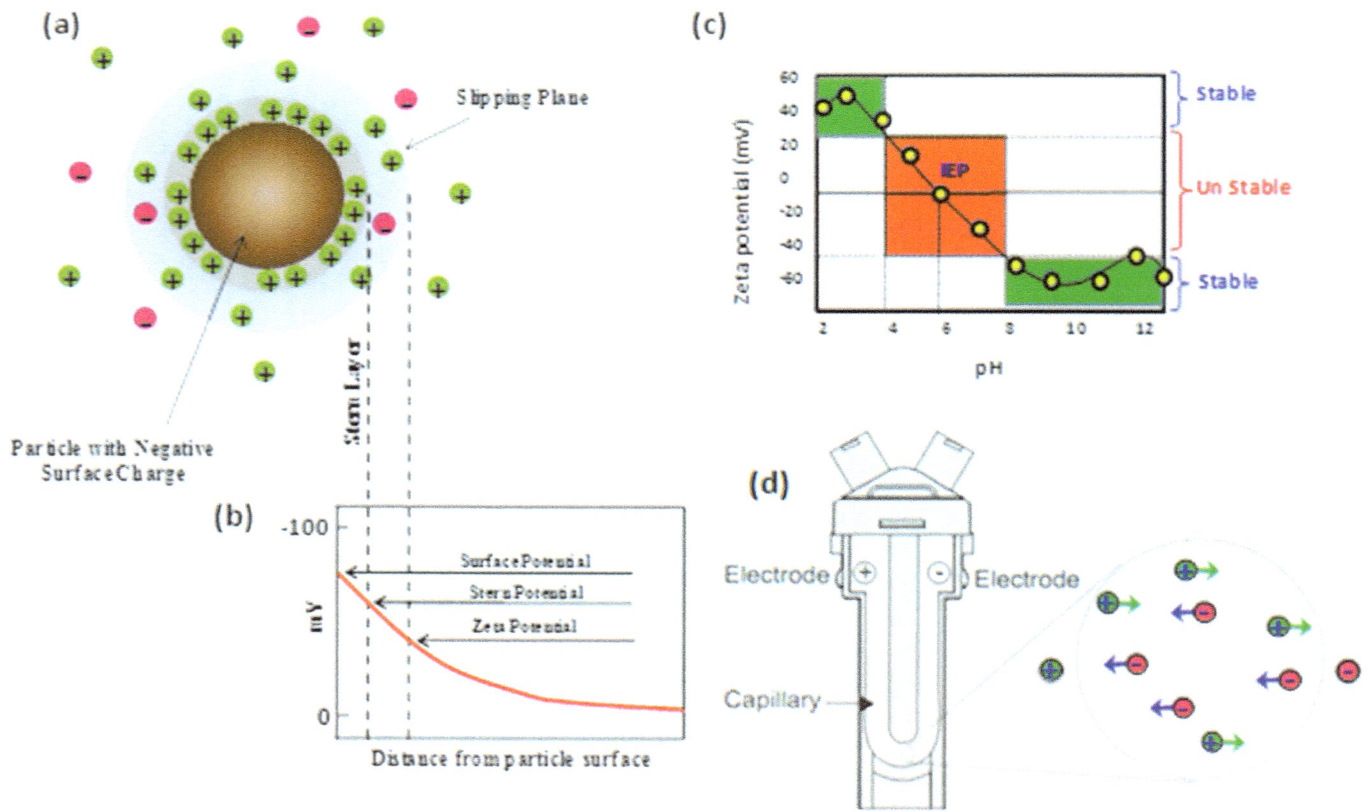

Figure 1.8 Schematic representation of (a) a charged particle with an electric double layer, (b) the distance from the particle surface and the corresponding potentials, (c) a typical plot of the zeta potential *versus* pH showing the position of the isoelectric point and the pH values where the dispersion is stable and unstable, and the (d) electrophoretic mobility in a folded capillary cell.

where, U_E is the electrophoretic mobility, ζ is the zeta potential, ε is the dielectric constant, η is the viscosity and $f(ka)$ is Henry's function ($\approx$1.5 by *Smoluchowski* approximation for a particle larger than 0.2 μm dispersed in electrolyte containing more than 10^{-3} molar salt). For small particles in low dielectric constant media $f(ka)$ becomes 1 *via* the Huckel approximation for non-aqueous measurements. κ is the inverse Debye length, and κ^{-1} is often taken as being a measure of the "thickness" of the electrical double layer.

For zeta potential measurements, the laser light source is split to provide an incident and reference beam. The incident laser beam passes through the center of the sample cell, and the scattered light is detected at an angle of 17°. When an electric field is applied to the cell, any particles moving through the measurement volume will cause the intensity of light detected to fluctuate with a frequency proportional to the particle speed, and this information is passed to a digital signal processor and then to a computer. The Zetasizer Nano software produces a frequency spectrum from which the electrophoretic mobility and hence zeta potential is calculated. The intensity of the detected scattered light must be within a specific range for the detector to successfully measure it, which is achieved using an attenuator. To correct for any differences in the cell wall thickness and dispersant refraction, compensation optics are installed to maintain optimum alignment.[239]

1.5.5 Phase Contrast Optical Microscopy

Phase contrast microscopy is a very useful technique that is used to look at the aggregation of nanoparticles in nanofluids. However, the limitation of this technique is that it can only measure aggregates of fairly larger sizes (>200 nm). In phase contrast optical microscopy, a phase object causes a small phase shift in the light passing through a transparent specimen, which is then converted into amplitude or contrast changes in the image. As light travels through a medium other than a vacuum, interaction with the medium causes amplitude and phase changes that depend on the properties of the medium. The basic principle to make phase changes visible in phase contrast microscopy is to separate the illuminating background light from the specimen scattered light, which makes up the foreground details, and to manipulate these differently. The ring-shaped illuminating light that passes through the condenser annulus is focused on the specimen by the condenser. Some of the illuminating light is scattered by the specimen. The remaining light is unaffected by the specimen and forms the background light. When observing an unstained biological specimen, the scattered light is weak and typically phase-shifted by −90°, relative to the background light. This leads to the foreground and the background having almost the same intensity, resulting in a low image contrast. For a phase contrast microscope, the image contrast can be improved in two steps. The background light is phase-shifted by −90° by passing it through a phase shift ring. This eliminates the phase difference between the background and the scattered light, leading to an increased intensity difference between the foreground and background. To further increase the contrast, the background is dimmed by a gray filter ring. Some of the scattered light will be phase-shifted and dimmed by the rings. However, the background light is affected to a much greater extent, which leads to a phase contrast effect. The above describes negative phase contrast. In its positive form, the background light is instead phase-shifted by $+90^\circ$. The background light will thus be 180° out of phase relative to the scattered light. This results in the scattered light being subtracted from the background light to form an image where the foreground is darker than the background. The performance of modern phase contrast microscopes is so refined that it enables specimens containing very small internal structures, or even just a few protein molecules, to be detected when the technology is coupled to electronic enhancement and post-acquisition image processing instruments.

1.5.6 Fourier-transform Infrared (FT-IR) Spectroscopy

FT-IR spectroscopy is a technique that is used to obtain insight into the nature of adsorption of functional groups at the particle interface. In FT-IR spectroscopy, the preferred IR spectroscopy method, IR radiation is passed through a sample. Some of the IR radiation is absorbed by the sample and some of it is transmitted. The resulting spectrum represents the molecular absorption and transmission, creating a molecular fingerprint of the sample. Like a human fingerprint, no two unique molecular structures produce the same IR spectrum, making IR spectroscopy useful for several types of analysis. The total internal energy of a molecule in a first approximation is the sum of its rotational, vibrational and electronic energy levels. In IR spectroscopy the interactions between matter and electromagnetic (EM) fields in the IR region is studied, where the EM waves mainly couple with the molecular vibrations. In other words, the excitation of molecules to a higher vibrational state occurs upon absorbing IR radiation. In FTIR spectroscopy studies, a small amount of solid sample is mixed thoroughly with potassium bromide in a 1 : 100 ratio (by weight) and the mixture is then compressed into a thin transparent pellet using a hydraulic press. These pellets are transparent to IR radiation and are used for analysis. If the photon energy coincides with the vibrational energy levels of the molecule the frequency is absorbed by the molecule. Hence, IR spectroscopy is a very powerful technique that provides fingerprint information on the chemical composition of the sample.

1.5.7 Thermogravimetric Analysis (TGA) and Differential Scanning Calorimetry (DSC)

TGA is a technique widely used to evaluate the amount of surfactant adsorbed on a particle surface. Specific heat capacity is a materials property that describes the energy required to induce a change in the temperature of a unit mass of the material. DSC is used to measure this quantity. The specific heat capacity, C_p is measured using DSC by heating a sample and measuring the temperature difference between the sample and a reference. The sample material is subjected to a linear temperature program, and the heat flow rate into the sample is continuously measured, which is proportional to the instantaneous specific heat of the sample. Reference and sample crucibles are placed on a sample holder inside a furnace, which generates heat radially toward the center. Thermocouples in contact with each crucible measure the temperature. One thermoelement is shared between the crucibles allowing the temperature difference to be measured as a voltage.

1.6 Conclusion

This chapter gives an overview of nanoparticles, nanofluids and the different techniques used to prepare nanoparticles, which include top-down approaches such as mechanical milling, etching, laser ablation, electro explosion, and bottom-up approaches such as precipitation, solvothermal, flow injection, decomposition of metal carbonyls, reverse micelle, electrodeposition, sol–gel, mechanochemical, hydrothermal, and bacterial synthesis techniques. This is followed by techniques used for the preparation of nanofluids *via* two-step and single-step methods. The direct evaporation technique, chemical reduction method, submerged arc nanoparticle synthesis method, laser ablation, microwave irradiation, polyol process, phase-transfer method, wet mechanochemical technique, sol–gel method (hydrolysis), emulsion-polymerization, and green synthesis are presented in detail. The stability issues of nanofluids are discussed, along with approaches used to improve the stability of nanofluids. Methods to enhance the stability of nanofluids using chemical treatment methods (surfactant addition and surface functionalization) and physical treatment methods (ultrasonication and ball milling) are discussed in detail. Various experimental techniques used for the characterization of nanomaterials and nanofluids are also detailed. Details of XRD, DLS, TEM, zeta potential measurements, phase contrast microscopy, FT-IR spectroscopy, TGA and DSC are presented for freshers who intend to begin research on this topic.

Acknowledgements

J. P. thanks Dr R. Divakar for his support.

References

1. G. Schmid, *Nanoparticles: From Theory to Application*, Wiely-VCH Verlag GmbH & Co., Weinheim, 2004.
2. B. Bhushan, *Springer Handbook of Nanotechnology*, Spinger, New York, 2004.

3. J. Philip, P. D. Shima and B. Raj, *Nanotechnology*, 2008, **19**, 305706.
4. N. R. Karthikeyan, J. Philip and B. Raj, *Mater. Chem. Phys.*, 2008, **109**, 50.
5. P. E. Gharagozloo, J. K. Eaton and K. E. Goodson, *Appl. Phys. Lett.*, 2008, **93**, 103110.
6. J. Garg, B. R. Poudel, M. Chiesa, J. B. Gordon, J. J. Ma, J. B. Wang, Z. F. Ren, Y. T. Kang, H. Ohtani, J. Nanda, G. H. McKinley and G. Chen, *J. Appl. Phys.*, 2008, **103**, 074301.
7. B. Wright, D. Thomas, H. Hong, L. Groven, J. Puszynski, E. Duke, X. Ye and S. Jin, *Appl. Phys. Lett.*, 2007, **91**, 173116.
8. E. V. Timofeeva, A. N. Gavrilov, J. M. McCloskey, Y. V. Tolmachev, S. Sprunt, L. M. Lopatina and J. V. Selinger, *Phys. Rev. E*, 2007, **76**, 061203.
9. J. Philip, P. D. Shima and B. Raj, *Appl. Phys. Lett.*, 2007, **91**, 203108.
10. C. H. Li and G. P. Peterson, *J. Appl. Phys.*, 2007, **101**, 044312.
11. S. P. Jang, J. H. Lee, K. S. Hwang and S. U. S. Choi, *Appl. Phys. Lett.*, 2007, **91**, 243112.
12. J. Eapen, W. C. Williams, J. Buongiorno, L. W. Hu, S. Yip, R. Rusconi and R. Piazza, *Phys. Rev. Lett.*, 2007, **99**, 095901.
13. X. Zhang, H. Gu and M. Fujii, *J. Appl. Phys.*, 2006, **100**, 044325.
14. B. Yang and Z. H. Han, *Appl. Phys. Lett.*, 2006, **89**, 083111.
15. R. Rusconi, E. Rodari and R. Piazza, *Appl. Phys. Lett.*, 2006, **89**, 261916.
16. S. M. S. Murshed, K. C. Leong and C. Yang, *J. Phys. D: Appl. Phys.*, 2006, **39**, 5316.
17. S. J. Kim, I. C. Bang, J. Buongiorno and L. W. Hu, *Appl. Phys. Lett.*, 2006, **89**, 153107.
18. K. S. Hong, T. K. Hong and H. S. Yang, *Appl. Phys. Lett.*, 2006, **88**, 031901.
19. W. Evans, J. Fish and P. Keblinski, *Appl. Phys. Lett.*, 2006, **88**, 093116.
20. J. Buongiorno, *ASME J. Heat Transfer*, 2006, **128**, 240.
21. P. B. Abdallah, *Appl. Phys. Lett.*, 2006, **89**, 113117.
22. R. Prasher, P. Bhattacharya and P. E. Phelan, *Phys. Rev. Lett.*, 2005, **94**, 025901.
23. S. M. S. Murshed, K. C. Leong and C. Yang, *Int. J. Therm. Sci.*, 2005, **44**, 367.
24. T. K. Hong, H. S. Yang and C. J. Choi, *J. Appl. Phys.*, 2005, **97**, 064311.
25. D. W. Zhou, *Int. J. Heat Mass Transfer*, 2004, **47**, 3109.
26. D. Wen and Y. Ding, *Int. J. Heat Mass Transfer*, 2004, **47**, 5181.
27. S. P. Jang and S. U. S. Choi, *Appl. Phys. Lett.*, 2004, **84**, 4316.
28. P. Bhattacharya, S. K. Saha, A. Yadav, P. E. Phelana and R. S. Prasher, *J. Appl. Phys.*, 2004, **95**, 6492.
29. Q. Z. Xue, *Phys. Lett. A*, 2003, **307**, 313.
30. H. E. Patel, S. K. Das, T. Sundararajan, A. S. Nair, B. George and T. Pradeep, *Appl. Phys. Lett.*, 2003, **83**, 2931.
31. S. K. Das, N. Putra, P. Thiesen and W. Roetzel, *ASME J. Heat Transfer*, 2003, **125**, 567.

32. J. A. Eastman, S. U. S. Choi, S. Li, W. Yu and L. J. Thompson, *Appl. Phys. Lett.*, 2001, **78**, 718.
33. Y. Xuan and W. Roetzel, *Int. J. Heat Mass Transfer*, 2000, **43**, 3701.
34. H. Younes, G. Christensen, X. Luan, H. Hong and P. Smith, *J. Appl. Phys.*, 2012, **111**, 064308.
35. S. Lee, S. U. S. Choi, S. Li and J. A. Eastman, *ASME J. Heat Transfer*, 1999, **121**, 280.
36. S. U. S. Choi, Z. G. Zhang, W. Yu, F. E. Lockwood and E. A. Grulke, *Appl. Phys. Lett.*, 2001, **79**, 2252.
37. H. Xie, H. Lee, W. Youn and M. Choi, *J. Appl. Phys.*, 2003, **94**, 4967.
38. P. D. Shima, J. Philip and B. Raj, *Appl. Phys. Lett.*, 2010, **97**, 153113.
39. P. Keblinski, S. R. Phillpot, S. U. S. Choi and J. A. Eastman, *Int. J. Heat Mass Transfer*, 2002, **45**, 855.
40. J. Philip and P. D. Shima, *Adv. Colloid Interface Sci.*, 2012, **183–184**, 30.
41. R. Taylor, S. Coulombe, T. Otanicar, P. Phelan, A. Gunawan, W. Lv, G. Rosengarten, R. Prasher and H. Tyagi, *J. Appl. Phys.*, 2013, **113**, 011301.
42. R. E. Rosenzweig, *Ferrohydrodynamics*, Cambridge University Press, Cambridge, 1985.
43. G. Puliti, S. Paolucci and M. Sen, *Appl. Mech. Rev.*, 2011, **64**, 030803.
44. J. Philip, P. D. Shima and B. Raj, *Appl. Phys. Lett.*, 2008, **92**, 043108.
45. M. J. Kao, C. H. Lo, T. T. Tsung, Y. Y. Wu, C. S. Jwo and H. M. Lin, *J. Alloys Compd.*, 2007, **134–435**, 672.
46. W. Yu and H. Xie, *J. Nanomater.*, 2012, **2012**, 1.
47. B. Dudda and D. Shin, *Int. J. Therm. Sci.*, 2013, **69**, 37.
48. O. Mahian, A. Kianifar, S. A. Kalogirou, I. Pop and S. Wongwises, *Int. J. Heat Mass Transfer*, 2013, **57**, 582.
49. D. Yang, F. Yang, J. Hu, J. Long, C. Wang, D. Fu and Q. Ni, *Chem. Commun.*, 2009, 4447.
50. V. Mahendran and J. Philip, *NDT&E Int.*, 2013, **60**, 100.
51. R. A. Taylor, T. Otanicar and G. Rosengarten, *Light Sci. Appl.*, 2012, **1**, 1.
52. R. Hernández, in *POLYSOLVAT-9, 9th International IUPAC Conference on Polymer-Solvent Complexes & Intercalates*, Ukraine, 2012.
53. K. Raj and R. Moskowitz, *J. Magn. Magn. Mater.*, 1990, **85**, 233.
54. P. D. Shima, J. Philip and B. Raj, *J. Phys. Chem. C*, 2010, **114**, 18825.
55. S. Vinod and J. Philip, *J. Mol. Liq.*, 2018, **257**, 1.
56. V. Mahendran and J. Philip, *Appl. Phys. Lett.*, 2012, **100**, 073104.
57. B. B. Lahiri, S. Ranoo, A. W. Zaibudeen and J. Philip, *J. Magn. Magn. Mater.*, 2017, **441**, 310.
58. V. Mahendran and J. Philip, *Sens. Actuators, B*, 2013, **185**, 488.
59. M. Nandy, B. B. Lahiri and J. Philip, *Sens. Actuators, A*, 2020, **314**, 112220.
60. A. W. Zaibudeen and J. Philip, *Sens. Actuators, B*, 2018, **255**, 720.
61. B. B. Lahiri, S. Ranoo and J. Philip, *Infrared Phys. Thecnol.*, 2017, **80**, 71.
62. S. Ranoo, B. B. Lahiri and J. Philip, *J. Magn. Magn. Mater.*, 2020, **498**, 166138.

63. C. Anushree and J. Philip, *Colloids Surf., A*, 2019, **567**, 193.
64. T. Muthukumaran and J. Philip, *Colloids Surf., A*, 2021, **610**, 125755.
65. F. Bozon-Verduraz, F. Fiévet, J.-Y. Piquemal, R. Brayner, K. E. Kabouss, Y. Soumare, G. Viau and G. Shafeev, *Braz. J. Phys.*, 2009, **39**, 134.
66. J. S. Benjamin, *Metall. Trans. B*, 1970, **1**, 2943.
67. J. F. d. Carvalho, S. N. d. Medeiros, M. A. Morales, A. L. Dantas and A. S. Carriço, *Appl. Surf. Sci.*, 2013, **275**, 84.
68. N. Salah, S. S. Habib, Z. H. Khan, A. Memic, A. Azam, E. Alarfaj, N. Zahed and S. Al-Hamedi, *Int. J. Nanomed.*, 2011, **6**, 863.
69. A. E. D. Mahmoud, A. Stolle and M. Stelter, *ACS Sustain. Chem. Eng.*, 2018, **6**, 6358.
70. Q. Li, M. Rellán-Piñeiro, N. Almora-Barrios, M. Garcia-Ratés, I. N. Remediakis and N. López, *Nanoscale*, 2017, **9**, 13089.
71. W. Cheng, W. Zhang, L. Hu, W. Ding, F. Wu and J. Li, *RSC Adv.*, 2016, **6**, 15900.
72. J. S. Duque, M. B. Madrigal, H. Riascos and Y. P. Avila, *Colloids Interfaces*, 2019, **3**, 25.
73. V. Amendola and M. Meneghetti, *Phys. Chem. Chem. Phys.*, 2009, **11**, 3805.
74. R. A. Ismail, G. M. Sulaiman, S. A. Abdulrahman and T. R. Marzoog, *Mater. Sci. Eng., C*, 2015, **53**, 286.
75. P. Sen, J. Ghosh, A. Abdullah and K. Prashant, *Chem. Sci.*, 2003, **115**, 499.
76. J. Berasategi, A. Gomez, M. M. Bou-Ali, J. Gutiérrez, J. M. Barandiarán, I. V. Beketov, A. P. Safronov and G. V. Kurlyandskaya, *Struct. Smart Mater. Struct.*, 2018, **27**, 045011.
77. J. W. Mullin and M. M. Osman, *Cryst. Res. Technol.*, 1973, **8**, 471.
78. A.-H. Lu, E. L. Salabas and F. Schuth, *Angew. Chem., Int. Ed.*, 2007, **46**, 1222.
79. M. A. Willard, L. K. Kurihara, E. E. Carpenter, S. Calvin and V. G. Harris, *Encyclopedia of Nanoscience and Nanotechnology*, American Scientific Publishers, Valencia, CA, 2004, vol. 1.
80. S. Sun and H. Zeng, *J. Am. Chem. Soc.*, 2002, **124**, 8204.
81. S. Sun, H. Zeng, D. B. Robinson, S. Raoux, P. M. Rice, S. X. Wang and G. Li, *J. Am. Chem. Soc.*, 2004, **126**, 273.
82. D. Chen and R. Xu, *Mat. Res. Bull.*, 1998, **33**, 1015.
83. Z. Li, Q. Sun and M. Gao, *Angew. Chem., Int. Ed.*, 2005, **44**, 123.
84. F. Q. Hu, L. Wei, Z. Zhou, Y. L. Ran, Z. Li and M. Y. Gao, *Adv. Mater.*, 2006, **18**, 2553.
85. G. S. Alvarez, M. Muhammed and A. A. Zagorodni, *Chem. Eng. Sci.*, 2006, **61**, 4625.
86. R. S. Chaurasiy and H. U. Hebbar, in *Nanoscience in Food and Agriculture 4*, Springer, Cham, 2017, p. 181.
87. S. Rana, J. Philip and B. Raj, *Mater. Chem. Phys.*, 2010, **124**, 264.
88. M. Ueda, H. Dietz, A. Anders, H. Kneppe, A. Meixner and W. Plieth, *Electrochim. Acta*, 2002, **48**, 377.

89. S. Banerjee, S. Roy, J. W. Chen and D. Chakravorty, *J. Magn. Magn. Mater.*, 2000, **219**, 45.
90. C. Burda, X. Chen, R. Narayanan and M. A. El-Sayed, *Chem. Rev.*, 2005, **105**, 1025.
91. B. L. Cushing, V. L. Kolesnichenko and C. J. O'Connor, *Chem. Rev.*, 2004, **104**, 3893.
92. J. W. Long, M. S. Logan, C. P. Rhodes, E. E. Carpenter, R. M. Stroud and D. R. Rolison, *J. Am. Chem. Soc.*, 2004, **126**, 16879.
93. J. Jin, S. Ohkoshi and K. Hashimoto, *Adv. Mater.*, 2004, **16**, 48.
94. T. Tsuzki and P. G. Mccormick, *J. Mater. Sci.*, 2004, **9**, 5143.
95. J. Ding, T. Tsuzuki and P. G. M. Cormick, *J. Magn. Magn. Mater.*, 1998, **177–181**, 931.
96. P. Druska, U. Steinike and V. Šepelák, *J. Solid State Chem.*, 1999, **146**, 13.
97. H. Yang, X. Zhang, W. Ao and G. Qiu, *Mater. Res. Bull.*, 2004, **39**, 833.
98. H. Yang, X. Zhang, C. Huang, W. Yang and G. Qiu, *J. Phys. Chem. Solids*, 2004, **65**, 1329.
99. J. Wang, C. Zeng, Z. Peng and Q. Chen, *Phys. B*, 2004, **349**, 124.
100. D. E. Zhang, X. J. Zhang, X. M. Ni, H. G. Zheng and D. D. Yang, *J. Magn. Magn. Mater.*, 2005, **292**, 79.
101. A. J. Barker, B. Cage, S. Russek, R. Garg, R. Shandas and C. R. Stoldt, American Society of Mechanical Engineers, 2005, p. 581.
102. S. Chaianansutcharit, O. Mekasuwandumrong and P. Praserthdam, *Cryst. Growth Des.*, 2006, **6**, 40.
103. R. Blackmore, *Science*, 1975, **190**, 377.
104. R. B. Frankel, R. P. Blakemore and R. S. Wolfe, *Science*, 1979, **203**, 1355.
105. D. L. V. Hyning, W. G. Klemperer and C. F. Zukoski, *Langmuir*, 2001, **17**, 3128.
106. Y. Chen, X. Luo, G.-H. Yue, X. Luo and D.-L. Peng, *Mat. Chem. Phys.*, 2009, **113**, 412.
107. S. S. Pati, V. Mahendran and J. Philip, *J. Nanofluids*, 2013, **2**, 94.
108. G. Gnanaprakash, S. Mahadevan, T. Jayakumar, P. K. Sundaram, J. Philip and B. Raj, *Mater. Chem. Phys.*, 2007, **103**, 168.
109. R. Prasher, P. E. Phelan and P. Bhattacharya, *Nano Lett.*, 2006, **6**, 1529.
110. H. Bonnemann and R. M. Richards, *Eur. J. Inorg. Chem.*, 2001, 2455.
111. Y. Xuan and Q. Li, *Int. J. Heat Fluid Flow*, 2000, **21**, 58.
112. K. Kwak and C. Kim, *Korean-Aust. Rheol. J.*, 2005, **17**, 35.
113. P. Keblinski, J. A. Eastman and D. G. Cahill, *Mater. Today*, 2005, **8**, 36.
114. M. S. Liu, M. C. C. Lin, I. T. Huang and C. C. Wang, *Int. Commun. Heat Mass Transfer*, 2005, **32**, 1202.
115. X. B. Wang, Z. M. Liu, P. A. Hu, Y. Q. Liu, B. X. Han and D. B. Zhu, *Appl. Phys. A: Mater. Sci. Process.*, 2005, **80**, 637.
116. D. Bom, R. Andrews, D. Jacques, J. Anthony, B. Chen, M. S. Meie and J. P. Selegue, *Nano Lett.*, 2002, **2**, 615.
117. S. Suresh, K. P. Venkitaraj, P. Selvakumar and M. Chandrasekar, *Colloids Surf., A*, 2011, **388**, 41.

118. M. M. Ghosh, S. Ghosh and S. K. Pabi, *Nanosci. Nanotechnol. Lett.*, 2012, **4**, 1.
119. Y. Hwang, J. K. Lee, J. K. Lee, Y. M. Jeong, S. Cheong, Y. C. Ahn and S. H. Kim, *Powder Technol.*, 2008, **186**, 145.
120. M. Moosavi, E. K. Goharshadi and A. Youssefi, *Int. J. Heat Fluid Flow*, 2010, **31**, 599.
121. C. Choi, H. S. Yoo and J. M. Oh, *Curr. Appl. Phys.*, 2007, **8**, 710.
122. D. Wen and Y. Ding, *IEEE Trans. Nanotech.*, 2006, **5**, 220.
123. J. M. Salehi, M. M. Heyhat and A. Rajabpour, *Appl. Phys. Lett.*, 2013, **102**, 231907.
124. Y. Li, J. e. Zhou, S. Tung, E. Schneider and S. Xi, *Powder Technol.*, 2009, **196**, 89.
125. Y. Chen, D. L. Peng, D. Lin and X. Luo, *Nanotechnology*, 2007, **18**, 505703.
126. J. Zhu, D. Li, H. Chen, X. Yang, L. Lu and X. Wang, *Mater. Lett.*, 2004, **58**, 3324.
127. S. U. Sandhya and S. A. Nityananda, *Nanomater. Nanotechnol.*, 2013, **3**, 1.
128. S. A. Kumar, K. S. Meenakshi, B. R. V. Narashimhan, S. Srikanth and G. Arthanareeswaranc, *Mater. Chem. Phys.*, 2009, **113**, 57.
129. K. Patel, S. Kapoor, D. P. Dave and T. Mukherjee, *J. Chem. Sci.*, 2005, **117**, 53.
130. H. Akoh, Y. Tsukasaki, S. Yatsuya and A. Tasaki, *J. Cryst. Growth*, 1978, **45**, 495.
131. V. Sridhara and L. N. Satapathy, *Nanoscale Res. Lett.*, 2011, **6**, 1.
132. S. U. S. Choi and J. A. Eastman, *US Pat.*, 6221275B1, 2001.
133. A. G. Kanaras, F. S. Kamounah, K. Schaumburg, C. J. Kiely and M. Brust, *Chem. Commun.*, 2002, **20**, 2294.
134. M. S. Liu, M. C. C. Lin, C. Y. Tsai and C. C. Wang, *Int. J. Heat Mass Transfer*, 2006, **49**, 3028.
135. U. S. Shenoy and A. N. Shetty, *Appl. Nanosci.*, 2014, **4**, 47.
136. M.-S. Liu, M. C.-C. Lin, C. Y. Tsai and C.-C. Wang, *Int. J. Heat Mass Transfer*, 2006, **49**, 3028.
137. F. Xun, Y. Wei, L. Yusheng, W. Debao, S. Huaqiang and Y. Fengyuan, *J. Dispers. Sci. Technol.*, 2005, **26**, 575.
138. Z. Said, M. A. Abdelkareem, H. Rezk, M. A. Nassef and H. Z. Atwany, *Powder Technol.*, 2020, **364**, 795.
139. D. Cabaleiro, P. Estellé, H. Navas, A. Desforges and B. Vigolo, *J. Nanofluids*, 2018, 7, 1081.
140. C. C. D. Santosa, W. R. Viali, E. D. S. N. Vialib Jr and M. Jafelicci Jr., *J. Mol. Liq.*, 2020, **113**, 113391.
141. C. H. Lo, T. T. Tsung and L. C. Chen, *J. Cryst. Growth*, 2005, **277**, 636.
142. C. H. Lo, T. T. Tsung and H. M. Lin, *J. Alloys Compd.*, 2007, **434–435**, 659.
143. H. J. Kim, I. C. Bang and J. Onoe, *Opt. Lasers Eng.*, 2009, **47**, 532.

144. S. W. Lee, S. D. Park and I. C. Bang, *Int. J. Heat Mass Transfer*, 2012, **55**, 6908.
145. T. X. Phuoc, Y. Soong and M. K. Chyu, *Opt. Lasers in Eng.*, 2007, **45**, 1099.
146. A. R. Sadrolhosseini, S. A. Rashid, A. Zakaria and K. Shameli, *J. Nanomater.*, 2016, **2016**, 1.
147. P. X. Tran and S. Yee, In *ASME Applied Mechanics and Materails Conference*, 2007.
148. V. Piriyawong, V. Thongpool, P. Asanithi and P. Limsuwan, *J. Nanomater.*, 2012, **2012**, 1.
149. M. C. Mbambo, S. Khamlich, T. Khamliche, M. K. Moodley, K. Kaviyarasu, I. G. Madiba, M. J. Madito, M. Khenfouch, J. Kennedy, M. Henini, E. Manikandan and M. Maaza, *Sci. Rep.*, 2020, **10**, 1.
150. P. V. Kazakevich, A. V. Simakin, V. V. Voronov and G. A. Shafeev, *Appl. Surf. Sci.*, 2006, **252**, 4373.
151. A. Riahi, S. Khamlich, M. Balghouthi, T. Khamliche, T. B. Doyle, W. Dimassi, A. Guizani and M. Maaza, *J. Mol. Liq.*, 2020, **304**, 112694.
152. W. K. Woo, Y. M. Hung and X. Wang, *Case Stud. Therm. Eng.*, 2021, **25**, 100993.
153. M. Rafique, M. S. Rafique, U. Kalsoom, A. Afzal, S. H. Butt and A. Usman, *Opt. Quantum Electron.*, 2019, **51**, 1.
154. J. Taha-Tijerina, S. Shaji, S. S. Kanakkillam, M. I. M. Palma and K. Aviña, *Appl. Sci.*, 2020, **10**, 1779.
155. H. Wang, J. Z. Xu, J. J. Zhu and H. Y. Chen, *J. Cryst. Growth*, 2002, **244**, 88.
156. H. T. Zhu, Y. S. Lin and Y. S. Yin, *J. Colloid Interface Sci.*, 2004, **277**, 100.
157. H. Bönnemann, S. S. Botha, B. Bladergroen and V. M. Linkov, *Appl. Organomet. Chem.*, 2005, **19**, 768.
158. A. K. Singh and V. S. Raykar, *Colloid Polym. Sci.*, 2008, **286**, 1667.
159. F. Seifikar and S. Azizian, *Energy Convers. Manage.*, 2021, **228**, 113675.
160. A. V. Nikam and A. H. Dadwal, *Adv. Powder Technol.*, 2019, **30**, 13.
161. F. Bonet, K. Tekaia-Elhsissen and K. V. Sarathy, *Bull. Mater. Sci.*, 2000, **23**, 165.
162. Y. Sun, B. Mayers, T. Herricks and Y. Xia, *Nano Lett.*, 2003, **3**, 955.
163. Y. L. Hsin, K. C. Hwang, F. R. Chen and J. J. Kai, *Adv. Mat.*, 2001, **13**, 830.
164. S. Zeroual, P. Estellé, D. Cabaleiro, B. Vigolo, M. Emo, W. Halim and S. Ouaskit, *J. Mol. Liq.*, 2020, **310**, 113229.
165. L. R. d. Oliveira, S. R. F. L. Ribeiro, M. H. M. Reis, V. L. Cardoso and E. P. B. Filho, *Diam. Relat. Mater.*, 2019, **96**, 216.
166. C. Wanga, J. Yanga and Y. Ding, *Prog. Nat. Sci.*, 2013, **23**, 338.
167. J. Yang, E. Sargent, S. Kelley and J. Y. Ying, *Nat. Mat.*, 2010, **8**, 683.
168. A. Kumar, H. M. Joshi, A. B. Mandale, R. Srivastava, S. D. Adyanthaya, R. Pasricha and M. Sastry, *J. Chem. Sci.*, 2004, **116**, 293.
169. X. Wang, S. Xu, J. Zhou and W. Xu, *J. Colloid Interf. Sci.*, 2010, **348**, 24.
170. W. Yu, H. Xie, X. Wang and X. Wang, *Nanoscale Res. Lett.*, 2011, **6**, 1.

171. H. Zhu, C. Zhang, Y. Tang, J. Wang, B. Ren and Y. Yin, *Carbon*, 2007, **45**, 226.
172. T. Muthukumaran, G. Gnanaprakash and J. Philip, *J. Nanofluids*, 2012, **1**, 85.
173. M. M. Can, S. Ozcan, A. Ceylan and T. Firat, *Mater. Sci. Eng., C*, 2010, **172**, 72.
174. M. Inkyo, Y. Tokunaga, T. Tahara, T. Iwaki, F. Iskandar, C. J. Hogan and K. Okuyama, *Ind. Eng. Chem.*, 2008, **47**, 2597.
175. S. Harjanto, H. Sutanto, R. Setiaji, A. H. Yuwono and D. Ferdian, *AIP Conf. Proc.*, 2011, **1415**, 110.
176. M. J. Nine, B. Munkhbayar, M. S. Rahman, H. Chung and H. Jeong, *Mater. Chem. Phys.*, 2013, **141**, 636.
177. L. Almásy, D. Creanga, C. Nadejde, L. Rosta, E. Pomjakushina and M. Ursache-Oprisan, *J. Serbian Chem. Soc.*, 2014, **80**, 367.
178. S. Tanimoto, Y. Nakamura, H. Yamaoka and Y. Hirokawa, *Int. J. Poly. Sci.*, 2010, 294790.
179. W. g. Kim, H. U. Kang, K.-m. Jung and S. H. Kim, *Sep. Sci. Technol.*, 2008, **43**, 3036.
180. S. M. Abdel-Samad, A. A. Fahmy, A. A. Massoud and A. M. Elbedwehy, *Curr. Nanosci.*, 2017, **13**, 586.
181. D. Jing, Y. Hu, M. Liu, J. Wei and L. Guo, *Sol. Energy*, 2015, **112**, 30.
182. V. M. Goud, V. Vaisakh, M. Joseph and V. Sajith, *Appl. Therm. Eng.*, 2019, **168**, 114862.
183. Z. H. Han, F. Y. Cao and B. Yang, *Appl. Phys. Lett.*, 2008, **92**, 243104.
184. G. Kim, A. Sousa, D. Meyers and M. Libera, *Microsc. Microanal.*, 2008, **14**, 459.
185. M. Joseph and V. Sajith, *Appl. Therm. Eng.*, 2019, **147**, 756.
186. E. V. Jitheesh, M. Joseph and V. Sajith, *Int. Commun. Heat Mass Transfer*, 2021, **127**, 105541.
187. B. A. Bhanvase, S. D. Sayankar, A. Kapre, P. Fule and S. H. Sonawane, *Appl. Therm. Eng.*, 2018, **128**, 134.
188. D. Yao, T. Li, Y. Zheng and Z. Zhang, *React. Funct. Polym.*, 2019, **136**, 131.
189. M. Sarafraz and F. Hormozi, *Exp. Therm. Fluid Sci.*, 2015, **66**, 279.
190. R. Kumar, J. N. Sharma and J. Sood, *Mater. Today: Proc.*, 2020, **28**, 1781.
191. M. S. Jameel, A. A. Aziz and M. A. Dheyab, *Nano-Struct. Nano-Objects*, 2020, **23**, 100484.
192. A. Rufus, N. Sreeju, V. Vilas and D. Philip, *J. Mol. Liq.*, 2017, **242**, 537.
193. G. B. Jegadeesan, K. Srimathi, N. S. Srinivas, S. Manishkanna and D. Vignesh, *Biocatal. Agric. Biotechnol.*, 2019, **21**, 101354.
194. R. Ranjbarzadeh, A. Moradikazerouni, R. Bakhtiari, A. Asadi and M. Afrand, *J. Clean. Prod.*, 2019, **206**, 1089.
195. E. C. Okonkwo, E. A. Essien, E. Akhayere, M. Abid, D. Kavaz and T. A. H. Ratlamwala, *Sol. Energy*, 2018, **170**, 658.
196. S. Omiddezyani, S. Gharehkhani, V. Yousefi-Asli, I. Khazaee, M. Ashjaee, R. Nayebi, F. Shemirani and E. Houshfard, *Colloids Surf., A*, 2021, **610**, 125923.

197. B. T. Sone, A. Diallo, X. G. Fuku, A. Gurib-Fakim and M. Maaza, *Arab. J. Chem.*, 2020, **13**, 160.
198. G. Sharmila, C. Muthukumaran, E. Sangeetha, H. Saraswathi, S. Soundarya and N. M. Kumar, *Nano-Struct. Nano-Objects*, 2019, **20**, 100380.
199. J. L. Luna-Sánchez, J. L. Jiménez-Pérez, R. Carbajal-Valdez, G. Lopez-Gamboa, M. Pérez-González and Z. N. Correa-Pacheco, *Thermochim. Acta*, 2019, **678**, 178314.
200. R. Sadri, M. Hosseini, S. N. Kazi, S. Bagheri, S. M. Ahmed, G. Ahmadi, N. Zubir, M. Sayuti and M. Dahari, *Energy Convers. Manage.*, 2017, **150**, 26.
201. R. Sadri, M. Hosseini, S. N. Kazi, S. Bagheri, N. Zubir, K. H. Solangi, T. Zaharinie and A. Badarudin, *J. Colloid Interface Sci.*, 2017, **504**, 115.
202. S. F. S. Shirazi, S. Gharehkhani, H. Yarmand, A. Badarudin, H. S. C. Metselaar and S. N. Kazi, *Mater. Lett.*, 2015, **152**, 192.
203. N. Pauzi, N. M. Nurul and A. A. Yusof, *J. Environ. Chem. Eng.*, 2020, **8**, 103331.
204. M. Hosseini, H. A. Abdelrazek, R. Sadri, A. R. Mallah, S. N. Kazi, B. T. Chew, S. Rozali and N. Yusoff, *Int. J. Heat Mass Transfer*, 2018, **127**, 403.
205. H. R. Kulkarni, C. Dhanasekaran, P. Rathnakumar and S. Sivaganesan, *Mater. Today: Proc.*, 2021, **42**, 1037.
206. Y. Mostafa Yusefi, S. Kamyar, S. Y. Ong, T. Sin-Yeang, H. Ziba, J. Hossein, W. Thomas J and K. Kuca, *Int. J. Nanomedicine*, 2020, **16**, 2515.
207. S. N. M. Zainon and W. H. Azmi, *Int. Commun. Heat Mass Transfer*, 2021, **126**, 105402.
208. M. M. Sarafraz, A. Arya, V. Nikkhah and F. Hormozi, *Chem. Biochem. Eng. Q.*, 2016, **30**, 489.
209. E. K. Goharshadi, S. Samiee and P. Nancarrow, *J. Colloid Interface Sci.*, 2011, **356**, 473.
210. M. K. Alam, K. Shah, E. Doroodchi, R. Azizian and B. Moghtaderi, *Int. J. Heat Mass Transfer*, 2016, **98**, 778.
211. S. A. Angayarkanni and J. Philip, *J. Phys. Chem. C*, 2013, **117**, 9009.
212. X. Zhang, H. Gu and M. Fujii, *Exp. Therm. Fluid Sci.*, 2007, **31**, 593.
213. M. Bloemen, W. Brullot, T. T. Luong, N. Geukens, A. Gils and T. Verbiest, *J. Nanopart. Res.*, 2012, **14**, 1.
214. S. Gyergyek, D. Makovec and M. Drofenik, *J. Colloid Interface Sci.*, 2011, **354**, 498.
215. Q. Lan, C. Liu, F. Yang, S. Liu, J. Xu and D. Sun, *J. Colloid Interface Sci.*, 2007, **310**, 260.
216. X.-M. Liu and L.-S. Wang, *Biomaterials*, 2004, **25**, 1929.
217. C. Y. Wang, J. M. Hong, G. Chen, Y. Zhang and N. Gu, *Chin. Chem. Lett.*, 2010, **21**, 179.
218. K. Yang, H. Peng, Y. Wen and N. Li, *Appl. Surf. Sci.*, 2010, **256**, 3093.
219. J. Wang, G. Li, T. Li and M. Zeng, *J. Therm. Anal. Calorim.*, 2020, **143**, 4057.

220. I. W. Almanassra, A. D. Manasrah, U. A. Al-Mubaiyedh, T. Al-Ansari, Z. O. Malaibari and M. A. Atieh, *J. Mol. Liq.*, 2020, **304**, 111025.
221. T. J. Choi, S. P. Jang and M. A. Kedzierski, *Int. J. Heat Mass Transfer*, 2018, **122**, 483.
222. L. Vekas, D. BICA and O. Marinica, *Rom. Rep. Phys.*, 2006, **58**, 257.
223. P. K. Das, A. K. Mallik, R. Ganguly and A. K. Santra, *J. Mol. Liq.*, 2018, **254**, 98.
224. S. A. Angayarkanni and J. Philip, *J. Appl. Phys.*, 2015, **118**, 094306.
225. B. Ruan and A. M. Jacobi, *Nanoscale Res. Lett.*, 2012, **7**, 127.
226. I. M. Mahbubul, R. Saidur, M. A. Amalina and M. E. Niza, *Int. Commun. Heat Mass Transfer*, 2016, **76**, 33.
227. P. Garg, J. L. Alvarado, C. Marsh, A. T. Carlson, A. D. Kessler and K. Annamalai, *Int. J. Heat Mass Transfer*, 2009, **52**, 5090.
228. K. Kwak and C. Kim, *Korea Aust. Rheol. J.*, 2005, **17**, 35.
229. H. Akoh, Y. Tsukasaki, S. Yatsuya and A. Tasaki, *J. Cryst. Growth*, 1978, **45**, 495.
230. K. Nemade and S. Waghuley, *Appl. Therm. Eng.*, 2016, **95**, 271.
231. E.-O.-L. Ettefaghi, H. Ahmadi, A. Rashidi, S. S. Mohtasebi and M. Alaei, *Int. J. Ind. Chem.*, 2013, **4**, 28.
232. M. Farbod, R. K. Asl and A. R. N. Abadi, *Colloids Surf., A*, 2015, **474**, 71.
233. B. D. Cullity and S. R. Stock, *Elements of X-Ray Diffraction*, Prentice Hall, London, 2001.
234. B. J. Berne and R. Pecora, *Dynamic light scattering: with applications to biology, chemistry and physics*, John Wiley, New York, 1976.
235. D. B. Williams and C. B. Carter, *Transmission electron microscopy: a textbook for materials science*, Springer, 1996.
236. F. Owens and C. P. Poole Jr., *The physics and chemistry of nanosolids*, John Wiley & Sons, New Jersy, 2008.
237. J. N. Israelachvili, *Intermolecular and surface forces*, Academic Press, San Diego, 1985.
238. V. Gupta and P. Trivedi, in *Lipid Nanocarriers for Drug Targeting*, William Andrew Applied Science, 2018, p. 563.
239. R. J. Hunter, *Zeta potential in colloid science: Principles and applications*, Academic Press, London, 1981.

CHAPTER 2

Stability of Nanofluids

M. IQBAL,* A. SERGIS AND Y. HARDALUPAS

Department of Mechanical Engineering, Imperial College London, Exhibition Road, South Kensington, London, SW7 2AZ, UK
*Email: mi4517@ic.ac.uk

2.1 Introduction

Nanofluids are constituted of typical heat transfer fluids (water, ethylene glycol) into which nano-sized particles are dispersed to enhance their heat transfer.[1] Nanoparticles are resistant to gravitational sedimentation due to their enhanced Brownian motion compared to micron-sized particles. However, nanoparticles in suspension are still subject to aggregation over time due to interparticle collisions, which degrade the stability of nanofluids.[2] In order to use nanofluids in industrial processes, long-term stability is required. In previous research, sedimentation and agglomeration have been observed during investigation of a dilute alumina–water nanofluid in a natural convection cell.[3] These issues were resolved through the manipulation of the nanofluid pH, decreasing it to a value far from the isoelectric point (IEP) of the nanopowder. This resulted in a net surface charge on the dispersed phase, enhancing the stability of the nanofluid through interparticle repulsion due to electrostatic forces. For one type of alumina nanoparticles, this was achieved through the addition of acetic acid. For another type of alumina nanoparticles, this reduction in pH was achieved through extensive sonication in an ultrasonic bath alone.[4]

Recent experiments conducted at the Instituto Superior Técnico (IST) in Lisbon (2019) investigated the thermal conductivity and forced convection performance of alumina–water nanofluids. During the preparation of the

Soft Matter Series No. 16
Fundamentals and Transport Properties of Nanofluids
Edited by S. M. Sohel Murshed

Published by the Royal Society of Chemistry, www.rsc.org

nanofluids for these experiments, the optimisation of their stability was conducted by controlling the applied ultrasonic time and power. The experimental setup for this nanofluid preparation is shown in Figure 2.1(b), in addition to the setup used at Imperial College London in Figure 2.1(a) (used for the experiments described in the previous paragraph). The key differences between the two setups are the different base fluids (deionised (DI) water *versus* distilled water) and the ultrasonication devices (low-power bath *versus* high-power probe).

During optimisation of the probe-sonication process, it was observed that excessive sonication (in excess of 5 min, 50% amplitude, volumetric concentration, $\phi = 0.1$ vol%) resulted in severe aggregation and reduction in nanofluid stability (Figure 2.2(b)); Figure 2.2(a) shows a stable nanofluid prepared *via* an ultrasonic bath after three weeks from previous work[3]). Additionally, temperamental changes in nanofluid pH depending on the applied ultrasonication and nanoparticle concentration were also noted. These observations raised the following critical questions:

1. Why, how and when does ultrasonication lead to re-agglomeration/aggregation of nanoparticle colloids? Is there a bound and finite optimum ultrasonication time and power density to create stable nanofluids?

   Ultrasonication initially breaks up aggregates/agglomerates, reducing their hydrodynamic size; however, after a certain specific energy input, the hydrodynamic size begins to increase. In this latter phase, rapid sedimentation (within minutes) is observed with the formation of large visible agglomerates. Additional sonication has no effect or can even worsen the stability. Why is it that continued sonication forms these micron flocs/aggregates but cannot redisperse them?
2. Why/how does ultrasonication change pH?

   Experiments with highly dilute nanofluids ($\phi < 0.01$ vol%) showed that sonication for around 5 h in a low power density ultrasonic bath, yielded a pH reduction from 7 to 5.7. At the same time, high power sonication of denser nanofluids ($\phi > 0.1$ vol%) led to an even greater pH reduction (yielding pH values of as low as 4.88). Interestingly, for sonication times longer than the optimal time (*i.e.*, after which significant nanoparticle agglomeration occurred), the pH value would then gradually restore to its initial value (neutral, 7). Linking this to the increase in specific surface area (SSA), could the formation/collapse (and modification during changing pH) of the electrical double layers (EDL) affect the nanofluid pH? Why is it not possible to affect pH change with dilute nanofluids in high-power applications?

The following sections serve to explore and answer the above questions. Section 2.2 considers the first question and Section 2.3 discusses the second question. Concluding remarks are added at the end of these sections to summarise the findings and propose the best approach for delivering optimum nanofluids. Finally, Section 2.4 provides final remarks.

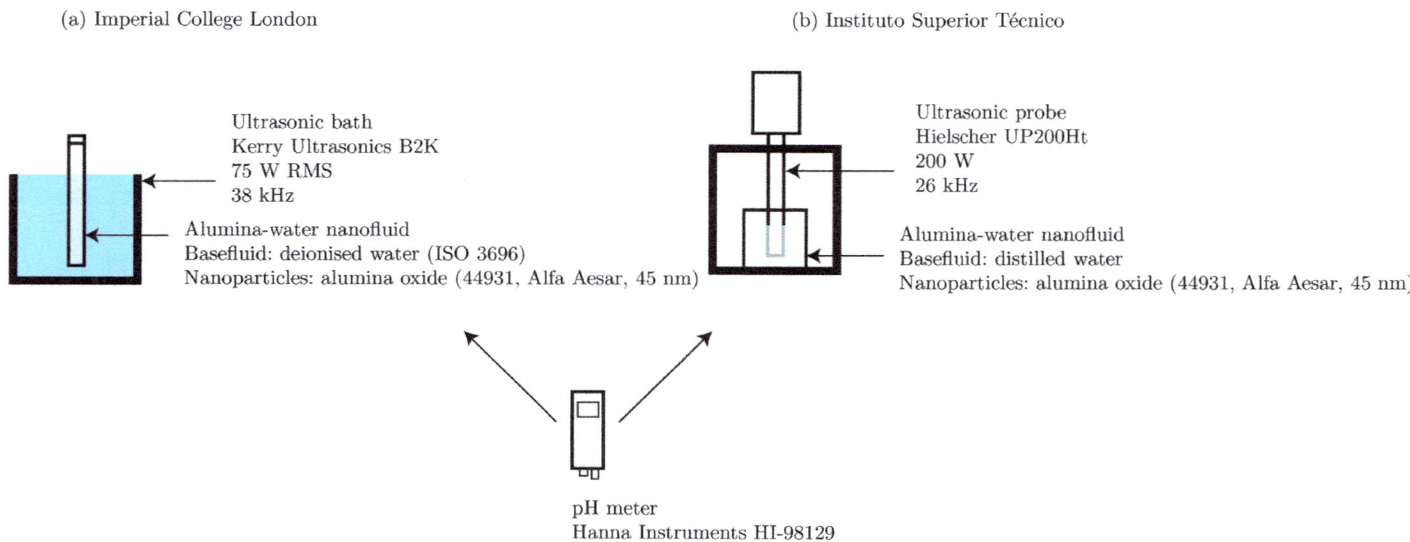

Figure 2.1 Experimental setup of nanofluid preparation at (a) Imperial College London and (b) Instituto Superior Técnico.

(a) (b)

Figure 2.2 (a) Stability of an alumina–water nanofluid prepared *via* ultrasonic bath after 3 weeks ($\phi = 0.06$ vol%; at Imperial College London).[3] (b) An example of an unstable alumina–water nanofluid sample after excessive probe-sonication ($\phi = 0.1$ vol%; at Instituto Superior Técnico). Adapted from ref. 3, DOI: 10.1016/j.ijheatmasstransfer.2016.05.113, under the terms of the CC BY 4.0 license, https://creativecommons.org/licenses/by/4.0/.

2.2 Ultrasonic Dispersion, Particle Breakage and Reaggregation

In this section, the working principles of ultrasonic dispersion are explained. The two different types of particle breakage mechanisms are introduced and considered in the context of ultrasonication, and the phenomenon of reaggregation is investigated and explained. This leads to considerations related to question 1 in the introductory Section 2.1.

2.2.1 Introduction

Ultrasonication refers to the exposure of a medium to ultrasonic waves (frequency, $f > 20$ kHz). In the context of nanofluids, the ultrasonication of the base fluid leads to rapid local pressure oscillations, which, with sufficient intensity/amplitude, can rarefy the medium to form voids (cavitation bubbles). It is the unstable collapse of these bubbles (occurring at high enough amplitudes, known as transient cavitation), which yields the major effects of ultrasonication. For reference, temperatures and pressures in these collapsing bubbles peak at around 5000 °C and 200 MPa.[5]

Ultrasonic devices come in two forms, namely baths (typically used in such applications) and probes. To disperse the colloids, in ultrasonic baths, the "bath" is filled up with water and the nanoparticle and base fluid solution (nanofluid) is placed within its own container (usually glass test tubes/containers). The nanofluid containers are subsequently submerged in the water bath of the device. The ultrasound path starts originally from the ultrasonic transducers that are usually attached to the external of the bath walls and follows a course through the bath wall, the water in the bath itself, the nanofluid container and finally into the nanofluid itself. For ultrasonic probes the ultrasound path is much shorter with fewer propagation media changes; the probe is directly immersed in the nanofluid and hence the ultrasound wave originally starts from the transducer, which is usually attached to the inner of the probe and then through the probe wall and into the nanofluid. Baths typically deliver ~20 W L^{-1} to the contained fluid (*i.e.*, the nanofluid), whereas ultrasonic probes typically deliver power densities that are 1000 times higher (~20 kW L^{-1}).[6] This is because the energy losses due to the increased ultrasound pathway and number of propagation media changes for ultrasonic baths are significant.[5]

Ultrasound frequency is also an important factor, with higher frequency devices requiring more power (amplitude) to induce cavitation, as at higher frequencies, the rarefaction cycles are shorter, thus suppressing cavitation.[5] The ultrasonic bath used during the current research had a frequency of 38 kHz and a power of 75 W (RMS), whilst the ultrasonic probe operated at a frequency of 26 kHz and a power of 200 W (DC). The effects of frequency change between the type of US baths used in the range reported have been stated to be negligible.[7]

Homogeneity of ultrasonication is quite easily achieved in probe applications through the sufficient submersion of the probe and minimisation of the distance between the probe and container walls. For bath devices, inhomogeneity of the ultrasonication is common, with local maxima and minima power regions forming inside the water bath volume. These can be visually identified through the immersion of aluminium foil inside the bath. Regions of high power in the bath appear as perforations on the aluminium foil due to the strong cavitation shear effects induced. The typical experimental setup for this test is shown in Figure 2.3, with the resulting perforation damage shown in Figure 2.4.[5] Ideally, in a bath one would locate the nanofluid in the region of maximal cavitation activity (*i.e.*, where the maximum damage to the aluminium foil occurs), however, this is not always practically possible. This has implications, for example, when sonicating multiple nanofluid samples simultaneously in an ultrasonic bath or when placing a sample at different locations within the bath. In these cases, the power delivery inhomogeneity results in differing rates of absorbed power by each sample, and hence differing rates of dispersion properties are achieved.

A key observation of the research study was the inability to reduce the pH of nanofluids with the use of an ultrasonic probe when they were below a certain concentration ($\phi < 0.1$ vol%), regardless of the duration/intensity of

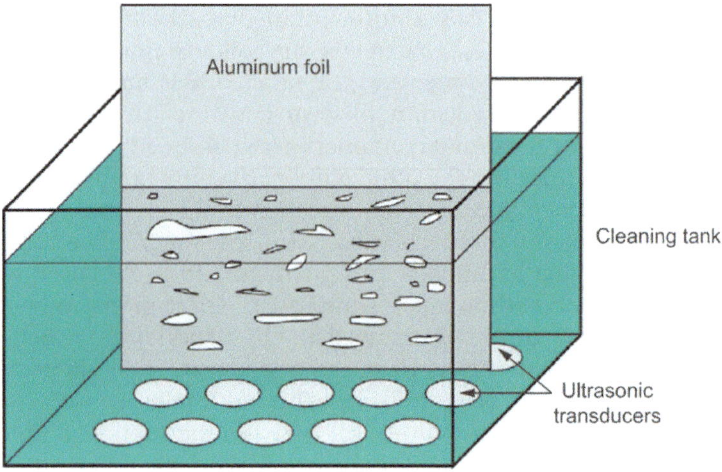

Figure 2.3 Schematic of the aluminium foil test in which a sheet of aluminium foil is placed vertically in an ultrasonic bath, causing perforation. Reproduced from ref. 54 with permission from Elsevier, Copyright 2015.

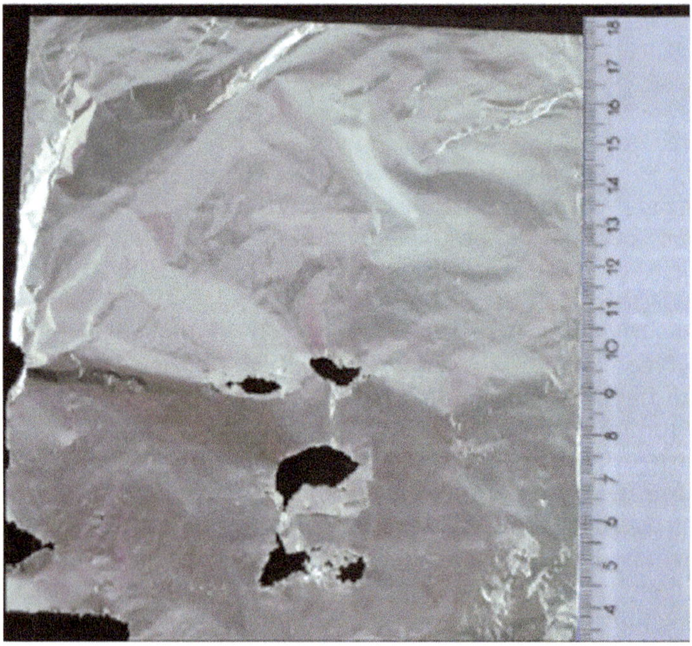

Figure 2.4 Aluminium foil perforation due to local maxima of power in an ultrasonic bath. Reproduced from ref. 5 with permission from John Wiley & Sons – Books, Copyright © 2009 Wiley-VCH Verlag GmbH & Co. KGaA.

ultrasonication (durations of up to 30 min at 100% amplitude). On the contrary, a pH reduction of a significantly more dilute nanofluid ($\phi = 0.00026$ vol%) was repeatedly achievable (pH reduction from 7 to 5.7 after 5 h of ultrasonication) when an ultrasonic bath was used. Opposed to the high quality DI water used in the ultrasonic bath,[4] distilled water was used for the ultrasonic probe experiments. This introduced ionic contaminants that may have affected the pH (discussed further in Section 2.3), but nevertheless, the major difference between the two experiments was the delivered power (~1000 times higher with the probe). It could hence be that lower delivered power (*i.e.*, the employment of a bath opposed to a probe) could be favourable in the formulation of stable nanofluids. As such, the analysis of the ultrasonic cavitation characteristics might provide an explanation for the dissimilar stability of the produced nanofluids.

To explore the above hypothesis, one must consider the interaction between ultrasonic cavitation bubbles and nanoparticles in their differing forms, *i.e.*, singlets, doublets, triplets, hard aggregates, soft agglomerates, *etc.*

2.2.2 Ultrasonic-induced Cavitation

During ultrasonication, induced local pressure fluctuations cause the formation of cavitation bubbles, which collapse. The induced shear causes mixing of the nanoparticles with the base fluid. When the cavitating bubbles in a sonicated fluid collapse near solid surfaces, they do so asymmetrically, resulting in the formation of a fast (~100 m s^{-1}) liquid jet,[8] which impacts the solid boundary (as shown in Figure 2.5), *i.e.*, Rayleigh–Taylor instability.[9] These jets can be expected to directly impact the primary (nano) particles in the suspension, shearing them apart. However, in order for these microjets

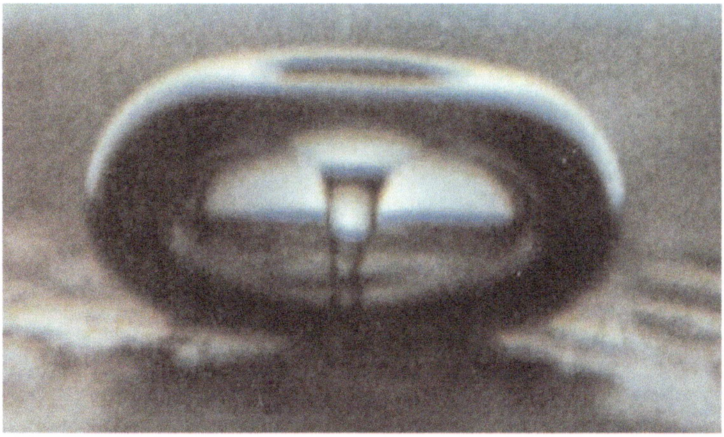

Figure 2.5 Picture of collapsing cavitation bubble at a solid surface, showing the formation of a liquid jet in the centre for a bubble size of ~1 mm. Reproduced from ref. 55 with permission from The Author, Copyright 1982.

to form, the solid boundary must be larger than the resonance bubble size[8] (in order to provide sufficient disturbance to the ultrasonic field[10]). This resonance (or critical) bubble size represents the size at which the transfer of ultrasound energy to the cavitation bubble is most efficient, enabling significant growth within one pressure cycle, after which the bubble is too big to support itself, and collapses violently (the aforementioned transient cavitation).[11] This critical bubble size is dependent on frequency and is given as ≈200 μm for $f = 20$ kHz.[8] Agglomerates smaller than this (and the primary nanoparticles in the nanofluid case) are thus only affected by the shockwaves emitted from collapsing cavitation bubbles.[10] Increasing ultrasonic power has been shown to reduce the size of the cavitation bubbles (~10 μm for power, $P > 150$ W),[12] hence, high-power sonication (probe applications) will form smaller cavitation bubbles, and thus microjet formation should occur and impact on smaller agglomerates, enhancing the effectiveness of the dispersion process. Regardless, both microjets and shockwaves serve to comminute the particles, with the chief criterion being the presence of base fluid cavitation.[13]

2.2.3 Fusing through Interparticle Collisions

The shockwaves emitted by the collapsing bubbles in the medium will have velocities in excess of the sound speed of the medium (>1400 m s^{-1} in water). From modelling underwater bomb blasts, laser-induced cavitation was found to yield shockwave pressures of up to 300 MPa, which, for infinitesimally small particles, allowed prediction of particle velocities of up to 160 m s^{-1}.[14] For the sonication of 10 μm particles, the impact velocity of interparticle collisions can be estimated to be between 100 and 500 m s^{-1}.[10] The high energy of collision of particles at these velocities results in their melting together, yielding a neck, as shown in Figure 2.6 for 5 μm zinc particles. The melting point of zinc is around 2000 °C, hence the temperatures reached during the collision of the two particles in Figure 2.6 must have exceeded this value (the authors[10] estimated this to have occurred between 2600 and 3400 °C, corresponding to the melting points of molybdenum and tungsten, materials for which significant necking is not observed). This effect is greater for smaller particles, with larger particles (>150 μm) relatively unaffected by this acceleration and fusing issue.[10] In the context of the employed nanoparticles in the performed experiment, alumina has a similar melting point to zinc, and the size reduction from micro to nano was expected to exacerbate this fusing phenomenon. Although smaller particles increased Brownian motion velocity, it was not expected for these velocities to exceed their convective velocity (*i.e.*, Peclet number, $Pe \geq 1$). These high velocity collisions are also abrasive, when glancing impacts occur, which are characterised by the smoothing of particle surfaces and removal of their oxide layers,[8,15] shown in Figure 2.7 – a feature of importance in Section 2.3.

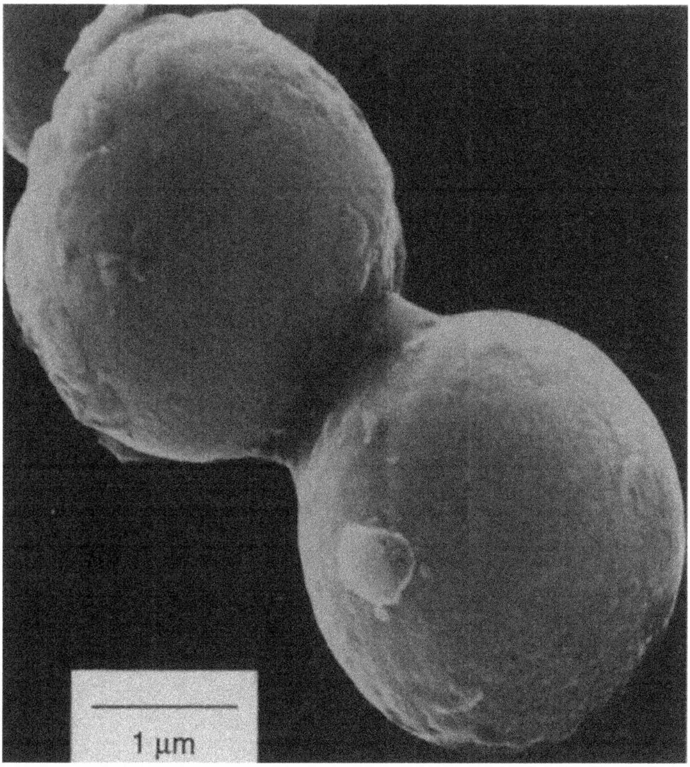

Figure 2.6 SEM of 5 µm Zn powder, exhibiting the formation of a neck in a doublet due to high-velocity collision (localised melting and cooling). Reproduced from ref. 10 with permission from AAAS, Copyright 1990.

2.2.4 Aggregates, Flocs and Derjaguin, Landau, Verwey, Overbeek (DLVO) Theory

The (hard) aggregation of nanoparticles in suspension due to partial melting on interparticle collision has been demonstrated in molecular dynamics studies.[16] Experimentally, it is more difficult to obtain a transmission electron microscope (TEM) image akin to Figure 2.6 for nanoparticles due to the smaller size of the suspended particles (nano *versus* micron-scale). However, TEM images of nanoparticles before and after sonication can still be studied, and do show the presence of hard aggregates (*e.g.*, triplets) which appear to form necks, and do not fragment any further.[17,18]

The fused agglomerate in Figure 2.6 is better defined as a 'hard' aggregate, requiring significant energy to redisperse (*i.e.*, ultrasonic milling or other comminution processes). These differ significantly from 'soft' agglomerates, also known as flocs, which are held together only by intermolecular forces such as van der Waals (vdW) interactions.[19] The differences between the two can be further understood from the characteristic Derjaguin, Landau, Verwey and Overbeek (DLVO) curve shown in Figure 2.8, with aggregates 'sitting' in a

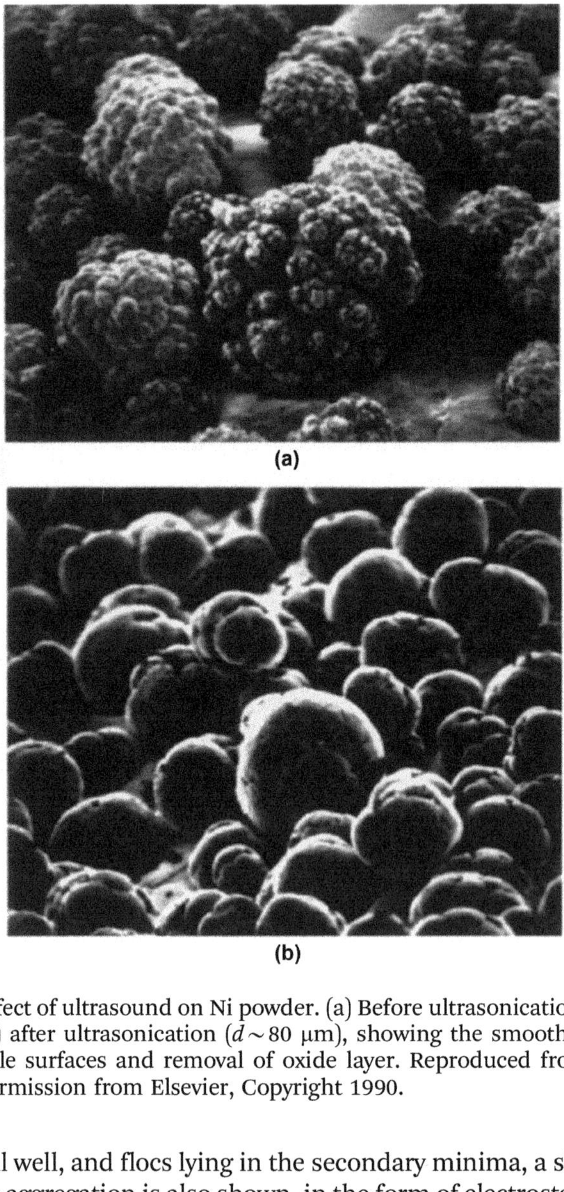

Figure 2.7 Effect of ultrasound on Ni powder. (a) Before ultrasonication ($d \sim 160$ μm), (b) after ultrasonication ($d \sim 80$ μm), showing the smoothing of the particle surfaces and removal of oxide layer. Reproduced from ref. 15 with permission from Elsevier, Copyright 1990.

deep potential well, and flocs lying in the secondary minima, a shallower well. The barrier to aggregation is also shown, in the form of electrostatic repulsion, Φ_m, with $\Phi_m > 15\ kT$ representing a moderate barrier to aggregation. However, the above-mentioned fused aggregates easily overcome this, as their kinetic energy is $E_k \approx 1000\ kT$ at 300 K with a particle diameter of $d = 50$ nm.[20] The formation of flocs is generally a reversible process, with the initial state of the colloid easily restored by stirring.[9] However, aggregate formation is known to be irreversible, due to the significant energy barrier, requiring immense amounts of energy to undo;[9] essentially this is akin to the energy delivered when the original comminution processes take place that yield the individual

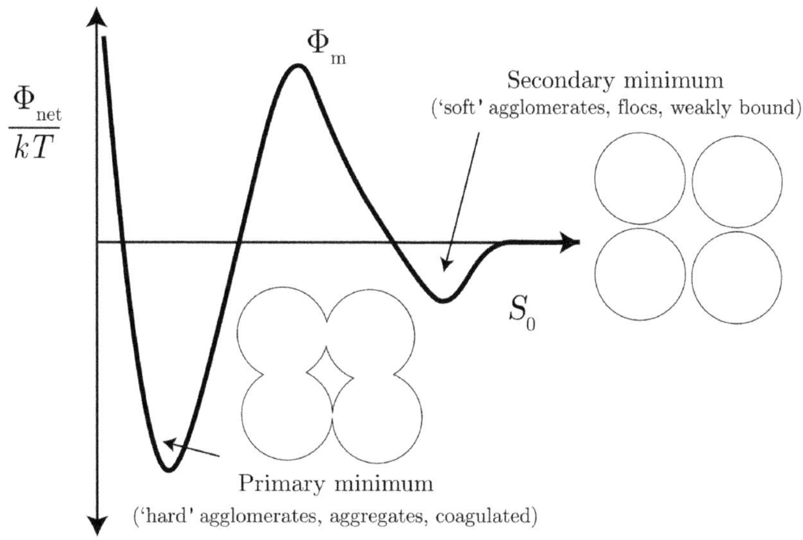

Figure 2.8 Characteristic DLVO curve (pairwise interaction between two spheres, Φ_{net}, normalised by thermal energy, kT). Depicts the primary and secondary minima and their characteristic agglomerate forms, in addition to the electrostatic barrier to aggregation (Φ_m). Solid–solid bridges/fusing are shown in the aggregates, applicable only when the nanoparticles are in a dry form and high velocity collisions occur.

primary particles.[19,21] The undesirable formation of aggregates might be encouraged by the employment of a "dry powder". In fact, the concept of a dry 'nanopowder' (*i.e.*, individualised nanoparticles stacked against each other) does not exist in reality. This is due to the formation of solid–solid diffusion bridges that occur in less than a few seconds when dry nanoparticles come into contact (an example of sintering/necking of silica nanoparticles due to this solid–solid diffusion is shown in Figure 2.9), *i.e.*, in the dry form provided by the manufacturer.[19,22]

2.2.5 The Implications of Dry Powder

For nanoparticles suspended in a liquid, the presence of a solvent layer (in addition to possibly an EDL/other stabilising agents) essentially eliminates the issue of solid–solid bridging, and slightly reduces the depth of the primary well, due to the presence of a hydration force (more simply, the volume occupancy of the solvent layer limits the minimum of the interparticle distance, S_0). The above makes it possible to restore stable colloidal conditions without significant energy input.[9] Hence, it is recommended to purchase nanoparticles in aqueous solution for all uses (as long as they are also kept wet during the entire manufacturing process, *e.g.*, sol–gel[23] opposed to pyrolysis[24,25]), so as to avoid needing to essentially ultrasonically

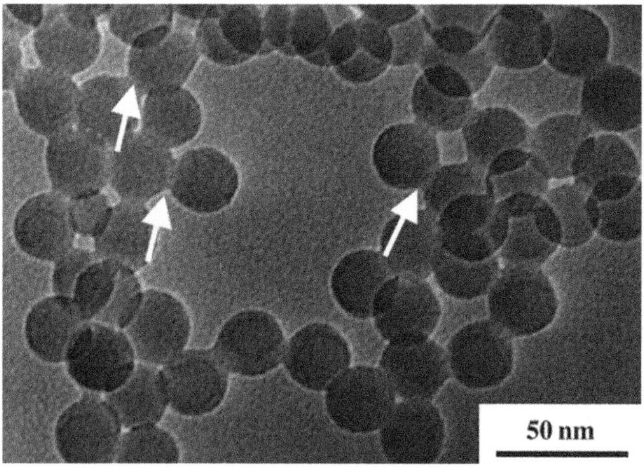

Figure 2.9 TEM of silica nanoparticles, exhibiting sintering and necking (no ultrasonication was used, hence these effects are due to solid–solid diffusion alone). Reproduced from ref. 56 with permission from John Wiley and Sons, Copyright 2006 The American Ceramic Society.

mill the nanopowder. For nanoparticles manufactured *via* the electric arc technique (as is the case for presently employed nanoparticles), an (aluminium) anode is vaporised and then quenched by a surrounding medium (water) to form the solid nanoparticles.[26] It may be that additional drying/separation processes are needed to yield the pure product. In any case, it is best to obtain this product in the wet form. This can help to explain the variation reported in ultrasonication studies,[6] as depending on whether one begins with a dispersed colloid or dry powder, the energy requirements to obtain the desired dispersion will differ greatly. The above also implicates nanofluid characterisation, namely agglomerate shape/size by TEM, as during this process the sample will be dried and is thus in contact with air for at least a few seconds, enabling the formation of these solid–solid bridges – *i.e.*, suggesting that larger/more agglomerates exist than those actually present *in situ*.

2.2.6 Energy Requirements for the Size Reduction of Colloids

The fact that most manufacturers use bottom-up methods in nanoparticle synthesis should indicate the significant energy demands required in top-down processes. For the comminution of polystyrene spheres to their primary size (0.91 μm) by ultrasonication, around 0.1 GJ m^{-3} is required[27] – requiring 500 s of ultrasonication in an ultrasonic bath (assuming a power delivery rate of 20 W L^{-1} to the colloid[6]). Reducing to a smaller primary size (∼nm) can be expected to require more energy, given the formation of an additional surface area and increase in adhesion forces *vs.* inertial forces (van der Waals forces scale with d whereas inertial forces scale with d^3).[28]

Generally, halving the agglomerate size requires 1000 times more energy. The result is that, in order to get to the primary size (*e.g.*, 45 nm in the presently employed alumina), around 5 GJ m^{-3} of deposited energy is required. In ultrasonic probe applications, this is infeasible due to the cooling required (this amount of energy is enough to evaporate a significant amount of the base fluid, *e.g.*, in a test tube) and probe erosion (contaminating the sample).[29] Both issues can be solved with the employment of an ultrasonic bath. However, then the processing time becomes impractically long, attaining a volumetric deposited energy of $E_v = 5$ GJ m^{-3}, neglecting power losses due to inefficiencies, would require around 3 days of continuous sonication (which would most likely damage the equipment).

2.2.7 Mechanisms of Agglomerate Breakup

Considering next the nature of agglomerate breakup, two distinct mechanisms have been observed, namely erosion and fragmentation. In the erosion breakup mechanism, individual particles from the exterior of the agglomerate are removed one at a time, whereas in the fragmentation breakup mechanism, the agglomerate splits into two smaller agglomerates. This is better understood visually in Figure 2.10, based on the work of Higashitani *et al.*[27] In their work they used polystyrene spheres of diameter 0.55–1.27 μm and investigated dispersion by comparing the effect of ultrasonication to that of a flow through a contracting orifice. The evidence to support the aforementioned mechanisms can be understood from the agglomerate size distributions in Figure 2.11 (from the same work[27]). Ultrasonication yields a significant increase in the proportion of primary particles ($i = 1$) whilst a reasonable proportion of large agglomerates remains, whereas the orifice contractile flow reduces the maximum agglomerate size (no agglomerates of >20 particles were observed) and yields a lower proportion of primary particles (20% *vs.* the 30% yielded by ultrasonication). The explanation of this result is as follows. In a contracting orifice flow, agglomerates align themselves with the flow, and are then elongated (and pulled apart) along the flow direction as the fluid accelerates through the contraction, proving adept at breaking up agglomerates.[30] Although ultrasonication differs to a hydrodynamic flow, a hydrodynamic flow is still induced in ultrasonic applications due to the shockwaves from collapsing cavitation bubbles (inducing fluid motion and shear).

Both erosion and fragmentation occur in ultrasonication applications, with the dominant mechanism dependant on the ultrasonic power and materials properties.[31] For example, Kusters *et al.*[13] observed titania powder to fragment, whilst silica powder eroded.

2.2.8 Optimising Ultrasonication Dispersion

Optimisation of the ultrasonication step in nanofluid preparation is critical to obtaining the best possible thermophysical performance and stability.

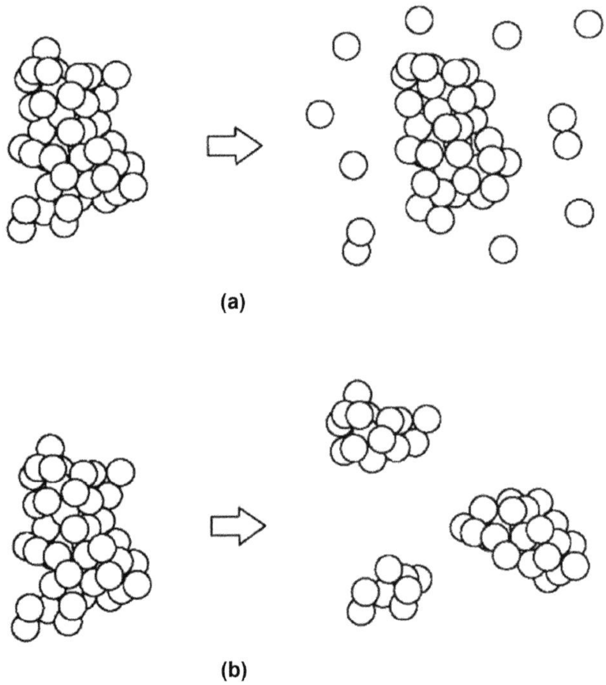

Figure 2.10 Mechanisms of (a) erosion and (b) fragmentation in agglomerate breakup. Reproduced from ref. 27 with permission from Elsevier, Copyright 1993.

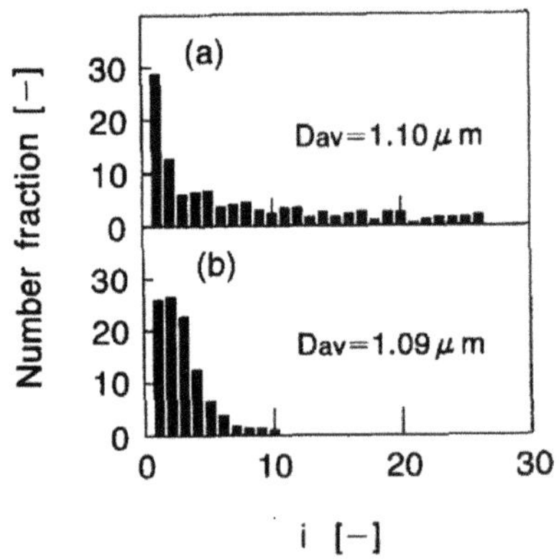

Figure 2.11 Agglomerate size distribution for micron sized polystyrene spheres subjected to (a) ultrasonication and (b) orifice contractile flow. Reproduced from ref. 27 with permission from Elsevier, Copyright 1993.

To compare different results of ultrasonic dispersion, they need to be standardised. To do this, the delivered energy is given specific to mass (E_m)[31] or volume (E_v).[27,32] However, the nature of the power delivery is important. Pulsed ultrasonication operation allows for the cooling of the sample, and for the escape of dissolved gases in the liquid (gases may become trapped in continuous operation), where it is known that the presence of dissolved gas significantly reduces the ultrasonic energy needed to form cavitation bubbles in the liquid.[5] The heating and consequential temperature rise of the sample will 'cushion' the collapse of cavitation bubbles, due to the increased vapour pressure, reducing their effects,[33] although, this increased vapour pressure also makes it easier to induce cavitation (less negative pressure required).[34] Additionally, the effect of ultrasonic power on cavitation bubble size has been mentioned,[12] with higher power applications yielding smaller cavitation bubbles, which would be able to form and impact microjets on smaller agglomerates.[10] The aforementioned thus further quantifies the challenges to the replicability of nanofluid studies. However, these issues can be overcome by focusing on replicating the particle size distribution (PSD) and stability characteristics (*e.g.*, zeta potential, pH, surfactant concentration) of the nanofluid. For example, it is expected that different investigators will need to use differing amounts of specific energy to yield sufficiently similar nanofluid suspensions.

A weakness of normalising by sample volume is that it does not account for the concentration (*i.e.*, number/size/mass) of particles present. Normalising by sample mass (E_m) somewhat alleviates this issue. From previous work, sonicating for 5 h in an ultrasonic bath (assuming a specific power of 20 W L^{-1}),[6] for $\phi = 0.00026$ vol% (15 mL sample, hence mass, $m = 1.5$ mg), one obtains $E_m = 360$ MJ kg^{-1}. Considering a 0.1 vol% 50 mL sample prepared using a high-power probe, (final pH of 5.9) after 1 min of sonication at 20% amplitude (*i.e.*, 40 W) one obtains $E_m = 13$ MJ kg^{-1}, showing that with the lower power approach, one is able to deposit over 20 times as much energy without negatively impacting stability (depositing more than 13 MJ kg^{-1} in the high-power sample resulted in significant aggregation and the restoration of the initial pH value). However, this can also be attributed to the concentration difference, as shear-induced (or otherwise known as orthokinetic) aggregation is more significant for denser nanofluids.[9] The larger nanoparticle concentration would also favour greater ultrasonication agglomeration breakdown efficiency, as with more particles in the system it is more likely those to be in the vicinity of collapsing cavitation bubbles (with a first order dependence,[13] *i.e.*, the probability of particle–cavity interaction $\propto \phi$).[31]

What is more interesting to consider is the difference in delivered power. Thoma *et al.*[35] found agglomeration to occur above a critical ultrasonication power for two different types of titania powder, however the same phenomena were not observed with silica.[13] As mentioned before, the titania powder exhibited fragmentation whilst the silica powder exhibited erosion. This result can be understood by recalling the third order dependence of orthokinetic (shear-induced) aggregation on particle size (assuming laminar flow conditions).[9] Therefore, larger aggregates subject to a shear stress are significantly more likely to aggregate further compared to smaller ones (according to orthokinetic aggregation rate $\propto d^3$, hence a doubling in agglomerate size will result in an eight-fold enhancement in orthokinetic aggregation). In other words, when agglomerates break up by erosion (single particles breaking off), they are so small that they are not subject to significant orthokinetic aggregation. When agglomerates break up by fragmentation, large agglomerates are generated, which conversely are subject to significant orthokinetic aggregation, which only increases more as aggregation proceeds.

Hence, the eroded primary particles in the silica were effectively insensitive to the applied shear, even when this was increased with higher ultrasonication powers. However, titania powder, upon ultrasonication, yield fragmented agglomerates, which would still be substantially larger than the primary particle size, and thus would be subject to significant orthokinetic aggregation. For the presently investigated alumina nanofluid, no significant agglomeration was observed when an ultrasonic bath was employed, suggesting that the nanofluid exhibited the eroding breakage mechanism in this case. Conversely, for the same nanofluid prepared using an ultrasonic probe, significant agglomeration was observed after a certain length of time, suggesting that the fragmentation breakage mechanism began to dominate. Hence, it appears that the applied power has some effect on the dominant breakage mechanism, where it is also clear that the erosion breakage mechanism is favourable.

Whether net coagulation or comminution is observed depends on whichever process dominates. The rate of change in particle number density (N_P) due to these processes was expressed by Aoki *et al.*[20] as:

$$\frac{dN_P}{dt} = k_1 N_C N_P - k_2 N_P^2 \tag{2.1}$$

where N_C is the cavity number density (the number of collapsing cavitation bubbles per unit volume) and k_1 is a rate constant quantifying the probability that a particle–cavity interaction will yield a comminution event. The first term on the right hand side (RHS) represents the rate of comminution, which depends on the probability of particle-cavity interaction (hence particle and cavity concentration – with cavity concentration increasing with applied power and particle number density, the latter offering additional nucleation sites for cavitation *via* gases trapped within a greater number of agglomerates). k_1 thus represents a materials property (the strength with which the agglomerated particles are bound, which accommodates for the differing strengths of adhesion dependant on particle/agglomerate shape/size). The second term on the RHS represents the loss of particles due to coagulation, which is proportional to $N_P{}^2$ as it is a particle–particle interaction,[9] and hence the rate constant k_2 represents the probability of aggregation on interaction (reduced by the presence of an electrostatic barrier, solvent layer, *etc.*).

From eqn (2.1) one can chiefly observe the dominance of coagulation for higher particle concentrations. Depending on the particle size distribution (*i.e.*, mean diameter, d_m), both rate constants will change, which can be understood as follows. The comminution rate constant (k_1) will reduce as aggregate size decreases (due to the increased strength of vdW forces in relation to inertial forces,[28] and the inability of microjets to form with smaller aggregates, ~100 μm[8]), *i.e.*, particle–cavity interactions will be less likely to yield comminution. This rate constant is also reduced as aggregates are 'reformed' by ultrasonication,[20] *i.e.*, those produced by high velocity interparticle collisions (fusing), with the reformed aggregates being more tightly bound than those in the initial dispersion. This behaviour was confirmed experimentally by Aoki *et al.*[20] and Park *et al.*,[36] with the latter publication finding that these aggregates were strong enough to withstand extensive high power sonication. The coagulation rate constant (k_2) will depend on solution pH and additive concentration, which can be easily controlled. However, as the mean particle size reduces, the probability of a particle–particle interaction yielding coagulation increases, as it becomes more favourable (decreases surface energy, and from DVLO theory, the coulombic barrier, Φ_m also decreases with agglomerate size[9]). The mechanism of breakage is also important, as the probability of particle–particle interaction resulting in coagulation is greater for larger agglomerates (third order dependence of orthokinetic aggregation), severely implicating the large aggregates yielded from fragmentation. This also ties in with sonication power, as high power applications have been reported to favour fragmentation,[36] as mentioned previously.

The rate constants k_1 and k_2 are obtained empirically. In their simplified heuristic model (eqn (2.1)), Aoki *et al.*[20] considered only the change in k_1 (*i.e.*, assuming k_2 to be constant). This was implemented in the form of two different values of k_1, denoted here as k_1 and k_1^*. This approach can be better understood from Figure 2.12, which shows how in region (a), the initial value of k_1 is used, leading to an exponential approach to a minimum agglomerate size after a

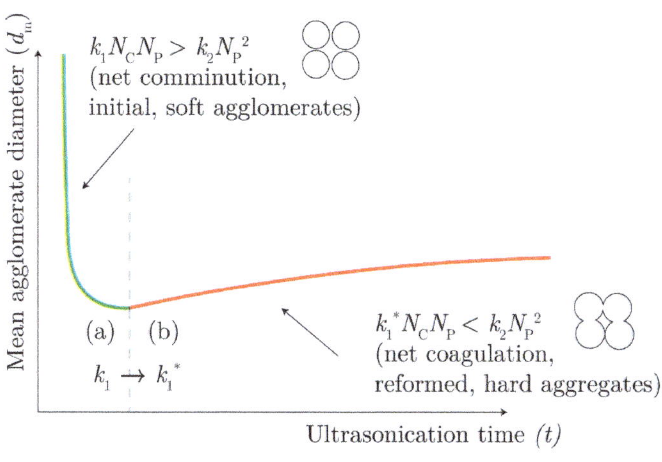

Figure 2.12 Effect of ultrasonication time on the mean agglomerate diameter, showing the transition from (a) net comminution to (b) net coagulation due to the formation of hard aggregates. Adapted from ref. 20 with permission from The Author.

certain duration of ultrasonication time. After this point, k_1 is substituted for k_1^* in the equation of $d_m(t)$ (eqn (3) in ref. 20), representing the dominance of the reformed, hard aggregates and hence the shift from net comminution to net coagulation. In region (b), the particle size distribution thus moves to a new equilibrium value that is greater than the minimum.

From the preceding description, it can thus be understood how high-power applications (yielding greater cavity concentrations, N_C), yield a greater rate of comminution. However, they also accelerate the rate of aggregate reformation, which results in a reduction in the comminution rate constant (k_1), outweighing the increase in cavity concentration (N_C). The coagulation rate constant (k_2) is also negatively implicated by high-power applications, if one assumes a fragmenting powder, as these larger fragmented aggregates (relative to eroded primary particles) will be subject to significant orthokinetic aggregation. Hence, after a critical deposited energy, the rate of coagulation will outweigh the rate of comminution, and net coagulation will be observed with time up to an equilibrium particle size distribution. This explains the inability of additional ultrasonication to restore the stability/reduce the agglomerate size of excessively sonicated nanofluids, with additional ultrasonication instead worsening the suspension stability and increasing the aggregate size. In addition to being reported by Aoki *et al.*,[20] these phenomena were also observed by Kole and Dey,[17] who studied a zinc oxide–ethylene glycol nanofluid, finding that after 60 h of probe sonication the mean particle size began to increase and the enhancement in thermal conductivity decreased.

Additional insight is offered by consideration of the dominant breakup mechanism (*i.e.*, erosion *vs.* fragmentation). Park *et al.*[36] deconvoluted multimodal particle size distributions to ascertain the dominance or either the erosion or fragmentation mechanisms in a variety of oxide powders, finding erosion to dominate at lower powers whilst fragmentation dominated at higher powers. This can be understood from the following. Increasing the applied power serves to increase the occurrence of acoustic cavitation in the sample – *i.e.*, the number of cavitation bubbles collapsing within the sample at a given moment – up to a maximum value (after which additional power reduces cavitation activity due to effects such as acoustic streaming and shielding/scattering).[37] This culminates in more cavitation bubbles impacting on/imploding near the dispersed phase, resulting in an increased 'cavitation pressure' applied to the dispersed phase.[38] Insufficient cavitation pressure exists at low powers to fragment aggregates, leaving erosion as the dominant mechanism, postulated to be caused by low angle impacts of agglomerates.[39] However, the exact nature of this mechanism remains contested (Kusters *et al.*[13] claim that there is no experimental evidence that interparticle collisions cause comminution).

The dominance of erosion at lower powers, combined with the insignificant effect of orthokinetic aggregation on these small, eroded primary particles, is expected to favour net comminution, and is indeed, what has been observed in the authors' lab using an ultrasonic bath.[3] One could argue that impractical sonication times are faced. However, this can be alleviated by reducing particle concentrations, which is the case for practical applications of nanofluids ($\phi < 0.1$ vol%). It should be noted that with sufficient time, fusing of primary particles may occur, limiting the smallest dispersible units obtainable. However, if one starts with a dry powder (as is the present case), significant numbers of unbreakable fused aggregates will already exist.

The lack of aggregation, regardless of sonication time/power for silica particles reported by Kusters *et al.*,[13] warrants additional consideration, as it would mean that one would not be required to reduce the applied power, saving on time in preparing stable suspensions. This behaviour, from the preceding discussion, is expected to be due to the invariant dominance of the erosion breakup mechanism. Indeed, the findings of Kusters *et al.*[13] have been confirmed by other experiments. For example, Retamal Marín *et al.*[40] deposited up to 1.4 GJ m^{-3} of ultrasonic energy (*via* a probe) into silica nanofluids of differing aggregate strengths, observing no aggregation. Vikash and Kumar,[41] through the use of a cooling system, were able to deposit ~5 GJ m^{-3} and observed no aggregation (steady state particle size distribution was reached after 60 min of high power sonication) in silica nanofluids of varying concentration. Their primary particle size was 12 nm, and from their dynamic light scattering (DLS) measurements the steady state hydrodynamic particle size distribution maximum was around 50 nm – hence it is expected that they achieved true size reduction to the minimum dispersible units (*i.e.*, fused doublets/triplets), with this deposited energy similar to what is expected[29] for such a case. Although silica appears promising from a dispersion standpoint, it has a lower macroscopic thermal conductivity compared to other oxides, and hence its suspension benefits may be outweighed by a lack of thermal enhancement, although at these small length scales the mechanisms of thermal conductivity differ, namely becoming phonon transfer dominant. This incentivises the discovery of eroding powders with better thermal properties, however the conclusive answer as to what determines a powder to be eroding or fragmenting remains to be seen.

2.2.9 Conclusion

To summarise the key points considering the first question initially posed in the introduction section, ultrasonication acts to both reduce the size of aggregates/flocs through the breakup mechanisms of erosion and fragmentation, in addition to forming them through interparticle collisions – yielding strongly-bound fused aggregates. It is expected for the presently employed alumina powder to predominantly exhibit the fragmentation breakup mechanism, yielding agglomerates subject to significant orthokinetic aggregation, which has a third order dependence on particle/agglomerate size, a second order dependence on concentration and a first order dependence on shear rate.[9] The rate of comminution decreases with increased deposited energy, as smaller aggregates are harder to breakup. This can be understood from the perspective of surface energy, with the increased dominance of vdW forces[28] and damage to the particle surface – negatively impacting the EDL.[8,15] The result is the observation of net coagulation after a critical deposited energy.

The aforementioned is also exacerbated by increases in applied ultrasonic power, as fragmentation has been shown to dominate at the high powers experienced in probe applications, whilst erosion dominates in low power bath applications. These observations were verified by the study of Park *et al.*[36] who used titania particles, although further experiments with different materials would help to affirm this hypothesis.

What is more interesting are eroding powders, such as silica, that exhibit no aggregation, regardless of sonication power and time (see *e.g.* ref. 40) – as this would enable the rapid preparation of stable nanofluids without the worries of an optimum time/deposited energy. The insensitivity of these eroding powders is hypothesised to be caused by the small sizes of the eroded primary particles being effectively insensitive to orthokinetic aggregation – recalling the third order dependence on size. Further research is warranted into why certain powders are eroding as opposed to fragmenting.

2.3 Ultrasonication and Nanofluid pH

This section considers the effects of ultrasonication on nanofluid pH. The importance of pH in the context of nanofluid stability is explained, and the mechanisms of pH change in nanofluid applications (including ultrasonication) are considered. This section discusses question 2 of the introductory Section 2.1.

2.3.1 Introduction

It is known that nanofluid pH is related to the zeta potential, which is used as a metric for nanofluid stability ($|\zeta| > 30$ mV). For example, in a previous study by the present authors, sedimentation of an alumina–water nanofluid in a natural convection cell was observed when the pH of the nanofluid was near to the IEP. As mentioned earlier, this sedimentation issue was rectified by reducing the pH of the nanofluid (from 8 to around 5), after which significantly less sedimentation was observed and the reduction in heat transfer performance reduced.[3,4] As was also mentioned, for one of the alumina powders employed (Sigma Aldrich, 544833, 50 nm, γ-Al_2O_3), the pH change was achieved by the addition of acid. For the second alumina powder (Alfa Aesar, 44931, 45 nm, 70 : 30 $\delta:\gamma$-Al_2O_3), the pH change was achieved by sonication alone. The following section expands on the underlying theory, so as to help understand how and why nanofluid pH changes with sonication (if at all).

2.3.2 DLVO Theory and the Electrical Double Layer

DLVO theory has already been introduced in Section 2.2. However, further elaboration is required to consider issues of pH alteration (and its consequences).

DLVO theory quantifies the net potential between two interacting bodies, constituting an attractive vdW component and a repulsive electrostatic component. For the pairwise interaction between two spheres,[9] the net potential, Φ_{net} is:

$$\Phi_{\text{net}} = -\frac{Aa}{12S_0} + \frac{64\pi CkT}{\kappa^2}\tanh^2\left(\frac{2ez\psi_\delta}{4kT}\right)\exp(-\kappa S_0) \tag{2.2}$$

Concerning the right hand side (RHS) of eqn 2.2, the first term represents the vdW attraction, which increases linearly with particle radius (a). A is the Hamaker constant, which is dependent on the particle/medium material, and quantifies the strength of the vdW interactions. S_0 is the closest distance of approach between two particles.

The second RHS term represents the coulombic repulsion, where C represents the concentration of salt (indifferent electrolyte, *e.g.*, NaCl[42]) in the bulk, of valency z. k, T and e are the Boltzmann constant, temperature and elementary charge, respectively. κ is the Debye parameter (nm^{-1}), defined more commonly as the Debye length κ^{-1} (nm), which, for water at room temperature with a single, symmetrical ($z-z$) electrolyte (*e.g.*, NaCl) is:

$$\kappa^{-1} = \frac{0.304}{|z|\sqrt{C}} \tag{2.3}$$

It is important to note that this situation is complicated by the presence of other, polyvalent ions, as together a multitude of ionic complexes can be formed with differing concentrations and valencies, with high cation valencies favoured under low pH conditions. This signifies the importance of using high quality DI (affirming deductions made in previous research[43,44]), as, depending on the quality of water used, differing (poly)valent ions of different concentrations will be present, thus determining the ionic strength (coulombic screening) within the nanofluid, ultimately having a tangible effect on the stability.

Returning to the final undefined term in eqn (2.2), ψ_δ is the potential (mV) in the Stern layer, which can be understood by the following.

Starting with an alumina nanoparticle in air (the presently employed dry nanopowder), it is known that the moisture in the air is sufficient to form hydroxyl groups from the exposed O^- ions on the surface, *i.e.*, Al–OH.[45] When dispersed in water, these reactive sites can either acquire or lose a proton depending on the pH, favouring protonation at pH values less than the IEP and deprotonation for higher pH values, in the form:[9]

$$\text{Al–OH}_2{}^+ \rightleftharpoons \text{Al–OH} \rightleftharpoons \text{Al–O}^- \tag{2.4}$$

The surface charge density is defined as:[9]

$$\bar{\sigma}_0 = F(\Gamma_{H^+} - \Gamma_{OH^-}) \tag{2.5}$$

where F is Faraday's constant and Γ is the molar adsorption of the ions (hydrogen and hydroxide, as they are the potential determining ions in the case of pure water[42]). Given that the neutral pH value of DI water (7) is less than the IEP of alumina ($pH_{IEP} \approx 8$–9.5, with the variation due to the co-ordination of oxygen to differing numbers of aluminium ions, caused by different crystallographic orientations and imperfections[46]), the molarity of adsorbed hydrogen ions will exceed that of the hydroxide ions, such that $\bar{\sigma}_0 > 0$, *viz.* a positive surface charge will exist.[47] This positive surface charge has an associated surface potential, ψ_0, which, for water at room temperature, can be expressed in terms of pH as:

$$\psi_0 = 59.2(pH_{PZC} - pH) \tag{2.6}$$

Eqn (2.6) also quantifies how alkaline pH values ($pH > pH_{PZC}$) will yield a negative potential, and thus negative surface charge. For oxides, the point of zero charge (PZC) usually corresponds to the IEP,[9] *i.e.*, $pH_{PZC} = pH_{IEP}$, hence justifying the assumption that nanofluids with pH values sufficiently far from the IEP will be stable (manifesting in a net surface charge, quantified by eqn (2.6)).

This positive charge/potential will correspondingly be neutralised by counterions in the solution, namely OH^- in the present case of water as a base fluid, with some tightly adsorbing to the particle surface (in the Stern plane) and the remaining charge neutralisation achieved by a diffuse layer of counterions. This arrangement is shown in Figure 2.13, which also shows the slip plane, the position at which the potential corresponds to the measured zeta potential.[9] The concentration of these counterions in the resulting EDL will exceed their values in the bulk, as shown in Figure 2.14. Likewise, the concentration of the co-ions (in this case H^+) will also be depleted in the EDL. This does not, however, affect the bulk concentrations of the ions, as displayed in Figure 2.14. This is due to the fact that an insignificant amount of the bulk solute inventory presides at any present interfaces,[9] *i.e.*, the formation of EDLs is not sufficient to affect bulk ion concentrations, and by extension, the solution pH.

2.3.3 The Effect of Autoionisation on pH

pH is, by definition, determined by the concentration of hydrogen ions in a solution (in pure water, these hydrogen ions exist as hydronium ions, H_3O^+, however these are written as H^+ for brevity). Pure water self-dissociates, in a process known as autoionisation:[48]

$$H_2O \leftrightharpoons H^+ + OH^- \tag{2.7}$$

Water dissociates to a very small extent, with the equilibrium concentrations of the ions on the RHS of eqn (2.7), $[H]_{eq} = [OH^-]_{eq} = 10^{-7}$ M, and thus acts as an incredibly weak acid. This equilibrium proton concentration, when plugged into the pH equation, yields the pH of pure water (7). These low ion

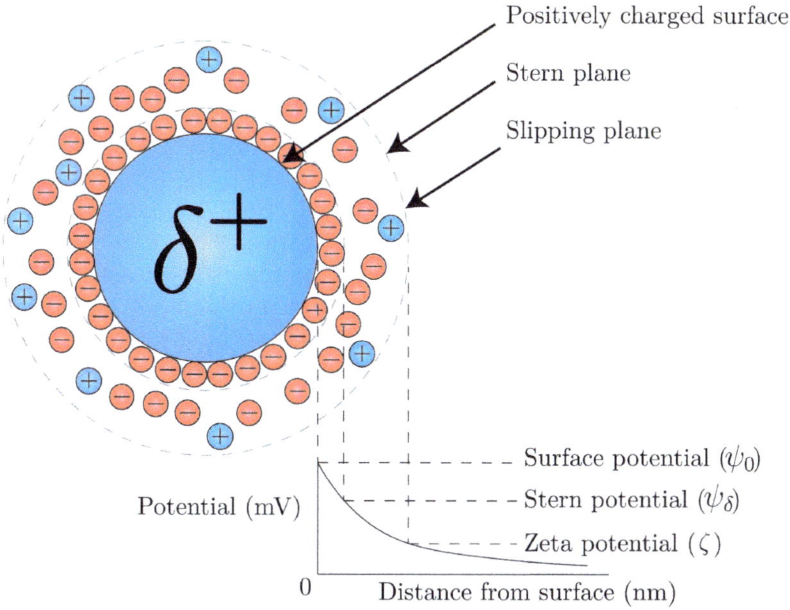

Figure 2.13 Schematic of the electrical double layer for a positively charged particle, with the different surface planes and their associated potentials (surface, Stern, zeta).

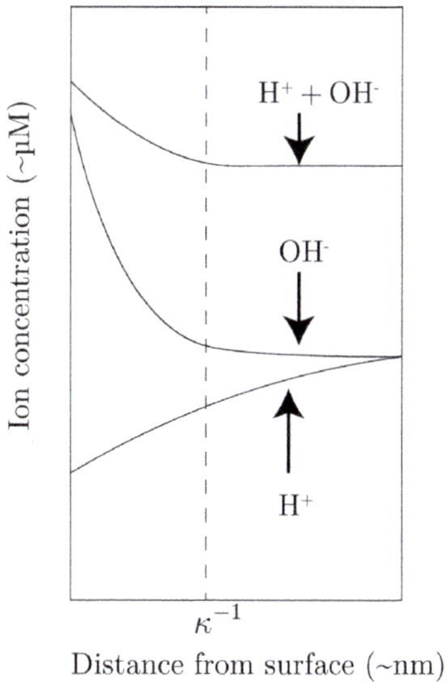

Figure 2.14 Typical concentration profiles within the electrical double layer (thickness κ^{-1}) showing the enrichment and depletion of the counter and co-ions. Adapted from ref. 9 with permission from World Scientific, Copyright 2009.

concentrations yield the low electrical conductivities observed in DI water, $\sigma = 0.055$ μS. Distilled water is less pure, and contains some ions, and hence has a slightly higher electrical conductivity of $\sigma = 0.5$ μS. Increasing the concentration of ions in the solutions serves to enhance the electrical conductivity, corresponding to greater deviations from neutral pH values. Some ions (*e.g.*, F^- for γ-Al_2O_3)[49] can specifically adsorb to the dispersed nanoparticles, yielding more effective charge neutralisation, decreasing the Stern potential, and thus the coulombic repulsion between particles. Dispersed ions that do not specifically adsorb may be denoted as salts, and act as an indifferent electrolyte, screening the Coulombic repulsion between particles, thus reducing the suspension stability (accounted for by the parameter C in eqn (2.2) and (2.3)).[9] The culmination of these two detrimental factors is a reduction in suspension stability with increased ion concentration. This explains the increased stability afforded by using higher quality DI water (reducing ion concentration), observed in previous work.[43] One can thus also understand how the increased ion concentrations (corresponding to increased electrical conductivity) of distilled water would be expected to yield less stable nanofluids, in principle (the deductions made above are complicated by the formation of polyvalent complexes – a process known as speciation[9]).

The autoionisation of water is mentioned as it can potentially determine the pH of a solution, applicable for pure water and for aqueous solutions of very dilute/weak acids ($C_a < 10^{-6}$ M, where C_a is the monoprotic acid concentration). Such a situation can be expected with DI and distilled water. However, as mentioned in Section 2.3.2, the changes in bulk ion concentrations due to EDL formation are negligible (if anything, the initial charging of the surface from eqn (2.4) consumes protons from the bulk). Hence, the bulk ion concentration changes required to effect pH change do not arise directly from the formation/collapse of EDLs. If this was the case, given that a pH reduction from 7 to 5 would require a 10 000 fold increase in the concentration of protons in the bulk, it is clear that the release of such a large number of protons cannot arise from an interfacial effect. Further, the presence of impurities (*e.g.*, dissolved CO_2), discussed in the following sections, would render the pH effect of autoionisation negligible ($C_a > 10^{-6}$ M).

Hence, it can be concluded that the formation/collapse of EDLs, as instigated through a change in specific surface area by ultrasonic dispersion, is not expected to affect the nanofluid pH. Further, the quality of the base fluid used is important, insofar as the ionic contaminants it contains – as these will affect the nanofluid pH and thus stability, in addition to effecting charge neutralisation at the surface of the dispersed phase.

2.3.4 Effect of Dissolved CO_2 on pH

A major source of acidity in any aqueous solution exposed to atmosphere is CO_2, which favourably dissolves into water until an equilibrium concentration is reached. A very small amount of the dissolved carbon dioxide (1%) reacts with water to form carbonic acid, which then effectively completely dissociates, yielding the net reaction:

$$CO_2 + H_2O \rightarrow H^+ + HCO_3^- \tag{2.8}$$

where the yielded protons serve to reduce the pH of the solution. For pure water at 25 °C, this can effect a pH reduction of as low as 5.7,[50] which explains why DI/distilled water, left exposed, exhibits a reduction in pH over time. However, for the presently used nanoparticles, pH values of as low as 4.5 have been observed after extensive low power sonication of very dilute nanofluids ($\phi < 0.0005$ vol%).[4] Under high-power sonication, as mentioned previously, it was not possible to replicate this result for very dilute nanofluids. However, pH reduction to as low as 4.88 was obtained with a 0.1 vol% nanofluid. The key here is to recognise that these pH values are lower than those expected from the equilibrium dissolution of CO_2 (<5 *vs.* the 5.7 expected from dissolved CO_2) – *i.e.*, dissolved CO_2 alone cannot have caused these pH reductions.

This hypothesis can be further justified by the negligible pH change of another alumina nanofluid (constituted of alumina nanoparticles from a different supplier) in previous research,[4] where sonication yielded a pH increase from 7 to 8. Further, ultrasonication is known to degas liquids[12,51] – as the growth of cavitation bubbles is fed by dissolved gases, after which they rise due to buoyancy and escape the system (although gas bubbles can become trapped during continuous high power sonication[5]). This degassing effect is expected to be diminished in the low power application, due to the reduced frequency of cavitation within the sample, meaning that significant amounts of CO_2 may dissolve and remain in solution, such that the reaction in eqn (2.8) proceeds. Hence, it can be deduced that the observed pH changes during sonication are not caused by the effects of dissolved CO_2.

2.3.5 The Effect of Impurities on pH

With proton generation from EDL formation (insignificant bulk ion concentration changes) and dissolved CO_2 (pH reduction of more than what is predicted by CO_2 saturation) refuted, impurities residing within/between the nanoparticles, released upon dispersion, are expected to be the cause of the observed pH change on ultrasonication. Indeed, the calcination (dehydration of surface by heating) of some alumina powders (comminuted by grinding) has been reported to have significant effects on their electrokinetic behaviour – due to the removal of an acidic surface contaminant.[52] Other authors have used the process of elution (washing with a solvent, such as NaOH)[49] to remove ionic impurities (for chromatography purposes). TEM spectral data of the presently employed powder[4] showed the presence of some impurities (C, Cu, Si and P), which may form complexes and contribute to the alteration of the nanofluid pH. Although the powder is very pure (>99%), even these trace amounts could significantly impact the pH (*e.g.*, from before, the dissociation of 1% of the dissolved CO_2 was sufficient to reduce the pH from 7 to 5.7). Another important factor to consider is the abrasion and disintegration of the oxide surface itself (as seen in Figure 2.7), hence alumina oxide complexes could be floating freely in the solution. However they are only soluble at extreme pH values,[53] and hence would not affect the pH. The fact that another nanofluid – formulated with alumina powder from another supplier – exhibited a pH increase on sonication[4] suggests the differing synthesis methods/treatments unique to each nanoparticle product line to be the cause of the pH change during sonication.

Considering the above in the context of experimental observations, namely, the failure of achieving pH reduction in dilute nanofluids ($\phi < 0.1$ vol%) in high-power applications, it is clear that even for trace amounts of nanoparticles ($\phi = 0.00026$ vol%) the amount of impurities is sufficient to effect significant pH reduction (7 to 4.5). This could be linked to the effective dispersion of the particles, releasing the impurities, whereas instead in the high-power application, such a dispersion and thus release of impurities was not achieved. For example, it may be that the erosion breakage mechanism, favoured at lower powers, could better release the impurities from between the nanoparticles themselves, whereas the impurities remain bound in the fragmented aggregates in high-power applications.

What remains to be explained is the observed restoration of pH by additional sonication. This observation coincides with the worsening of the suspension stability – the formation of large agglomerates, seemingly redispersed on stirring/sonication however re-agglomerating and settling shortly thereafter. In this case, the oxide layers may also be eroded – explaining the significant loss of stability (even a neutral pH should yield some repulsion due to the EDL), with these free alumina complexes reacting with the impurities to remove them from the solution, restoring the initial pH, or forming other complexes that effectively yield the initial pH.

2.3.6 Conclusion

Considering question 2 posed initially in the introduction Section 2.1, it can be concluded that the observed pH change during sonication of the presently investigated nanofluids is not due to the formation of EDLs or dissolved CO_2. Instead, impurities originating from the alumina powder itself, released into the solution on ultrasonication, are expected to be the cause. This is supported by the differing pH changes observed when using alumina nanoparticles from different manufacturers in previous work,[4] as different impurities arise from the different nanoparticle synthesis and processing methods used by different manufacturers. The inability of high-power applications to effect pH reduction in dilute nanofluids can be attributed to the dominance of fragmentation at high powers, which was less effective at releasing impurities than the erosion breakup mechanism that dominates in lower power applications (ultrasonic baths).

Uncertainties remain on the restoration of the initial pH value observed after extensive sonication, warranting further investigation. Some postulations have been made regarding damage to the oxide layer, as the suspension stability became worse than the initial condition, even though both have neutral pH values, which should enable the formation of some net surface charge, EDL and interparticle repulsion. However, given the incredibly short shelf life of these excessively sonicated nanofluids, it is clear that this was not the case. The resulting pH change can thus be attributed to the damage to the oxide layer.

2.4 Concluding Remarks

This chapter posed two key questions pertaining to nanofluid stability that required answering. Firstly, what are the reasons and mechanisms for ultrasonication leading to re-aggregation of nanoparticle colloids, and does there exist an optimum application to maximise nanofluid stability? Secondly, how does ultrasonication of a nanofluid lead to a change in pH?

With regards to the first question, it was shown that excessive high-power ultrasonication of nanofluids can lead to net aggregation due to the irreversible fusion of particles as a consequence of high-velocity interparticle collisions and the dominance of orthokinetic aggregation applicable to the larger, fragmented agglomerates that are yielded from high-power ultrasonication. For low-power applications and certain 'eroding' powders (*e.g.*, silica), there is no point of diminishing returns with respect to stability and applied ultrasonication (a reduced, steady-state particle size distribution is reached). In high-power applications, researchers should thus reduce the amplitude and duration so as to avoid significant coagulation (except if eroding powders such as silica are used, as these will not exhibit net coagulation). The optimum ultrasonication parameters (time, amplitude) in high-power applications will depend on the base fluid, nanoparticles (material, shape, size), pH and nanofluid concentration. In the present experiments using alumina–water nanofluids (distilled water at pH 7), for a concentration of $\phi = 0.01$ vol%, coagulation was observed after 4 min of ultrasonication at 100% amplitude and after 1 h at 20% amplitude – demonstrating the interplay between these two ultrasonic parameters for an identical nanofluid. For the case of 100% amplitude, coagulation was not observed at an applied ultrasonication time of 2 min, hence this represents the optimum time for this amplitude and concentration. For a higher concentration of $\phi = 0.1$ vol%, coagulation was observed after 5 min of ultrasonication at 50% amplitude. Reducing the amplitude to 20% and the applied ultrasonication time to 1 min yielded the best result (as quantified by the pH change of the nanofluid, from 7 to 5.9, yielding excellent stability). Note that in both cases the system volumes were 50 mL, with this variable affecting the optimum parameters.

Considering the second question with regards to ultrasonication and pH change, it was deduced that this effect is caused by the release of impurities contained within the dispersed phase. At low powers, erosion dominates, enabling the release of these impurities in dilute alumina nanofluids ($\phi < 0.1$ vol%). At high power, fragmentation of denser alumina nanofluids ($\phi > 0.1$ vol%) releases additional impurities, resulting in a greater change in pH. The restoration of the initial pH value for dense nanofluids under high-power ultrasonication is still not fully understood. However, it is postulated that damage to the oxide layer due to particle collisions contributes to this effect.

To conclude, in order to improve nanofluid stability, one should begin with an aqueous dispersion of nanoparticles (*i.e.*, avoid the use of dry powders) – so long as the given nanoparticles are kept in solution throughout the entirety of their manufacturing process (to avoid the formation of solid–solid bridges). If using an ultrasonic bath, sufficient time (several hours) should be allocated to the dispersion, as these devices have a much lower power delivery to the nanofluid. If using an ultrasonic probe and an eroding powder (such as silica), there is no need to reduce the applied power and time. However, for other fragmenting powders (such as alumina), amplitude and time should be reduced as to avoid net coagulation and reduced stability. pH should be measured before and after ultrasonication in all cases to check for the release of impurities into the solution and adjusted if necessary to obtain a sensible zeta potential.

Acknowledgements

The authors would like to acknowledge the EPSRC (ICO Nuclear CDT) and UKAEA for supporting this work. Additional acknowledgement is given to Professor Murshed for inviting the present authors to contribute to this book, and for hosting the corresponding author at their institution under a short-term scientific mission as part of the Nanouptake COST Action.

References

1. S. U. S. Choi, *Am. Soc. Mech. Eng., Fluids Eng. Div., FED*, 1995, **231**, 99–105.
2. R. Choudhary, D. Khurana, A. Kumar and S. Subudhi, *J. Exp. Nanosci.*, 2017, **12**, 1–12.
3. K. Kouloulias, A. Sergis and Y. Hardalupas, *Int. J. Heat Mass Transfer*, 2016, **101**, 1193–1203.
4. K. Kouloulias, A. Sergis and Y. Hardalupas, in *12th International Conference on Heat Transfer, Fluid Mechanics and Thermodynamics*, 2016, pp. 1465–1472.
5. H. M. Santos, C. Lodeiro and J. L. Capelo-Martínez, *The Power of Ultrasound*, 2009.
6. A. Asadi, F. Pourfattah, I. Miklós Szilágyi, M. Afrand, G. Żyła, H. Seon Ahn, S. Wongwises, H. Minh Nguyen, A. Arabkoohsar and O. Mahian, *Ultrason. Sonochem.*, 2019, **58**, 104701.

7. M. Pohl and H. Schubert, in *PARTEC*, NurnbergMesse GmbH, Nuremberg, 2004.
8. K. S. Suslick, in *Kirk-Othmer Encyclopedia of Chemical Technology*, John Wiley & Sons, Inc., Hoboken, NJ, USA, 2000.
9. J. C. Berg, *An Introduction to Interfaces and Colloids*, World Scientific, 2009.
10. S. J. Doktycz and K. S. Suslick, *Science*, 1990, **247**, 1067–1069.
11. K. S. Suslick, *Sci. Am.*, 1989, **260**, 80–86.
12. N. A. Tsochatzidis, P. Guiraud, A. M. Wilhelm and H. Delmas, *Chem. Eng. Sci.*, 2001, **56**, 1831–1840.
13. K. A. Kusters, S. E. Pratsinis, S. G. Thoma and D. M. Smith, *Chem. Eng. Sci.*, 1993, **48**, 4119–4127.
14. A. Vogel and W. Lauterborn, *J. Acoust. Soc. Am.*, 1988, **84**, 719–731.
15. K. S. Suslick and S. J. Doktycz, *Adv. Sonochem.*, 1990, **1**, 197–230.
16. Z. Yao, H. Zhang, Y. Hu, J. Bian, G. Wang, J. Lu and X. Niu, *Extreme Mech. Lett.*, 2016, **7**, 71–77.
17. M. Kole and T. K. Dey, *Thermochim. Acta*, 2012, **535**, 58–65.
18. K. Nemade and S. Waghuley, *Appl. Therm. Eng.*, 2016, **95**, 271–274.
19. A. F. Rawle, *Powder Technol.*, 2007, **174**, 6–9.
20. M. Aoki, T. A. Ring and J. S. Haggerty, *Adv. Ceram. Mater.*, 1987, **2**, 209–212.
21. E. M. Hotze, T. Phenrat and G. V Lowry, *J. Environ. Qual.*, 2010, **39**, 1909–1924.
22. M. R. Zachariah and M. J. Carrier, *J. Aerosol Sci.*, 1999, **30**, 1139–1151.
23. S. Said, S. Mikhail and M. Riad, *Mater. Sci. Energy Technol.*, 2020, **3**, 344–363.
24. P. Manivasakan, A. Karthik and V. Rajendran, *Powder Technol.*, 2013, **234**, 84–90.
25. C. Sauter and H. P. Schuchmann, *Chem. Eng. Technol.*, 2007, **30**, 1401–1405.
26. B. Yanık, H. Ağustos, Y. İpek, A. Koyun and D. Uzunsoy, *Arabian J. Sci. Eng.*, 2013, **38**, 3587–3592.
27. K. Higashitani, K. Yoshida, N. Tanise and H. Murata, *Colloids Surf., A*, 1993, **81**, 167–175.
28. M. B. Ranade, *Aerosol Sci. Technol.*, 1987, **7**, 161–176.
29. R. R. Retamal Marín, F. Babick and M. Stintz, *Powder Technol.*, 2017, **318**, 451–458.
30. K. Higashitani, N. Inada and T. Ochi, *Colloids Surf.*, 1991, **56**, 13–23.
31. K. A. Kusters, S. E. Pratsinis, S. G. Thoma and D. M. Smith, *Powder Technol.*, 1994, **80**, 253–263.
32. K. Hielscher, *MRS Proc.*, 2012, **1479**, 21–26.
33. T. J. Mason, J. P. Lorimer and D. M. Bates, *Ultrasonics*, 1992, **30**, 40–42.
34. S. B. Awad and N. F. Awad, *Part. Adhes. Removal*, 2015, 201–241.
35. S. G. Thoma, M. Ciftcioglu and D. M. Smith, *Powder Technol.*, 1991, **68**, 71–78.
36. B. S. Park, D. M. Smith and S. G. Thoma, *Powder Technol.*, 1993, **76**, 125–133.

37. I. Tzanakis, G. S. B. Lebon, D. G. Eskin and K. Pericleous, *Trans. Indian Inst. Met.*, 2015, **68**, 1023–1026.
38. S. G. Thoma, D. M. Smith and M. Ciftcioglu, *Powder Technol.*, 1991, **68**, 63–69.
39. M. A. Mullier, J. P. K. Seville and M. J. Adams, *Powder Technol.*, 1991, **65**, 321–333.
40. R. R. Retamal Marín, F. Babick, G. G. Lindner, M. Wiemann and M. Stintz, *Nanomaterials*, 2018, **8**, 454.
41. Vikash and V. Kumar, *Ultrason. Sonochem.*, 2020, **65**, 105061.
42. C. P. Huang and W. Stumm, *J. Colloid Interface Sci.*, 1973, **43**, 409–420.
43. K. Kouloulias, PhD Thesis, Imperial College London, 2018.
44. K. Kouloulias, A. Sergis and Y. Hardalupas, *Int. J. Heat Mass Transfer*, 2016, **101**, 1193–1203.
45. E. Tombácz, *Period. Polytech., Chem. Eng.*, 2009, **53**, 77.
46. J. Webber, J. E. Zorzi, C. A. Perottoni, S. Moura e Silva and R. C. D. Cruz, *J. Mater. Sci.*, 2016, **51**, 5170–5184.
47. A. Ghadimi, R. Saidur and H. S. C. Metselaar, *Int. J. Heat Mass Transfer*, 2011, **54**, 4051–4068.
48. B. G. Cox, *Acids and Bases*, Oxford University Press, 2013.
49. W. Szczepaniak and H. Kościelna, *Anal. Chim. Acta*, 2002, **470**, 263–276.
50. C. E. Boyd, in *Water Quality*, Springer, US, Boston, MA, 2000, pp. 105–122.
51. T. Hielscher, in *ENS 2005*, Paris, 2007, pp. 138–143.
52. A. S. Rao, *Adv. Ceram.*, 1987, **21**, 517.
53. G. Lefèvre, M. Duc and M. Fédoroff, *J. Colloid Interface Sci.*, 2004, **269**, 274–282.
54. F. J. Fuchs, in *Power Ultrasonics*, Elsevier, 2015, pp. 577–609.
55. L. A. Crum, in *1982 Ultrasonics Symposium*, IEEE, 1982, vol. 1, pp. 1–11.
56. J. Wang, W. B. White and J. H. Adair, *J. Am. Ceram. Soc.*, 2006, **89**, 2359–2363.

CHAPTER 3

Nanofluids Long-term Stability Challenges and Guidelines

S. CHAKRABORTY,[a] P. KUMAR[b] AND S. CHAKRABORTY*[b]

[a] School of Chemical Engineering, Vellore Institute of Technology, Vellore 632014, Tamil Nadu, India; [b] Department of Chemical Engineering, Indian Institute of Technology Kharagpur, Kharagpur 721302, West Bengal, India
*Email: sc@che.iitkgp.ernet.in

3.1 Introduction

Advancement in nanotechnology has led to its application in a plethora of domains, *i.e.*, catalysis,[1,2] water treatment,[3,4] electronics,[5,6] composite materials,[7,8] CO_2 sequestration,[9,10] thermal engineering,[11–15] lubricants,[16–18] medicine,[19–21] medical diagnostics,[22–24] supercapacitors,[25,26] and so on. The utilization of colloidal suspensions of nanoparticles (nanofluids) has gained popularity in several domains, as coolant,[27–31] nano lubricants,[32–34] medical diagnostics,[35–37] use in CO_2 sequestration,[38–40] and so on. The invention of nanofluids was inspired by the limitations of micron-sized particle suspensions, which have the following shortcomings: poor suspension stability, particle deposition tendency, pipeline erosion, higher particle loading requirements, augmented viscosity, and increased pumping costs, which prevent their widespread industrial application despite their moderate thermal conductivity.[13,41] With the development of nanofluids in the Argonne National Laboratory back in the 1990s by Stefan Choi,[42] most of the aforementioned problems were eliminated. The beneficial attributes of nanofluids, such as high thermal conductivity, low viscosity increment, low particle loading, altered specific heat, and ease of application in micro channels, have increased

Soft Matter Series No. 16
Fundamentals and Transport Properties of Nanofluids
Edited by S. M. Sohel Murshed

Published by the Royal Society of Chemistry, www.rsc.org

their usability in various domains.[30,41,43–51] However, issues with the stability of nanofluids largely prevent industry from utilizing them.[52–56] The current chapter focuses specifically on the stability aspects of nanofluids to explore the underlying mechanism and identify the major factors responsible for their destabilization. The recognition of key factors related to the stability of nanofluids, such as the role of nanoparticles, base fluids, synthesis techniques, operating conditions, external forces, and stabilization techniques, is essential for maintaining their long-term stability. In addition to the above factors, compatibility between base fluid–nanoparticle–stabilizers and the applicational suitability of these three components are crucial for stabilization purposes. This chapter provides a detailed insight into Derjaguin-Landau-Verwey-Overbeek (DLVO)[57,58] and non-DLVO[59,60] theory and how they play a critical role in colloidal destabilization. As the chapter primarily focuses on aspects of the stability of nanofluids, stabilization techniques are the center of attention. The stabilization techniques covered in this chapter include a discussion on both mechanical and chemical stabilization.[52,55,61] Mechanical stabilization can be achieved using ultrasonic, magnetic stirring, vortex shaking, and ball milling techniques, whereas chemical stabilization can be broadly classified into three distinctive classifications, *i.e.*, electrostatic, steric, and electrosteric stabilization. All of these techniques are discussed in detail in the chapter. However, all of these features have also been discussed extensively in the literature. This chapter also summarizes the operating principles of different stability evaluation methods, such as zeta potential analysis, dynamic light scattering (DLS), centrifugation, UV–visible spectroscopy, 3ω methods, *etc.*[52,55,62] To provide a clear understanding for readers on what the ideal properties of nanofluids are and the guidelines necessary for maintaining their long-term stability, parameters such as particle size–shape, zeta potential, nanoparticle–base fluid compatibility, thermal and oxidation stability of the nanoparticles and base fluid, the density difference between the base fluid and nanoparticles, high dielectric constant, boiling and freezing points of the base fluid, and viscosity of the base fluid are also discussed. It should be pointed out that discussion on the challenges of long-term nanofluid stabilization remains relatively unexplored, which is extensively addressed in this chapter. The long-term stability of nanofluids is highly important to ensure their reusability in industrial, environmental, and medical applications. Instability in colloidal suspension makes them unsuitable for any applications. Therefore, discussion on this topic is necessary to offer researchers and industry professionals, adequate ideas to assess and mitigate the complications in attaining long-term nanofluid stability. Guidelines have been chalked out to provide a possible roadmap to achieve the long-term stability of nanofluids. Stabilizer addition is considered to be one such route to this goal. However, several limitations exist in this approach as the addition of additives may also lead to a decline in several important parameters such as thermophysical characteristics (specific heat, thermal conductivity, and viscosity) and give rise to other problems such as foaming tendency, thermal degradation, poor freezing point and even lead to destabilization in a saline environment. Therefore, to ensure maximum benefits of stabilizer addition, the correct type of stabilizer needs to be selected based on nanoparticle type, operating conditions, and base fluid properties, which is also discussed herein. This chapter concludes with a discussion on future research directions.

3.2 Key Objectives of the Current Review

The motivation of the current chapter arises from the limited use of nanofluids in industrial/real-world applications despite their plethora of positive attributes such as enhanced thermophysical, electrical, optical, magnetic properties, low pumping costs, micro channel compatibility, low particle loading requirements, potential in drug delivery, medical diagnostics, CO_2 sequestration, and enhanced oil recovery applications. However, the biggest bottleneck preventing their widespread uses lies in their long-term stability and reusability. Therefore, this chapter aims to identify the sweeping problems associated with nanofluid stability and reusability and provide definite answers to mitigate such issues. The key objectives of the present chapter are listed herewith:

- This chapter provides a clear understanding of the physics involved in colloidal instability through classical theory and models. This information has been used to explain the stabilization/destabilization mechanism and to help in identifying key factors responsible for nanofluid destabilization.
- Moreover, it also identifies desired properties and difficulties that need to be overcome to ensure long-term nanofluid stability.
- This chapter provides strict guidelines that need to be followed to ensure long-term stability of nanofluids and mitigate the chances of particle aggregation during their applications.
- The selection of correct nanoparticle, base fluid, and stabilizer (stabilization techniques) combinations is the first step towards addressing the stability issues. To obtain prolonged stabilization of a nanofluid without additives, hydrophilic nanoparticles should be dispersed in polar solvents and hydrophobic particles in a non-polar medium. Nanoparticles with an intrinsic high surface charge (zeta potential) are desirable for the same reason. Favorable particle size, shape, thermal, and oxidation stability are also required to ensure long-term nanofluid stability.
- Understanding the role of density gradient (between the nanoparticles and base fluid), base fluid viscosity, boiling and freezing point, nanoparticle size, and shapes on colloidal destabilization.
- Suitable stabilization techniques selection should be based on the operating conditions, *i.e.*, temperature, pressure, salinity, electric field, magnetic field, and pH, which play a major part in determining the effectiveness of stabilization during real-time applications. If the nanofluid is exposed to a high temperature for a long time, surfactant and polymer-based stabilization become ineffective, as additives degradation reduces their stabilization

effect. Dispersant/stabilizers should be compatible with both base fluid and nanoparticles to ensure longer colloidal stability.
- Future researchers should explore specific aspects of nanofluid stability where information is missing to attain long-term nanofluid stability.

The following section discusses the fundamental theories necessary to understand the mechanism involved in colloidal destabilization, which are also useful for deciphering nanofluid instability aspects.

3.3 Fundamentals of Colloidal Stability/Instability

In the realm of colloidal science, colloidal dispersions can be classified into two categories; (a) lyophobic and (b) lyophilic. According to Freundlich,[63] lyophilic colloidal suspensions are thermodynamically stable, whereas their lyophobic counterparts are considered unstable from a thermodynamic viewpoint. When water is used as the dispersion media, lyophilic and lyophobic colloids can be termed hydrophilic and hydrophobic, respectively. From a stability standpoint, lyophilic colloid particles can be readily dispersed to prepare a stable colloidal suspension. However, lyophobic particles require additional assistance, such as mechanical (sonication, homogenization) and chemical (electrostatic, steric, and electrosteric) stabilization techniques, for their dispersal.

Kruyt[64] classified colloidal suspensions into two categories: (a) reversible and (b) irreversible colloids. Reversible colloids are spontaneously distributed in the dispersion medium due to a temperature gradient, whereas irreversible colloids, due to their thermodynamic instability, are difficult to disperse in a base fluid. The secret behind the better dispersion stability of lyophilic colloids lies in their compatible interaction with the dispersion medium. The two factors that govern lyophilic colloid stability are colloid electric charge and solvation energy. For a suspension to remain thermodynamically stable, the Gibbs's free energy for the system should be less than zero. The suspension remains stable even after the addition of electrolytes in small quantities. However, electrolyte addition may alter physical properties such as viscosity. Lyophobic colloid suspensions, which are naturally unstable, are made stable through external means (chemical and mechanical), and their stability solely depends on surface electric charge. Therefore, even at lower electrolyte concentrations, the stability of lyophobic colloids can be compromised.

Even though individual colloidal particles are electrically charged due to electrical double layer (EDL) formation, this EDL is an amalgamation of surface charge, Stern layer, and diffuse layer, where the net electrical charge is zero. To make the overall system surface charge and nanoparticles electrically neutral, equivalent opposite charges need to be present in the dispersion medium to compensate for the charge of the colloidal particles. The surface charge and the ionic atmosphere system are represented by the EDL shown in the Figure 3.1. Details of EDL are discussed along with the concepts of surface, Stern layer, and zeta potential in the subsequent sections. A schematic of the EDL surrounding a charged particle is shown in Figure 3.1.

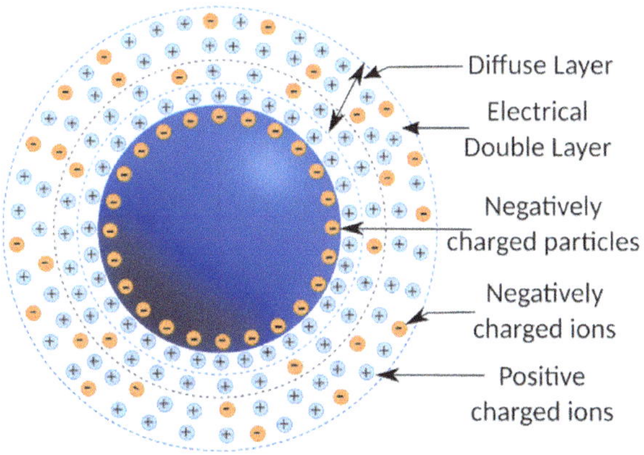

Figure 3.1 Schematic depiction of the EDL.

In a colloidal dispersion, each particle, despite carrying an EDL, maintains individual identity through EDL repulsion. Once particles approach each other, this ultimately leads to entropically unfavorable compression and confinement of the diffuse layer; thus, it prevents particle aggregation or ensures interparticle separation. Therefore, for a colloid to remain stable, a strong double layer repulsion is necessary, which can be achieved *via* a high particle surface charge or additional stabilizers.

Dispersion forces that govern the stability of colloidal particles are the most prevalent constituent force of van der Waals attractions and can work up to a separation distance of ~10 nm. With the possibility of carrying one or several charges, the colloidal particles influence the neighbouring particles; thus, the nature of the colloidal particle dictates its stability and instability characteristics. Lyophobic particles in colloidal form are considered thermodynamically unstable in the continuous phase, leading to particle aggregation, which minimizes interfacial area. Therefore, different stabilization techniques need to be implemented for stable colloidal dispersion in a continuous phase.

3.3.1 DLVO Theory

The first-ever quantitative colloidal stability theory for lyophobic colloids was proposed by Derjaguin, Landau,[65,66] Verwey, and Overbeek,[67] referred to as the DLVO theory. The DLVO theory states that the energy changes during particle–particle interactions due to Brownian motion[68] can be calculated in terms of total interaction potential, V_T, which is the sum of potential energies of attraction (London–van der Waals dispersion, V_A), and repulsion (electrostatic repulsive potential, V_R), using eqn (3.1).

$$V_T = V_A + V_R \tag{3.1}$$

Total interaction potential varies with changing particle separation distance. However, several assumptions need to be considered while applying the

DLVO theory. These assumptions are as follows: (a) it should be valid for low particle concentration, (b) gravitational force and density difference between the particle and base fluid is negligible, whereas van der Waals attractions and repulsive force (owing to EDL) are taken into account, (c) the colloidal suspension is uniform, and (d) distribution of the ionic charge in the dispersion medium is due to Brownian motion, electrostatic forces and entropic dispersion.

The dimension and morphology of the dispersed particles in the base fluid significantly influence the resultant attractive and repulsive potential of the colloidal suspension. The role of size and shape on attractive and repulsive potential is shown in Table 3.1. A closer look into the van der Waals attractive potential equation reveals the significance of the Hamekar constant (A) value on the particle aggregation tendency. The attractive potential value (V_A) is proportional to the Hamekar constant; therefore, its low value is desirable from a stability viewpoint.[52,69,70]

Suspension stability and instability are established based on attractive and repulsive forces; when attractive forces govern over repulsive forces, particle–particle collisions lead to aggregation and settling of particles, while with the prevalence of high repulsive forces, the suspension remains stable. Therefore, dominant repulsive forces between particles are required for stable suspension formation, which is achievable using any steric, electrostatic, and electrostatic stabilization techniques. The total interaction potential (V_T) responsible for colloidal stabilization varies with particle separation distance, and a high V_T value is a pre-requisite to maintaining prolonged stability, which is the manifestation of a higher repulsive force, whereas a low V_T value signifies nanofluid instability, which is shown in Figure 3.2 (a and b).

3.3.2 Non-DLVO Theory

The continuum-based DLVO theory fails to justify interparticle interactions at a reduced separation distance below the nanometric range. As a result, non-DLVO forces arise due to the non-functionality of either one or both continuum-based attractive van der Waals and repulsive EDL forces, and these additional forces can have different natures: monotonically repulsive, monotonically attractive, or oscillatory. The strength of these forces for influencing dispersion stability is much stronger than the forces considered in the DLVO theory.[71] Several non-DLVO-based theories have been studied by several researchers over the years, namely ion–ion corelation,[72–75] protrusion forces (repulsive forces between amphiphilic particles),[76] attractive hydrophobic forces (applicable to hydrophobic particles),[77] and repulsive hydration forces (applicable to hydrophilic particles).[76–81] In the pioneering study on ion–ion interaction forces, Kjellander *et al.*[73] established that ion–ion interactions can be neglected when surface charge density and electrolyte density are very low. However, at the increased magnitude of the abovesaid factors, attractive interactions between EDLs are applicable at a small separation distance (3–5 Å).

Table 3.1 Summary of the different attractive and repulsive potential equations for different particle sizes, shapes, and orientations[ab] [52,82].

Geometry	Attractive potential (V_A)	Boundary condition	Geometry	Repulsive potential (V_R)	Boundary condition
Sphere/sphere	$V_A = -\frac{A}{6H} \times \left[\frac{r_1 r_2}{r_1 + r_2}\right]$ (3.2)	$r_1, r_2 \gg H$	Sphere/sphere, $r = r_1 = r_2$ $\varphi = \varphi_1 = \varphi_2$	$V_R = 2\pi\varepsilon\varepsilon_o r\varphi^2 \ln(1 + \exp(-\kappa H))$ (3.3)	$\kappa r \gg 1$ $\varphi \leq 25$ mV Potential constant
Sphere/sphere	$V_A = -\frac{16Ar_1^3 r_2^3}{9H^6}$ (3.4)	$r_1, r_2 \ll H$	Sphere/sphere $r = r_1 = r_2$	$V_R = 2\pi\varepsilon\varepsilon_o r\varphi^2 \exp(-\kappa H)$ (3.5)	$\kappa r < 5$ $\varphi \leq 25$ mV Potential constant
Plate/plate (semi-infinite)	$V_A = -\frac{A}{12\pi H^2} \times A_{plate}$ (3.6)	NA	Plate/plate $\varphi = \varphi_1 = \varphi_2$	$V_{R,plate} = \varepsilon\varepsilon_o \kappa\varphi^2 \ln\left(1 - \tan h\left(-\frac{\kappa H}{2}\right)\right)$ (3.7)	$\varphi \leq 25$ mV Potential constant
Cylinder/ cylinder (parallel arrangement)	$V_{vdw} = V_A = -\frac{A}{12\sqrt{2}H^{3/2}} \times \left[\frac{r_1 r_2}{r_1 + r_2}\right] \times L$ (3.8)	$r_1, r_2 \gg H$	Cylinder/ cylinder (parallel arrangement) $r_1 \neq r_2$	$V_R = \frac{\sqrt{r_1 r_2}}{\sqrt{r_1 + r_2}} \left(\int_H^\infty \frac{V_{R,plate}}{\sqrt{x - H}} dx\right) \times L$ (3.9)	Use eqn (3.7) for $V_{R,plate}$ $\kappa r \gg 1$; $r > H$ Potential constant
Cylinder/ cylinder (parallel arrangement)	$V_A \propto \frac{1}{H^5} \times L$ (3.10)	$r_1, r_2 \ll H$	Cylinder/ cylinder (crossed arrangement)	$V_R = 2\pi\sqrt{r_1 r_2} \int_H^\infty V_{R,plate} dx$ (3.11)	Use eqn (3.7) for $V_{R,plate}$ $\kappa r \gg 1$; $r > H$ Potential constant

Table 3.1 (*Continued*)

Geometry	Attractive potential (V_A)	Boundary condition	Geometry	Repulsive potential (V_R)	Boundary condition
Cylinder/cylinder (crossed arrangement)	$V_A = -\frac{A\sqrt{r_1 r_2}}{6H}$ (3.12)	$r_1, r_2 \gg H$	Cylinder/sphere	$V_R = \frac{2\pi r_s \sqrt{r_c}}{\sqrt{r_s + r_c}} \int_H^\infty V_{R,plate} dx$ (3.13)	Use eqn (3.7) for $V_{R,plate}$ $\kappa r \gg 1$; $r > H$ Potential constant
Plate/sphere	$V_A = -\frac{Ar}{6H}\left(1 + \frac{H}{2r+H} + \frac{H}{r}\ln\left(\frac{H}{2r+H}\right)\right)$ (3.14)	$r \gg H$	Plate/sphere	$V_R = 2\pi r \int_H^\infty V_{R,plate} dx$ (3.15)	Use eqn (3.7) for $V_{R,plate}$ $\kappa r \gg 1$; $r > H$ Potential constant
Plate/sphere	$V_A \propto -\frac{1}{H^3}$ (3.16)	$r \ll H$			

[a] H = separation distance between interacting particles (nm), $r_1 = r_2 = r$ = characteristic length (nm), L = cylinder length (nm), A_{plate} = area of plate (m^2), r_s = radius of sphere (nm), r_c = radius of cylinder (nm), φ = surface potential (mV), ε = dielectric constant of dispersion medium, ε_o = permittivity of vacuum ($F\,m^{-1}$).

[b] κ^{-1} = Debye length (nm), where κ is the Debye constant, $\kappa = \sqrt{\left(n\sum_i (Z_i E_{electron})^2/(\varepsilon k_B T)\right)}$, k_B, T, $E_{electron}$, n, and Z_i refer to the Boltzmann constant ($J\,K^{-1}$), absolute temperature (K), the charge of a single electron, ionic species number density in the colloidal suspension, and valency of ions; A = Hamekar constant (J) = $A = \left(A_P^{\frac{1}{2}} - A_{BF}^{\frac{1}{2}}\right)^2$, where A_P and A_{BF} represent the particle and the base fluid Hamekar constant.

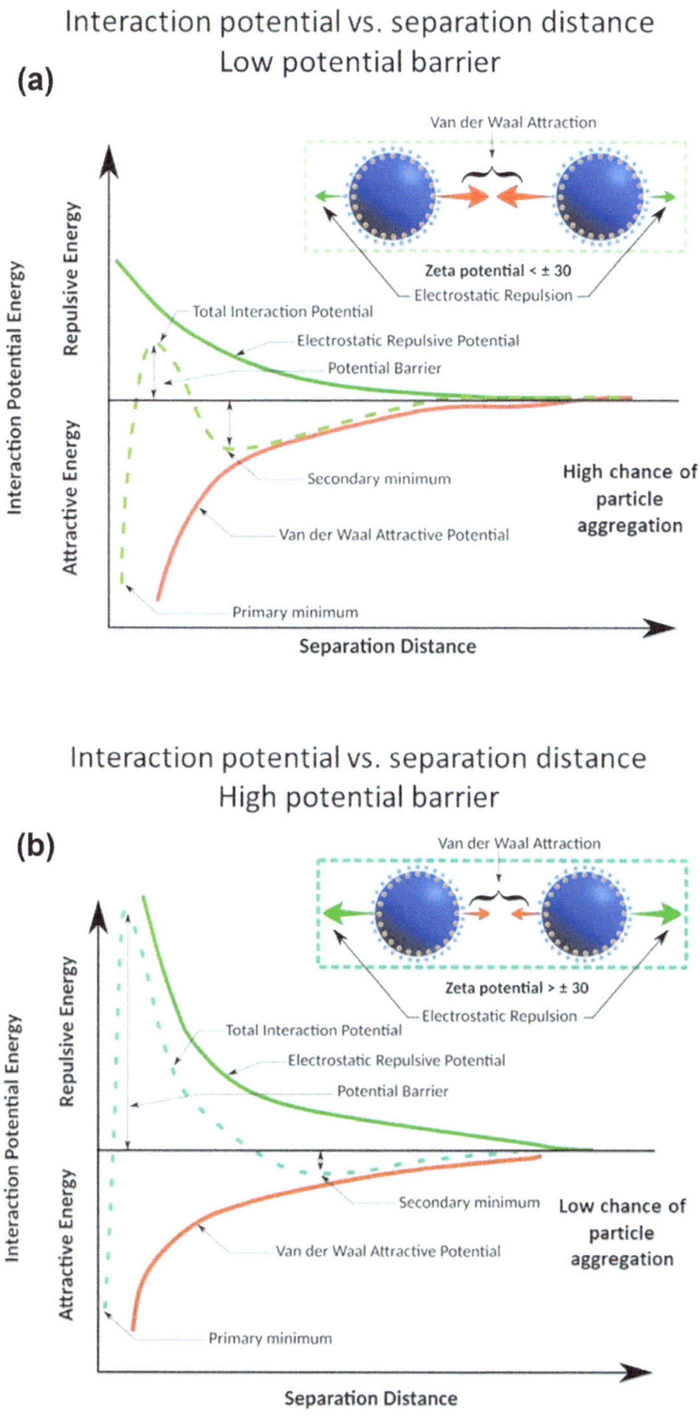

Figure 3.2 (a and b) Effect of (a) low and (b) high potential barriers on particle aggregation tendency. Adapted from ref. 52 with permission from Elsevier, Copyright 2020.

In addition to the theories mentioned above, the colloidal stability (nanofluid) also depends on a plethora of factors, such as dispersed phase material (size, shape, nature, *etc.*), the dispersion medium (compatibility, boiling point, freezing point, dielectric constant, viscosity, *etc.*), and stabilizer type (ionic and non-ionic surfactant, ionic and non-ionic polymer), which are discussed in the subsequent section.

3.4 Factors Influencing Nanofluid Stability

Nanofluid stability hinges on several aspects: the nature of the nanoparticles and base fluid, nanofluid preparation methods, operating conditions (temperature, pressure, pH, and salinity), external forces (gravity, magnetic and electrical field), and correct stabilizer selection. The aforementioned factors are summarized in Figure 3.3 and are broadly illustrated in the following section.

3.4.1 Role of Nanoparticles

Nanofluid stability is dependent on several factors associated with nanoparticles, such as nanoparticle type, particle size, particle shape, and nanoparticle concentration. These effects are discussed in the next section in detail.

3.4.1.1 Particle Type

Nanofluids can be categorized based on the types of nanoparticles used to prepare them. These types are: (1) metal-based, (2) metal oxide-based, (c) carbon-based, and (d) mixed metal/hybrid (*i.e.*, metal–carbon) based. Figure 3.1 shows different types of nanofluids based on the nature of the nanoparticles. Compatibility between nanoparticles and base fluid is crucial in accomplishing long-term nanofluid stabilization. When making an aqueous nanofluid, hydrophilic nanomaterials (metal oxides, metal hydroxide, graphene oxide, and hydrophilic hybrid particle) should be considered, but only hydrophobic nanoparticles should be considered when making oil or organic solvent-based nanofluids. Therefore, the selection of the correct nanoparticle and base fluid combination is critical to achieving long-term nanofluid stability. Metal-based nanoparticles like Ag,[83–85] Au,[86,87] Cu[88] are generally easily dispersed in an aqueous or polar medium, however, they can also be dispersed in non-polar solvents (oil) with the help of additional stabilizers.[89–91] In general, metal nanoparticles are more prone to oxidation and show relatively poor stability in base fluid compared to their metal oxide counterparts. A few researchers, however, have been able to produce metal-based nanofluids that exhibit a high zeta potential value (−43 mV) and also maintain stability for a long period (30 days).[87,91] As mentioned earlier, metal oxide nanofluids such as CuO,[92–96] TiO_2,[30,97–100] Fe_3O_4,[101–103] ZnO,[104–106] and Al_2O_3[96,107,108] are easier to stabilize in water/polar solvents compared to nonpolar solvents (silicone oil, transformer oil, kerosene, *etc.*).[109–112]

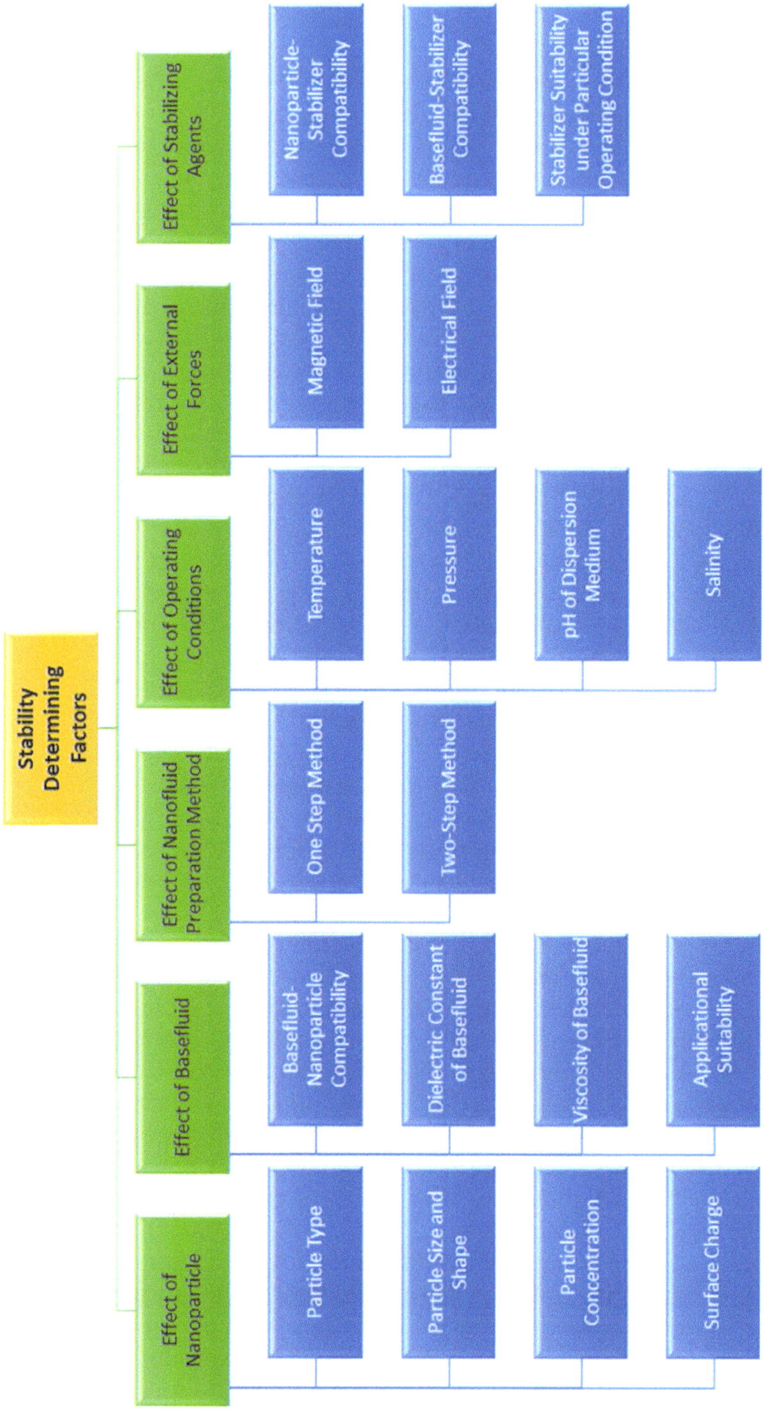

Figure 3.3 Identification of the various factors influencing the stability of nanofluids.

Polar solvent-based metal oxide nanofluids also show stability over a longer duration[92,93,95,96,98,100,103,113] and possess high surface potential[96,102,103,107,114] both with and without the addition of a stabilizer. In addition to stabilization aspects, metal oxide nanoparticles are cheaper than metal-based nanofluids, which is an additional reason behind their popularity.

In contrast to metal and metal oxide-based nanofluids, carbon-based nanoparticles (carbon nanotubes (CNTs), graphene) are easily dispersed due to their hydrophobic characteristics in non-polar/non-aqueous solvents such as decane, kerosene, pentane, hexane, toluene, benzene, acetic acid, and chloroform, *etc.* Despite their excellent thermophysical[115–119] and electrical[117,119–122] properties, long-term stabilization of these nanofluids remains a challenge. Xie and coworkers[118] reported the synthesis of a stable CNTs/decane nanofluid using oleylamine as a dispersant. The nanofluid was reported to remain stable for up to 60 days, which is considered remarkably stable. Carbon-based nanofluids in an aqueous/polar medium can also be stabilized using a plethora of techniques, such as electrostatic[123,124] and steric[125,126] stabilization. It should be clearly stated that without assistance from ionic and non-ionic surfactants or polymers, carbon-based nanofluids are impossible to stabilize in polar solvents. According to our literature review, the highest reported stability duration for carbon-based nanomaterials is three months (in an aqueous medium). Several researchers have achieved long-term stability for single-wall carbon nanotube (SWCNT)–water and multi-wall carbon nanotube (MWCNT)–water nanofluids using gum rrabic (GA, steric),[127] sodium dodecylbenzene sulphonate (SDBS, electrostatic),[128] and sodium dodecyl sulphate (SDS, electrostatic)[129] as stabilizers. Yazid and coworkers[130] recently published a comprehensive literature review on the stability aspects of CNT-based nanofluids, wherein the authors discussed the effects of chemical (both non-covalent and covalent surface modification) and mechanical treatment on the long-term stability of nanofluids.

There are very few research articles available in the open literature that discuss the long-duration stability of hybrid nanofluids (combination of metal–carbon, metal oxide–carbon, mixed metal oxide, metal alloy, or mixed metal hydroxides) in various dispersion media. Since carbon-based nanomaterials are hydrophobic in nature and metal/metal oxide type nanoparticles are regarded as hydrophilic, the hybrid nanofluid can be dispersed in both polar and non-polar dispersion media with the help of electrostatic or steric stabilization techniques. Song and colleagues[131] reported the moderate stability of a stainless steel–water nanofluid using cetyl trimethyl ammonium bromide (CTAB, cationic), and SDBS (anionic) as stabilizers. The researchers achieved nanofluid stability of up to 10 days, and the zeta potential value was also reported to be above 60 mV (absolute value). Without stabilizer addition, Nine *et al.*[132] and Syam Sundar *et al.*[133] stabilized Al_2O_3–MWCNT–water and nanodiamond–Fe_3O_4–water nanofluids for 5 and 3 days, respectively. However, more studies on stability need to be conducted on hybrid nanofluids to better understand and enhance their

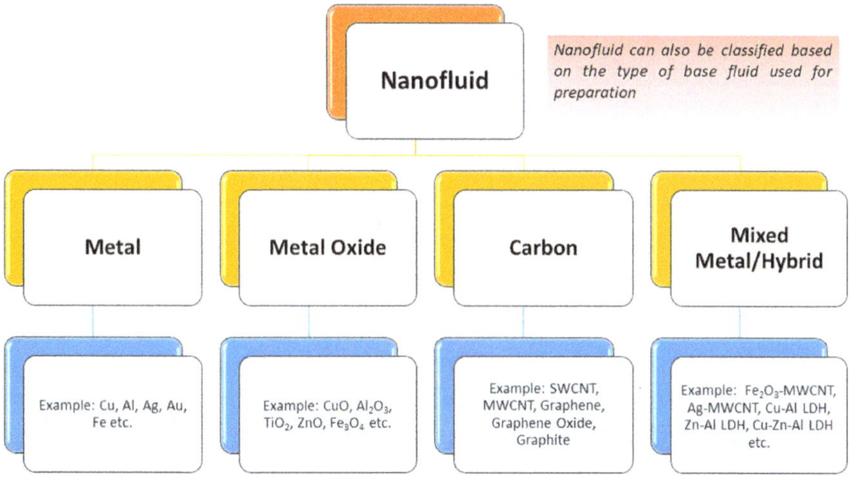

Figure 3.4 Different types of nanoparticles used for nanofluid preparation.

stability. Different nanoparticles used for nanofluid preparation are shown in Figure 3.4.

3.4.1.2 *Effect of Particle Size and Shape*

The role of nanoparticle size and shape on the stability of nanofluids has been verified by several researchers.[82,134,135] As discussed in the mechanism section, both electrostatic repulsive forces (owing to EDL) and van der Waals attractive forces are highly reliant on the size of the suspended nanoparticles. The reason behind such behaviour is due to the fact that small-sized particles possess a greater amount of surface atoms which increases the density of active sites[136] and significantly contributes to particle aggregation tendency.

He and coworkers[134] reported the effects of decreasing particle size on the aggregation tendency of spherical iron oxide nanoparticles. The study revealed that with a decrease in particle size, nanoparticle (hematite, α-Fe_2O_3) aggregation also increased. Under similar operating conditions (pH and ionic strength), greater aggregation of nanoparticles was reported for a small particle size. Table 3.1 shows the influence of nanoparticle size and shapes on van der Waals attractive (V_A) and electrostatic repulsive (V_R) potential. This clearly shows the impact of particle size, shape, and particle separation distance on total interaction, attraction, and repulsion potential. He *et al.*[134] demonstrated the effect of iron oxide particle size and separation distance on attractive, repulsive and total potential by plotting these potential values against interparticle separation distance. The authors predicted the interaction potential of hematite nanoparticles through DLVO theory and concluded that a decline in particle size leads to augmented attractive potential, whereas the repulsive and total interaction potentials decrease. Similar conclusions were also drawn by Chakraborty *et al.*[52] in their recent review on nanofluid stability. The article also discussed the effects on the particle size of aggregation tendency. In addition to the attractive and repulsive potential values, at a small particle size, the critical coagulation concentration (CCC) of the colloidal suspension[137,138] also reduces. The particle size also impacts the isoelectric point (IEP), which is directly related to the pH and zeta potential value of nanofluids[139,140] and thereby indirectly affects suspension stability.

In addition to particle size, particle shape and orientation also play a crucial role in nanofluid stability (particle aggregation tendency). The effect of particle shape (spherical, cylindrical, and plate/sheet), the interaction between different shapes (plate/sphere, plate/cylinder, cylinder/sphere), and particle orientations on attractive and repulsive potential have been demonstrated by Walker and colleagues.[82] The role of nanoparticle shapes on the stability of a water-based alumina nanofluid was reaffirmed by Kim *et al.*,[141] who used nanoparticles of three shapes (blade, brick, and plate) in an aqueous medium. Among the three types, the blade-shaped nanoparticles showed the fastest sedimentation, followed by the platelet-shaped particles, whereas brick-shaped nanoparticles remained stable for the longest duration. The role of particle shape on colloidal stability has also been discussed by McNew *et al.*[142]

The classical DLVO hypothesis covers only spherical particles. The prediction of electrostatic repulsive and van der Waal attractive potential for non-spherical shapes such as ellipses, platelets, rod, rhombohedral, and triangles[69,136] is difficult. In the case of non-isometric shapes, the relationship between the separation distance and attractive/repulsive potential is more complex than for the spherical counterpart. Due to the presence of more surface atoms, the attractive potential is greater than the repulsive potential in anisometric particles (compared to spherical particles), which is responsible for particle aggregation.[143] The effect of nanoparticle shape on DLVO theory was also reported by Flores *et al.*[144] using silica particles with spherical and bullet-shaped structures. The terminal settling velocity of nanoparticles also depends significantly on the particle shape as it contributes to drag forces that act on the particles during sedimentation.[145,146] The higher the drag force value, the lower the settling velocity, which is valid for particles with a high aspect ratio.

3.4.1.3 Particle Concentration

The particle loading in the base fluid plays a vital role in defining the stability of a nanofluid. Table 3.1 clearly shows the role of particle separation distance on attractive and repulsive potential. As the particle loading in the base fluid increases, the interparticle gap (separation distance) decreases. At a smaller separation distance, repulsive potential diminishes, whereas attractive potential increases significantly. As attractive potential overpowers the electrostatic repulsive barrier, particle aggregation begins to occur.

High particle loading also makes electrostatic stabilization techniques ineffective due to a very low interparticle gap. Thus, steric and electrosteric stabilization remain the only options to maintain/improve the stability of nanofluids. Several researchers have reported the effects of particle concentration on nanofluid stability.[15,45,147,148] At a high particle loading, the mean cluster size of an Fe–EG nanofluid increased from 1.2 μm to 2.3 μm, which was clear indication of clustering.[149] Chakraborty and coworkers[15,30,45] reported clear evidence of particle clustering at high particle loading using a variety of aqueous Cu–Al LDH, Cu–Zn–Al LDH, and TiO_2 nanofluids. Researchers have used multiple instruments such as DLS, transmission electron microscopy (TEM), zeta potential analysis, and sedimentation photography to observe the role of particle concentration on nanofluid stability.

3.4.1.4 Surface Charge of Nanoparticles

To ensure better nanofluid stability without stabilizer addition, nanoparticles used to prepare nanofluids should exhibit high surface charge. A high surface charge is also manifested by a high zeta potential value, which is vital to maintaining the long-term stability (except steric stabilization) of nanofluid. According to several researchers,[52,54,55,150] it is desirable to have a zeta potential of greater than ±30 mV from a colloidal stability standpoint. However, a zeta potential greater than ±45 mV is considered to be good, and beyond ±60 mV is regarded as excellent in the same regard (see Table 3.2). Particles with a high zeta potential value are expected to display greater electrostatic repulsive potential, which leads to better nanofluid stability. Mixed metal hydroxide nanoparticles such as Cu–Al layered double hydroxide (LDH), Zn–Al LDH, Cu–Zn–Al LDH exhibit high surface charge (±35 mV) and can be easily dispersed in base fluid (water) without the need for additional stabilizers.[12,15,45,151] However, metal oxide and carbon-based nanofluids do not exhibit the high surface charge necessary for stabilization. Therefore, additional stabilizers, pH adjustment, and surface functionalization are necessary to achieve higher stability in most metal oxide and carbon-based nanofluids.[52,55]

In addition to the above criteria, minimum density differences between nanoparticles and base fluid and compatibility of nanoparticles with base fluid and stabilizers are also crucial to ensuring the long-term stability of nanofluids.

Table 3.2 Interpretation of various zeta potential values from a stability standpoint.[55,108]

Zeta potential value (±mV)	Interpretation
0	No stability and high settling
15	Low stability and moderate settling
30	Moderate stability
45	Good stability with low settling
60	Extremely high stability, negligible settling

3.4.2 Role of Nanofluid Preparation Techniques

The preparation methodology adopted for nanofluid synthesis plays a crucial role in determining the stability of the resulting nanofluid. In general, nanofluids are prepared *via* two techniques, *i.e.*, (a) one-step and (b) two-step methods. A concise overview of the aforementioned methodology is discussed in the next section.

3.4.2.1 One-step Methodology

In one-step nanofluid synthesis techniques, an intermediate particle drying step is avoided and nanoparticle preparation and dispersion are performed in a single stage. By ensuring single-step *in situ* synthesis and dispersion in the base fluid, the risk of nanoparticle agglomeration occurring during the particle drying stage is eliminated. Vacuum Evaporation on Running Oil Substrate (VEROS) and Submerged Arc Nanoparticle Synthesis System (SANSS) are prominent one-step nanofluid synthesis techniques. VEROS is also known as physical vapour deposition (PVD), where nanoparticles are produced *via* deposition or condensation from the vapour state of a raw material into a flowing dispersion medium that has a low vapour pressure. No intermediate drying step is needed as nanoparticles are produced and dispersed simultaneously. Non-polar ferromagnetic (Co, Fe, and Ni) nanofluids have been prepared *via* VEROS techniques,[152] whereas different types of dielectric liquid-based copper (CuO, CuO_2, and Cu) nanofluids have been prepared using the SANSS method.[153] The second method provides exceptional control over nanoparticle shape with negligible aggregation tendency. The shape of the nanoparticles plays a critical role in determining thermal properties and aggregation tendency. Different shapes of nanoparticles formed using these techniques was studied by TEM analysis, which also revealed uniform particle size distribution. Choice of the dielectric fluid plays an important role in determining nanoparticle shape. Different researchers also modified VEROS to improve product quality, gain greater control over nanoparticle size, and achieve better stability without the use of additional stabilizers.[154,155] Eastman and colleagues modified the existing VEROS method to produce a Cu–EG nanofluid *via* the vapor condensation of Cu into the low-pressure base fluid (EG). Chang and Kao[156] described a novel approach for making nanofluids that further improved existing SANSS techniques. In comparison to the older SANSS methodology, the new method enabled an increase in the amount of nanofluid produced and improved the quality of the nanoparticles. A primary benefit of the single-step method is low particle agglomeration tendency, resulting in increased nanofluid stability. This approach eliminates the costs of drying and dispersion. One-step laser ablation methods have also been used to prepare Al and Ag nanofluids.[157] However, these methods of nanofluid preparation are less utilized due to their scalability and production cost-related issues.[158] The major drawbacks of this method include the inability to scale up the process due to high manufacturing costs and its compatibility with only low vapour pressure base fluids.

To overcome the aforementioned problems, a few chemical routes have also been explored by researchers working on a comparatively less expensive one-step nanofluid synthesis scheme (chemical). An EG-based copper nanofluid was created by chemically reducing sodium hypophosphite monohydrate and copper sulphate in an EG medium using microwave irradiation, according to a study by Zhu *et al.*[159] As a stabilizer, polyvinylpyrrolidone (PVP-K30) was utilized. This approach resulted in a uniformly dispersed Cu nanofluid that was extremely stable. The stability of the nanofluid was evaluated over a period of 14 days at elevated temperatures in a stationary environment, and it exhibited no signs of sedimentation. The method also produced a nanofluid with a low particle size. Microwave irradiation techniques were also used to prepare a silver nanofluid from $AgNO_3$ and PVP in an ethanol base fluid. The stability of an Ag nanofluid synthesized *via* this technology was maintained for 30 days under normal gravity (at room temperature) and over 10 h under centrifugation (at 3000 rpm) without showing any signs of sedimentation.[160] Yu *et al.*[161] used a phase transfer approach to create a stable homogenous nonpolar solvent-based graphene oxide (GO) nanofluid. In the presence of oxygen-bearing functional groups, *i.e.*, C=O (carbonyl), –OH (hydroxyl), and –COOH (carboxyl), the GO nanoparticles showed affinity towards water molecules and could readily be dispersed in an aqueous medium. To prepare GO nanoparticles with favourable compatibility in a nonpolar dispersion medium, oleylamine was utilized as a stabilizer to transform them into hydrophobic nanoparticles. Ayyappan and coworkers[162] described a single-step method for preparing various metal-based nanofluids (silver, gold, copper, and palladium) in an alcoholic medium utilizing PVP as the dispersant. The study used salts of the aforementioned metals to produce nanofluids *via* a chemical reduction method. However, it failed to produce uniformly-dispersed Cu nanoparticles even with an additional catalyst (Mg). In general, this method produced metal nanofluids (Ag, Au, Pd) with a particle size in the range of 5–35 nm, except for a Cu nanofluid.

In conclusion, one-step nanofluid synthesis involves the simultaneous synthesis and dispersion of nanoparticles in base fluid in a one-step process (VEROS/PVD, SANSS, laser ablation, microwave irradiation, phase transfer, *etc.*) without the need for a particle drying stage. The major benefits of one-step methods are superior control over particle size and shape, ideal for long duration and high-temperature stability, no intermediate particle drying requirements, and methods such as VEROS and SANSS do not require an additional stabilizer. However, the bottleneck of this technique includes (a) dependency on the batch process and low production output, (b) difficulty in separating the unreacted reactant post nanofluid synthesis, and (c) the high cost of production (except for a few chemical methods).

3.4.2.2 Two-step Methodology

The most extensively used approach for preparing nanofluids is a two-step procedure. This strategy is both inexpensive and simple to implement. The major disadvantage of this approach is the inability to control particle aggregation. The most popular strategy for controlling agglomeration is to use considerable ultrasonication and regular stirring. Another approach that can be used to control particle aggregation is the addition of different stabilizers, *i.e.*, surfactants (anionic, cationic, and non-ionic type), and polymers (hydrophilic and hydrophobic type) during nanofluid production. The use of stabilizers may vary depending on the nature of the base fluid and the type of nanomaterial used. Several studies have also employed the two-step process to create stable nanofluid dispersions without the addition of surfactants and polymers.

Nanofluid preparation using a two-step methodology can be subdivided into two distinct stages. **Stage 1**: chemical and mechanical (planetary ball mill) techniques are used to prepare different types of nanoparticles in the dried form, and **Stage 2**: the nanoparticle dispersion in the base fluid also requires the use of different mechanical and chemical routes. The mechanical/physical route utilizes a magnetic stirrer, bath, probe sonicator, high-pressure homogenizer, and other such items. The chemical route can be further sub-divided into three methods, *i.e.*, electrostatic, steric, and electrosteric stabilization. Electrostatic and steric stabilization use different surfactants and polymers to improve nanofluid stability, whereas electrosteric stabilization utilizes ionic polymers and ionic liquids for improving stability. Among these techniques, electrostatic and steric stabilization are the most widely used for stability enhancement. It should also be mentioned that two-step methods are widely used for dispersing metal oxide nanoparticles in a base fluid compared to a metal nanofluid.[163] The bulk of the research on the two-step methodology has been conducted[92–94,97,101,164,165] on different metal oxide (CuO, TiO_2, Al_2O_3, ZnO, SiO_2, Fe_3O_4, *etc.*) based nanofluids compared to any other nanofluids. However, in the recent past, focus has also been on the synthesis of carbon-based (graphene and MWCNTs)[124,125,166,167] and hybrid (Al_2O_3–MWCNT, Ag–MWCNT) nanofluids[132,168] using a two-step method.

However, the key disadvantages of two-step methods lie in their stability-related issues where particle aggregation post-synthesis is difficult to avoid. Chemical stabilizers (polymer and surfactant) become ineffective under high-temperature conditions due to thermal degradation. Thus, the implementation of frequent ultrasonication or homogenization is the only option. These chemical and mechanical stabilization techniques are also energy-intensive and expensive.

3.4.3 Effect of Base Fluid Properties

The role of base fluid properties on nanofluid stability is multifaceted. The influence of base fluid can be classified into four categories: (A) base fluid dielectric constant, (B) base fluid–nanoparticle compatibility, (C) viscosity of the base fluid, and (D) applicational suitability.

3.4.3.1 Base Fluid Dielectric Constant

The repulsive potential is proportional to the base fluid dielectric constant; therefore the higher the dielectric constant, the better the stability. To understand the necessity of the dielectric constant value, we need to look at the following equations for electrostatic repulsive interaction potential. It should also be mentioned that the electrostatic repulsive potential depends on the shape of the nanoparticles.

In the case of sphere/sphere interactions, the electrostatic repulsive interaction potential (see Table 3.1, eqn (3.3)) is

$$V_R = 2\pi\varepsilon\varepsilon_o r\varphi^2 \ln(1+\exp(-\kappa H)) \tag{3.3}$$

The boundary condition is that the characteristics radius (r) is equal for both spheres ($r = r_1 = r_2$), and the surface potential for both spheres ($\varphi = \varphi_1 = \varphi_2$) is also equal. The size and surface potential criteria are thus: $\kappa r \gg 1$ and $\varphi \leq 25$ mV.[82]

Similarly, in the case of plate/plate interactions, the electrostatic repulsive interaction potential equation is (see Table 3.1, eqn (3.7)):

$$V_{R,plate} = \varepsilon\varepsilon_o\kappa\varphi^2 \ln\left(1 - \tan h\left(-\frac{\kappa H}{2}\right)\right) \tag{3.7}$$

Here, the boundary condition is that the surface potential for both plates ($\varphi = \varphi_1 = \varphi_2$) is also equal. The size and surface potential criteria are as thus: $\kappa r \gg 1$ and $\varphi \leq 25$ mV.[82]

where, H, r, ε, ε_o, φ, and κ^{-1} represent the particle separation distance, characteristic radius, base fluid dielectric constant, vacuum permittivity, and Debye length, respectively (See Table 3.1). The aforementioned equation represents the effect that the surface potential has on the electrostatic repulsive interaction potential. The dielectric constants of acetone, benzene, ethanol, ethylene glycol, methanol, hexane, and water are 20.7 (25 °C), 2.28 (20 °C), 25.3 (20 °C), 37 (20 °C), 33 (20 °C), 1.89 (20 °C), and 80.1 (20 °C), respectively.[52,169] Water, among other base fluids, has the highest dielectric constant, indicating its tremendous repulsive potential.

3.4.3.2 Base Fluid–nanoparticle Compatibility

The selection of the nanoparticle and base fluid should be made based on their compatibility, which plays a crucial role in determining the stability of the nanofluid. Metal oxide-based nanoparticles such as CuO, TiO_2, Al_2O_3, ZnO, SiO_2, Fe_3O_4, *etc.*, are more compatible with aqueous medium or water-miscible solvents[92–94,96,97,101,164,165,170,171] whereas carbon-based nanoparticles are more compatible with organic solvent or hydrocarbon-based liquids.[172–175] Carbon-based nanoparticles can also be suspended in aqueous or water-miscible media but only through surface functionalization or *via* additive (surfactant/polymer) based stabilization.[124,125,166,167] In a nutshell, it can be concluded that the nature (hydrophilicity/hydrophobicity and oleophilicity/oleophobicity) of individual nanoparticles should be known before selecting their base fluid.

3.4.3.3 Viscosity of the Base Fluid

The viscosity of the nanofluid also plays a critical role in maintaining and improving nanofluid stability. A base fluid with high viscosity is highly undesirable from pressure drop, pumping costs, power, and energy economics perspectives. However, concerning the stability of the nanofluid, a base fluid with high viscosity helps to reduce the settling velocity of the suspended nanoparticles. The terminal settling velocity (V_{settling}) of a spherical, smooth nanoparticle under the effect of gravity can be calculated using the following expression,[176] which considers the Stokes law regime ($\text{Re}_{\text{p}} \ll 1$, laminar regime).

$$V_{\text{settling}} = \frac{\left(\rho_{\text{p}} - \rho_{\text{o}}\right) g d_{\text{p}}^2}{18\mu} \tag{3.17}$$

where, d_{p}, g, ρ_{p}, and μ denote the diameter of the nanoparticle, gravitational constant, the density of the nanoparticle, the density of the base fluid, and the viscosity of the base fluid, respectively. This equation clearly shows that attaining increased nanofluid stability requires a small particle size, a base fluid with high viscosity, and a nominal density differential between the nanoparticle and the base fluid. The higher the settling velocity, the lower the nanofluid stability. It is also worth noting that small particles exhibit more surface energy, which can contribute to particle clustering.

3.4.3.4 Applicational Suitability

The base fluid should be chosen according to the final application area. An oil/molten salt-based nanofluid is more stable than a water-based nanofluid in high-temperature and insulation-related applications.[177–181] Poor nanofluid stability in heat transfer applications can cause several problems, *i.e.*, deterioration in thermal conductivity, particle deposition in heat transfer surfaces/pipelines, increased pressure drop in flow channels, pipeline erosion, *etc.* Low freezing point coolants, such as various dielectric liquids (FC-72, FC-87, R-134a, HFE-7000, HFE-7100, poly alpha-olefin) or water–EG solutions, can be employed in sub-freezing point applications. For military-grade electronics applications, the operating temperature range varies between −65 °C to 225 °C. Even automobile engine coolants must perform over a large temperature range between −67 °C (in Siberia) to 200 °C (under the hood in desert areas). The freezing of coolant inside electronics may cause severe damage to the internal parts of electronics devices (microchannel) and car engines. Oil/ionic liquid-based nanofluids are more attractive for lubrication because they can diminish adhesion between two interacting surfaces and prevent suspended nanoparticles from depositing on the substrate.[182–184] Excessive particle deposition during lubrication may lead to a rise in friction, thus negating the benefits of the lubricant. For carbon capture or gas/clathrate hydrate promoter-related applications, aqueous nanofluids are thus preferred.[185–187] Particle aggregation reduces the gas and particle interaction, increasing aggregate size and leading to eventual sedimentation. These effects severely undermine the carbon-capturing potential of nanofluids. Therefore, it is well-understood that the selection of the base fluid is essential from a stability viewpoint and for maintaining the performance of nanofluids in different fields of application.

3.4.4 Effect of Operating Conditions

This section discusses the effect of different operating factors such as pressure, salinity, temperature, and pH of the dispersion medium on the stability of synthesized nanofluids. The effect of these factors on long-term stability is discussed in the following section.

3.4.4.1 Temperature Effect

Nanofluids are predominantly designed and used for thermal engineering applications, where they are subjected to repeated heating and cooling cycles. Nanofluids that have poor temperature stability and are agglomerated under such conditions as high temperature promotes increased Brownian motion due to a high particle diffusion coefficient. High Brownian motion leads to more collisions and an increased chance of particle aggregation. Solvent loss is also considered to be another factor for nanoparticle aggregation in the base fluid. In addition to the above reasons, the poor thermal stability of stabilizers (polymer or surfactants) also plays a key part in nanofluid destabilization.[188] During the cooling cycle of nanofluids, a significant reduction in Brownian motion may also lead to particle aggregation. Ideal nanofluids should maintain their stability over a wide range of temperatures. However, in real-world applications, nanofluids tend to agglomerate and deteriorate quickly due to a plethora of factors such as operating conditions (pH, temperature, pressure, and salinity) and external forces (gravity, magnetic and electrical field). Among these factors, temperature is considered to be one of the most critical factors responsible for nanofluid destabilization. The agglomeration of nanofluids has been widely reported by several researchers using different nanofluids.[189–191] Researchers have reported the aggregation of CuO nanoparticles and a reduction in nanofluid stability at elevated temperatures using DLS analysis (average cluster size increase) and zeta potential measurements. The study reported a reduction in zeta potential value from 22 mV to 13 mV using the nanofluid for over 1 h at 60 °C. A more practical approach to attaining high-temperature stability is to avoid surfactant and polymer and use mechanical stabilization such as high-pressure homogenization and probe sonication or chemical stabilization methods such as acid treatment and covalent surface functionalization (plasma treatment). Plasma treatment has been found to stabilize nanofluids under high-temperature conditions.[189] Tavares and coworkers[189] reported the high-temperature stabilization of a Cu–EG nanofluid using plasma treatment up to 150 °C. However, as the operating temperature appraoched the boiling point of EG (at 200 °C), the nanofluid agglomerated due to solvent loss, increased particle Brownian motion, and interparticle interaction. In contrast to surfactant/polymer-based stabilization, where the aggregation is permanent, ultrasonication can disintegrate the particle aggregates formed due to high temperature. Hordy and coworkers[192] also used plasma-based surface functionalization to prepare a MWCNT nanofluid with long-term and thermal stability in different base fluids (EG, propylene glycol (PG), Therminol VP-1, and water). Except for Therminol VP1 based MWCNT nanofluid, nanofluids made using the other base fluids remained stable for up to 8 months. The nanofluids remained stable up to 85 °C (at atmospheric pressure the boiling point is 100 °C) for water-based nanofluids and 170 °C for glycol-based (EG boiling point: 197 °C; PG boiling point: 188.2 °C) nanofluid, whereas the stability of the Therminol-based MWCNT nanofluid was compromised at 220 °C (Therminol VP-1 boiling point: 257°), which was much lower than its boiling point and maximum operating temperature limit of 400 °C. The poor dispersion stability of the MWCNT nanofluid in Therminol VP-1 medium is that it is non-polar in nature, and plasma-based functionalization works better in polar media such as water, EG, and PG. The oxygen-bearing functional groups (carboxylic and hydroxyl) incorporated by plasma treatment are more compatible in a polar medium. The study also shows the impact of multiple heating cycles on the stability of the Therminol-based MWCNT nanofluid by measuring the nanoparticle concentration after each heating cycle. In a recent study, the high-temperature stability of an indium tin oxide (ITO)–therminol 66 nanofluid[193] was reported up to 345 °C by capping ITO nanoparticles with oleylamine. Besides plasma treatment, long-term stability (180 days) of carbon-based nanofluids can also be attained using high pressure (1000 bar) homogenization.[194] This technique has been used to prepare a carbon nanohorn nanofluid in EG, which remained stable up to 150 °C. Carboxylic acid group-based functionalization of CNT nanoparticles in an aqueous medium showed improvement in stability (up to 21 days), thermal conductivity (56%), and thermal performance index (44%). The nanofluid remained stable in a temperature range of 50–70 °C under laminar and turbulent flow regimes. In this study, nonylphenol ethoxylate was used as a stabilizer. The limitations of conventional stabilizers such as oleic acid for high-temperature applications has been reported by Li *et al.*[91] The authors observed the rapid reduction in stabilizing effect at a higher temperature using an oleic acid-coated Ag nanofluid in an n-heptane medium. Under room temperature conditions, the Ag nanofluid remained stable over 30 days, whereas at elevated temperature (160 °C), stability reduced rapidly to 2 h. Similarly, a coconut oil-based silver nanofluid under convective boiling conditions showed a negligible fouling effect (nanoparticle deposition) up to 185 °C over 17 h and a minimal particle clustering effect.

Poor nanofluid stability at elevated temperature is considered to be highly detrimental for thermal engineering applications based on several aspects, *i.e.*, poor thermal conductivity, increased pipeline erosion, enhanced viscosity, increased pressure drop, high pumping costs, and augmented heat transfer resistance caused by nanoparticle deposition (fouling factor).[14,94,97,98,195,196] The detrimental effects of particle deposition on the thermal performance of thermosyphon (a TiO_2–water nanofluid) were reported by Sarafraz and coworkers.[98] The authors observed better thermal performance at the initial stages of TiO_2 nanofluid usage. However, prolonged exposure (10 days) to high temperature increased particle deposition and led to a poor heat transfer coefficient value. In this study, increased nanoparticle deposition was due to high heat load, high particle loading, and the inclination angle in the thermosyphon. Sarafraz *et al.*[197] also observed the negative effect of particle deposition on heat transfer coefficient attainment in flow boiling using a Therminol 66-based MgO nanofluid. Wei and coworkers[198] also observed moderate stability (10 days) of an oil-based TiO_2 nanofluid at elevated temperatures (20–50 °C) without the need for additional dispersants.

Despite all the disadvantages of the high-temperature destabilization of nanofluids, particle deposition on a heat transfer surface to an optimum level is desirable from a nucleate boiling viewpoint. Particle deposition on heat transfer surface acts as additional nucleation sites and helps in enhancing heat flux and heat transfer coefficient.[15,199]

3.4.4.2 *Effect of Pressure*

The operating pressure plays a vital role in nanofluid stability and performance in various applications; studies conducted over the years have accounted their role in pool boiling, thermosyphon, flat-pipe evaporator, and heat pipes. The impact of high pressure on colloidal stability is manifested by a lower interparticle gap that induces particle aggregation and destabilization.

In a comparative study dealing with atmospheric and sub-atmospheric pressure with functionalized silica nanofluid stability in pool boiling, Yang *et al.*[200] observed that a functionalized nanofluid was relatively stable, devoid of deposition, and performed better in heat transfer applications under sub-atmospheric conditions. In a study dealing with a copper wire-bonded flat pipe, Chen *et al.*[201] observed that the operating pressure directly influenced the existing binding forces between the nanoparticles. This resulted in pronounced Brownian motion and increased interaction with the system thermal boundary layer, and, consequently, thermal boundary layer fragmentation ensured uniform temperature and nanofluid stability in the heat transfer enhancement process. In thermosyphon, Liu *et al.*[202] used a 1 wt% optimal CuO nanofluid, which showed performance improvement with reduction in operating pressure; above the optimum concentration, nanoparticle agglomeration likely destabilized the nanofluid, which was unaffected by operating pressure. Liu *et al.*[203] concluded that in the micro-grooved flat-pipe evaporator, the heat transfer performance of the CuO nanofluid improved up till the optimum nanofluid concentration, beyond which the thermal performance reduced. Such reduction in thermal performance was due to CuO micro-groove surface coating, indicating challenges associated with nanofluid stability. For heat transfer performance assessment of a micro-grooved heat-pipe, Yang *et al.*[204] studied the effect of varying operating pressure on heat transfer. Here, the authors reported that under low-pressure conditions and optimal particle loading (1 wt%), improvement in heat performance was observed compared to plain water, which was possibly due to reduction in particle aggregation at optimal concentration. Few researchers have investigated the effects of operating pressure on nanofluid stability; therefore, more systematic studies need to be conducted to elucidate the mechanism involved in pressure-based colloidal instability.

3.4.4.3 Effect of Salinity

Other than their potential in thermal engineering application, nanofluids have also been used for enhanced oil recovery (EOR) due to several factors such as the ability to alter the wettability of reservoir rock, decrease the interfacial tension of oil–water medium, alter disjoining pressure, and help oil displacement inside the reservoir by increasing the viscosity of injected water.[205–208] However, to attain the favorable attributes of nanofluids in enhanced oil recovery, it is necessary to maintain their stability in saline medium (in an oil reservoir), which has been regarded as highly challenging by several researchers.[205,209–211]

In the presence of salt, the zeta potential of different nanofluids alters, which plays a critical role in defining nanofluid stability. When a nanofluid is used in the presence of high salt concentration, the electrostatic repulsive potential rapidly decreases, and the diffuse double layer (also known as the Gouy–Chapman double layer) also shrinks to a point where the van der Waals attractive forces overpower the electrostatic repulsive potential. This effect is known as the "salting out" phenomenon, where a sufficient amount of salt is added to the colloidal suspension to destabilize an electrostatically stable nanofluid. When the concentration of the free salt ions rises, the zeta potential falls. Therefore, it is pretty clear that naked (despite having a high surface charge) or even electrostatically stabilized nanoparticles are more likely to agglomerate and destabilize in a saline environment. Al-Anssari *et al.*[210] reported a rapid reduction in the zeta potential of a SiO_2 nanofluid from -45 mV to 0 mV with an increase in salt (NaCl) concentration, which indicated nanofluid destabilization. The authors used both cationic (CTAB) and anionic (SDS) surfactants to improve the nanofluid stability at a high salt concentration (up to 5 wt%) to stabilize the nanofluid and found that anionic surfactants (SDS) worked better under low surfactant concentration (below the critical micelle concentration). However, the cationic surfactant destabilized the nanofluid irrespective of the salt concentration as it neutralized the negative surface charge on the SiO_2 nanoparticles in the presence of salt.

Jayezi and coworkers[206] reported the effectiveness of a polymer coating on the stability of nanofluids for EOR application. The study reported improved stability, solubility, and ease of transport of the polymer-coated nanofluid in the porous medium. The polymer coating of nanoparticles also contributed to higher oil recovery as it increased the viscosity of the injected fluid, resulting in the generation of a piston-like driving force within the oil reservoir. Application of surfactant and polymer-coated nanoparticles in EOR can reduce the interfacial tension and alteration in the liquid wettability characteristics of reservoir rock (oil–wet reservoir rock to water–wet reservoir rock) which is crucial from an oil recovery viewpoint. Without the additional stabilization, it is challenging to maintain nanofluid stability inside the oil reservoir (saline environment) as excessive particle deposition on the rock surface can block the pore spaces, reducing rock permeability and affecting the oil recovery process. Optimal nanoparticle deposition on reservoir rock is considered beneficial from a wettability alteration viewpoint. An identical study was performed by Bayat *et al.*[212] on a silica and titanium dioxide nanofluid, where nanoparticle deposition on limestone rock surface (a common type of reservoir rock) favorably altered the wettability behavior of the rock.

It is to be clearly stated that although both surfactant and polymer-based coating of nanoparticles can be used to stabilize nanofluids in a saline medium, and that the effectiveness of polymer-based steric stabilization and ionic polymer or ionic liquid-based electrosteric stabilization works better for stabilization purposes. Among the ionic surfactants, zwitterionic or amphoteric surfactants work best in a saline environment. The stabilizing effects of various highly acidic polyelectrolytes, *i.e.*, polyacrylic acid (PLA), poly(2-acrylamido-2-methyl-1-propanesulfonic acid) (poly(AMPS)), polystyrene sulphonate, amphoteric surfactant, *etc.* in a saline environment have been reported by several researchers.[206,213,214] In addition to the above options, ionic copolymers like poly(2-acrylamido-3-methylpropanesulfonate-*co*-acrylic acid) have shown excellent and long-term stabilization of nanofluids in a saline environment.[103,215]

The presence of salinity is not restricted only to EOR but extends to many industrial applications, such as petroleum refinery, petrochemical plant, food industry, nuclear power plant, *etc.*, where brine solution is used, or salt is created due to chemical reaction. In the aforementioned sectors where nanofluids can be utilized, their stability becomes a critical issue to be addressed, further iterating the necessity of this section.

3.4.4.4 pH of Dispersion Medium

As discussed in the previous section, the zeta potential value of different nanofluids is directly dependent on the pH of the dispersion medium.[216–220] Therefore, in the presence of salt, acid, or alkali, the pH of the nanofluid suspension alters significantly, which also affects the zeta potential of

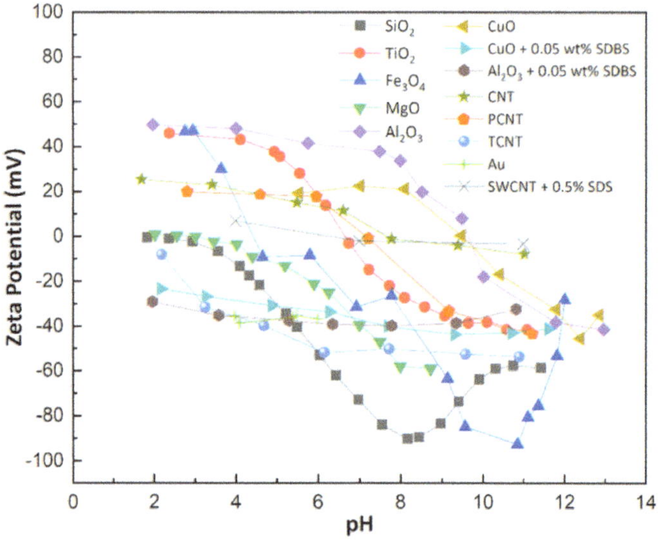

Figure 3.5 Zeta potential variation with pH of different nanofluids.[52,118,163,221,224–227]

nanofluids. The zeta potential of a nanofluid must be more than 30 mV to be deemed stable.[55] However, to attain excellent stability, the absolute value of the zeta potential should be 60 mV and above. The variation in the pH of the nanofluid suspension significantly affects the density of surface electron charge density, which is key to maintaining the high electrostatic repulsive force responsible for nanofluid stabilization. The variation in zeta potential due to change in pH has been reported in several articles, which is shown in Figure 3.5.[62,221] The effect of the various zeta potential ranges on the stability aspects of nanofluids is shown in Table 3.2.[55] One of the crucial challenges of maintaining nanofluid stability under varying pH conditions is preventing the colloidal suspension from reaching the IEP. At the IEP, suspended colloidal particles possess no net charges; henceforth, the zeta potential value becomes zero. Table 3.2 depicts the effects of zero zeta potential value on nanofluid stability. Umar *et al.*[216] reported the effects of pH alteration (by addition of (acetic acid, sodium hydroxide (NaOH), and ammonium hydroxide (NH_4OH)) on the clustering of alumina and copper oxide nanofluids. The effects of unfavourable pH alteration led to the aggregation of the nanoparticles (observed by TEM analysis) and reduced surface charge in acidic medium (measured by zeta potential analysis). The effect of pH on nanoparticle size and shape has also been observed by several researchers, which has direct or indirect implications on nanofluid stability.[222,223] An unfavourable pH of a colloidal suspension may also lead to non-uniform particle size and shape (spherical to cylindrical) distribution, which has been reported by Li *et al.*[91] using a silver nanofluid in basic medium (pH = 7).

3.4.5 Effect of External Forces

This section discusses the role of different external forces, *i.e.*, electrical field and magnetic field, on the long-term stability of nanofluids.

3.4.5.1 Effect of Electrical Field

The presence of an electric field affects the stability of nanofluids and can alter their thermophysical properties,[228–230] electrical insulation,[231,232] and wettability[233,234] features. Plausible reasons for heat transfer enhancement under the effect of an electric field are electrophoretic, dielectrophoretic, and electrostriction forces.[235] Other factors such as nanoparticle volume fraction, gravity, pressure, Reynolds number, Rayleigh number (ratio between buoyancy and viscous force, considered important when the mode of heat transfer is free convection) also influence heat transfer enhancement in the existence of an electric field. Though relatively stable, the nanofluid faces stability issues under the effect of an electric field under certain conditions. Many scholars have studied and reported stability-associated problems that occur under the influence of an electrical field.

Chand *et al.*[236] studied instability issues in a nanofluid horizontal layer under vertical electric field and rotation. Instability in nanofluids is primarily caused by buoyancy and is independent of Brownian motion and thermophoresis. Under the influence of an electrical field, if dielectric liquids are used as the base fluid, the addition of nanoparticles, even at low concentration, leads to significant augmentation in electrical insulation properties.[231,232,237] However, this also leads to the accumulation of positive ionic charges on the nanoparticle surface, which ultimately leads to a decrease in the number of free electrons in the dielectric nanofluid under electrical stress.[231,232,237]

Goudarzi *et al.*[238] observed that the application of a SDS stabilized Al_2O_3–water nanofluid under electrical field (inside car radiator) is only favorable up to an optimum nanoparticle concentration, beyond that limit heat transfer coefficient declines. Kong and coworkers[184] discussed the effect of a tribo-film mechanism, which is responsible for anti-wear and anti-friction properties of nanofluids for tribological purposes. In the presence of an electric field, exoelectron accumulation on the rubbing surface leads to nanofluid instability due to surface accumulation of nanoparticles.[184] In an electrical-field induced enhanced oil recovery (EEOR) application using NiO and CuO nanofluids,[239] around 9–22% improvement in oil recovery was reported compared to regular water flooding. This enhancement is due to the electrorheological effects resulting from orientational polarization and induced dielectric polarization, leading to an increase in viscosity in the resulting network of polarized molecules.[240] The nanofluid viscosity enhancement in EEOR could also lead to desirable stability enhancement due to increased inter-particle disjoining pressure[241]. Li *et al.*[242] studied the role of tapping depth mechanism in maximum potential creation, which has significance in stabilizing surfactant-coated nanoparticles under the influence of an electric field. The authors reported polarization of surfactant molecules attached to Fe_3O_4 nanoparticles (base fluid: vegetable oil) in the presence of a high electric field, which is considered beneficial for nanofluid stability. To enhance the dispersion stability of a clay-based material (sepiolite) in electrorheological fluid, *i.e.*, silicone oil, under the effect of an electric field, the development of a solid–gel-like structure is desired.[243] The formation of the solid-gel structure is beneficial for increased sedimentation resistance.

3.4.5.2 *Effect of Magnetic Field*

Nanofluids are utilized in several applications such as drug delivery,[244] patterning,[245,246] photonics,[247] cooling,[248–250] medical diagonistics,[251] *etc.*, where they are subjected to a magnetic field. Therefore, maintaining their stability is of paramount importance from an application viewpoint. However, under the influence of an external magnetic field, magnetic nanoparticles tend to agglomerate, which can have both positive and negative impacts on the use of nanofluids depending on the application area.[246,247,252–256] The beneficial attributes of nanoparticle aggregation under the influence of an magnetic field can be observed in drug delivery, medical diagnostics, photonics, patterning, and to a certain extent aggregation in heat transfer applications.[235,246,247,252,254–256] The positive attributes of magnetic nanoparticle clustering under a magnetic field (for heat transfer application) are due to enhanced particle–particle, particle–wall, and particle–liquid interactions/collisions, which augments the heat transfer rate. Shakiba *et al.*[257] reported improvement in the Nusselt number under the effect of the magnetic field. The beneficial attributes of magnetic field on nanofluid-based heat transfer have also been reported by other research groups.[258–260] A few studies have also[261] reported that beyond the optimal magnetic field strength, the heat transfer value is negatively affected. However, excessive aggregation of nanoparticles is undesired in most of the applications due to flowability issues, micro channel blockage, deposition/fouling effects, reduction in the surface-to-volume ratio of nanoparticles, reduced thermal properties, and increase in viscosity and pressure drop, *etc.* The negative impact of magnetic field on the stability and heat transfer performance of nanofluids has been reported by several researchers.[262–264]

Several researchers have reported the destabilizing role of magnetic fields on the stability of both magnetic and non-magnetic nanofluids. Chang and coworkers[265] studied the role of magnetic field on a copper oxide nanofluid and observed that the nanofluid lost stability rapidly. When the nanofluid was subjected to a magnetic field, the electrostatic repulsive potential diminished, the attractive potential took over, the particle size increased, and the zeta potential decreased, indicating particle aggregation and eventual nanofluid destabilization. Aggregation and rapid sedimentation of paramagnetic nanoparticles such as MnO, Mn_2O_3, $Mn_2P_2O_7$, and Cr_2O_3 under the effect of a strong vertical magnetic field have also reported by Parker *et al.*[266]

The consequence of magnetic field on the formation of the different colloidal assemblies (1D, 2D, and 3D) has also been observed by Gareev *et al.*[267]

Challenges associated with the maintenance of nanofluid stability after the removal of the magnetic field can be attained using electrostatic or steric stabilizing techniques. Nanofluid stability is also vital to ensure the reusability of magnetic nanofluids when subjected to a strong magnetic field. In comparison to other stabilizers, oleic acid has gained more popularity in ensuring the stability of nanofluids under a magnetic field.[251,268,269] The long polymeric chains of oleic acid create a protective barrier that prevents magnetic nanoparticles from forming permanent aggregates so that they can be easily auto dispensed once the magnetic field is removed.

3.4.6 Effect of Stabilizing Agents

Understanding the suitability of different stabilizers depending on nanoparticle type and loading, base fluid type, and operating conditions is crucial for attaining the prolonged stability of nanofluids. Electrostatic stabilization using ionic surfactants (SDS, CTAB, SDS, ALS, PLS, *etc.*) are unsuitable for stabilizing nanofluids at high particle concentration, which can be achieved through steric stabilization.[52,55] For high-temperature applications, surfactant and polymer-based stabilization techniques are highly ineffective as elevated temperature causes stabilizers to degenerate, agglomerate, and cause excessive foaming.[270,271] All of the aforementioned outcomes are highly undesirable for promoting the stability and thermophysical properties of nanofluids. However, surface functionalization, pH adjustment, and acid treatment are considered more suitable methods of high-temperature nanofluid stabilization.[126,272–274] The compatibility between suspended nanoparticles and stabilizers is considered to be most important for stabilization purposes. To enhance the stability of charged nanoparticles in the base fluid, the surface charge (positive/cationic or negative/anionic) of the used surfactant plays a key role in stabilization. For negatively charged particles, the anionic surfactant should be used for nanofluid stabilization or *vice versa*. The mechanism behind electrostatic stabilization using ionic surfactants is due to the existence of both hydrophobic and hydrophilic sections, which effectively interact with polar solvent molecules and can be used to disperse both hydrophobic and hydrophilic nanoparticles in the base fluid. Adsorption of surfactant molecules on the surface of the nanoparticles also leads to the formation of an EDL (in the case of metal oxide nanoparticles), which is responsible for the electrostatic repulsive barrier. Electrostatic stabilization using ionic surfactant is not suitable for electrolyte-sensitive systems. In the presence of salt, nanofluids need to be stabilized using steric or electrosteric stabilization (ionic copolymer). It should also be mentioned that surface functionalization or acid treatment (electrostatic) may also ensure stability in saline medium.

In the case of polymer-based stabilization, polymeric chain length, the affinity between the anchor section of the polymer with the nanoparticle, and favorable interaction between long polymeric chains with solvent molecules are crucial for nanofluid stability.[275] Steric stabilization can be attained using homopolymer, copolymer, and non-ionic surfactants. Copolymers can provide better steric stabilization than homopolymers as they are made of monomers with higher affinity towards nanoparticles (anchor part) and other monomers with a greater affinity towards solvent molecules, thereby providing enhanced stability. Copolymers compatible in polar solvents (polyvinyl chloride–polyvinyl pyrrolidone) are unsuitable for non-polar media (polyvinyl chloride–polyisobutylene and polyacrylonitrile–polystyrene).[52,276] Ionic polymers (polyacrylic acid–block–poly butyl acetate, polyvinyl sulfonic acid, polystyrene sulfonic acid) are used for electrosteric stabilization of nanofluids, where the adsorbed ionic (charged) polymer provides a combination of steric hindrance and electrostatic repulsive potential.[277,278] Electrosteric stabilization is capable of providing modest stability in a saline environment. A detailed discussion on different stabilization methods using surfactant, polymer, and acid treatment is discussed in the following section.

3.5 Different Methods of Nanofluid Stabilization

This section discusses the various stabilization approaches for improving nanofluid stability in dispersion media. The classification of several stabilization routes is shown in Figure 3.6, which is divided into mechanical and

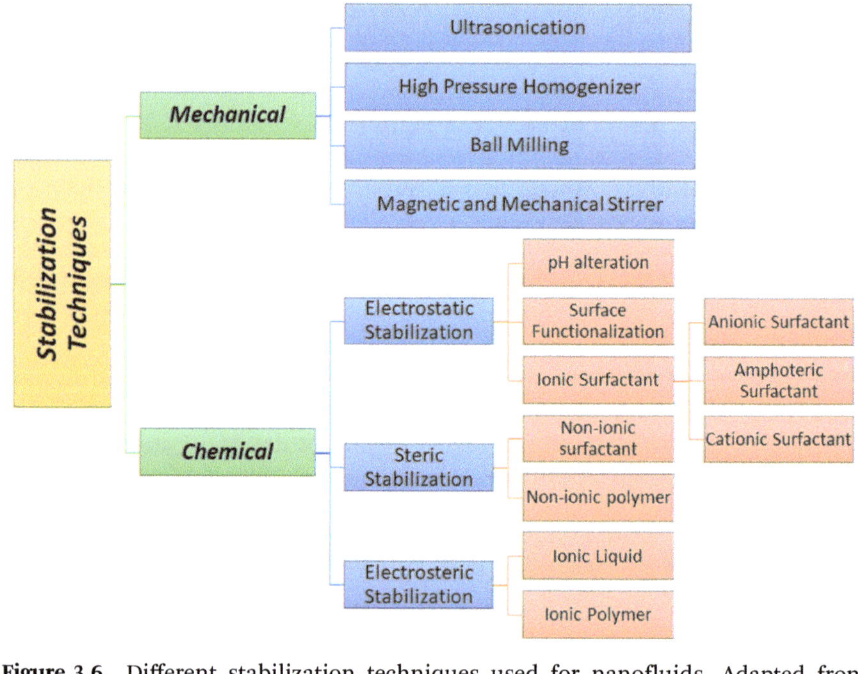

Figure 3.6 Different stabilization techniques used for nanofluids. Adapted from ref. 52 with permission from Elsevier, Copyright 2020.

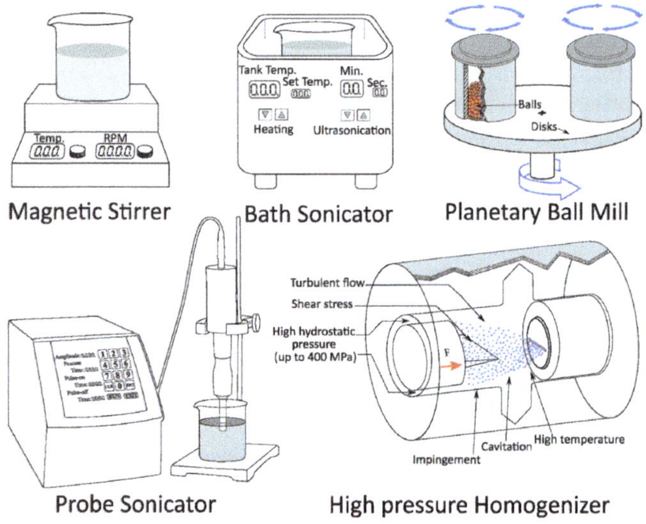

Figure 3.7 Schematic representation of mechanical stabilization instruments. Adapted from ref. 52 with permission from Elsevier, Copyright 2020.

chemical techniques. In the following section, we will go through these strategies in more detail.

3.5.1 Mechanical Stabilization

Several mechanical stabilization methods are used to reduce the cluster size of nanoparticles and maintain uniform particle dispersion, techniques such as magnetic stirring, ultrasonication, ball milling, and high-pressure homogenization. The following section provides an overview of these various approaches. Figure 3.7 shows a schematic representation of the different instruments used for mechanical stabilization, *i.e.*, a high-pressure homogenizer, planetary ball mill, bath sonicator, probe sonicator, and magnetic stirrer.

3.5.1.1 High-shear and High-pressure Homogenization

A high shear homogenizer consists of a rotor and a stator that rotate at a high rate. The aggregated particle clusters present in the base fluid are broken down into the primary particle size using this device.[279] As a substitute option, high-pressure homogenization is used, which drives nanoparticles through fine holes by applying high pressure. When compared with other mechanical stabilization techniques, high-pressure homogenization may be the best option. Hwang and coworkers[280] implemented a high-pressure homogenizer (at 1241 bar) to synthesize a carbon black–water nanofluid, wherein SDS was used as a stabilizer to improve the dispersion and ensure long-term stability. The study also proved that high-pressure homogenization helps disintegrate the particle aggregates present in the base fluid, based on the combined effects of cavitation and high shear force. Fedele and colleagues[281] implemented ultrasonication and high-pressure homogenization to synthesize CuO, TiO_2, and SWNT nanofluids in aqueous media. The study further established (*via* particle size measurements) the effectiveness of high-pressure homogenization over ultrasonication.

A high-shear homogenizer has also been used for preparing different nanofluids, such as an Ag/DI water nanofluid.[282] The nanofluid synthesized using this technique was found to be stable for more than 3 months. This type of homogenization was performed under a high pressure of 400 bar. The homogenizer, as mentioned earlier, is made up of two feeding microchannels that converge in a mixing chamber. Since the microchannel diameter is very small, this leads to a rise in velocity within the mixing chamber, and while the liquid flows through the microchannel, it experiences high shear due to wall effects. Once the two-fluid stream releases into the mixing chamber due to high-velocity impacts on the chamber walls, it leads to the formation of powerful cavitations within the nanofluid. High shearing, high impact velocity, and powerful cavitation cause nanoparticle clusters to disintegrate into a uniform particle suspension and stabilized nanofluid. Oil-based nanofluids of MWCNTs and diamond have also been prepared *via* high-pressure homogenization, with the results found to be favorable for stability enhancement.[55,283]

However, it is to be noted that a combination of multifaceted stabilization strategies (both chemical and mechanical stabilization) is necessary for attaining long-term stability. In a similar study by Drzazga and coworkers,[284] an amalgamation of nonionic surfactant and mechanical stabilization (ultrasonication, and high-pressure homogenization) was used to produce stable CuO–water nanofluids.

3.5.1.2 *Ultrasonic Vibration Technique*

The second-best mechanical method for nanofluid stabilization is the dispersion of nanoparticles *via* ultrasonic vibration. The ultrasonic vibration method can be broadly classified into two types; (a) direct sonication *via* probe sonication and (b) indirect sonication *via* bath sonication. Compared to the indirect sonication technique, direct sonication enables better size reduction and enhances suspension stability. Ultrasonication is considered to be the most popular mechanical stabilization method due to its low operating costs, better cluster size reduction, and improved dispersion stability without the need for additional stabilizers. Ultrasonic treatment of nanofluids is also advantageous for enhancement of thermophysical properties, as reported by several researchers.[285–290] Sadeghi and coworkers[291] used zeta potential measurements and DLS to determine the stability of an Al_2O_3 nanofluid. They found that when the duration of ultrasonication (0–180 min) increased, the zeta potential of the nanofluid doubled (25 to 50 mV), regardless of the concentration of the nanoparticles. Similarly, with increasing vibration time, the DLS data revealed a decrease in particle cluster size. The efficiency of ultrasonication techniques in preparing a monodisperse alumina nanofluid was also reported by Nguyen and coworkers;[292] who used ultrasonication for 500 s at a vibrational amplitude of 30% to lower the mean cluster size of the nanofluid from 230 nm to 130 nm. Chung and colleagues[293] discovered the effectiveness of an ultrasonic horn (probe sonicator) in preparing a stable zinc oxide/water nanofluid. Probe sonicators are more efficient for size reduction and lowering sedimentation rate. However, the optimum duration of sonication varies for various nanoparticles and base fluids. also It has also been discovered that a longer sonication time does not always lead to a small particle size. Kole and Dey[294] prepared a 1 vol% ZnO/EG nanofluid using ultrasonic vibration, with the vibration time was varied in the range of 4–100 h. Particle size analyses were conducted to measure the influence of vibration time on nanoparticle size. The study revealed that particle cluster size reduced up to 60 h, then increased again, indicating this to be the optimal sonication time.

Therefore, it is well understood that the length of ultrasonication should be optimized as too much ultrasonication might lead to structural flaws in nanoparticles and can diminish the benefits of size reduction. When surfactants are used as additives in nanofluids, excessive ultrasonication can damage the surfactant molecules due to heat generation. According to Walvekar and coworkers,[295] regardless of the concentration of the stabilizer utilized, found that there was an optimal sonication period to obtain a stable nanofluid. CNT nanofluids were shown to be stable after 4 h of sonication in their study. Amrollahi and colleagues[296] have also investigated the effects of ultrasonication on MWCNT/EG nanofluid sedimentation and thermal conductivity and found that stability and thermal conductivity were improved. As mentioned earlier, in the case of graphene or CNT-based nanofluids, excessive ultrasonication might cause flaws in their structure, resulting in CNT length shortening. The optimization of sonication time is required to avoid such ultrasonication-induced flaws.[297] Sadri *et al.*[287] looked at the thermophysical aspects of ultrasonication (heat conductivity, viscosity, and density). Ultrasonication causes a rise in the temperature of the suspension, which has an impact on its thermophysical properties. When using a heat-sensitive nanofluid, ultrasonication implementation in a programmable pulse and auto-tuning mode is instead suggested to control the temperature augmentation.

3.5.1.3 Ball Milling

Planetary ball milling (PBM) is considered one of the most effective mechanical techniques by which to produce nanoparticles without the need for chemical and other mechanical routes such as ultrasonication and high-pressure homogenization. However, to produce a stable nanofluid suspension, the PBM techniques should be complemented with other mechanical dispersion methods such as ultrasonication and magnetic/mechanical stirring. Boopathy and coworkers.[298] used a combination of three mechanical techniques, *i.e.*, PBM (wet milling), ultrasonication, and magnetic stirring to synthesize an Al and Cu nanofluid in a two-step approach. Munkhbayar *et al.*[299] used both PBM and ultrasonication methods to make an aqueous MWCNT nanofluid. The use of PBM reduces the average cluster size, making smaller-sized particles easy to disperse and improves the resulting stability of the nanofluid. The ball milling process for nanoparticle dispersion, in general, can be divided into two types: (a) stirred bead milling[300] and (b) dry and wet ball milling.[301]

In stirred bead milling, submicron level dispersion of nanoparticle aggregates in the base fluid is achieved without the need for additional stabilizers. The overall process consists of two stages: coarse milling (6 h) and fine milling (6 h), spanning over 12 h. Post milling operation, nanoparticles are passed through a high shear homogenization process to further lower the particle size and improve the stability. The process is also highly scalable for bulk production. In the dry and wet milling process, a planetary ball mill is utilized. In this process, zirconia balls (ZrO_2) are used for grinding, and the duration of the operation is 1 h. Two cylindrical pots are arranged symmetrically at opposite corners on a spinning disc. Here, 80% of the volume is occupied with particles and balls in a 1 : 210 weight ratio. The spinning disc (200–500 rpm) rotates in the opposite direction compared to the pots. For the wet milling process, 20 mL of deionized water is added into each pot in addition to all the dry milling components. Compared to wet milling, the size reduction and cluster breakage are better using this method. Post dry and wet milling process, surfactant addition, and high shear homogenization can further improve nanoparticle dispersion in the base fluid.[301]

In addition to the aforementioned ball milling techniques, surfactant-assisted ball milling can also be used to functionalize the nanoparticle surfaces and to increase their stability in base fluid.[302,303] Surfactant aided ball milling has several beneficial attributes such as preventing aggregate formation, better size reduction, increase in settling time, surface modification improves dispersion, increase in surface lubricity, prevents milling ball weight loss and contamination from milling balls.

3.5.1.4 Magnetic/Mechanical Stirring

Other than the aforementioned mechanical stabilization techniques, stirring (both magnetic and mechanical), shaking, and vortexing are also implemented to synthesize stable uniformly-dispersed nanofluids. For nanofluid preparation, a magnetic stirrer is a must-have tool. To prevent the formation of the bubbles during nanofluid synthesis, the magnetic stirring speed must be optimized. However, this approach is not capable of maintaining nanofluid stability over long periods on its merit and has to be complemented by high-pressure homogenization, ultrasonication, or chemical stabilization methods.[15,30,151] Moreover, this method is considered to be ineffective at higher nanoparticle loading.

3.5.2 Chemical Stabilization

It is unnecessary to use chemical stabilization techniques when base fluid and nanoparticles are compatible with each other; metal oxides, metal hydroxides, and graphene oxide are easily suspended in polar solvents with the help of ultrasonication high pressure/shear homogenization. Nonpolar solvents like oil, carbon tetrachloride, benzene, hexane can be used to disperse nanomaterials such as CNTs, graphene, and carbon black without the need for additional stabilizers. However, chemical stabilization strategies are necessary to generate stable nanofluids when hydrophilic nanoparticles need to be suspended in non-polar solvents and hydrophobic nanoparticles in polar solvents. The chemical stabilization techniques can be categorized into three types, (a) electrostatic, (b) steric, and (c) electrosteric stabilization. Electrostatic stabilization can be attained *via* three routes, (a) addition of ionic surfactant, (b) pH alteration, and (c) nanoparticle surface functionalization. Steric stabilization can be achieved *via* two pathways, (a) addition of non-ionic surfactant and (b) use of non-ionic polymer. Eletrostreric stabilization is a combination of the aforementioned methods, which can be attained using ionic polymers or ionic liquids. The classification of several chemical-based stabilizing techniques is depicted in Figure 3.6. Table 3.4 summarises the use of different chemical-based stabilizing strategies (electrostatic, steric, and electrosteric) for enhancing nanofluid stability.

3.5.2.1 Electrostatic Stabilization

When the van der Waals attractive potential is counterbalanced by an electrostatic repulsive potential acting between two similarly charged particles, then electrostatic stabilization is attained. The surface charge necessary for electrostatic stabilization can be obtained *via* surface group ionization or ionic surfactant adsorption on the nanoparticle surface. Ion adsorption on the nanoparticle surface occurs with ions acquired from the dispersion media. As the surface charge is acquired, the counterions also surround the nanoparticle surface, which leads to the formation of an EDL. In the case of negatively charged particles, a positively charged layer surrounds the nanomaterials while providing an electrostatic repulsive barrier against particle aggregation, thus enhancing the stability of the nanofluid. The aforementioned zeta potential value, which is regarded as the major indicator of nanofluid stability, is also obtained from measuring the electrical potential of the diffuse layer, which is a part of the EDL. Electrostatic stabilization is achieved by increasing the electrostatic potential barrier between two interacting nanoparticles preventing them from forming aggregates and leading to sedimentation. Figure 3.8(a) summarizes the mechanism involved in the electrostatic stabilization of the nanofluid for

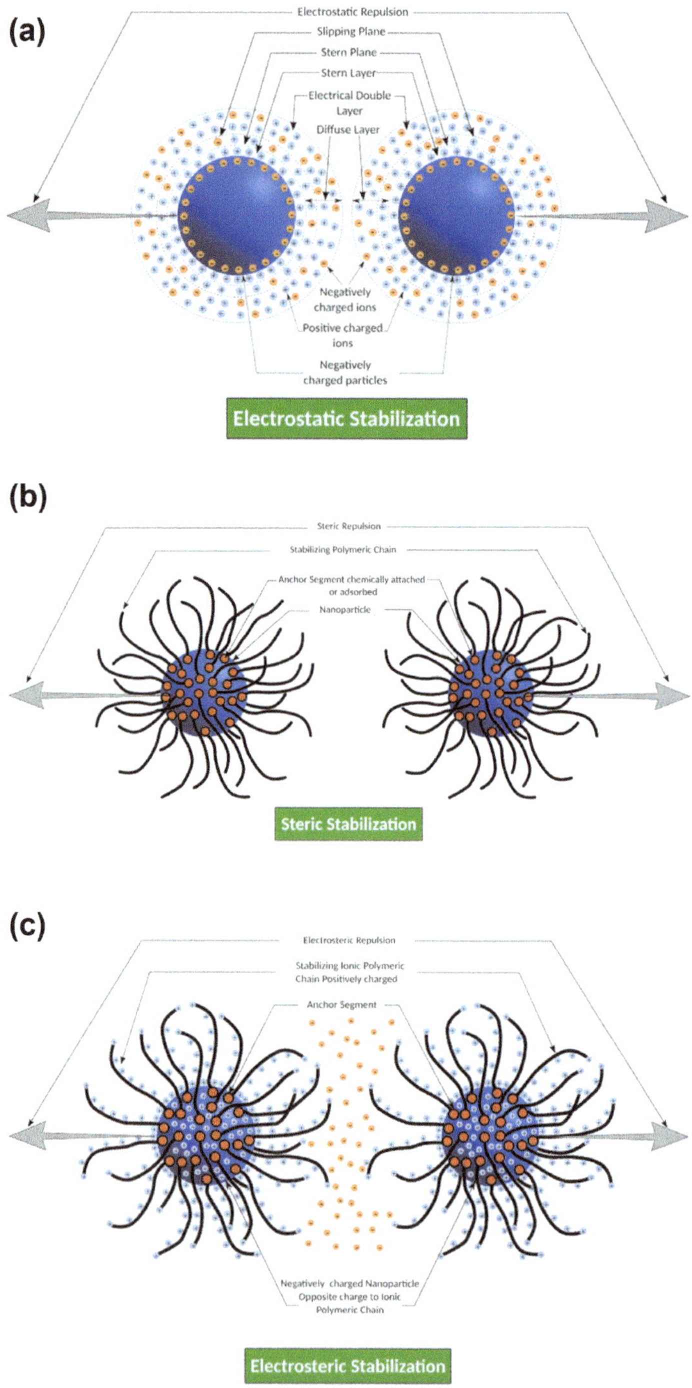

Figure 3.8 (a–c) Schematic representation of different chemical stabilization techniques: (a) electrostatic, (b) steric, and (c) electrosteric stabilization. Adapted from ref. 52 with permission from Elsevier, Copyright 2020.

augmenting its stability. As mentioned earlier, electrostatic stabilization can be accomplished by three routes, namely, the addition of ionic surfactants, pH alteration, and surface functionalization, which are discussed in detail in the following section.

3.5.2.1.1 Application of Ionic Surfactant. Ionic surfactants utilized for nanofluid stabilization can be categorized into three types, *i.e.*, anionic, cationic, and amphoteric.[52,54,55,61,163,304–306] The effectiveness of ionic surfactant in stability enhancement is due to two factors: (a) the presence of a hydrophobic tail and hydrophilic head, which are necessary to disperse hydrophobic nanoparticles in polar solvents or *vice versa*, and (b) an increase in the surface charge of nanoparticles due to surfactant adsorption, which thereby increases the electrostatic repulsive potential. Among the aforementioned surfactants, anionic surfactants are negative, whereas cationic surfactants are positively charged. The third type of surfactant, amphoteric surfactants, have both cationic and anionic groups in their chemical structure. The unique feature of these surfactants is that they can form an anion or cation depending on the pH of the base fluid or dispersion medium.[307] Compared to the previous two types of surfactants, they are also less toxic, resistant to water hardness, and can be used in combination with other anionic or cationic surfactants. The use of pyrene-derived stabilizers, which fall under the anionic surfactant category, is considered a noncovalent functionalization approach. However, it is to be noted that non-covalent stabilization of nanofluids can also be attained using other additives, such as nonionic surfactants and polymeric stabilizers, *etc.*[52,308–310] For carbon-based nanofluid stabilization (such as graphene and CNTs), non-covalent functionalization is highly effective, where additives such as SDBS, SDS, gum arabic, and Triton X-100 are used. Among the aforementioned stabilizers, for stability enhancement of a carbon-based nanofluid in an aqueous medium, SDBS shows the best performance, which is due to the robust π–π interaction between carbon-based nanomaterials. The main disadvantage of ionic surfactant-based stabilization lies in its limitation under high-temperature conditions as most ionic surfactants are temperature sensitive and degrade at elevated temperatures causing nanoparticles to agglomerate and destabilize. The use of ionic surfactant in an excess amount has a negative effect on thermophysical characteristics (lowers thermal conductivity and specific heat, increases suspension viscosity) and may also lead to foaming, which reduces the heat transfer area and heat transfer rate/coefficient. Commonly used ionic surfactants are listed in Table 3.3 based on their type.

3.5.2.1.2 pH Alteration. The IEP is defined as the pH value at which a particle bears no net electric charge and minimal hydration forces. Due to the absence of repulsive forces, nanofluids become unstable when their pH is in the vicinity of the IEP. The zeta potential of nanofluids is zero at the isoelectric point. Therefore, the repulsive forces between interacting nanoparticles diminish, and clustering takes place.[55,311] As the pH value deviates from the IEP, the zeta potential value improves, which is desired for nanofluid stabilization. Higher hydration forces are also an indication of improved nanofluid stability and are desirable for long-term stability. It should be noted that the surface charges on the nanoparticles tend to fluctuate as the pH of the base fluid is altered, thereby impacting the particle–particle interaction, which may lead to nanofluid destabilization. In the presence of anionic SDBS surfactant, the effect of pH alteration plays a critical role in instability for both Cu/water[312] and CNT/water[313] nanofluids. A Cu–water nanofluid at pH 9.5 has been shown to display excellent stability, due to the accumulation of surface charge as a result of the presence of an anionic stabilizer. Similarly, Ju *et al.*[313] investigated the role of pH on the agglomeration of CNT nanoparticles in aqueous media in the presence of SDBS. This study also concluded that the aggregation kinetics of CNTs is pH dependent. At a higher pH value (pH = 10), the functional group created by the SDBS surfactant alters, causing the surface to acquire more negative charges. The authors also reported the shrinking of the hydrodynamic radius of the CNTs as the suspension pH was changed from 3 to 10. Wang and coworker[314] studied the influence of pH on the particle aggregation tendency of Cu–water and Al_2O_3–water nanofluid. Both the nanofluids showed destabilization tendency below pH 7, whereas they showed improved stability above a pH value of 7.5. This enhancement in stabilization is due to higher electrostatic repulsion forces and better mobility of nanoparticles in the base fluid. The nanoparticles acquire high surface charge density at the optimal pH level, which is essential for maintaining nanofluid stability.[52,315] To disperse hydrophobic CNTs in an aqueous medium, acid treatment of nanoparticles is performed to incorporate hydroxyl groups for surface modification; this modification enables CNT dispersion in a hydrophilic medium.[118] For any thermal application, it is highly desirable to have a pH value of around 7 as higher or lower pH leads to risk of corrosion of heat transfer surfaces.[279] Figure 3.5 summarizes the different studies conducted on the effect of pH variation on the zeta potential value of various nanofluids, which provides a clear indication of its stability. The effect of the IEP point (pH value at which surface charge becomes zero) on the zeta potential values of various nanofluids is obvious from the aforementioned figure.

3.5.2.1.3 Surface Functionalization. The alternate route for attaining electrostatic stabilization is by altering or introducing polar or non-polar functional groups depending on the nature of the base fluid used. The surface functionalization of the nanoparticle is achieved using alkaline, acid, or plasma treatment.

The selection of functional groups is dictated by the nature of the base fluid in which nanoparticles are to be dispersed. Therefore, polar functional groups such as hydroxyl, carbonyl, carboxylic, amine, sulphate, and phosphate can be used to improve surface compatibility or dispersion of nanoparticles in the polar base fluid. Such type of treatment is referred to as

Table 3.3 Details of commonly used ionic surfactants for nanofluid stabilization.[52,55]

Surfactant name	Chemical formula	Molecular weight ($g\,mol^{-1}$)	Surfactant type
Sodium dodecyl sulphate, SDS	$CH_3(CH_2)_{11}SO_4Na$	288.38	Anionic
Sodium dodecyl benzene sulfonate, SDBS	$CH_3(CH_2)_{11}C_6H_4SO_3Na$	348.48	Anionic
Ammonium lauryl sulphate, ALS	$C_{12}H_{29}NO_4S$	283.43	Anionic
Potassium lauryl sulphate, PLS	$C_{12}H_{25}KO_4S$	304.49	Anionic
8-Hydroxypyrene-1,3,6-trisulfonic acid trisodium salt	$C_{16}H_7Na_3O_{10}S_3$	524.39	Anionic
Tetrasodium 1,3,6,8-pyrene tetrasulfonic acid	$C_{16}H_6Na_4O_{12}S_4$	610.43	Anionic
1-Pyrenesulfonic acid sodium salt	$C_{16}H_9NaO_3S$	304.30	Anionic
1-Pyrenebutanol	$C_{20}H_{18}O$	274.4	Anionic
Cetyl trimethyl ammonium bromide, CTAB	$C_{19}H_{42}BrN$	364.45	Cationic
Benzalkonium chloride	$C_{27}H_{42}NO_2.Cl$	448.1	Cationic
Cetrimonium chloride	$C_{19}H_{42}ClN$	320	Cationic
Distearyl dimethylammonium chloride	$C_{38}H_{80}ClN$	586.52	Cationic
Lecithin	$C_{35}H_{66}NO_7P$	643.9	Amphoteric
Sodium lauroamphoacetate	$C_{18}H_{34}N_2NaO_3$	349.471	Amphoteric
Cocamidopropyl betaine	$C_{19}H_{38}N_2O_3$	342.288	Amphoteric

covalent functionalization. Many types of covalent functionalization exist, including carbonyl, carboxyl, hydroxyl, and amine-based functionalization.[52,130] Amiri and coworkers[316] investigated the influence of both covalent and non-covalent functionalization on the stability of a CNT–water nanofluid using cysteine (covalent), Ag (covalent), and GA (noncovalent), respectively. Cyanamide was used to improve dispersion stability *via* the amine-based surface modification of crumpled nitrogen-doped graphene nanosheets.[317] This functionalization helped in the dispersion of the graphene nanosheets in a water/EG medium. Long-term stabilization of a water–quantum dot nanofluid was achieved *via* amine functionalization without the need for additional stabilizers. Amine functionalization was also found to be effective at high temperature as the nanofluid maintained its stability and thermal performance, and no fouling was observed on the heat transfer surface.[318] Kayhani *et al.*[319] used a combination of mechanical (ultrasonication, rotary evaporation) and chemical treatment (1,1,1,3,3,3 hexamethyldisilazan, $C_6H_{19}NSi_2$) to stabilize TiO_2–water nanofluids, with the nanofluids remaining stable for several days without any sign of particle aggregation or sedimentation. According to the authors, the introduction of the hydrophilic ammonium group was the reason behind enhanced nanofluid stability. Similarly, through functionalization of SiO_2–water nanofluids using (3-glycidoxylpropyl) trimethoxysilane, Yang and coworker[200] were able to maintain nanofluid stability for 12 months even with very high particle loading (10 wt%). The functionalization also prevented nanoparticles from depositing on heat transfer surfaces to cause a fouling effect and also improved Heat Transfer Coefficient (HTC) compared to a pristine nanofluid. Tang *et al.*[320] reported the surface functionalization of ZnO nanoparticles with polymethacrylic acid to improve their stability in an aqueous medium. The hydroxyl groups present on the surface of ZnO nanoparticles intermingled with the carboxyl groups of poly(methacrylic acid) (PMAA) to create a poly(zinc methacrylate) complex on the nano ZnO surface. Grafting of the aforementioned complex on the ZnO surface helped to prevent particle aggregation and resulted in improved stability in the aqueous medium. Yu and coworkers[321] employed plasma treatment with a blend of methane and oxygen to introduce various polar groups on nanodiamond surfaces to improve their stability in water. Ettefaghi *et al.*[322] used citric acid and urea as precursors to make a low-cost, long-lasting water-graphene quantum dot nanofluid. They employed a hydrothermal approach for creating a stable nanofluid that was both cost-effective and environmentally benign.

Wepasnick and coworkers[274] used a surface modification method to improve the dispersion stability of MWCNTs by incorporating polar functional groups using several oxidizing agents such as HNO_3, H_2SO_4, and $KMnO_4$. Similar studies on the acid-based functionalization of carbon-based nanofluids were also conducted by Smith *et al.*[272,323] Esumi *et al.*[273] investigated the consequence of acid treatment (HNO_3, H_2SO_4) on the dispersion stability of CNTs in a plethora of solvents. The study revealed that compared to untreated CNTs, the treated CNTs showed better stability in water, ethanol, and hexane. Xie *et al.*[175] reported that CNTs treated with HNO_3 or a mixture of HNO_3/H_2SO_4 showed long-term stability of up to 2 months in water and EG medium. Alkaline-based functionalization of graphene nanoparticles also showed favorable potential in enhancing their stability in an aqueous medium. Here, the authors used potassium persulfate ($K_2S_2O_8$) as an oxidizing agent and potassium hydroxide (KOH) for pH alteration to prepare water-soluble graphene nanosheets. The water-soluble nature of graphene was due to the incorporation of polar functional hydroxyl and carboxyl groups on the boundary of the graphene sheets.[124]

3.5.2.2 Steric Stabilization

Steric stabilization of nanofluids can be attained by two means, (a) by addition of polymeric molecules on a nanoparticle surface and (b) *via* non-ionic surfactant addition.[52,306,316,324,325] When polymers are added to enhance suspension stability, the polymer becomes attached to the particle surface, forming a protective layer surrounding it. This layer prevents neighboring nanoparticles from adhering to each other, therefore preventing particle aggregation and eventual sedimentation. The prevention of particle aggregation due to attached polymeric molecules is termed as steric hindrance. This type of stabilization is also known as non-covalent functionalization. This stabilization is highly effective under high particle loading, which is impossible *via* the electrostatic stabilization route. As mentioned earlier, the non-covalent functionalization of nanoparticles can be done using both ionic and non-ionic surfactants and polymers.

The mechanism responsible for attaining steric stabilization is due to the restricted movement of nanoparticles in the dispersion medium due to the presence of attached polymer and surfactant.[277,326] Several factors majorly affect the performance of the steric stabilization process, *i.e.*, (a) polymeric chain length, (b) adsorption capacity, and (c) polymer and nanoparticle compatibility.[275] Shorter chain length and poor adsorption capacity are highly undesirable for creating sufficient steric hindrance responsible for preventing particle aggregation. An amphiphilic polymer containing no net charges can be used for steric stabilization purposes. These polymers contain an anchor and tail section. The anchor section is adhered to the nanoparticle surface, whereas the polymeric tail part intermingles with the base fluid molecules to stop particle agglomeration. To minimize the strong van der Waals attractive potential responsible for the destabilization, it is necessary to have powerful enthalpic interactions between the attached polymeric tail and base fluid molecules. In addition to the above criteria, a long polymer chain length and high density of the attached polymer play a crucial role in steric stabilization.[52,65,327] The steric stabilization mechanism is schematically depicted in Figure 3.8(b).

As mentioned earlier, steric stabilization can be achieved using both polymer and non-ionic surfactants. The commonly used nonionic surfactants for steric stabilization are GA, oleylamine, oleic acid, Rokanol K7, Rokacet O7, Span 80, Triton X 100, Tween 20, Tween 80, and PVP, *etc.*[55,95,127,328,329] In addition to the above, aromatic surfactants, *i.e.*, polycyclic aromatic hydrocarbons such as perylenebisimide-based bolaamphiphile surfactant and the tetra potassium salt of coronene tetracarboxylic acid, are also used for steric stabilization. Due to the amphiphilic nature of non-ionic surfactants, they can adhere to the surface of nanoparticles easily and provide better stability in both polar and non-polar base fluids. For carbon-based nanofluids, there are several challenges associated with their steric stabilization as these surfactants tend to thermally degrade at elevated temperature, resulting in excessive foam formation, which not only reduces heat transfer area but also affects thermophysical properties. The long-term steric stabilization (3 months) of a water-based SWCNT nanofluid was achieved using GA as a stabilizer.[127] Tang *et al.*[329] reported the preparation of a highly-stable (2 months) water-based MWCNT nanofluid using PVP and Triton X-100 as dispersants. The effectiveness of oleic acid as a steric stabilizer has been reported by Hwang *et al.*,[280] where authors were able to maintain nanofluid stability for up to 2 months by suspending Ag nanoparticles in a non-polar (silicone oil) medium. Li *et al.*[91] also prepared a highly stable Ag nanofluid using oleic acid and *N*-butylamine as dispersants in heptane medium.

In the case of polymer-based steric stabilization, the type of polymer used can either be homopolymer or copolymer. Homopolymers are made out of identical monomers, whereas copolymers are made of different monomers combined into a single polymeric chain. Copolymers can be categorized into three types: random, graft, and block type. Compared to random copolymers, block and graft copolymers perform better in terms of steric stabilization. However, the selection of a copolymer is dictated by two factors, *i.e.*, (a) compatibility and affinity between the copolymer with dispersed nanoparticles where the polymers are attached *via* chemisorption or physisorption and (b) favorable interaction between the chemisorbed polymer and the dispersion medium.[275] For ideal steric stabilization, the presence of two types of monomer (as in a copolymer), one monomer should display affinity towards the suspended nanoparticles, whereas the other should show affinity towards solvent molecules. Due to the presence of such a comonomer in the copolymer chain, better stabilizing effects are observed. However, it is to be pointed out that copolymers that are compatible with polar solvents are not suitable for non-polar solvents. Copolymers such as polyvinyl chloride–polyvinyl pyrrolidone (PVC–PVA), polyacrylonitrile–polymethacrylic acid (PAN–PMAA), polystyrene–polyethylene oxide (PS–PEO), polyvinyl acetate–polyvinyl alcohol (anchor polymer-stabilizing moieties), and polypropylene–polyvinyl methyl ether (PP–PVME) are more compatible with polar solvents (water), whereas in the case of non-polar solvents, copolymers such as PVC–polyisobutylene (PVC–PIB), PAN–PS), PEO–polylauryl methacrylate (PLMA), and PP–polydimethylsiloxane (PDMS), *etc.*,[95,127,280,330] are implemented. In the above-mentioned list of copolymers, the first monomers act as the anchor polymer, whereas the second monomers are responsible for interaction with solvent molecules and contribute majorly to suspension stability. Similar to copolymers, non-ionic surfactants with long chains also provide effective stabilization in both polar and non-polar solvents. Triton X-100 is a prime example of this behavior, which shows effectiveness in stabilizing both hydrophilic and hydrophobic nanoparticles in polar and non-polar solvents.[330,331]

It should be mentioned that quantification of steric stabilization is not possible through zeta potential measurements as polymers and surfactants used for stabilization are of neutral charge. To evaluate the suspension stability, stability evaluation methods such as sedimentation, centrifugation, absorbance–transmittance, and particle size measurements are the preferred options. As discussed earlier, the primary disadvantage of this stabilization lies in its limitations at high temperatures and its adverse effect on thermophysical properties. However, there are several advantages of steric stabilization techniques that overcome its limitations. (a) Unlike electrostatic stabilization *via* steric stabilization, aggregated particle clusters can be dispersed/redispersed in base fluids, (b) nanofluid stability can be maintained even with high particle loading, (c) they are useful for electrolyte sensitive systems, and (d) they are ideal for multiphase systems. Table 3.4 shows several examples of nanofluids stabilized *via* the steric stabilization route.

Table 3.4 Summary of the performance of different implemented nanofluid stabilization techniques.

Authors	Nanofluid	Dispersants	Zeta potential and duration of stability
Mechanical stabilization			
Srinivas *et al.*[335]	Water–CNTs	None	20.5 mV
Fedele *et al.*[281]	Water–single walled carbon nanohorn (SWCNH)	None	35 mV
Choudhury *et al.*[108]	Water–Al_2O_3	None	14 mV
Chakraborty *et al.*[165]	Water–TiO_2	None	2 h
Sandhu *et al.*[96]	Water–EG (50 : 50) –Al_2O_3	None	57 mV, 28 days
Pang *et al.*[164]	Methanol–Al_2O_3	None	60 mV
Chakraborty *et al.*[15]	Water–Cu–Zn–Al LDH	None	38.6 mV, 18 h
Nine *et al.*[132]	Water–MWCNTs–Al_2O_3	None	5 days
Sundar *et al.*[336]	Water–nanodiamond–Fe_3O_4	None	−34.6 mV, 3 days
Electrostatic stabilization			
Srinivas *et al.*[335]	Carboxylated Water–CNTs	Ethylexanoic acid, sebacic acid	38.2 mV
Wu *et al.*[337]	Water–SWCNTs	Humic acid	10 days
Islam *et al.*[338]	Water–MWCNTs	SDBS	3 months
Yu *et al.*[129]	Water–MWCNTs	SDS	3 months
Kim *et al.*[331]	Water–MWCNTs	SDBS	1 day
	Water–MWCNTs	CTAB	1 day
Wusiman *et al.*[123]	Water–MWCNTs	SDS	1 month
	Water–MWCNT	SDBS	1 month
Fedele *et al.*[281]	Water–SWCNH	SDS	−40 mV, −26.9 mV, 15 days
Hwang *et al.*[280]	Water–carbon Black	SDS	−26.9 mV
Li *et al.*[91]	Water–Cu	CTAB	28.1 mV, 7 days
	Water–Cu	SDBS	−43.8 mV, 7 days
Wang *et al.*[220]	Water–Cu	SDBS	−43.8 mV
Sandhu *et al.*[96]	Water–CuO	SDS	40 mV, 2 days
	Water–EG (50 : 50)–CuO	SDS	47 mV, 4 days
Sarafraz *et al.*[94]	EG–CuO	SDS + HCl + NaOH	45 days
Kamalgharibi *et al.*[92]	EG–CuO	SDS	75 days
Choudhury *et al.*[108]	Water–Al_2O_3	SDS	−30 mV, 16 days

Table 3.4 (*Continued*)

Authors	Nanofluid	Dispersants	Zeta potential and duration of stability
Wang *et al.*[131]	Water–Al_2O_3	SDBS	−40.1 mV
Mo *et al.*[100]	Water–TiO_2	SDS	12 days
Ghadimi *et al.*[339]	Water–TiO_2	SDS	−55 mV, 7 days
Salari *et al.*[97]	Water–TiO_2	SDS	11 days
Song *et al.*[131]	Water–stainless steel	SDBS	−70 mV, 10 days
	Water–stainless steel	CTAB	60.1 mV, 10 days
Chakraborty *et al.*[340]	Water–Cu–Zn–Al LDH	SDS	−50.6 mV, >1 day
Suresh *et al.*[341]	Water–Al_2O_3–Cu	SDS	+46 mV
Steric stabilization			
Bandyopadhyaya *et al.*[127]	Water–SWCNTs	GA	3 months
Tang *et al.*[330]	Water–MWCNTs	PVP	2 months
	Water–MWCNTs	Triton X-100	2 months
Indhuja *et al.*[125]	Water–MWCNTs	GA	−29.1 mV, 30 days
Xie *et al.*[118]	Water/EG/decane–pristine CNTs (PCNTs)	Oleylamine	5 min
	Water/EG/decane-treated CNTs (TCNTs)	Oleylamine	2 months
Sarafraz *et al.*[334]	Water–COOH–CNTs	Nonylphenol ethoxylate	21 days
Li *et al.*[91]	Water–Cu	Triton X-100	−8.3 mV, 7 days
Hwang *et al.*[280]	Silicon oil–Ag	Oleic acid	2 months
Li *et al.*[91]	*n*-heptane–Ag	Oleic acid and *N*-butylamine	1 month
Sahooli *et al.*[95]	Water–CuO	PVP	32.3 mV, 7 days
Salari *et al.*[101]	Water–Fe_3O_4	Nonylphenol ethoxylate	−32 mV, 21 days
Chakraborty *et al.*[165]	Water–TiO_2	PVP	8 h
Chakraborty *et al.*[340]	Water–Cu–Zn–Al LDH	Tween 20	24.3 mV, 12 h
Electrosteric stabilization			
Bagaria *et al.*[102]	Water–Fe_3O_4 (NaCl, $CalCl_2$ salt-containing water)	Poly(2-acrylamido-3 methylpropanesulfonate-*co*-acrylic acid)	−45 mV
Iqbal *et al.*[333]	Water–Fe_3O_4 (NaCl, $CalCl_2$ salt-containing water)	Poly(AMPS-*co*-AA)	−53.6 mV, 1 month

3.5.2.3 Electrosteric Stabilization

Electrosteric stabilization is the least explored method of stabilization compared to its counterparts. Electrosteric stabilization combines the features of electrostatic and steric stabilization. This type of stabilization generally uses ionic polymer and ionic liquids to stabilize suspended nanoparticles. Here, similar to steric stabilization, the ionic polymers are attached to the nanoparticles surface, which restrict particles from coming close to each other and thereby preventing agglomeration and colloidal destabilization. However, in the case of electrosteric stabilization, due to the presence of charge in the adsorbed polymeric chain, an additional electrostatic potential barrier is also formed, which provides two layers of protection against nanoparticle destabilization. Ionic polymers are also compatible with nanoparticles that have different surface charges. Similar to the stabilizers used for steric stabilization, ionic polymers also contain two distinct segments: (a) an anchor section of the polymer is adsorbed on the nanoparticle surface and (b) an ionic polymeric chain to act as an additional electrostatic repulsion barrier against particle aggregation and to provide strong enthalpic interaction with the surrounding base fluid molecules. Ionic polymers are also known as polyelectrolytes. Polyelectrolytes are made of repeated polymeric blocks containing ionizable groups such as –COOH, –SO_3H, and –PO_3H_2, *etc.* Electrosteric stabilizers include polyacrylic acid-block-polybutyl acetate, polyvinyl sulfonic acid, polystyrene sulfonic acid, poly(1-vinylpyrrolidone-*co*-acrylic acid), and poly(methacrylic acid), among others.[277,278] An electrically charged polymer is formed when the ionizable part of the polymer is dissociated in the presence of the dispersion medium. To facilitate electrosteric stabilization, both the nanoparticle and the polyelectrolyte must possess opposite charges to facilitate efficient polymer adsorption on the nanoparticle surface. Polymers containing both cationic and anionic groups are referred to as polyampholytes and can be used to stabilize positively and negatively charged nanoparticles in different base fluids. Implementation of ionic liquids is an alternate approach to attain the electrosteric stabilization of nanofluids. Ionic liquids (hexafluorophosphate, imidazolium, pyridinium, tetrafluoroborate, *etc.*) also work similarly to ionic polymers as they provide both steric and electrostatic stabilization due to high surface charge and strong steric hindrance by providing an aggregation preventive outer layer.[277,332] Electrosteric stabilization has also been shown to be effective in a saline environment to produce a stable iron oxide (Fe_3O_4) nanofluid. Several researchers have used poly(2-acrylamido-3 methylpropanesulfonate-*co*-acrylic acid) (PDMS-*co*-AA) as a stabilizer to prepare a stable iron oxide nanofluid.[102,333] The stability of an Fe_3O_4 nanofluid was tested using sedimentation (up to 1 month) and zeta potential analysis (above −45 mV), which proved its long-term stability. Figure 3.8(c) shows the underlying mechanism involved in nanofluid stabilization *via* the electrosteric stabilization route.

Table 3.4 summarizes commonly used stabilization techniques (mechanical and chemical), which clearly shows the stabilization effectiveness of the different techniques, as reported by several researchers.[15,91,92,94,95,97,100–102,108,118,123,125,127,132,164,165,281,330,333,334]

3.6 Stability Evaluation Methods

For the determination of nanofluid stability, numerous methods have been reported in the literature, such as sedimentation and centrifugation, absorbance and transmittance measurements, DLS, zeta potential measurements, transmission, and scanning electron microscopy (SEM), and 3ω methods. Figure 3.9 schematically represents the different stability evaluation methods. The following section provides a quick overview of the various strategies.

3.6.1 Sedimentation and Centrifugation

Sedimentation is the most versatile and widely used method for stability determination. This method is reliant on the settling of suspended particles under the influence of gravity. The density difference between the dispersed phase and dispersion medium, particle size, and medium viscosity play a crucial role in determining settling/sedimentation velocity. Here, in this method, the time required to form the sedimentation layer at the bottom of the container is an indication of nanofluid stability. The higher the sedimentation time, the better the stability of the nanofluid suspension. The sedimentation velocity, derived from Stokes law,[342] of dispersed species in the suspension can be calculated using eqn (3.7).

The sedimentation velocity equation is derived by taking into account drag, buoyancy, and gravitational forces. However, the equation does not consider the presence of neighbouring particles and particle–particle repulsive forces (electrostatic repulsive forces). Here, particle size is also large compared to the mean free path of the fluid molecule. Many researchers[15,165,196,343,344] have employed the sedimentation method for nanofluid stability evaluation. The stability of a chitosan-based MWCNT nanofluid[345] was reported using sedimentation techniques where the authors achieved nanofluid stability over months compared to a nanofluid prepared without chitosan, which settled within 30 min. It is to be pointed out that the major limitation of this technique lies in its highly time-consuming nature.

The centrifugation technique is a quicker alternative to the time-consuming sedimentation-based stability evaluation technique. Sedimentation techniques are also more feasible for particle sizes for as small as 1 μm. However, when dealing with submicron particles or nanoparticles (<100 nm), centrifugation is a more practical route as it significantly lowers the time required for stability evaluation compared to normal gravity-based sedimentation techniques. Under the influence of high centrifugal force, the mixing tendency that arises from diffusion and convection subsides. Therefore, the sedimentation

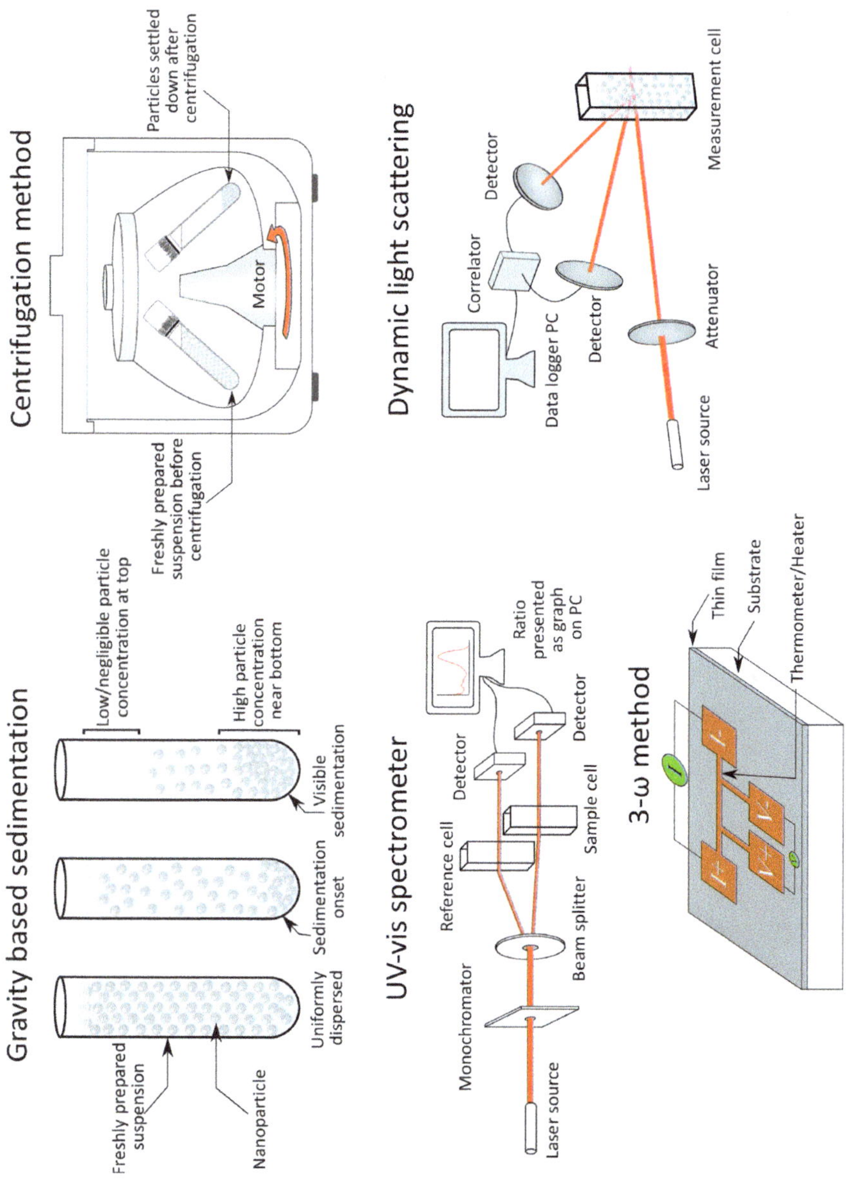

Figure 3.9 Schematic depiction of the different stability evaluation methods.

process is accelerated, which reduces sedimentation time. A closer look at eqn (3.18) reveals that gravitation constant (g) is replaced by $\omega^2 X$, where ω denotes the angular velocity of the centrifuge and X represents the separation distance between the centrifuge tube and the rotation axis of the centrifuge. A normal laboratory centrifuge can exert gravitational acceleration a thousand times powerful than normal "g", which is why settling time drastically reduces.

For spherical nanoparticles that obey the Stokes law regime ($\mathrm{Re_p} < 1$), terminal velocity for centrifugation is given by eqn (3.18):

$$V_c = \frac{d_p^2 \left(\rho_p - \rho_o\right) \omega^2 X}{18\mu} \tag{3.18}$$

where V_c, ω, and X are the terminal settling/sedimentation velocity under centrifugation, centrifuge angular velocity ($= 2 \times \pi \times$revolution per second), and the separation distance between the centrifuge tube and the rotation axis of the centrifuge, respectively. The remaining terms are the same as in the previous equation. The equation shows the importance of particle size, density difference, angular velocity, and medium viscosity on settling velocity. Moderate viscosity, minimal density difference, and small particle size are desirable to ensure reduced settling velocity and improve suspension stability. A few researchers[160,346] have also employed centrifugation techniques for stability evaluation. Sharma *et al.*[346] tested the stability of a SDBS-stabilized CNT nanofluid using both sedimentation and centrifugation techniques. The authors reported both instantaneous and long-term stability (up to 15 months) of a CNT–SDBS nanofluid. Similarly, Singh *et al.*[160] used a centrifugation method at 3000 rpm for 10 h to evaluate the suspension stability of an ethanol-based Ag nanofluid (dispersant: polyvinyl pyrrolidone). The authors reported no sedimentation of the nanofluid within the duration of the centrifugation. This result was cross-verified by a normal sedimentation study where the nanofluid remained stable for over a month, which reaffirmed the stability and reliability of centrifugation as a stability evaluation method.

3.6.2 Absorbance and Transmittance Measurements

Determination of spectral absorbance is a popular technique by which to evaluate the stability of nanofluids. In this method, the reduction in spectral absorbance with time is regarded as an indication of particle sedimentation. There exists a linear relationship between spectral absorbance and particle loading. The UV–vis spectrophotometer is a widely used instrument for stability determination. However, the main limitation of this technique lies in its absorbance wavelength range of 190–1100 nm.[56,62,347] If a nanoparticle has a characteristic absorbance peak in the aforementioned wavelength range, its stability in the continuous phase can be easily measured using a UV–vis spectrophotometer. In addition to stability evaluation, this technique can also be used to measure the concentration of nanoparticles in base fluid

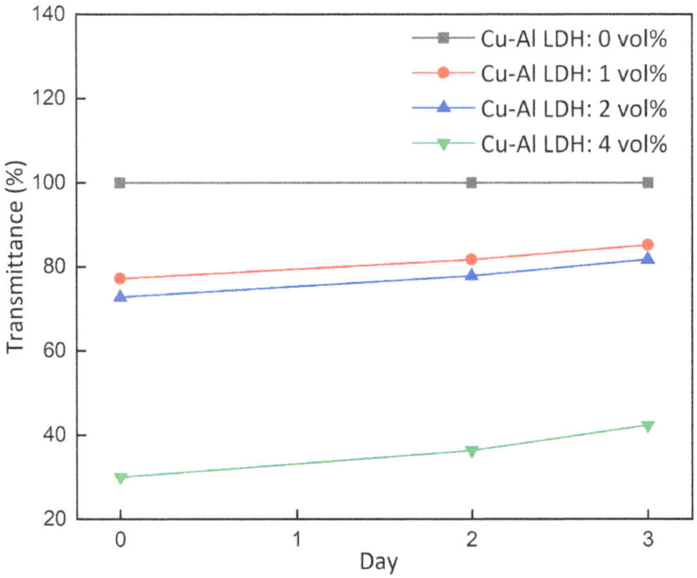

Figure 3.10 Evaluation of nanofluid stability through transmittance measurements at different particle loading and time intervals. Adapted from ref. 52 with permission from Elsevier, Copyright 2020.

(supernatant liquid) at various time intervals, which is not possible using other methods. Many researchers have used UV–vis spectrophotometers to evaluate the stability of different nanoparticles (CNTs, Cu, CuO, SiO_2, TiO_2, Al_2O_3 Ag, MXenes, Fullerenes, *etc.*)[62,291,348–352] in different media (water, paraffin oil, lubricants, propylene glycol).[62,291,347,348] A UV–vis spectrometer is more suitable for faint-colored and less concentrated colloidal suspensions.[150,353] The characteristic absorption band alters over time as particle sedimentation and aggregation take place, which is continuously monitored using a UV–vis spectrophotometer to determine nanofluid stability. As mentioned earlier, the spectral absorbance (A) is directly proportional to particle concentration in the base fluid. Therefore, the reduction in absorbance over time indicates the instability/sedimentation of a nanofluid suspension.

Similar to spectral absorbance, transmittance can also be used to determine nanofluid stability. Since transmittance (T) and absorbance (A) are inversely proportional to each other, an increase in transmittance therefore indicates poor nanofluid stability and particle aggregation. Chakraborty *et al.*[52] assessed the stability of a Cu–Al LDH nanofluid using a transmittance method, with the r revealing that at high particle concentration, the transmittance increases with time, which is a distinct representation of stability deterioration. Figure 3.10 shows the transmittance value of an LDH nanofluid at different time intervals and particle loading. The significant alteration in the transmittance value is only visible at 4 vol% nanofluid concentration.

3.6.3 DLS

The DLS method is generally employed for particle size, cluster size, and particle size distribution assessment for different colloidal suspensions. The instrument used in DLS determines particle size based on the Stokes–Einstein equation[354] and uses laser illumination to detect nanoparticles suspended in the base fluid. The particle detection process is accomplished using a photon detector that observes scattered light fluctuation as the laser encounters random particle motion (Brownian motion) in the base fluid. The intensity of light scattering is directly related to the particle size, which is determined from the Stokes–Einstein equation.[354] The equation used for particle size determination is as follows:

$$R_{\mathrm{H}} = \frac{k_{\mathrm{B}}T}{6\pi\mu D_t} \tag{3.19}$$

where the Boltzmann constant, hydrodynamic radius, temperature, translational diffusion coefficient, and viscosity of the base fluid are represented by k_{B}, R_{H}, T, D_t, and μ, respectively. To determine nanofluid stability using DLS analysis, it is necessary to measure particle size at regular intervals. Such measurements would allow a measure of particle aggregation tendency, which eventually leads to sedimentation. Several researchers have used DLS measurements to detect instability in a plethora of nanofluids such as CuO,[281,355] TiO_2,[281] Al_2O_3,[291] SiO_2,[356] SWCNH,[281] *etc.*, by measuring the increase in cluster size.

3.6.4 Zeta Potential Measurement

Zeta potential measurement is an effective and widely accepted method for colloidal stability determination. This method is based on the electrophoretic characteristics of colloidal suspension. The aforementioned characteristics provide a clear indication of the ability of suspended nanoparticles to create a stable colloidal suspension in the base fluid. As colloid particles are innately ionic, surface charge is generated when they are dispersed in the base fluid and attracts the opposite charge surrounding the nanoparticle, termed as the stern layer. Another layer also forms adjacent to the Stern layer, commonly known as the diffuse layer, wherein both positively and negatively charged ions are present. However, the stern layer is relatively loosely packed. The surface, stern, and diffuse layers combine to form an EDL, which is electrically neutral. The zeta potential value does not represent surface potential or stern potential rather the potential difference between the stern layer and continuous medium. The zeta potential value is measured in millivolts, and its value is much lower than the surface and stern potential. Figure 3.11 shows all three potentials that constitute the EDL. The interpretation of various zeta potential values from a stability standpoint is shown in Table 3.2. In a nutshell, it can be summarized that a high zeta potential value (absolute) is necessary to maintain long-term nanofluid stability and prevent particle aggregation as it manifests a

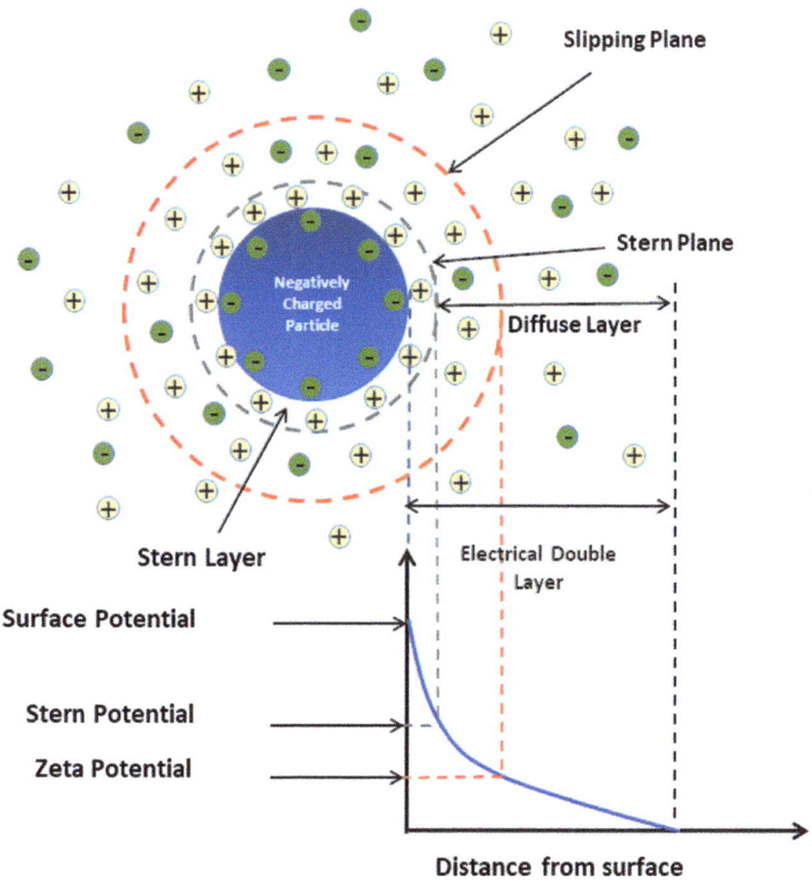

Figure 3.11 Schematic representation of the surface, stern, and zeta potential of charged particles.

higher repulsive potential. To promote better stability of a nanofluid, the zeta potential value should be greater than ±45 mV. A zeta potential of lower than ±30 mV reflectspoor nanofluid stability.

3.6.5 TEM and SEM

Similar to DLS analysis, TEM can also be used to measure particle shape, size distribution, and assessment of particle aggregation tendency. This information can be utilized as an alternative route to evaluate nanofluid stability. Numerous researchers have utilized TEM to report nanoparticle aggregation tendency, size, and shape.[15,62,165,340,357,358] Das *et al.*[359] reported the stabilization of an Al_2O_3–water nanofluid using CTAB, SDS, and SBBS as stabilizers. The authors reported better stability in the presence of stabilizers compared to a pristine Al_2O_3 nanofluid. The lowest aggregation of nanoparticles was obtained using SDBS as a stabilizer. Kim and coworkers[358] used TEM analysis to investigate the aggregation tendency of ethanol and EG-based Cu and Ni nanofluids. Compared to ethanol, An EG-based nanofluid showed minimal aggregation tendency, which was observed *via* TEM analysis. It should be mentioned that TEM analysis for nanofluid stability evaluation can only be conducted under low particle loading conditions as at high particle concentration aggregation is likely to happen due to evaporation of base fluid from the TEM grid.

Like TEM analysis, SEM can also be used to determine particle agglomeration tendency.[360] This test is performed by casting a nanofluid drop on the SEM grid (carbon tape) and then allowing it to dry before measuring the aggregation tendency of the sample.

3.6.6 3ω Method

The 3ω technique is a relatively less explored method for assessing the dispersion stability of different nanofluids. This method uses a combination of sedimentation and thermal conductivity measurements to evaluate stability. In this method, the thermal conductivity of the nanofluid is measured at the onset of sedimentation and particle aggregation.[361] The alteration in thermal conductivity is due to instability in the nanofluid. The advantage of this method lies in its applicability for a wide range of particle concentrations. Recently, several studies have utilized this technique to evaluate the stability of different nanofluids such as CuO,[362] Al_2O_3,[363] and TiO_2.[364] Figure 3.9 shows the experimental setup used to measure nanofluid stability using the 3ω method. The effect of particle aggregation can have both incremental and detrimental impacts on the thermal conductivity of nanofluids, as reported by several researchers,[362,363,365] which can be detected using the 3ω method. Oh *et al.*[363] and Karthikeyan *et al.*[362] reported a rise in thermal conductivity due to particle aggregation, whereas Evans and Prasher *et al.*[366,367] concluded that thermal conductivity increased only up to an optimal aggregation level, beyond which the thermal conductivity declined again.

Table 3.5 shows a summary of different stability evaluation methods for working principles, parameters, and interpretation of stability data.

3.7 Properties to Attain for Long-term Nanofluid Stability

Despite extensive research on nanofluid synthesis, property measurements (thermal, optical, magnetic, electrical, *etc.*), and their applications, the long-term stability and reusability aspects of nanofluids have remained elusive. Many researchers have reported difficulties associated with the long-term stabilization of nanofluids, but only a few have been able to ensure stability over a considerable time duration. All the desirable properties necessary for attaining long-term nanofluid stability are listed below:

(a) Nanoparticle size should be >20 nm to minimize the number of atoms on the surface of the particles. The greater the number of surface

Table 3.5 Summary of different nanofluid stability evaluation methods.

Name of process	Parameter	Working principle	Interpretation
Sedimentation	Settling velocity, particle size, particle and solution density, liquid viscosity, gravitational force	Gravitational force, Stokes law	• High settling velocity indicates unstable nanoparticles suspension. • Large density difference of dispersed particles and the dispersion medium. • Particles size is above 1 μm.
Centrifugation	Settling velocity, particle and solution density, angular velocity, centrifuge distance from the rotation axis, liquid viscosity	Centrifugal force, Stokes law	• Higher settling velocity indicates unstable nanoparticle dispersion. • Very high solid and liquid density difference.
Zeta potential	Surface potential, stern potential, zeta potential (mV), distance from the surface	Electrical diffuse layer	• Zeta Potential greater than (±30 mV) indicates the moderately stable suspension. • Zeta Potential value below (±30 mV) indicates instability in suspension.
Spectral absorbance and transmittance	Incident laser beam intensity, passing laser beam intensity, absorptivity, light path length, particle concentration	Beer–Lambert law	• Decrease in absorbance indicates particle sedimentation. • Not suitable for dark-colored suspensions.
3ω method	Thermal conductivity	Thermal conductivity, sedimentation	• Thermal conductivity measurement due to particle aggregation and sedimentation for instability evaluation.
TEM	Particle shape and size	Electron beam focusing	• Changes in particle shape and size due to agglomeration.
DLS	Particle hydrodynamic radius, translational diffusion coefficient, absolute temperature, viscosity	Stokes–Einstein equation	• Particle size determination over time indicates its agglomeration tendency.

atoms available for interaction, the greater the risk of particle agglomeration due to higher attractive forces between neighbouring particles. However, it should be noted that a particle size of <20 nm is desirable from a thermal properties viewpoint.[14]

(b) Nanoparticles should be spherical so that the surface area available for interaction is minimal. Therefore, the chances of agglomeration are also less compared with nanoparticles that have a high aspect ratio. However, should be mentioned that a high aspect ratio of nanoparticles is desirable from a thermal properties (thermal conductivity and specific heat) standpoint.

(c) A minimal density difference between the nanofluid and base fluid is desirable from a stability standpoint.

(d) Moderate viscosity of the base fluid is required to prevent high settling velocity in the Stokes law regime. However, too high a viscosity leads to higher frictional loss, elevated pumping power requirements, channel erosion, *etc.*[14,52,196]

(e) Compatibility between the base fluid and nanoparticles is highly desirable. Hydrophilic nanoparticles (metal oxide, metal hydroxide, *etc.*) display better dispersion in polar solvents such as water compared to non-polar solvents. However, for hydrophobic nanoparticles such as CNTs, graphene, and carbon black, it is desirable to use non-polar solvents like transformer oil or silicone oil, *etc.*

(f) A high dielectric constant of the base fluid is also considered to be favorable from a repulsion potential standpoint, which prevents nanoparticles from adhering to each other.[52]

(g) For high-temperature applications, oil-based nanofluids are more preferrable than water or alcohol-based nanofluids due to less chance of solvent loss. The vapor pressure of the base fluid should be low for high-temperature applications, or else overpressure can occur inside the heat exchanger. However, for low-temperature applications, the base fluid freezing point should be well below 0 °C to ensure that the nanofluid does not freeze in the microchannel or heat exchangers. For the aforementioned applications, refrigerants such as FC-72, FC-87, HFE-7000, HFE-7200, Polyalphaolefins, Syltherm XLT, Coolanol 25R, water–EG, ethanol, and methanol should be used as a base fluid to maintain nanofluid stability or to ensure glitch-free operation of the nanofluid under sub-zero conditions.

(h) A high surface charge (zeta potential $> \pm 45$ mV) of the nanoparticles is essential to maintain the long-term stability of the nanofluid without the need for additional stabilizers.

(i) Excessive particle concentration in the base fluid is highly undesirable from a stability viewpoint. With an increase in particle concentration, the interparticle gap diminishes, which leads to stronger van der Waals attractive forces and accelerated particle aggregation and sedimentation. Therefore, a low particle concentration of the nanofluid is desirable from a stability viewpoint.

(j) For electrolyte-sensitive systems (saline environment), *i.e.*, enhanced oil recovery, stabilization of the nanofluid should be achieved using amphoteric surfactants, steric stabilization, surface functionalization, or acid/alkaline treatment. Commonly used surfactants such as SDS, CTAB, ALS, and SDBS are ineffective under such conditions.

(k) To ensure redispersion/redistribution of magnetic nanoparticles in base fluid when the magnetic field is removed, an oleic acid coating is necessary to attain the desired outcome.

3.8 Difficulties Associated with Long-term Nanofluid Stabilization

Several roadblocks exist that prevent nanofluids from attaining highly desired long-duration stability. The major difficulties are as follows.

a) The colloidal suspension (nanofluid) is unstable in nature. When nanoparticles are suspended in a base fluid, they tend to aggregate due to their high surface area, strong van der Waals attractive forces, greater Brownian motion, and high collision between each other and the container/channel wall. Due to these factors, nanofluids are more likely to destabilize over a long period.

b) Loss of solvent is also a major factor while considering long-term stability due to base fluid evaporation. Suspended particles tend to agglomerate easily, causing the nanofluid to destabilize.

c) The difference in density between the prepared nanofluid and dispersion medium plays a huge role in determining the uniform dispersion and long-term stability of a nanofluid.

d) The most important factor that determines the long-term stability of nanofluids lies in the compatibility issues concerning the nanofluid, stabilizer, and base fluid. The selection of base fluid should be made by keeping in mind the nature of the nanoparticles and stabilizers used. In a polar medium (water), hydrophilic nanoparticles and ionic surfactants with similar charges should be used, and *vice versa*.

e) With applications that require repeated heating-cooling cycles, many commonly used stabilizers such as ionic surfactants (SDS, SDBS, CTAB, and ALS) and polymers (GA, PVP, PVA) can degrade due to thermal exposure, shearing effects, salinity, pH, *etc.*, reducing dispersion capability and also detrimentally impacting the thermal and optical properties, among others.

f) After each cycle of use, irrespective of application area, nanoparticles suspended in base fluid tend to deposit or sediment, which introduces several other problems, such as reduced heat transfer rate, pore, and channel blockage, increase in pressure drop, *etc.* Therefore, the evaluation of stability should be done at regular intervals.

g) Due to particle aggregation and solvent loss, the viscosity of nanofluids increases, which is directly linked to increased pressure drop, pumping

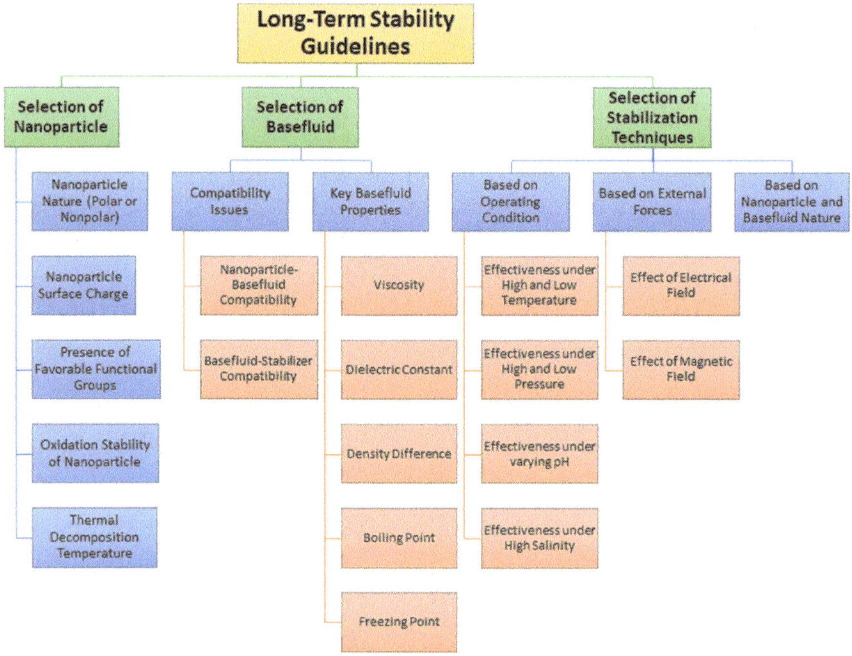

Figure 3.12 Summary of the guidelines that should be followed to ensure long-term nanofluid stability.

costs/power, increased resistance to heat transfer due to fouling, and channel erosion. Such aggregation has serious effects on the thermal, optical, electrical, and magnetic properties of nanofluids, irrespective of their type.

h) Under the effects of external forces such as magnetic and electrical fields, even stable nanofluids tend to destabilize with time. These factors can induce particle aggregation by shrinking the EDL thickness surrounding the nanoparticles, thereby reducing the zeta potential value responsible for stability.

3.9 Guidelines to Attain Long-term Stabilization of Nanofluids

This section details clear guidelines to attain long-term nanofluid stability. These guidelines include several factors, such as nanoparticle size, shape, nanoparticle–base fluid compatibility, nanoparticle–stabilizer compatibility, base fluid–stabilizer compatibility, choice of base fluid, stabilizers, and stabilization techniques based on the operating conditions, effect of external forces, *etc.* Figure 3.12 shows the guidelines that need to be followed to ensure long-term nanofluid stability.

For attaining the long-term stability of nanofluids, the guidelines detailed in the next section need to be followed. The guidelines can be classified into three general categories, namely: (a) the selection of nanoparticles, (b) selection of the base fluid, and (c) the selection of stabilizers or stabilization techniques.

3.9.1 Selection of Nanoparticles

The selection of nanoparticles is dependent on multiple factors, *i.e.*, nanoparticle nature, the surface charge of nanoparticles, the presence of favorable functional groups, the oxidation stability of nanoparticles, and the thermal stability of nanoparticles. It is thus of paramount importance to determine the nature of nanoparticles (hydrophilic/polar or hydrophobic/non-polar) before selecting a base fluid and stabilization techniques. Selection of the wrong base fluid or stabilization methods can do more harm to nanofluid stability than good. Polar nanoparticles should be dispersed in a polar solvent (water), and accordingly, the correct stabilizers should be chosen based on compatibility between the nanoparticles, base fluid, and stabilizer.

It is necessary to attain a high surface charge to maintain the stability of a nanofluid in dispersion media without the need for additional dispersants. However, a high surface charge alone cannot ensure long-term stability. In a saline environment, the EDL tends to shrink, which causes the repulsive potential to diminish and attractive potential to dominate, therefore resulting in particle aggregation and destabilization. Under the effect of high temperature, the Brownian motion in a nanofluid also increases, which induces more particle–wall and particle–particle collision interactions, thereby increasing particle aggregation tendency. Therefore, acid treatment and surface functionalization can be used to ensure long-term stability at a high temperature or in a saline environment. It is also necessary to check the oxidation and thermal degradation tendency of nanoparticles, as such tendency can hinder nanofluid stability significantly, as graphene is hydrophobic in nature and compatible in a non-polar solvent, whereas graphene oxide is hydrophilic in nature and easily dispersible in polar media. Thermal degradation can induce more particle aggregation, and is hence considered to be undesirable from a stability viewpoint. The presence of favorable functional groups is another important factor to consider when selecting the correct nanoparticles. To uniformly disperse polar nanoparticles (hydrophilic) in a polar (aqueous) medium, the presence of polar functional groups such as amine, carbonyl, carboxylic, hydroxyl, phosphate, and sulphate groups is necessary. Similarly, to be compatible with a non-polar medium, a non-polar functional group such as a methyl group should be present.

3.9.2 Selection of a Base Fluid

Selection of a base fluid in nanofluid preparation should be made based on two factors: (a) compatibility issues and (b) key properties associated with stability. The compatibility issues can be classified into two categories: base fluid compatibility with both nanoparticle and stabilizer. To the best of our understanding, polar nanoparticles (metal oxide, graphene oxide, *etc.*) should be suspended in polar solvents (alcohol, water, ethylene glycol, *etc.*), and non-polar nanoparticles (SWCNT, MWCNT, graphene, *etc.*) should be dispersed in non-polar solvents (silicone oil, transformer oil, decane, n-heptane, *etc.*). These nanoparticles contain favorable functional groups compatible with the functional group present in the base fluid. Similarly, the compatibility between base fluid molecules and stabilizers (surfactant or polymer) is equally important to minimize the strong van der Waals attractive potential, which is responsible for destabilization. The favorable enthalpic interaction between adsorbed surfactant and the polymeric tail with the base fluid molecules is essential to achieve greater electrostatic repulsive potential, as mentioned in an earlier section.[52,65]

In addition to compatibility issues, it is also important to look at a few important properties of the base fluid, *i.e.*, boiling point, freezing point, dielectric constant, viscosity, and density difference, *etc.* Information on the boiling and freezing points of the base fluid is highly necessary despite using the best possible stabilization techniques. When a nanofluid is used close to its boiling and freezing points, solvent loss takes place due to evaporation or freezing, which is highly detrimental to its stability. Therefore, the base fluid used for high-temperature applications should have a high boiling point, and when applied under low-temperature conditions, the base fluid must have low pour and freezing points. The stability of nanofluids becomes a serious concern when they are used under stationary conditions, whereas when used in laminar or turbulent regimes due to fluid velocity, their stability can be maintained for a relatively longer duration. Under stationary conditions or under the Stokes law regime, the particle settling velocity (indication of stability) is inversely proportional to the viscosity of the medium. Therefore, high viscosity is desired from a stability viewpoint, but at the same time, it can also increase pressure drop and enhance pumping costs. Hence, moderate viscosity is needed to maintain the stability of a colloidal suspension (nanofluid). In addition to the above three parameters, the dielectric constant and density difference between the nanoparticles and base fluid can also impact the long-term nanofluid stability. The high dielectric constant of the base fluid (to ensure high electrostatic repulsive force) and minimal density difference are highly necessary to ensure long-term stability, which has already been discussed in the previous section.

3.9.3 Selection of Stabilizers or Stabilization Techniques

The selection of stabilizers and stabilization techniques depends on three major factors, *i.e.*, operating conditions, external forces, and compatibility of stabilizers with nanoparticles and base fluids.

The influence of operating conditions on the nanofluid stability can be subdivided into four factors: temperature, pressure, pH, and salinity. Temperature, pressure, and pH are the most critical working parameters while considering the operational stability of synthesized nanofluids. The effect of high and low temperatures on nanofluid stability has already been discussed in the previous section. The majority of the polymers and surfactants used for stabilization purposes are thermally unstable. When these surfactants are subjected to high temperatures, they tend to form aggregates and lose dispersion potential. In low temperatures (freezing point), many surfactants and polymers also lose their dispersion potential. The effect of pressure variation also detrimentally impacts the stabilizing potential of nanofluids by affecting the surfactant and polymer structure. High pressure induces particle aggregation, as the interparticle distance reduces significantly, which increases the van der Waals attractive forces. However, low pressure (sub-atmospheric) can lead to solvent loss, which detrimentally affects the performance of stabilizers. pH variation of the dispersion medium is equally damaging to stabilization aspects as it can reduce the zeta potential of the nanofluid significantly. Near the IEP, a nanofluid is considered unstable, and under such circumstances, electrostatic stabilization is considered ineffective. Steric or electrosteric stabilization along with mechanical treatment is necessary to maintain stability under these conditions. For enhanced oil recovery applications, the nanofluid is subjected to a saline environment that detrimentally affects the zeta potential value. In a saline environment, mechanical stabilization along with steric or electrosteric stabilization is needed to maintain stability.

In addition to operating conditions, magnetic and electric fields also severely affect the stability of the nanofluid, as they increase particle aggregation. Under an electrical field, charged stabilization techniques such as electrostatic and electrosteric stabilization become ineffective if the nanoparticle surface charge is positive. However, when the particles are negatively charged, ionic stabilization is more effective. Similarly, to stabilize magnetic nanoparticles in a base fluid, non-ionic surfactants (such as oleic acid) are widely used, as their presence prevents magnetic nanoparticles from forming aggregates once the magnetic field is removed. Nanoparticles coated with oleic acid are compatible both with polar and non-polar media, depending on the number of coating layers.

As mentioned in an earlier section, the compatibility of stabilizers/dispersants with base fluid molecules and nanoparticle surfaces is also highly critical to ensuring long-term nanofluid stability.

3.10 Conclusion and Future Scope of Research

Due to their plethora of advantageous features, such as enhanced thermal properties, low viscosity, microchannel applicability, low pressure drops, and pumping costs, nanofluids have gained popularity in a variety of industrial applications. However, the large-scale industrial application of nanofluids is mainly hindered due to their uncertain stability. This chapter scrutinizes the numerous aspects of nanofluid stability, including the fundamentals of nanofluid stabilization/destabilization and the identification of several factors associated with their destabilization. Destabilization factors such as the role of nanoparticles, role of synthesis techniques, effect of the base fluid, influence of operating conditions and external forces, as well as the effect of stabilizing agents or stabilization methods, have been discussed in great detail. The effectiveness of different nanofluid stabilization techniques under various operating conditions has been given special attention in this chapter. Also, different stability evaluation methods have been discussed, such as sedimentation, centrifugation, zeta potential, DLS, *etc.* This chapter also identifies desirable properties that are necessary for long-term nanofluid stabilization and details definite guidelines to attain this.

The major points discussed in this chapter, which are essential to long-term nanofluid stabilization, are as follows.

a) To ensure long-term nanofluid stabilization, the selection of correct nanoparticles, base fluid, and dispersant or stabilization techniques are the most important factors.
b) Selection of nanoparticles is dictated by their nature (hydrophilic or hydrophobic), surface charge, presence of favourable functional groups, thermal and oxidation stability, whereas base fluid selection is attributed to nanoparticle compatibility and the five key properties of viscosity, density difference, boiling point, freezing point, and dielectric constant.
c) The selection of stabilizer and stabilization techniques are greatly dependent on operating conditions (temperature, pressure, pH, and salinity) and external forces (magnetic and electrical field) that affect the nanofluid stability. Additionally, the compatibility of the nanoparticles and base fluid with the stabilizer is also important.
d) As the majority of nanofluids are used in thermal engineering applications, the thermal and oxidation stability of stabilizers and nanoparticles is highly important. For high-temperature applications, the boiling point of the base fluid should be high. Therefore, oil-based nanofluids are preferred at high temperatures from a stability standpoint. In terms of stabilization techniques, mechanical stabilization, pH adjustment, and surface functionalization (non-surfactant/polymer) are the most effective. In the case of pH-dependent or saline

environments (EOR applications), steric and electrosteric stabilization are more useful from a stability standpoint.

It should also be pointed out that other than the above-mentioned operating parameters, the wall effect (specific to microchannel applications) and shearing effect (for tribological applications) can also affect nanofluid stability, which have not been discussed in this chapter. It has been observed through a literature review that compared to steric and electrostatic stabilization, electrosteric stabilization has been less explored. However, this method has immense potential in pH and electrolyte-sensitive media. The authors also strongly believe that a combination of mechanical and chemical stabilization techniques is critical to ensuring long-term nanofluid stability. Therefore, additional investigations on such a hybrid approach are of utmost necessity to ensure long-term stability under a plethora of operating conditions. Recently, many researchers have explored the effect of temperature and pH on nanofluid stability. However, pressure and salinity have been relatively less studied, which need to be further studied. Similarly, the effect of electrical field on nanofluid destabilization should also be investigated in greater detail. Compared to metal, metal-oxide, and carbon-based nanofluids, studies on the stability of hybrid nanofluids have been less reported. Thus, more studies on their stability enhancement need to be conducted to ensure proper utilization of hybrid nanofluids with enhanced thermophysical, optical, and electrical properties.

Conflicts of Interest

The current book chapter has no conflict of interest.

Acknowledgements

The authors would like to thank the School of Chemical Engineering, VIT Vellore, and Department of Chemical Engineering, IIT Kharagpur for their support.

References

1. D. Astruc, *Chem. Rev.*, 2020, **120**, 461–463.
2. R. Raja, S. Hermans, D. S. Shephard, B. F. G. Johnson, R. Raja, G. Sankar, S. Bromley and J. Meurig Thomas, *Chem. Commun.*, 1999, 1571–1572.
3. C. Martinez-Boubeta and K. Simeonidis, in *Nanoscale Materials in Water Purification*, eds. S. Thomas, D. Pasquini, S.-Y. Leu and D. A. Gopakumar, Elsevier, 2019, pp. 521–552.
4. T. Pradeep and Anshup, *Thin Solid Films*, 2009, **517**, 6441–6478.
5. J. S. Kang, H. S. Kim, J. Ryu, H. Thomas Hahn, S. Jang and J. W. Joung, *J. Mater. Sci.: Mater. Electron.*, 2010, **21**, 1213–1220.

6. J. Yan, G. Zou, A. Wu, J. Ren, J. Yan, A. Hu and Y. Zhou, *Scr. Mater.*, 2012, **66**, 582–585.
7. M. Kumar, S. Chakraborty, P. Upadhyaya and G. Pugazhenthi, *J. Appl. Polym. Sci.*, 2018, **135**, 1–11.
8. S. Chakraborty, M. Kumar, K. Suresh and G. Pugazhenthi, *Chin. J. Polym. Sci.*, 2016, **34**, 739–754.
9. S. Faridi, H. Bose and T. Satyanarayana, *Energy Fuels*, 2017, **31**, 3002–3009.
10. S. Seo, G. A. Perez, K. Tewari, X. Comas and M. Kim, *Sci. Rep.*, 2018, **8**, 11786.
11. S. V. Ravikumar, J. M. Jha, K. Haldar, S. K. Pal and S. Chakraborty, *J. Heat Transfer*, 2015, **137**, 051504.
12. A. M. Tiara, S. Chakraborty, I. Sarkar, S. K. Pal and S. Chakraborty, *Appl. Clay Sci.*, 2017, **143**, 241–249.
13. S. K. Das, N. Putra and W. Roetzel, *Int. J. Heat Mass Transfer*, 2003, **46**, 851–862.
14. S. K. Das, S. U. S. Choi and H. E. Patel, *Heat Transfer Eng.*, 2006, **27**, 3–19.
15. S. Chakraborty, I. Sarkar, A. Ashok, I. Sengupta, S. K. Pal and S. Chakraborty, *Appl. Therm. Eng.*, 2018, **141**, 339–351.
16. S. M. Alves, B. S. Barros, M. F. Trajano, K. S. B. Ribeiro and E. Moura, *Tribol. Int.*, 2013, **65**, 28–36.
17. A. Hernández Battez, R. González, J. L. Viesca, J. E. Fernández, J. M. Díaz Fernández, A. Machado, R. Chou and J. Riba, *Wear*, 2008, **265**, 422–428.
18. Y. Chen, P. Renner and H. Liang, *Lubricants*, 2019, 7.
19. S. K. Murthy, *Int. J. Nanomed.*, 2007, **2**, 129–141.
20. I. Yildiz, S. Shukla and N. F. Steinmetz, *Curr. Opin. Biotechnol*, 2011, **22**, 901–908.
21. J. Wolfram, M. Zhu, Y. Yang, J. Shen, E. Gentile, D. Paolino, M. Fresta, G. Nie, C. Chen, H. Shen, M. Ferrari and Y. Zhao, *Curr. Drug Targets*, 2015, **16**, 1671–1681.
22. S. K. Nune, P. Gunda, P. K. Thallapally, Y.-Y. Lin, M. L. Forrest and C. J. Berkland, *Expert Opin. Drug Delivery*, 2009, **6**, 1175–1194.
23. C. Fang and M. Zhang, *J. Mater. Chem.*, 2009, **19**, 6258–6266.
24. T. D. Schladt, K. Schneider, H. Schild and W. Tremel, *Dalton Trans.*, 2011, **40**, 6315–6343.
25. S. Nagamuthu, S. Vijayakumar and G. Muralidharan, *Energy Fuels*, 2013, **27**, 3508–3515.
26. D. Yang, in *Nanocomposites-New Trends and Developments*, IntechOpen, 2012, pp. 299–328.
27. D. Madhesh and S. Kalaiselvam, *Procedia Eng.*, 2014, **97**, 1667–1675.
28. S. Mitra, S. K. Saha, S. Chakraborty and S. Das, *Appl. Therm. Eng.*, 2012, **37**, 353–359.
29. S. K. Singh and J. Sarkar, *Int. Commun. Heat Mass Transfer*, 2018, **98**, 41–48.
30. S. Chakraborty, I. Sarkar, D. K. Behera, S. K. Pal and S. Chakraborty, *Powder Technol.*, 2017, **307**, 10–24.

31. J. M. Jha, I. Sarkar, S. Chakraborty, S. K. Pal and S. Chakraborty, *Exp. Heat Transfer.*, 2017, **30**, 500–516.
32. Ş. Şirin and T. Kıvak, *Tribol. Int.*, 2019, **137**, 180–192.
33. S. S. Rahman, M. Z. I. Ashraf, A. K. M. N. Amin, M. S. Bashar, M. F. K. Ashik and M. Kamruzzaman, *J. Cleaner Prod.*, 2019, **206**, 180–196.
34. H. Sadabadi and A. S. Nezhad, *Nanomaterials*, 2020, **10**, 1–13.
35. L. J. Habeeb, H. S. Majdi, in *Applications of Nanobiotechnology*, eds. C. A. Buckner, R. M. Lafrenie, J. A. Dénommée, J. M. Caswell, D. A. Want, G. G. Gan, Y. C. Leong, P. C. Bee, E. Chin, A. K. H. Teh, S. Picco, L. Villegas, F. Tonelli, M. Merlo, J. Rigau, D. Diaz, M. Masuelli, S. Korrapati, P. Kurra, S. Puttugunta, S. Picco, L. Villegas, F. Tonelli, M. Merlo, J. Rigau, D. Diaz, M. Masuelli, M. Tascilar, F. A. de Jong, J. Verweij and R. H. J. Mathijssen, IntechOpen, 2020, pp. 1–20.
36. N. A. Khan and F. Sultan, *Computational Approaches in Biomedical Nano-Engineering*, Wiley-VCH, Germany, 2018, pp. 227–253.
37. M. Sheikhpour, M. Arabi, A. Kasaeian, A. Rokn Rabei and Z. Taherian, *Nanotechnol., Sci. Appl.*, 2020, **13**, 47–59.
38. O. Nashed, S. M. Youssouf, K. M. Sabil, A. M. Shariff, S. Sufian and B. Lal, *IOP Conf. Ser. Mater. Sci. Eng.*, 2018, **458**, 1–7.
39. S. D. Zhou, Y. S. Yu, M. M. Zhao, S. L. Wang and G. Z. Zhang, *Energy Fuels*, 2014, **28**, 4694–4698.
40. S. Yan, W. Dai, S. Wang, Y. Rao and S. Zhou, *Energies*, 2018, **11**, 1–13.
41. S. K. Das, S. U. S. Choi and H. E. Patel, *Heat Transfer Eng.*, 2006, **27**, 3–19.
42. S. U. S. Choi, in *American Society of Mechanical Engineers, Fluids Engineering Division (Publication) FED*, 1995, vol. 231, pp. 99–105.
43. H. Riazi, T. Murphy, G. B. Webber, R. Atkin, S. S. M. Tehrani and R. A. Taylor, *Int. J. Therm. Sci.*, 2016, **107**, 25–38.
44. J. A. Eastman, S. U. S. Choi, S. Li, W. Yu and L. J. Thompson, *Appl. Phys. Lett.*, 2001, **78**, 718–720.
45. S. Chakraborty, I. Sarkar, K. Haldar, S. K. Pal and S. Chakraborty, *Appl. Clay Sci.*, 2015, **107**, 98–108.
46. H. A. Abdulbari and F. L. W. Ming, *J. Eng. Res.*, 2015, **12**, 60–67.
47. T.-H. Tsai and R. Chein, *Int. J. Heat Fluid Flow*, 2007, **28**, 1013–1026.
48. A. Abdollahi, H. A. Mohammed, S. M. Vanaki, A. Osia and M. R. Golbahar Haghighi, *Alexandria Eng. J.*, 2017, **56**, 161–170.
49. M. Zargartalebi and J. Azaiez, *Int. J. Heat Mass Transfer*, 2018, **127**, 1233–1242.
50. O. Yıldız, Ö. Açıkgöz, G. Yıldız, M. Bayrak, A. S. Dalkılıç and S. Wongwises, *Heat Mass Transfer*, 2020, **56**, 1–24.
51. I. M. Shahrul, I. M. Mahbubul, S. S. Khaleduzzaman, R. Saidur and M. F. M. Sabri, *Renewable Sustainable Energy Rev.*, 2014, **38**, 88–98.
52. S. Chakraborty and P. K. Panigrahi, *Appl. Therm. Eng.*, 2020, **174**, 115259.
53. A. Ghadimi, R. Saidur and H. S. C. Metselaar, *Int. J. Heat Mass Transfer*, 2011, **54**, 4051–4068.

54. D. Dey, P. Kumar and S. Samantaray, *Heat Transfer Res.*, 2017, **46**, 1413–1442.
55. Babita, S. K. Sharma and S. M. Gupta, *Exp. Therm. Fluid Sci.*, 2016, **79**, 202–212.
56. W. Yu and H. Xie, *J. Nanomater.*, 2012, **2012**, 1–17.
57. T. Missana and A. Adell, *J. Colloid Interface Sci.*, 2000, **230**, 150–156.
58. G. R. Wiese and T. W. Healy, *Trans. Faraday Soc.*, 1970, **66**, 490–499.
59. D. Grasso, K. Subramaniam, M. Butkus, K. Strevett and J. Bergendahl, *Rev. Environ. Sci. Bio/Technol.*, 2002, **1**, 17–38.
60. C. J. van Oss, R. F. Giese and P. M. Costanzo, *Clays Clay Miner.*, 1990, **38**, 151–159.
61. A. Ghadimi, R. Saidur and H. S. C. Metselaar, *Int. J. Heat Mass Transfer*, 2011, **54**, 4051–4068.
62. N. Ali, J. A. Teixeira and A. Addali, *J. Nanomater.*, 2018, **2018**, 6978130.
63. H. Freundlich, *Colloid and Capillary Chemistry*, Methuen, London, 1926.
64. H. R. Kruyt, *Colloid Science*, Elsevier, Amsterdam, 1952.
65. B. Derjaguin and L. Landau, *Prog. Surf. Sci.*, 1993, **43**, 30–59.
66. B. V. Derjaguin, N. V. Churaev and V. M. Miller, *Surface Forces*, Consultants Bureau, New York, 1987.
67. E. J. W. Verwey, N. Laboratorium and D. Waals, *J. Phys. Chem.*, 1947, **51**, 631–636.
68. I. Popa, G. Gillies, G. Papastavrou and M. Borkovec, *J. Phys. Chem. B*, 2010, **114**, 3170–3177.
69. E. M. Hotze, T. Phenrat and G. V. Lowry, *J. Environ. Qual.*, 2010, **39**, 1909–1924.
70. C. Rodríguez-Abreu, in *Nanocolloids: A Meeting Point for Scientists and Technologists*, eds. M. Sanchez Dominguez and C. R. Abreu, 2016, pp. 1–36.
71. J. Israelachvili, *Intermolecular and Surface Forces*, Elsevier Academic Press, London, 3rd edn, 2005.
72. F. Mange, P. Couchot, A. Foissy and A. Pierre, *J. Colloid Interface Sci.*, 1993, **159**, 58–67.
73. R. Kjellander, *Berichte der Bunsengesellschaft für physikalische Chemie*, 1996, **100**, 894–904.
74. G. Lagaly, in *Handbook of clay science*, eds. F. B. Bergaya, B. K. G. Theng and G. Lagaly, Elsevier, 2006, pp. 191–245.
75. K. Jasmund and G. Lagaly, *Clay minerals and clays – Structure, properties, applications and use in industry and the environment*, Steinkopff-Verlag Heidelberg, 1st edn, 1993.
76. J. N. Israelachvili and H. Wennerström, *Langmuir*, 1990, **6**, 873–876.
77. E. Ruckenstein, *J. Colloid Interface Sci.*, 1991, **147**, 1–4.
78. V. N. Paunov, R. I. Dimova, P. A. Kralchevsky, G. Broze and A. Mehreteab, *J. Colloid Interface Sci.*, 1996, **182**, 239–248.
79. S. Rohrsetzer, S. Bán, P. Kovács and I. Pászli, *Colloid Polym. Sci.*, 1995, **273**, 189–194.

80. C. F. Zukoski, J. L. Look and G. H. Bogush, *Colloid. Silica Fundam. Appl.*, 2005, 565–572.
81. J. L. Look and C. F. Zukosi, *J. Am. Ceram. Soc.*, 1995, **78**, 21–32.
82. D. A. Walker, B. Kowalczyk, M. O. de la Cruz and B. A. Grzybowski, *Nanoscale*, 2011, **3**, 1316–1344.
83. K. S. Pavithra, S. C. Gurumurthy, M. P. Yashoda, T. Mateti, K. Ramam, R. Nayak and M. S. Murari, *J. Therm. Anal. Calorim.*, 2021, **146**, 601–610.
84. M. Ramezanizadeh and M. Alhuyi Nazari, *Int. J. Low-Carbon Technol.*, 2019, **14**, 468–474.
85. M. Atashafrooz, *Alexandria Eng. J.*, 2018, **57**, 4277–4285.
86. T. Hayat, B. Ahmed, F. M. Abbasi and A. Alsaedi, *J. Mol. Liq.*, 2017, **234**, 324–329.
87. H. J. Kim, I. C. Bang and J. Onoe, *Opt. Lasers Eng.*, 2009, **47**, 532–538.
88. M. Khoshvaght-Aliabadi, S. Pazdar and O. Sartipzadeh, *Int. Commun. Heat Mass Transfer*, 2016, **70**, 84–92.
89. M. Rafiq, M. Shafique, A. Azam and M. Ateeq, *Ain Shams Eng. J.*, 2021, **12**, 555–576.
90. S. S. Botha, P. Ndungu and B. J. Bladergroen, *Ind. Eng. Chem. Res.*, 2011, **50**, 3071–3077.
91. D. Li, B. Hong, W. Fang, Y. Guo and R. Lin, *Ind. Eng. Chem. Res.*, 2010, **49**, 1697–1702.
92. M. Kamalgharibi, F. Hormozi, S. A. H. Zamzamian and M. M. Sarafraz, *Heat Mass Transfer*, 2016, **52**, 55–62.
93. V. Nikkhah, M. M. Sarafraz, F. Hormozi and S. M. Peyghambarzadeh, *Exp. Therm. Fluid Sci.*, 2015, **60**, 83–95.
94. M. M. Sarafraz and F. Hormozi, *Int. Commun. Heat Mass Transfer*, 2014, **53**, 116–123.
95. M. Sahooli, S. Sabbaghi and M. Shariaty Niassar, *Int. J. Nanosci. Nanotechnol.*, 2012, **8**, 27–34.
96. H. Sandhu, G. Dasaroju and M. Singh, in *4th Thermal and Fluids Engineering Conference*, Las Vegas, 2019, pp. 1289–1297.
97. E. Salari, S. M. Peyghambarzadeh, M. M. Sarafraz and F. Hormozi, *Period. Polytech., Chem. Eng.*, 2016, **60**, 106–122.
98. M. M. Sarafraz, F. Hormozi and S. M. Peyghambarzadeh, *Appl. Therm. Eng.*, 2015, **82**, 212–224.
99. A. Ghadimi and I. H. Metselaar, *Exp. Therm. Fluid Sci.*, 2013, **51**, 1–9.
100. S. Mo, Y. Chen, L. Jia and X. Luo, *Appl. Energy*, 2012, **93**, 65–70.
101. E. Salari, S. M. Peyghambarzadeh, M. M. Sarafraz, F. Hormozi and V. Nikkhah, *Heat Mass Transfer*, 2017, **53**, 265–275.
102. H. G. Bagaria, B. M. Neilson, A. J. Worthen, Z. Xue, K. Y. Yoon, V. Cheng, J. H. Lee, S. Velagala, C. Huh, S. L. Bryant, C. W. Bielawski and K. P. Johnston, *J. Colloid Interface Sci.*, 2013, **398**, 217–226.
103. M. Iqbal, B. A. Lyon, E. E. Ureña-Benavides, E. Moaseri, Y. Fei, C. McFadden, K. J. Javier, C. J. Ellison, K. D. Pennell and K. P. Johnston, *Colloids Surf., A*, 2017, **520**, 257–267.
104. K. S. Suganthi and K. S. Rajan, *Int. J. Heat Mass Transfer*, 2014, **71**, 653–663.

105. D. Cabaleiro, L. Colla, F. Agresti, L. Lugo and L. Fedele, *Int. J. Heat Mass Transfer*, 2015, **89**, 433–443.
106. K. S. Pavithra, M. P. Yashoda, S. Prasannakumar and S. Mutalik, *Iran. Polym. J.*, 2020, **29**, 185–196.
107. C. Pang, J. Y. Jung, J. W. Lee and Y. T. Kang, *Int. J. Heat Mass Transfer*, 2012, **55**, 5597–5602.
108. R. Choudhary, D. Khurana, A. Kumar and S. Subudhi, *J. Exp. Nanosci.*, 2017, **12**, 140–151.
109. T. T. Baby and R. Sundara, *J. Phys. Chem. C*, 2011, **115**, 8527–8533.
110. A. A. Minea, *Nanomaterials* 2019, **9**, 1–22.
111. A. Agarwala, N. Kaynan, S. Zaidiner and R. Yerushalmi, *Chem. Commun.*, 2014, **50**, 5397–5399.
112. W. Yu, H. Xie, L. Chen and Y. Li, *Colloids Surf., A*, 2010, **355**, 109–113.
113. A. Ghadimi and I. H. Metselaar, *Exp. Therm. Fluid Sci.*, 2013, **51**, 1–9.
114. B. LotfizadehDehkordi, S. N. Kazi, M. Hamdi, A. Ghadimi, E. Sadeghinezhad and H. S. C. Metselaar, *Heat and Mass Transfer*, 2013, **49**, 1109–1115.
115. E. Sadeghinezhad, M. Mehrali, R. Saidur, M. Mehrali, S. Tahan Latibari, A. R. Akhiani and H. S. C. Metselaar, *Energy Convers. Manage.*, 2016, **111**, 466–487.
116. W. Yu, H. Xie and W. Chen, *J. Appl. Phys.*, 2010, **107**, 94317.
117. A. Naddaf and S. Zeinali Heris, *Int. Commun. Heat Mass Transfer*, 2018, **95**, 116–122.
118. H. Xie, H. Lee, W. Youn and M. Choi, *J. Appl. Phys.*, 2003, **94**, 4967–4971.
119. Z. Said, M. A. Abdelkareem, H. Rezk, A. M. Nassef and H. Z. Atwany, *Powder Technol.*, 2020, **364**, 795–809.
120. C. Almeida, S. Paul, L. G. Asirvatham, S. Manova, R. Nimmagadda, J. R. Bose and S. Wongwises, *Fluids*, 2020, **5**, 1–13.
121. A. Al-Sharafi, A. Z. Sahin and B. S. Yilbas, *J. Heat Transfer*, 2016, **138**, 1–12.
122. H. Yarmand, S. Gharehkhani, S. F. S. Shirazi, A. Amiri, E. Montazer, H. K. Arzani, R. Sadri, M. Dahari and S. N. Kazi, *Int. Commun. Heat Mass Transfer*, 2016, **72**, 10–15.
123. K. Wusiman, H. Jeong, K. Tulugan, H. Afrianto and H. Chung, *Int. Commun. Heat Mass Transfer*, 2013, **41**, 28–33.
124. A. Ghozatloo, M. Shariaty-Niasar and A. M. Rashidi, *Int. Commun. Heat Mass Transfer*, 2013, **42**, 89–94.
125. A. Indhuja, K. S. Suganthi, S. Manikandan and K. S. Rajan, *J. Taiwan Inst. Chem. Eng.*, 2013, **44**, 474–479.
126. M. M. Sarafraz, F. Hormozi and V. Nikkhah, *Exp. Therm. Fluid Sci.*, 2016, **78**, 41–49.
127. R. Bandyopadhyaya, E. Nativ-Roth, O. Regev and R. Yerushalmi-Rozen, *Nano Lett.*, 2002, **2**, 25–28.
128. M. F. Islam, E. Rojas, D. M. Bergey, A. T. Johnson and A. G. Yodh, *Nano Lett.*, 2003, **3**, 269–273.

129. J. Yu, N. Grossiord, C. E. Koning and J. Loos, *Carbon N. Y.*, 2007, **45**, 618–623.
130. M. N. A. W. M. Yazid, N. A. C. Sidik, R. Mamat and G. Najafi, *Int. Commun. Heat Mass Transfer*, 2016, **78**, 253–263.
131. Y. Y. Song, H. K. D. H. Bhadeshia and D. W. Suh, *Powder Technol.*, 2015, **272**, 34–44.
132. M. J. Nine, M. Batmunkh, J. H. Kim, H. S. Chung and H. M. Jeong, *J. Nanosci. Nanotechnol.*, 2012, **12**, 4553–4559.
133. L. S. Sundar, E. Venkata Ramana, M. P. F. Graça, M. K. Singh and A. C. M. Sousa, *Int. Commun. Heat Mass Transfer*, 2016, **73**, 62–74.
134. Y. T. He, J. Wan and T. Tokunaga, *J. Nanopart. Res.*, 2008, **10**, 321–332.
135. T. Ambreen and M.-H. Kim, *Appl. Energy*, 2020, **264**, 114684.
136. C. Rodríguez-Abreu, *Nanocolloids: Some Basic Concepts and Principles of Their Stabilization*, 2016.
137. J.-P. Hsu and B.-T. Liu, *J. Phys. Chem. B*, 1998, **102**, 334–337.
138. J.-P. Hsu and B.-T. Liu, *J. Colloid Interface Sci.*, 1998, **198**, 186–189.
139. K. Suttiponparnit, J. Jiang, M. Sahu, S. Suvachittanont, T. Charinpanitkul and P. Biswas, *Nanoscale Res. Lett.*, 2011, **6**, 1–8.
140. K. H. Krishna, S. Neti, A. Oztekin and S. Mohapatra, *J. Appl. Phys.*, 2015, **117**, 1–9.
141. H. J. Kim, S. H. Lee, J. H. Lee and S. P. Jang, *Energy*, 2015, **90**, 1290–1297.
142. C. P. McNew, N. Kananizadeh, Y. Li and E. J. LeBoeuf, *Chemosphere*, 2017, **168**, 65–79.
143. M. J. Vold, *J. Colloid Sci.*, 1954, **9**, 451–459.
144. A. Gomez-Flores, S. A. Bradford, G. Hwang, S. Choi, M. Tong and H. Kim, *Colloids Surf., A*, 2020, **599**, 124921.
145. H. Başağaoğlu, S. Succi, D. Wyrick and J. Blount, *Sci. Rep.*, 2018, **8**, 8583.
146. D. D. Liyanage, R. J. K. A. Thamali, A. A. K. Kumbalatara, J. A. Weliwita and S. Witharana, *J. Nanomater.*, 2016, **2016**, 1–7.
147. S. Ponnada, T. Subrahmanyam, S. V. Naidu and K. V. Sharma, *Advances in Power Generation from Renewable Energy Sources (APGRES)*, 2019, 1–15.
148. T. Zhang, Q. Zou, Z. Cheng, Z. Chen, Y. Liu and Z. Jiang, *Powder Technol.*, 2021, **379**, 457–465.
149. K. S. Hong, T.-K. Hong and H.-S. Yang, *Appl. Phys. Lett.*, 2006, **88**, 31901.
150. A. Ghadimi, R. Saidur and H. S. C. Metselaar, *Int. J. Heat Mass Transfer*, 2011, **54**, 4051–4068.
151. I. Sarkar, S. Chakraborty, J. M. Jha, S. K. Pal and S. Chakraborty, *Int. J. Therm. Sci.*, 2017, **116**, 52–62.
152. H. Akoh, Y. Tsukasaki, S. Yatsuya and A. Tasaki, *J. Cryst. Growth*, 1978, **45**, 495–500.
153. C.-H. Lo, T.-T. Tsung and L.-C. Chen, *J. Cryst. Growth*, 2005, **277**, 636–642.

154. J. A. Eastman, S. U. S. Choi, S. Li, L. J. Thompson and S. Lee, *Enhanced thermal conductivity through the development of nanofluids, in 1996 Fall meeting of the Materials Research Society (MRS)*, USA, Boston, 1996.
155. S. U. S. Choi, W. Yu, J. R. Hull, Z. G. Zhang and F. E. Lockwood, *SAE Trans.*, 2002, **111**, 38–43.
156. H. Chang and M. Kao, *J. Chinese Soc. Mech. Eng.*, 2007, **28**, 187–194.
157. P. X. Tran and Y. Soong, Not published – presentation only, United States, 2007.
158. Y. Li, J. Zhou, S. Tung, E. Schneider and S. Xi, *Powder Technol.*, 2009, **196**, 89–101.
159. H. Zhu, Y. Lin and Y. Yin, *J. Colloid Interface Sci.*, 2004, **277**, 100–103.
160. A. K. Singh and V. S. Raykar, *Colloid Polym. Sci.*, 2008, **286**, 1667–1673.
161. W. Yu, H. Xie, X. Wang and X. Wang, *Nanoscale Res. Lett.*, 2010, **6**, 47.
162. S. Ayyappan, R. S. Gopalan, G. N. Subbanna and C. N. R. Rao, *J. Mater. Res.*, 1997, **12**, 398–401.
163. N. Ali, J. A. Teixeira and A. Addali, *J. Nanomater.*, 2018, **2018**, 1–33.
164. C. Pang, J.-Y. Jung, J. W. Lee and Y. T. Kang, *Int. J. Heat Mass Transfer*, 2012, **55**, 5597–5602.
165. S. Chakraborty, I. Sarkar, D. K. Behera, S. K. Pal and S. Chakraborty, *Powder Technol.*, 2017, **307**, 10–24.
166. K. Wusiman, H. Jeong, K. Tulugan, H. Afrianto and H. Chung, *Int. Commun. Heat Mass Transfer*, 2013, **41**, 28–33.
167. M. M. Sarafraz, M. R. Safaei, Z. Tian, M. Goodarzi, E. P. Bandarra Filho and M. Arjomandi, *Energies*, 2019, **12**, 1929.
168. B. Munkhbayar, M. R. Tanshen, J. Jeoun, H. Chung and H. Jeong, *Ceram. Int.*, 2013, **39**, 6415–6425.
169. D. C. Gregg, *J. Chem. Educ.*, 1952, **29**, 320.
170. A. M. Tiara, S. Chakraborty, I. Sarkar, S. K. Pal and S. Chakraborty, *Heat and Mass Transfer*, 2017, **53**, 2187–2197.
171. O. Pourmehran, M. M. Sarafraz, M. Rahimi-Gorji and D. D. Ganji, *J. Taiwan Inst. Chem. Eng.*, 2018, **88**, 37–48.
172. N. Ali, A. M. Bahman, N. F. Aljuwayhel, S. A. Ebrahim, S. Mukherjee and A. Alsayegh, *Nanomaterials*, 2021, **11**, 1–78.
173. C.-M. Cheng, S.-P. Yu and T.-P. Teng, *J. Nanomater.*, 2018, **2018**, 3264621.
174. N. T. Tam, N. V. Phuong, P. H. Khoi, P. N. Minh, M. Afrand, P. Van Trinh, B. H. Thang, G. Żyła and P. Estellé, *Nanomaterials*, 2020, **10**, 1–39.
175. H. Xie, H. Lee, W. Youn and M. Choi, *J. Appl. Phys.*, 2003, **94**, 4967–4971.
176. R. G. Harrison, P. Todd, S. R. Rudge and D. P. Petrides, *Bioseparations science and engineering*, 2015.
177. Y. Chen, X. Quan, Z. Wang, C. Lee, Z. Wang, P. Tao, C. Song, J. Wu, W. Shang and T. Deng, *J. Mater. Chem. A*, 2016, **4**, 17503–17511.
178. N. Navarrete, A. Gimeno-Furio, R. Mondragon, L. Hernandez, L. Cabedo, E. Cordoncillo and J. E. Julia, *Sci. Rep.*, 2017, **7**, 17580.

179. D. Shin, B. Jo, H. Kwak and D. Banerjee, 2010 14th International Heat Transfer Conference, 2010, 583–591.
180. S. N. Suhaimi, A. R. A. Rahman, M. F. M. Din, M. Z. Hassan, M. T. Ishak and M. T. bin Jusoh, *J. Nanomater.*, 2020, **2020**, 2061343.
181. N. A. Azizie and N. Hussin, *J. Phys. Conf. Ser.*, 2020, **1432**, 12025.
182. N. Saurín, T. Espinosa, J. Sanes, F.-J. Carrión and M.-D. Bermúdez, *Lubricants*, 2015, **3**, 650–663.
183. J. V. Jose, A. V. Menon, S. P. Kurup N, and S. Sasi, *Int. J. Innov. Res. Sci.*, 2013, **2**, 832–838.
184. L. Kong, J. Sun and Y. Bao, *RSC Adv.*, 2017, **7**, 12599–12609.
185. M. Arshadi, H. Taghvaei, M. K. Abdolmaleki, M. Lee, H. Eskandarloo and A. Abbaspourrad, *Appl. Energy*, 2019, **242**, 1562–1572.
186. Y. Zhang, B. Zhao, J. Jiang, Y. Zhuo and S. Wang, *Int. J. Greenh. Gas Control*, 2016, **50**, 49–56.
187. J. Yan, Y.-Y. Lu, D.-L. Zhong, Z.-L. Zou and J.-B. Li, *Energy*, 2019, **180**, 728–736.
188. V. O. Aseyev, H. Tenhu and F. M. Winnik, in *Conformation-Dependent Design of Sequences in Copolymers II*, Springer, 2006, pp. 1–85.
189. J. Tavares and S. Coulombe, *Powder Technol.*, 2011, **210**, 132–142.
190. R. A. Taylor, P. E. Phelan, T. P. Otanicar, C. A. Walker, M. Nguyen, S. Trimble and R. Prasher, *J. Renewable Sustainable Energy*, 2011, **3**, 23104.
191. M. Fischedick, J. Marzinkowski, P. Winzer and M. Weigel, *J. Cleaner Prod.*, 2014, **84**, 563–580.
192. N. Hordy, D. Rabilloud, J.-L. Meunier and S. Coulombe, *Sol. Energy*, 2014, **105**, 82–90.
193. J. Lin, X. Ju, C. Xu, Y. Yang and X. Du, *Opt. Mater.*, 2020, **109**, 110373.
194. E. Sani, L. Mercatelli, S. Barison, C. Pagura, F. Agresti, L. Colla and P. Sansoni, *Sol. Energy Mater. Sol. Cells*, 2011, **95**, 2994–3000.
195. V. Nikkhah, M. M. Sarafraz, F. Hormozi and S. M. Peyghambarzadeh, *Exp. Therm. Fluid Sci.*, 2015, **60**, 83–95.
196. S. Chakraborty, I. Sarkar, K. Haldar, S. K. Pal and S. Chakraborty, *Appl. Clay Sci.*, 2015, **107**, 98–108.
197. M. M. Sarafraz, H. Arya, M. Saeedi and D. Ahmadi, *Appl. Therm. Eng.*, 2018, **138**, 552–562.
198. B. Wei, C. Zou and X. Li, *Int. J. Heat Mass Transfer*, 2017, **104**, 537–543.
199. A. M. Tiara, S. Chakraborty, I. Sarkar, A. Ashok, S. K. Pal and S. Chakraborty, *Exp. Therm. Fluid Sci.*, 2017, **89**, 295–303.
200. X.-F. Yang and Z.-H. Liu, *Int. J. Therm. Sci.*, 2011, **50**, 2402–2412.
201. Y. J. Chen, P. Y. Wang, Z. H. Liu and Y. Y. Li, *Int. J. Heat Mass Transfer*, 2013, **67**, 548–559.
202. Z. H. Liu, X. F. Yang and G. L. Guo, *J. Appl. Phys.*, 2007, **102**, 013526.
203. Z. hua Liu, J. guo Xiong and R. Bao, *Int. J. Multiphase Flow*, 2007, **33**, 1284–1295.
204. X. F. Yang, Z. H. Liu and J. Zhao, *J. Micromech. Microeng.*, 2008, **18**, 035038.

205. S. Li, Y. H. Ng, H. C. Lau, O. Torsæter and L. P. Stubbs, *J. Nanomater.*, 2020, **10**.
206. H. Shamsijazeyi, C. A. Miller, M. S. Wong, J. M. Tour and R. Verduzco, *J. Appl. Polym. Sci.*, 2014, **131**, 1–13.
207. K. Li, D. Wang and S. Jiang, *Oil Gas Sci. Technol.*, DOI: 10.1051/ogst/2018052.
208. B. Moradi, P. Pourafshary, F. J. Farahani, M. Mohammadi and M. A. Emadi, *SPE Oil and Gas India Conference and Exhibition*, 2015, 8.
209. L. Hendraningrat, S. Li and O. Torsæter, *J. Pet. Sci. Eng.*, 2013, **111**, 128–138.
210. S. Al-Anssari, M. Arif, S. Wang, A. Barifcani and S. Iglauer, *J. Colloid Interface Sci.*, 2017, **508**, 222–229.
211. R. S. Kumar, K. R. Chaturvedi, S. Iglauer, J. Trivedi and T. Sharma, *J. Pet. Sci. Eng.*, 2020, **195**, 107634.
212. A. Esfandyari Bayat, R. Junin, A. Samsuri, A. Piroozian and M. Hokmabadi, *Energy Fuels*, 2014, **28**, 6255–6266.
213. T. Phenrat, N. Saleh, K. Sirk, H.-J. Kim, R. D. Tilton and G. V Lowry, *J. Nanopart. Res.*, 2008, **10**, 795–814.
214. E. L. Foster, Z. Xue, C. M. Roach, E. S. Larsen, C. W. Bielawski and K. P. Johnston, *ACS Macro Lett.*, 2014, **3**, 867–871.
215. H. G. Bagaria, B. M. Neilson, A. J. Worthen, Z. Xue, K. Y. Yoon, V. Cheng, J. H. Lee, S. Velagala, C. Huh, S. L. Bryant, C. W. Bielawski and K. P. Johnston, *J. Colloid Interface Sci.*, 2013, **398**, 217–226.
216. S. Umar, F. Sulaiman, N. Abdullah and S. N. Mohamad, *AIP Conference Proceedings*, 2018, **2031**, 020031.
217. J. Huang, X. Wang, Q. Long, X. Wen, Y. Zhou and L. Li, in *2009 Symposium on Photonics and Optoelectronics*, 2009, pp. 1–4.
218. K. Goudarzi, F. Nejati, E. Shojaeizadeh and S. K. Asadi Yousef-abad, *Exp. Therm. Fluid Sci.*, 2015, **60**, 20–27.
219. M. Zareei, H. Yoozbashizadeh and H. R. Madaah Hosseini, *J. Therm. Anal. Calorim.*, 2019, **135**, 1185–1196.
220. X. Wang, D. Zhu and S. yang, *Chem. Phys. Lett.*, 2009, **470**, 107–111.
221. P. I. P. Soares, C. A. T. Laia, A. Carvalho, L. C. J. Pereira, J. T. Coutinho, I. M. M. Ferreira, C. M. M. Novo and J. P. Borges, *Appl. Surf. Sci.*, 2016, **383**, 240–247.
222. H. Zhang, B. Chen and J. F. Banfield, *J. Phys. Chem. C*, 2010, **114**, 14876–14884.
223. H. Konakanchi, R. S. Vajjha, G. A. Chukwu and D. K. Das, *Heat Trans. Eng.*, 2015, **36**, 81–90.
224. U. Paik, J. Y. Kim and V. A. Hackley, *Mater. Chem. Phys.*, 2005, **91**, 205–211.
225. X. Wang, X. Li and S. Yang, *Energy Fuels*, 2009, **23**, 2684–2689.
226. A. Amrollahi, A. M. Rashidi, M. Emami Meibodi and K. Kashefi, *J. Exp. Nanosci.*, 2009, **4**, 347–363.
227. Y. Assef, D. Arab and P. Pourafshary, *J. Pet. Sci. Eng.*, 2014, **124**, 331–340.
228. A. Bateni, A. Amirfazli and A. W. Neumann, *Colloids Surf., A*, 2006, **289**, 25–38.

229. S. Zeinali Heris, F. Mohammadpur and A. Shakouri, *Mater. Res. Bull.*, 2014, **53**, 21–27.
230. G. Wang, Z. Zhang, R. Wang and Z. Zhu, *J. Nanomater.*, 2020, **10**, 1–24.
231. D. E. A. Mansour, A. M. Elsaeed and M. A. Izzularab, *IEEE Trans. Dielectr. Electr. Insul.*, 2016, **23**, 3364–3372.
232. J. Miao, M. Dong and L. P. Shen, *Proceedings of 2012 IEEE International Conference on Condition Monitoring and Diagnosis, CMD* 2012, 2012, pp. 1126–1129.
233. D. Zong, Z. Yang and Y. Duan, *Appl. Therm. Eng.*, 2017, **122**, 71–79.
234. C. Stamatopoulos, P. Bleuler, M. Pfeiffer, S. Hedtke, P. Rudolf von Rohr and C. M. Franck, *Langmuir*, 2019, **35**, 4876–4885.
235. G. Wang, Z. Zhang, R. Wang and Z. Zhu, *Nanomaterials*, 2020, **10**, 1–24.
236. R. Chand, D. Yadav and G. C. Rana, *Int. J. Nanopart.*, 2015, **8**, 241–261.
237. Y. F. Du, Y. Z. Lv, J. Q. Zhou, M. T. Chen, X. X. Li and C. R. Li, Proceedings of IEEE International Conference on Dielectric Liquids, 2011, pp. 12–15.
238. K. Goudarzi, H. Jamali and V. Kalaei, *Adv. Energy Convers. Mater.*, 2020, 15–24.
239. M. Haroun, A. Ansari, N. Al Kindy, N. A. Sayed, B. Ali and H. Sarma, *Soc. Pet. Eng. - Abu Dhabi Int. Pet. Exhib. Conf. 2012, ADIPEC 2012 - Sustain. Energy Growth People, Responsib. Innov.*, 2012, 5, pp. 3610–3622.
240. H. Soleimani, Z. Abbas, K. Khalid, N. Yahya and H. Soleimani, *Eur. J. Sci. Res.*, 2010, **39**, 105–110.
241. Y. Sun, D. Yang, L. Shi, H. Wu, Y. Cao, Y. He and T. Xie, *Energy Fuels*, 2020, **34**, 1202–1218.
242. J. Li, B. Du, F. Wang, W. Yao and S. Yao, *Phys. Lett. A*, 2016, **380**, 604–608.
243. E. Kutalkova, T. Plachy and M. Sedlacik, *J. Mol. Liq.*, 2020, **309**, 113120.
244. M. Chorny, I. Fishbein, B. B. Yellen, I. S. Alferiev, M. Bakay, S. Ganta, R. Adamo, M. Amiji, G. Friedman and R. J. Levy, *Proc. Natl. Acad. Sci. U. S. A.*, 2010, **107**, 8346 LP–8351.
245. K. H. Li and B. B. Yellen, *Appl. Phys. Lett.*, 2010, **97**, 83105.
246. S. K. Saroj and P. K. Panigrahi, *Colloids Surf., A*, 2019, **580**, 123672.
247. J. Ge, L. He, J. Goebl and Y. Yin, *J. Am. Chem. Soc.*, 2009, **131**, 3484–3486.
248. M. Rajnak, Z. Wu, B. Dolnik, K. Paulovicova, J. Tothova, R. Cimbala, J. Kurimsky, P. Kopcansky, B. Sunden, L. Wadso and M. Timko, *Energies*, 2019, **12**, 1–11.
249. A. Zakinyan, S. Kunikin, A. Chernyshov and V. Aitov, *Magnetochemistry*, 2021, **7**, 1–11.
250. S. Doganay, R. Alsangur and A. Turgut, *Mater. Res. Express*, 2019, **6**, 112003.
251. X. Chen, B. Wu, K. W. Jayawardana, N. Hao, H. S. N. Jayawardena, R. Langer, A. Jaklenec and M. Yan, *Adv. Healthcare Mater.*, 2016, **5**, 2007–2012.
252. J. S. Andreu, J. Camacho and J. Faraudo, *Soft Matter*, 2011, **7**, 2336–2339.

253. H. Ezzaier, J. A. Marins, C. Claudet, G. Hemery, O. Sandre and P. Kuzhir, *Nanomaterials*, 2018, **8**, 623.
254. M. Asfer, S. K. Saroj and P. K. Panigrahi, *J. Magn. Magn. Mater.*, 2017, **436**, 47–56.
255. G. Verma and P. A. Hassan, *Phys. Chem. Chem. Phys.*, 2013, **15**, 17016–17028.
256. M. Asfer, B. Mehta, A. Kumar, S. Khandekar and P. K. Panigrahi, *Int. J. Heat Fluid Flow*, 2016, **59**, 74–86.
257. A. Shakiba and K. Vahedi, *J. Magn. Magn. Mater.*, 2016, **402**, 131–142.
258. M. M. Larimi, A. Ghanaat, A. A. Ramiar and A. A. Ranjbar, *Int. J. Mech. Sci.*, 2016, **118**, 101–112.
259. H. K. Moghadam, S. S. Baghbani and H. Babazadeh, *J. Therm. Anal. Calorim.*, **143**, 3849–3866.
260. F. Fadaei, M. Shahrokhi, A. Molaei Dehkordi and Z. Abbasi, *J. Magn. Magn. Mater.*, 2019, **475**, 304–315.
261. M. Sheikholeslami, M. Gorji-Bandpay and D. D. Ganji, *Int. Commun. Heat Mass Transfer*, 2012, **39**, 978–986.
262. A. A. Alnaqi, S. Aghakhani, A. H. Pordanjani, R. Bakhtiari, A. Asadi and M. D. Tran, *Int. J. Heat Mass Transfer.*, 2019, **133**, 256–267.
263. M. Sheikholeslami, H. F. Öztop, N. Abu-Hamdeh and Z. Li, *Int. J. Numer. Methods Heat Fluid Flow*, 2019, **29**, 294–308.
264. M. Sheikholeslami and K. Vajravelu, *Chin. J. Phys.*, 2018, **56**, 1578–1587.
265. H. Chang, T. T. Tsung, C. R. Lin, H. M. Lin, C. K. Lin, C. H. Lo and H. T. Su, *Mater. Trans.*, 2004, **45**, 1375–1378.
266. M. R. Parker, R. P. A. R. van Kleef, H. W. Myron and P. Wyder, *J. Magn. Magn. Mater.*, 1982, **27**, 250–256.
267. K. G. Gareev, I. E. Kononova, V. S. Levitckii, V. A. Moshnikov and S. S. Nalimova, *J. Phys. Conf. Ser.*, 2014, **572**, 12027.
268. S. Masur, B. Zingsem, T. Marzi, R. Meckenstock and M. Farle, *J. Magn. Magn. Mater.*, 2016, **415**, 8–12.
269. X. Teng and H. Yang, *J. Mater. Chem.*, 2004, **14**, 774–779.
270. S. S. J. Aravind, P. Baskar, T. T. Baby, R. K. Sabareesh, S. Das and S. Ramaprabhu, *J. Phys. Chem. C*, 2011, **115**, 16737–16744.
271. H. S. Xue, J. R. Fan, Y. C. Hu, R. H. Hong and K. F. Cen, *J. Appl. Phys.*, 2006, **100**, 104909.
272. B. Smith, K. Wepasnick, K. E. Schrote, A. R. Bertele, W. P. Ball, C. O'Melia and D. H. Fairbrother, *Environ. Sci. Technol.*, 2009, **43**, 819–825.
273. K. Esumi, M. Ishigami, A. Nakajima, K. Sawada and H. Honda, *Carbon N. Y.*, 1996, **34**, 279–281.
274. K. A. Wepasnick, B. A. Smith, J. L. Bitter and D. Howard Fairbrother, *Anal. Bioanal. Chem.*, 2010, **396**, 1003–1014.
275. H. Zecha, *Acta Polym.*, 1981, **32**, 582.
276. R. Bandyopadhyaya, E. Nativ-Roth, O. Regev and R. Yerushalmi-Rozen, *Nano Lett.*, 2002, **2**, 25–28.
277. E. Piacenza, A. Presentato and R. J. Turner, *Crit. Rev. Biotechnol.*, 2018, **38**, 1137–1156.

278. E. González, M. Paulis and M. J. Barandiaran, *Eur. Polym. J.*, 2014, **59**, 122–128.
279. D. Wen and Y. Ding, *J. Nanopart. Res.*, 2005, **7**, 265–274.
280. Y. Hwang, J. K. Lee, J. K. Lee, Y. M. Jeong, S. ir Cheong, Y. C. Ahn and S. H. Kim, *Powder Technol.*, 2008, **186**, 145–153.
281. L. Fedele, L. Colla, S. Bobbo, S. Barison and F. Agresti, *Nanoscale Res. Lett.*, 2011, **6**, 300.
282. E. P. Bandarra Filho, O. S. H. Mendoza, C. L. L. Beicker, A. Menezes and D. Wen, *Energy Convers. Manage.*, 2014, **84**, 261–267.
283. D. H. Fontes, G. Ribatski and E. P. Bandarra Filho, *Diamond Relat. Mater.*, 2015, **58**, 115–121.
284. M. Drzazga, G. Dzido, M. Lemanowicz and A. Gierczycki, 14th European ConferenceMI3., 2012, vol. 10–13, pp. 89–94.
285. I. M. Mahbubul, R. Saidur, M. A. Amalina and M. E. Niza, *Int. Commun. Heat Mass Transfer*, 2016, **76**, 33–40.
286. J. Shah, M. Ranjan, S. K. Gupta and Y. Sonvane, *AIP Conf. Proc.*, 2018, **1951**, 20008.
287. R. Sadri, G. Ahmadi, H. Togun, M. Dahari, S. N. Kazi, E. Sadeghinezhad and N. Zubir, *Nanoscale Res. Lett.*, 2014, **9**, 151.
288. B. Ruan and A. M. Jacobi, *Nanoscale Res. Lett.*, 2012, **7**, 127.
289. M. F. Zawrah, R. M. Khattab, L. G. Girgis, H. El Daidamony and R. E. Abdel Aziz, *HBRC J.*, 2016, **12**, 227–234.
290. K. Wusiman, H. Jeong, K. Tulugan, H. Afrianto and H. Chung, *Int. Commun. Heat Mass Transfer*, 2013, **41**, 28–33.
291. R. Sadeghi, S. G. Etemad, E. Keshavarzi and M. Haghshenasfard, *Microfluid. Nanofluid.*, 2015, **18**, 1023–1030.
292. V. S. Nguyen, D. Rouxel, R. Hadji, B. Vincent and Y. Fort, *Ultrason. Sonochem.*, 2011, **18**, 382–388.
293. S. J. Chung, J. P. Leonard, I. Nettleship, J. K. Lee, Y. Soong, D. V Martello and M. K. Chyu, *Powder Technol.*, 2009, **194**, 75–80.
294. M. Kole and T. K. Dey, *Thermochim. Acta*, 2012, **535**, 58–65.
295. W. Rashmi, A. F. Ismail, I. Sopyan, A. T. Jameel, F. Yusof, M. Khalid and N. M. Mubarak, *J. Exp. Nanosci.*, 2011, **6**, 567–579.
296. A. Amrollahi, A. A. Hamidi and A. M. Rashidi, *Nanotechnology*, 2008, **19**, 315701.
297. A. Shahsavar, M. R. Salimpour, M. Saghafian and M. B. Shafii, *Thermochim. Acta*, 2015, **617**, 102–110.
298. J. Boopathy, R. Pari, M. Kavitha and P. C. Angelo, *Preparation of Nano Fluids by Mechanical Method*, 2012, vol. 1461.
299. B. Munkhbayar, M. R. Tanshen, J. Jeoun, H. Chung and H. Jeong, *Ceram. Int.*, 2013, **39**, 6415–6425.
300. M. Silambarasan, S. Manikandan and K. S. Rajan, *Int. J. Heat Mass Transfer*, 2012, **55**, 7991–8002.
301. B. Munkhbayar, M. J. Nine, J. Jeoun, M. Bat-Erdene, H. Chung and H. Jeong, *Powder Technol.*, 2013, **234**, 132–140.
302. Y. Wang, Y. Li, C. Rong and J. P. Liu, *Nanotechnology*, 2007, **18**, 465701.

303. V. M. Chakka, B. Altuncevahir, Z. Q. Jin, Y. Li and J. P. Liu, *J. Appl. Phys.*, 2006, **99**, 08E912.
304. W. Yu and H. Xie, *J. Nanomater.*, 2012, **2012**, 1–17.
305. S. Mukherjee, P. C. Mishra and P. Chaudhuri, *ChemBioEng Rev.*, 2018, **5**, 312–333.
306. A. Amiri, M. Naraghi, G. Ahmadi, M. Soleymaniha and M. Shanbedi, *FlatChem*, 2018, **8**, 40–71.
307. K. Kiatkittipong, M. A. A. M. Shukri, W. Kiatkittipong, J. W. Lim, P. L. Show, M. K. Lam and S. Assabumrungrat, *Processes*, 2020, **8**, 1–22.
308. D. Parviz, S. Das, H. S. T. Ahmed, F. Irin, S. Bhattacharia and M. J. Green, *ACS Nano*, 2012, **6**, 8857–8867.
309. X. An, T. Simmons, R. Shah, C. Wolfe, K. M. Lewis, M. Washington, S. K. Nayak, S. Talapatra and S. Kar, *Nano Lett.*, 2010, **10**, 4295–4301.
310. M. Zhang, R. R. Parajuli, D. Mastrogiovanni, B. Dai, P. Lo, W. Cheung, R. Brukh, P. L. Chiu, T. Zhou, Z. Liu, E. Garfunkel and H. He, *Small*, 2010, **6**, 1100–1107.
311. W. B. Russel, D. A. Saville and W. R. Schowalter, *Colloidal Dispersions*, Cambridge University Press, Cambridge, 1989.
312. X. Li, D. Zhu and X. Wang, *J. Colloid Interface Sci.*, 2007, **310**, 456–463.
313. L. Ju, W. Zhang, X. Wang, J.-D. Hu and Y. Zhang, *Colloids Surf., A*, 2012, **409**, 159–166.
314. W. Xian-Ju and L. Xin-Fang, *Chin. Phys. Lett.*, 2009, **26**, 056601.
315. D. Lee, J.-W. Kim and B. G. Kim, *J. Phys. Chem. B*, 2006, **110**, 4323–4328.
316. A. Amiri, M. Shanbedi, H. Amiri, S. Z. Heris, S. N. Kazi, B. T. Chew and H. Eshghi, *Appl. Therm. Eng.*, 2014, **71**, 450–459.
317. A. Amiri, M. Shanbedi, B. T. Chew, S. N. Kazi and K. H. Solangi, *Chem. Eng. J.*, 2016, **289**, 583–595.
318. M. Soleymaniha, A. Amiri, M. Shanbedi, B. T. Chew and S. Wongwises, *Int. Commun. Heat Mass Transfer*, 2018, **95**, 147–154.
319. M. H. Kayhani, H. Soltanzadeh, M. M. Heyhat, M. Nazari and F. Kowsary, *Int. Commun. Heat Mass Transfer*, 2012, **39**, 456–462.
320. E. Tang, G. Cheng, X. Ma, X. Pang and Q. Zhao, *Appl. Surf. Sci.*, 2006, **252**, 5227–5232.
321. Q. Yu, Y. J. Kim and H. Ma, *Appl. Phys. Lett.*, 2008, **92**, 2006–2009.
322. E. Ettefaghi, B. Ghobadian, A. Rashidi, G. Najafi, M. H. Khoshtaghaza and S. Pourhashem, *Energy Convers. Manage.*, 2017, **153**, 215–223.
323. B. Smith, K. Wepasnick, K. E. Schrote, H.-H. Cho, W. P. Ball and D. H. Fairbrother, *Langmuir*, 2009, **25**, 9767–9776.
324. M. Shanbedi, S. Zeinali Heris, M. Baniadam and A. Amiri, *Exp. Heat Transfer*, 2013, **26**, 26–40.
325. K. J. Park, D. Jung and S. E. Shim, *Int. J. Multiphase Flow*, 2009, **35**, 525–532.
326. G. Cao and Y. Wang, *Nanostructures and Nanomaterials*, World Scientific, 2011, vol. 2.

327. J.-Y. Choi, J. Y. Yoo, H.-S. Kwak, B. Uk Nam and J. Lee, *Curr. Appl. Phys.*, 2005, **5**, 472–474.
328. M. Shanbedi, S. Zeinali Heris and A. Maskooki, *J. Therm. Anal. Calorim.*, 2015, **120**, 1193–1201.
329. Q. Y. Tang, S. Ismathullakhan, Y. C. Chanl, N. B. Wong and R. Cheung, *J. Nanosci. Nanotechnol.*, 2010, **10**, 4967–4974.
330. Q. Y. Tang, I. Shafiq, Y. C. Chanl, N. B. Wong and R. Cheung, *J. Nanosci. Nanotechnol.*, 2010, **10**, 4967–4974.
331. H.-S. Kim, W.-I. Park, M. Kang and H.-J. Jin, *J. Phys. Chem. Solids*, 2008, **69**, 1209–1212.
332. K. Manojkumar, A. Sivaramakrishna and K. Vijayakrishna, *J. Nanopart. Res.*, 2016, **18**, 103.
333. M. Iqbal, B. A. Lyon, E. E. Ureña-Benavides, E. Moaseri, Y. Fei, C. McFadden, K. J. Javier, C. J. Ellison, K. D. Pennell and K. P. Johnston, *Colloids Surf., A*, 2017, **520**, 257–267.
334. M. M. Sarafraz, F. Hormozi and V. Nikkhah, *Exp. Therm. Fluid Sci.*, 2016, **78**, 41–49.
335. V. Srinivas, C. V. K. N. S. N. Moorthy, V. Dedeepya, P. V. Manikanta and V. Satish, *Heat and Mass Transfer*, 2016, **52**, 701–712.
336. L. S. Sundar, E. Venkata Ramana, M. P. F. Graça, M. K. Singh and A. C. M. Sousa, *Int. Commun. Heat Mass Transfer*, 2016, **73**, 62–74.
337. G. Wu, J. Yang, S. Ge, Y. Wang, M. Chen and Y. Chen, *Adv. Mater. Res.*, 2009, **60–61**, 394–398.
338. M. F. Islam, E. Rojas, D. M. Bergey, A. T. Johnson and A. G. Yodh, *Nano Lett.*, 2003, **3**, 269–273.
339. A. Ghadimi and I. H. Metselaar, *Exp. Therm. Fluid Sci.*, 2013, **51**, 1–9.
340. S. Chakraborty, I. Sengupta, I. Sarkar, S. K. Pal and S. Chakraborty, *Appl. Clay Sci.*, 2019, **168**, 43–55.
341. S. Suresh, K. P. Venkitaraj, P. Selvakumar and M. Chandrasekar, *Colloids Surf., A*, 2011, **388**, 41–48.
342. L. L. Schramm, *Emulsions, Foams, and Suspensions: Fundamentals and Applications*, Wiley, New York, 2005.
343. S. Chakraborty, I. Sarkar, A. Ashok, I. Sengupta, S. K. Pal and S. Chakraborty, *Powder Technol.*, 2018, **335**, 285–300.
344. L. Chen, H. Xie, Y. Li and W. Yu, *Thermochim. Acta*, 2008, **477**, 21–24.
345. T. X. Phuoc, M. Massoudi and R. H. Chen, *Int. J. Therm. Sci.*, 2011, **50**, 12–18.
346. B. Sharma, S. K. Sharma, S. M. Gupta and A. Kumar, *Arabian J. Sci. Eng.*, 2018, **43**, 6155–6163.
347. L. Jiang, L. Gao and J. Sun, *J. Colloid Interface Sci.*, 2003, **260**, 89–94.
348. D. A. Mansour and E. G. Atiya, in *2016 IEEE Conference on Electrical Insulation and Dielectric Phenomena (CEIDP)*, 2016, pp. 671–674.
349. S. Mohd Zaki, W. Azmi, A. A. M. Redhwan, N. N. M. Zawawi and R. Mamat, *J. Mech. Eng*, 2017, 233–247.

350. B. Bakthavatchalam, K. Habib, R. Saidur, N. Aslfattahi and A. Rashedi, *Appl. Sci.*, 2020, **10**, 1–20.
351. A. Ghadimi, H. Metselaar and B. LotfizadehDehkordi, Nanofluid stability optimization based on UV–vis spectrophotometer measurement, *J. Eng. Sci. Technol.*, 2015, **7**, 32–40.
352. E. Saeed, M. M. Piñeiro, C. Hermida-Merino and M. J. Pastoriza-Gallego, *Nanomaterials*, 2019, **9**, 252.
353. W. Chamsa-Ard, S. Brundavanam, C. C. Fung, D. Fawcett and G. Poinern, *Nanomaterials*, 2017, **7**, 131.
354. S. Bhattacharjee, *J. Controlled Release*, 2016, **235**, 337–351.
355. M. Kole and T. K. Dey, *Int. J. Therm. Sci.*, 2011, **50**, 1741–1747.
356. R. Ranjbarzadeh, A. Moradikazerouni, R. Bakhtiari, A. Asadi and M. Afrand, *J. Cleaner Prod.*, 2019, **206**, 1089–1100.
357. S. K. Sahoo, M. K. Das and P. Rath, *IEEE Trans. Compon., Packag., Manuf. Technol.*, 2018, **8**, 416–426.
358. H. S. Kim, F. Yilmaz, P. Dharmaiah, D. J. Lee, T. H. Lee and S. J. Hong, *Arch. Metall. Mater.*, 2017, **62**, 999–1004.
359. P. K. Das, N. Islam, A. K. Santra and R. Ganguly, *J. Mol. Liq.*, 2017, **237**, 304–312.
360. S. Rubalya Valantina, K. Arockia Jayalatha, D. R. Phebee Angeline, S. Uma and B. Ashvanth, *J. Mol. Liq.*, 2018, **256**, 256–266.
361. S. Mukherjee and S. Paria, *IOSR J. Mech. Civ. Eng.*, 2013, **9**, 63–69.
362. N. R. Karthikeyan, J. Philip and B. Raj, *Mater. Chem. Phys.*, 2008, **109**, 50–55.
363. D. W. Oh, A. Jain, J. K. Eaton, K. E. Goodson and J. S. Lee, *Int. J. Heat Fluid Flow*, 2008, **29**, 1456–1461.
364. V. A. Martínez, D. A. Vasco and C. M. García–Herrera, *Int. Commun. Heat Mass Transfer*, 2018, **91**, 234–238.
365. R. Prasher, W. Evans, P. Meakin, J. Fish, P. Phelan and P. Keblinski, *Appl. Phys. Lett.*, 2006, **89**, 143119.
366. J. Philip, P. D. Shima and B. Raj, *Appl. Phys. Lett.*, 2007, **91**, 1–4.
367. W. Evans, R. Prasher, J. Fish, P. Meakin, P. Phelan and P. Keblinski, *Int. J. Heat Mass Transfer*, 2008, **51**, 1431–1438.

Section II: Properties of Nanofluids

CHAPTER 4

Fundamental and Critical Aspects of the Rheological Behaviour of Nanofluids

P. ESTELLÉ*[a] AND G. ŻYŁA[b]

[a] Univ Rennes, LGCGM, F-35000 Rennes, France; [b] Department of Physics and Medical Engineering, Rzeszów University of Technology, Rzeszów, Poland
*Email: patrice.estelle@univ-rennes1.fr

4.1 Introduction

Nanofluids are colloidal suspensions containing solid particles with at least one dimension of <100 nm that can be made of metals, oxides, carbides, nitrides, or carbon dispersed in conventional heat transfer fluids (HTFs). Since the early 1990s, they have received great attention, which has risen exponentially over recent years, due to their potential applications in heat transfer devices and energy systems as a result of their enhanced thermal properties. Moreover, they appear to be promising candidates for conversion and management of energy for responding to global warming and reducing the temperature increase of our planet.

From a practical point of view, the rheological behaviour of nanofluids is an important issue for applications involving fluid flow and has been used to evaluate the required pumping power, pressure drop, convective heat transfer and natural convection. Thus, such flow properties are now the subject of many studies, as shown later in Table 4.1, just beyond thermal conductivity.[1] The dynamic viscosity of HTFs, and their rheological

Soft Matter Series No. 16
Fundamentals and Transport Properties of Nanofluids
Edited by S. M. Sohel Murshed

Published by the Royal Society of Chemistry, www.rsc.org

Table 4.1 Number of JCR-indexed publications extracted from Web of Science.[5]

Keywords	"nanofluid*" and "viscosi*"	"nanofluid*" and "viscoelasti*"	"nanofluid*" and "thixotrop*"	"nanofluid*" and "magnetorheolog*"
Number of publications			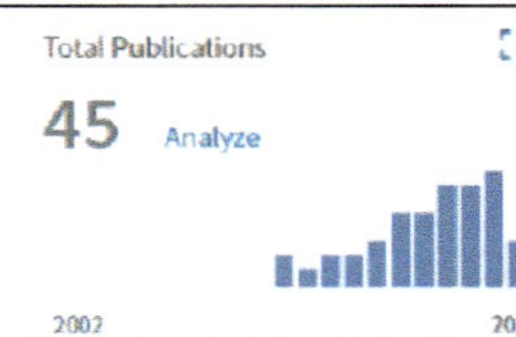	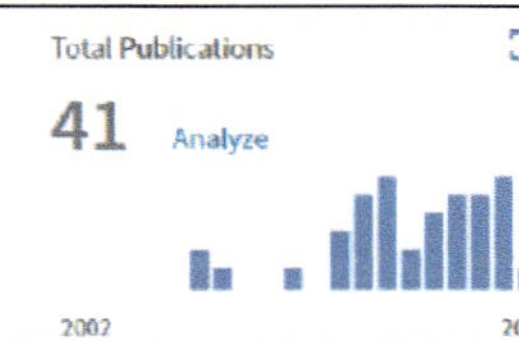

properties at large, can change due to the addition of solid nanoparticles and can result in increased pressure drops,[2] which affect the efficiency of energy systems. It is also closely related to the stability, the structure and interaction of solid nanoparticles, their dispersion state and possible presence of aggregates. Consequently, it can help in the understanding of physical and structural properties of nanofluids that can be related to thermal behaviour. Also, in many dimensionless numbers such as the Prandtl, Brinkman, Reynolds, and Rayleigh numbers, viscosity needs to be known. Viscosity also appears in figure of merits calculations that are generally used for the evaluation of nanofluid efficiency compared to base fluids.[3,4]

Rheology is the science related to the study of deformation and flow of matter. Viscosity is defined as the resistance of a liquid to flow. An understanding of rheology and viscosity of nanofluids is thus crucial to further their development and commercial usage. Two types of flow are of practical interest in the rheological measurements of nanofluids, namely steady state shearing and dynamic shearing (oscillations). Up to now, most of the studies have mainly focused on the first category, focusing on dynamic viscosity measurements of nanofluids with shear rate. The second category allows the evaluation of more complex rheological behavior, including viscoelasticity. In addition, time-dependant behaviour and the evaluation of electrorheological/magnetorheological properties have been less focused on. This can be illustrated by looking at Table 4.1, wherein the statistics, extracted from Web of Science,[5] of number of publications featuring specific keywords, including both "nanofluid*" and a specific rheological property, are reported. In addition to the data in Table 4.1, it should be noted that there are only 28 publications on the electrorheological properties of nanofluids.

So, the aim of this chapter is to highlight some critical rheological issues that are not generally addressed in the nanofluid field but appear to be essential for better understanding of the flow behaviour of nanofluids in relationship to their internal structure and nanoparticle organisation and arrangement. This chapter is organized as follows. First, some fundamental aspects and key issues about rheological properties and dynamic viscosity of nanofluids are discussed. Then, the main findings about the viscoelastic structure of nanofluids are introduced. Also, the time-dependant behaviour of nanofluids is presented in the third section. Finally, the influence of magnetic and electrical fields on the rheological behaviour of nanofluids is illustrated. As a conclusion, the last section summarizes our findings and some recommendations are proposed for future research. Without attempting an exhaustive survey, our intention is to raise awareness in the nanofluid community of crucial steps in nanofluid development and characterization from a rheological point of view.

4.2 Critical Aspects in Rheological Properties and the Dynamic Viscosity of Nanofluids

First of all, viscometers are still too often used to evaluate the viscosity of nanofluids. Unfortunately, they are not able to elucidate the possible shear dependence of nanofluid flow behaviour, in particular with low viscous base fluid based nanofluids, and other rheological features.

The rheological behaviour and shear flow properties of nanofluids can be more effectively evaluated using rheometers under well basic controlled conditions that must include:

- a selection of appropriate geometries that allow homogeneous shear within the sample gap, a cone and plate geometry is recommended, with a large diameter and small angle, in particular for low viscous materials,
- steady-state flow conditions when shear flow behaviour is inspected, and a shear rate range with at least 2 or 3 decades,
- control of possible slip along the geometries or surface tension and inertia effects, especially at low shear/strain rate
- right sample content within the geometries, right control of sample temperature and avoid sample evaporation
- qualification and proper uncertainty evaluation of the experimental procedure with fluids of well-known properties, repeatability of experiments,
- visual inspection of the samples after the measurements as an indication of stability under shear flow.[6,7]

The rheological properties of nanofluids cannot be reduced to one viscosity value, without a clear demonstration of Newtonian behaviour, and have to include a clear picture of at least shear flow behaviour. Actually, most usual HTFs behave in a Newtonian manner, with viscosity values of typically 1 mPa s for distilled water, 16.2 mPa s for ethylene glycol and around 200 mPa s for lubricating engine oil at 20 °C under atmospheric pressure. The addition of nanoparticles can drastically modify such a behaviour and the rheological behaviour of nanofluids cannot be categorized *a priori*. The transition from a Newtonian behaviour to a non-Newtonian one is strongly dependent on many parameters, such as the initial viscosity value of HTF, content, size and shape of nanoparticles, the addition (or not) of surfactant, their dispersion state and interaction, shear rate, and pH. Thus, a nanofluid behaves as a simple Newtonian fluid when steady-shear experimental results show a linear relationship between shear stress and shear rate (or a constant dynamic viscosity with shear rate; see Figures 4.1 and 4.2). This relationship is expressed by the following equation, where μ, τ and $\dot{\gamma}$ are the apparent dynamic viscosity, the shear stress and the shear rate, respectively.

$$\tau = \mu \cdot \dot{\gamma} \tag{4.1}$$

When the apparent dynamic viscosity changes with shear rate, this corresponds to a simple non-Newtonian behaviour. Shear-thinning behaviour is related to a decrease in apparent dynamic viscosity, while its increase leads to shear-thickening behaviour. Such effects can be attributed to particle interactions and ordering, as shown in Figure 4.1.[8]

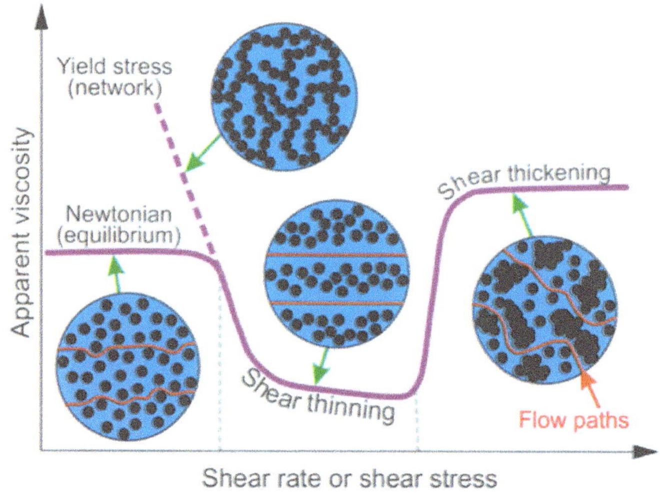

Figure 4.1 Steady shear flow behaviour of a nanofluid. Reproduced from ref. 8 with permission from Elsevier, Copyright 2020.

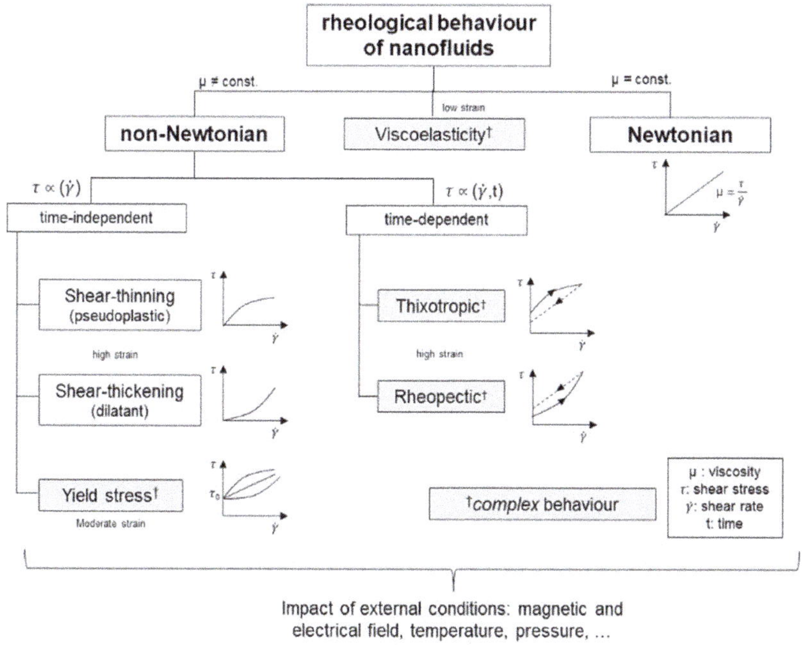

Figure 4.2 Classification of the rheological behaviour of nanofluids.

Shear-thinning is generally connected with the disentanglement, reorganization or alignment of nanoparticles (or their agglomerates/aggregates when present) under shear, resulting in less viscous forces. Inversely, shear-thickening arises from the structure of disordered nanoparticles.

The non-Newtonian nature of nanofluids, and materials in general, can be modelled by the following power law, as shown in eqn (4.2). In this equation, k refers to consistency, while n is the flow index. With shear-thinning materials, $n<1$; for shear-thickening materials, $n>1$. For $n=1$, this indicates Newtonian behaviour and eqn (4.1) is obtained. With regards to apparent dynamic viscosity, its expression is given by eqn (4.3)

$$\tau = k \cdot \dot{\gamma}^n \tag{4.2}$$

$$\mu = k \cdot \dot{\gamma}^{n-1} \tag{4.3}$$

Sometimes, nanofluids can also exhibit yield stress, *e.g.*, minimum stress must be applied to initiate the nanofluid flow and reach a liquid regime. Such behaviour is associated with the apparent dynamic viscosity evolution reported in Figure 4.1.[8] It should be mentioned that nanofluids can also behave as Newtonian or simple non-Newtonian nanofluids in the liquid regime above the yield stress, τ_0. In that case, the shear flow behaviour can thus be modelled using the Bingham or Herschel–Bulkley models, respectively. Those models are respectively written as follows when the yield stress is exceeded. Typical shear flow curves linked to this behaviour are shown in Figure 4.2.

$$\tau = \tau_0 + \mu \cdot \dot{\gamma} \tag{4.4}$$

$$\tau = \tau_0 + k \cdot \dot{\gamma}^n \tag{4.5}$$

Dynamic shearing or oscillatory tests can be used to determine both the viscous and elastic properties of nanofluids under low deformation from storage and loss moduli respectively, denoted as G' and G''. Such tests give information about the possible viscoelastic structure of the material linked to the solid regime. Crossover of the moduli obtained under fixed frequency oscillation (in the solid regime) and varying the stress amplitude is also related to the yield stress value. Such behaviour, including yield stress and viscoelastic structure, can be defined as complex behaviour. In addition, the time dependence of nanofluid flow can be included in this category. Actually, some nanofluids can be sensitive to shearing time. Such behaviour can be observed under up-and-down steady shear experiments when a hysteresis loop is obtained. One distinguishes thixotropic behaviour to rheopectic behaviour when the up steady shear curve is above the down curve. This behaviour can also be observed when the apparent dynamic viscosity decreases or increases during steady shear experiments, respectively, at constant shear rate, or if nanofluids are sensitive to preshearing and resting time.[9] Finally, a classification is proposed in Figure 4.2.

The type of HTF, content, size and shape of nanoparticles, their dispersion state and interaction, shear rate, *etc.*, also contribute to the modification of nanofluid dynamic viscosity. Over the last ten years, there have been several comprehensive reviews on the analysis and discussion of the impact of nanofluid formulation parameters on the steady-state shear flow behaviour of nanofluids.[2,10–15] It is clear from these different review papers that common viscosity models initially developed for suspensions containing microsized spherical particles, *e.g.*, the Einstein model[16] and its derivatives,[17,18] underpredict the viscosity of nanofluids in most cases.[2,10–14] Unfortunately, they are mostly largely and wrongly used in the literature as models for numerical simulations. It should be noted that the Maron–Pierce[19] and Krieger–Dougherty[20] models, which are extensions of the Einstein model for high particle content, can be relevant when nanoparticles mainly form aggregates and if the nanoparticle volume fraction is replaced by an effective volume fraction of aggregates.[21–23] It should be mentioned that this effective volume fraction of aggregates is also dependent on nanoparticle volume fraction and the radii of both primary nanoparticles and aggregates, with the geometry of the aggregates also being considered as a fractal-like structure with a fractal index.[21]

In addition, based on experimental data, many empirical viscosity correlations have also been proposed, as reviewed in different studies.[2,10–14] However, these correlations are generally only relevant and applicable for specific nanofluids, concentration and temperature ranges. This implies to systematically investigate the rheological behaviour and dynamic viscosity of newly produced nanofluids. It must also be mentioned that these viscosity values have to be associated with specific shear rate for non-Newtonian nanofluids. While the viscosity of nanofluids is generally increased with nanoparticle content,[2,10–15] a decrease is sometimes reported and attributed to lubrication effects of nanoparticles.[24,25] Such behaviour is promising in terms of applications if it is also accompanied by an enhancement in thermal properties.

Moreover, some efforts have been made to propose more generic models, but these contributions still remain limited. As an example, Pal[26] modelled the viscosity enhancement of different nanodispersions and nanoemulsions with nanoparticle and droplet content. The proposed model is an extension of the Oldroyd model based on the viscosity ratio, solvation and aggregation of nanoparticles/nanodroplets. Udawattha *et al.*[27] developed a viscosity model for nanofluids considering the effect of the interfacial layer on the nanoparticles as a solid-like layer between the base fluid and nanoparticle surface, and the Brownian motion of the nanoparticles. A general viscosity equation for graphene-based nanofluids was proposed by Vallejo *et al.*,[28] including the dependence on both the temperature and the graphene nanoplatelet volume concentration.

However, to date and to the best of our knowledge, there has been no universal correlation(s) based on physical insights able to clearly capture the nanofluid viscosity evolution with regards to all main composition parameters, whatever the nanoparticle size, dimension, shape, nature and concentration, and temperature.

Finally, a universal behaviour of nanofluid viscosity is its reduction with temperature increase.[2] Such a trend is well captured by the Vogel–Fulcher–Tammann–Hesse equation expressed in eqn (4.6),[29,30] as used in many studies, see ref. 7 and 28 as examples. In eqn (4.6), μ_0, D, and T_0 are fitting parameters, while μ and T are the dynamic viscosity and the temperature of the nanofluid, respectively.

$$\mu = \mu_0 \cdot e^{\frac{D \cdot T_0}{T - T_0}} \tag{4.6}$$

Vogel–Fulcher–Tammann–Hesse parameters also provide physical information regarding the energy associated with molecular packing in the liquid (μ_0), the dimensionless fragility of the liquid ($1/D$) or the ideal glass transition temperature (T_0).

4.3 Viscoelastic Structure

As it was presented in a previous subchapter, research studies have mainly focused on the dynamic viscosity evaluation of nanofluids. However, after simply checking the basic nature of the examined samples, most often the investigation is abandoned and no further examination is performed. So, even if the dynamic viscosity measurements were planned and performed correctly, a full picture of the rheological behaviour of the nanofluid is not obtained. After basic viscosity examination showing that the sample is a non-Newtonian fluid, a series of further experiments should be performed. One of these examinations is oscillatory measurements in which the sample is subjected to oscillation that can be applied by varying (a) frequency and the (b) deformation value.

Such experiments are generally conducted in two steps. First, the linear viscoelastic region (*i.e.*, the range in which oscillations are not large enough to irreversibly destroy sample structure) is defined by means of deformation tests at constant frequency. Then, storage and loss moduli are investigated by varying the oscillatory frequency and maintaining deformation within the linear viscoelastic range. In this type of examination, the viscoelastic structure of nanofluids can thus be examined and quantified from the storage modulus, G', and loss modulus, G'', and curves of the function of the frequency (noted ω or w) and deformation (noted γ). The most general G' and G'' responses of a structured fluid upon frequency scanning are depicted in Figure 4.3.[31] These data lead to important information about the influence that nanoparticles have on the rheological properties of liquids.

Depending on the trends in G' and G'' with oscillatory frequency, and which one of the two moduli prevails, it is possible to classify the viscoelastic response of the material under the investigated conditions. It must be pointed out that with the sensitivity of most available rheometers it is often not possible to observe the entire viscoelastic spectrum, only a portion.

Strong changes in the rheological nature of nanofluids and the appearance of a viscoelastic structure have been well described for ethylene glycol (EG) based nanofluids. EG itself is a well-known simple Newtonian fluid, but after adding even a small amount of nanoparticles, the produced nanofluids can exhibit complex non-Newtonian nature. Such an example is shown in Figure 4.4, which shows both shear flow and G' and G'' curves for carbon black EG nanofluids at 298.15 K.[32]

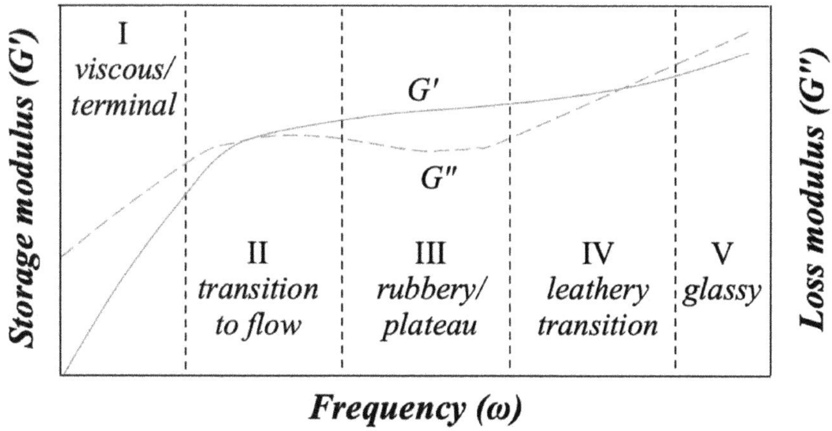

Figure 4.3 Typical viscoelastic plots for structured fluids upon frequency scanning.

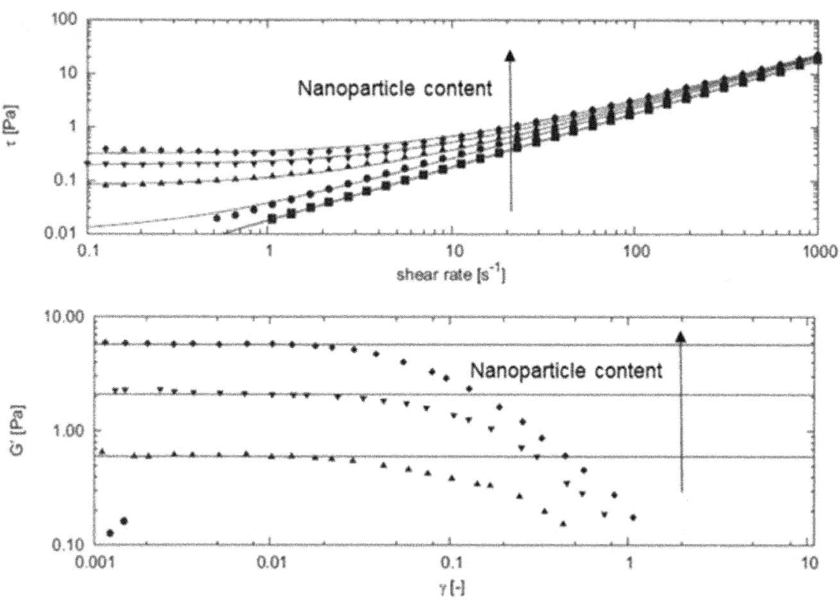

Figure 4.4 Steady shear flow curves of carbon black EG nanofluids at 298.15 K showing the impact of the varying of nanoparticle content from 0.1 to 1% in wt. (top) Storage (G') modulus as a function of deformation for various mass fractions of carbon black nanoparticles in EG nanofluids at 298.15 K and oscillation frequency of 1.0 Hz (bottom). Symbols represent different concentration, see ref. 32. Reproduced from ref. 32 with permission from Elsevier, Copyright 2020.

First, the change in shear flow behaviour with nanoparticles content, from 0.1 to 1 wt%, can be clearly evidenced with a change in the shear flow curve. For low nanoparticle content, the nanofluid is Newtonian, then mainly shear-thinning with the appearance of yield stress for higher nanoparticle concentration. The yield stress gradually increases with the nanoparticle concentration. Then, Figure 4.4. also shows the appearance of a viscoelastic structure from a nanoparticle content of 0.5 wt% and the value of G' is also increased for higher concentrations.

Another example can be found in a study by Hermida-Merino *et al.*,[33] who presented the complex nature of EG-based nanofluids containing a high fraction of graphite nanoparticles. After presenting shear flow and viscosity curves as a proof of non-Newtonian nature, the authors performed a series of oscillatory tests. First of all, oscillatory measurements with constant oscillation frequency and various deformations were performed to determine the values of the storage and loss moduli. In this study, the linear viscoelastic region, *e.g.*, the range of deformation in which these modules are constant as explained before, was determined. This allowed the evaluation of the value of deformation at which the sample was not damaged and the performing of a series of experiments in which the sample was "safely" deformed while the frequency of oscillation changed. From that, the authors also performed creep and recovery experiments. In this study, an interesting situation was observed. The authors described a local increase in the loss moduli value while increasing the deformation, and explained it as an overshoot phenomenon. Such an observation was also previously described for EG-based nanofluids. Actually, a few years earlier, the same group examined the rheological behaviour of an EG-based nanofluid with Fe_2O_3 nanoparticles.[34] This local increase was explained by the fact that the structure is damaged due to the disaggregation of nanoparticles during high deformations over the critical strain, and after that nanoparticles aligned with the flow direction. An overshoot phenomenon was later observed for EG-based nanofluids containing carbon black particles.[32,35] Viscoelastic behaviour of EG-based nanofluids was also reported for some other types of nanomaterials. The impact of SnO_2 nanoparticles on the rheological behaviour of EG was investigated in ref. 36. Typical viscoelastic behaviour was observed for these nanofluids from strain sweep and frequency sweep tests. As shown in Figure 4.5, the storage modulus is higher than the loss modulus in the studied frequency range. This means that the elastic properties are dominant over the viscous properties, both properties being increased with nanoparticle concentration.

Viscoelastic behavior was also reported for EG-based TiN nanofluids, which was also shown to be dependant on nanoparticle size.[37] At the same concentration, the lower the nanoparticle size, the better the viscoelastic structure. In addition to previous carbon structures, viscoelastic behaviour was also observed for EG nanofluids made using different carbon structures, namely nanodiamonds[38] and a graphite/diamond mixture.[39]

Finally, one must note that despite the fact that there have been many studies on the dynamic viscosity of EG-based nanofluids, specific research studies on the viscoelastic behaviour of these nanofluids are scarce, even though they make a huge contribution toward understanding the mechanism

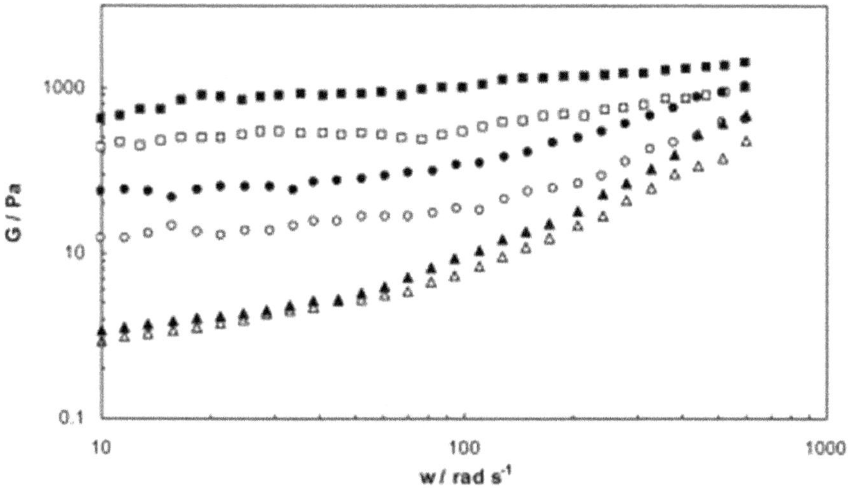

Figure 4.5 Storage modulus (G', filled symbols) and loss modulus (G'', empty symbols) as a function of frequency for SnO_2–EG nanofluids at a constant strain of 0.08%, triangles: 5 wt%, circles: 15 wt% and squares: 25 wt% in wt% of nanoparticles. Reproduced from ref. 36 with permission from Elsevier, Copyright 2013.

of the rheological properties of nanofluids. Also, it appears that studies on the viscoelastic properties of water-based nanofluids and oil-based nanofluids are even more limited.

4.4 Time Dependent Nature

At this point, it is already known that some nanoparticles dispersed in fluid can create a spatial structure that is damaged when the sample is deformed or under forced flow. This leads to the question of if time plays a role in the rheological properties evolution of nanofluids. To answer this, EG-based nanofluids are focused on as an example. The most encountered time-dependant behaviour is thixotropy. Thixotropy is simply a time-dependent shear behaviour of the fluid, and its detailed definition along with theoretical approach can be found in the literature.[40] An experiment in which the thixotropy of a fluid is examined is quite simple to perform. It requires conducting a steady-shear experiment while the shear rate is increasing, then holding the sample at a high constant shear rate and repeating the steady-shear experiment but instead decreasing the shear rate. With non-time-dependent nanofluids, both obtained flow curves cover each other, while thixotropic or rheopectic fluid show a hysteresis loop, as shown earlier in Figure 4.2. One can calculate the difference between the area under the flow curves when the shear rate is increased and decreased respectively and this value can be used as a "thixotropic index". In pure EG, the flow curves obtained with any type of measurement on a fresh or pre-examined sample will be the same and the value of the shear stress is not dependent on the shear

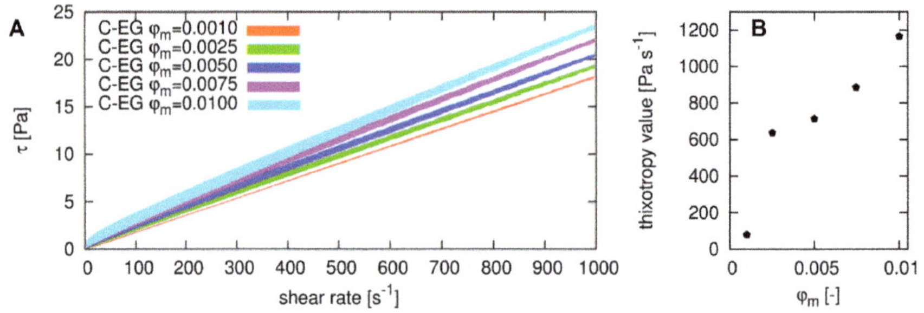

Figure 4.6 (A) Difference of areas under flow curves taken with increasing and decreasing shear rate for various mass fractions of particles. (B) Thixotropic index values at 298.15 K. Reproduced from ref. 32 with permission from Elsevier, Copyright 2020.

rate as it is for Newtonian fluids. However, thixotropic behaviour is observed for some nanofluids developed using EG. Shu *et al.*[41] showed that a high (35 wt%) concentration of TiO_2 nanoparticles makes an EG-based nanofluid thixotropic. They showed thixotropic loops at three temperatures. This type of behaviour was also presented by Żyła *et al.* for various types of carbon nanoparticles and low concentrations.[32,38] In these cases, besides the qualitative presentation of thixotropy loops, the authors calculated the area under flow curves derived under different conditions, and then the difference between these was quantitatively presented. In ref. 32 the authors showed that even a small amount (up to 1 wt%) of carbon black nanoparticles led to a nanofluid with a thixotropic nature. Figure 4.6 shows a summary of the main results, from which it can be noted that the differences between the flow curves with increasing and decreasing shear rate (presented as a coloured area) become increasingly greater upon an increase in the fraction of the carbon black particles in the nanofluid. After calculating the "thixotropy index", Figure 4.6 shows a non-linear increase with the mass fraction of the nanoparticles, denoted as φ_m. Similar enhancement of the thixotropic index was also observed for nanodiamond–EG-based nanofluids.[38]

Additionally, the fact that the thixotropy value is dependent on the size of nanoparticles in EG-based nanofluids was presented by Żyła *et al.*[37] The authors performed a study on the thixotropic behaviour of nanofluids containing TiN nanoparticles with two different average sizes, 20 nm and 50 nm, respectively. It was presented that for a low fraction of nanoparticles, nanofluids containing nanoparticles with a high diameter exhibited higher thixotropy values, while after a critical concentration value this situation changes. Previous studies on the time-dependent behaviour of nanofluids, and more specifically their thixotropic nature, were conducted using EG-based nanofluids. However, it can be noted that similar behaviour was also reported for water-based nanofluids containing multiwalled carbon nanotubes and alumina with a mass fraction of 1%[42] and functionalized carbon nanotubes with a COOH group with a mass fraction of 0.5 and 1 wt%,[43] for example.

4.5 Influence of External Fields on the Rheological Properties of Nanofluids

The influence of the electrical and magnetic field on the particles that have electric charge or magnetic moment are well understood and have been described in detail. However, the nanofluids as a relatively new group of materials can be made of such nanoparticles and at this point the influence of the fields on the rheological behaviour of nanofluids should be taken into account. Obviously, this type of analysis can only be performed with nanoparticles that are sensitive to electrical and magnetic fields, such as Fe_2O_3, Fe_2O_4, $CoFe_2O_4$, Fe and other conductive nanoparticles The base fluid should also be a non-electric and non-magnetic carrier liquid. Actually, electro/magneto nanofluids can be used in many applications by controlling/adjusting fluid motion and nanoparticle organization from external fields. In the following subchapters, the influence of external fields on the rheological behaviour of nanofluids is introduced, considering a few examples of case studies.

4.5.1 Electrorheology of Nanofluids

The electrorheological properties of colloidal suspensions were presented and described a long time before the term "nanofluids" appeared in the literature. In 1989, Gast and Zukowski[44] described the influence of the electrical field on the rheological behaviour of the suspension, with one of the conclusions being: "*Poorly understood areas such as the effect of charge carriers in the continuous phase and particle size, shape and chemistry are delineated to indicate areas deserving further research*". Unfortunately, after over thirty years from this conclusion, nanofluids with electro- and magneto-rheological properties are definitely out of mainstream research, despite the real-life application of nanofluids requiring the use of such materials. The basics of such measurements are quite simple – it is basically the classic rheological measurements, described earlier, but performed under the application of electric/magnetic field between the geometries of the rheometer. Figure 4.7 shows a scheme of the measuring geometry used in electro-rheological experiments.

Changes of the rheological behaviour of nanofluids under the application of electrical field were presented by Pereira and McIntyre[46] for oil-based nanofluids containing nanodiamonds (ND). They showed that an electric field dramatically changed the viscosity of nanofluids immediately after turning on the electric field. The authors showed that the ND–oil nanofluid exhibited a non-Newtonian nature even in the absence of the field, and when it was exposed to the field this nature was maintained, but the value of the viscosity increased significantly and the yield stress was proportional to the applied field. Interestingly, the authors pursued their investigation and performed oscillatory tests under an electric field. They presented that the viscoelastic structure also changed under such conditions and both the storage and loss modulus increased while the electric field increased.

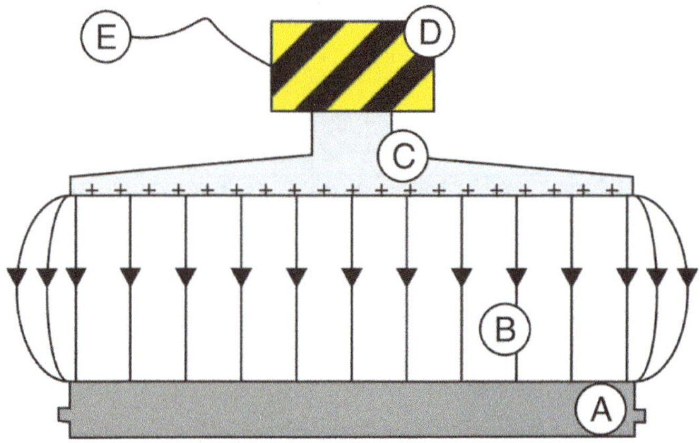

Figure 4.7 Diagram of an electric field in a mounted electrorheological system. (A) Stable lower plate, (B) electric field lines, (C) ER rotor, (D) ER adaptor, (E) positive electrode connected to a high-voltage power supply. Reproduced from ref. 45, DOI: 10.1186/1556-276X-9-170, under the terms of the CC BY 2.0 license, https://creativecommons.org/licenses/by/2.0.

Such behaviour was explained by the enhancement in the formation and strength of the ND structures in the base fluid under electrical field.[46] The influence of the applied electric field on the dynamic viscosity of $MgAl_2O_4$–DG nanofluids was investigated in ref. 45. It was shown in this study that the electric field did not significantly change the dynamic viscosity of nanofluids or the thixotropic behaviour and thixotropic index. The electrical rheological behaviour of Fe_2O_3 nanofluids was tested by Raykar and coworkers[47] under low DC electric fields. The strong effect of the electrical field on the viscosity of an Fe_2O_3 nanofluid of a high concentration was evidenced at 10 V mm^{-1}. This was attributed to the formation of a ring-like structure of Fe_2O_3 nanoparticles due to the current passing within the nanofluid and an electrohydrodynamic convection effect.

4.5.2 Magnetorheology of Nanofluids

Besides the aforementioned influence of electrical field, the application of a magnetic field can also change the rheological properties of nanofluids. Shima and Philip[48] evidenced the impact that a magnetic field has on the dynamic viscosity changes and increase of kerosene-based Fe_3O_4 nanofluids. Such behaviour was clearly linked to the appearance of chain-like structures due to strong interparticle interactions under the influence of a magnetic field, as evidenced from microscopy images. It was observed that the higher the magnetic field, the greater the nanoparticle interactions and resistance to flow. They also observed full reversible rheological behaviour upon turning on/turning off the magnetic field.

Shear and dynamic rheological measurements under a magnetic field were comprehensively performed by Felicia and Philip[49] using Fe_3O_4

dispersed in oil, with shear thinning behaviour with the presence of yield stress reported for the nanofluids. The viscosity was shown to increase with magnetic field due to the formation of heterogeneous linear structures. Magnetic dipolar interactions led to the creation of linear chains of nanoparticles under a moderate magnetic field, while dense columns were observed under higher fields. Dynamic measurements evidenced the elastic behaviour of the nanofluids due to the field-induced structure of the nanoparticles, with the elastic properties of the nanofluids being enhanced with magnetic field until saturation was reached under high-field strengths.

The rheological behaviour of Fe–Ni nanoparticles dispersed in a paraffin base oil was studied by Katiyar *et al.*[50] focusing on the effect of particle content on the shear flow behaviour of nanofluids under the influence of magnetic flux. It was mainly observed that viscosity increased with both nanoparticle content and magnetic field due to the orientation of the nanoparticles inducing a strong resistance to flow, as illustrated in Figure 4.8, with a significant change from 8 wt% in nanoparticles. It was further shown that at fixed concentration of 10 wt% of nanoparticles, the yield stress of the nanofluid increased in line with the magnetic field.

More recently, Yang *et al.*[51] also investigated how the viscosity of an oil-based Fe_3O_4 ferrofluid changed when a magnetic field was applied. An unexpected result was observed in this study. It was shown that the viscosity of these nanofluids decreased while the magnetic field induction increased. However, at a particular magnetic field value, the viscosity increased dramatically and after that began to decrease again. Such behaviour was related

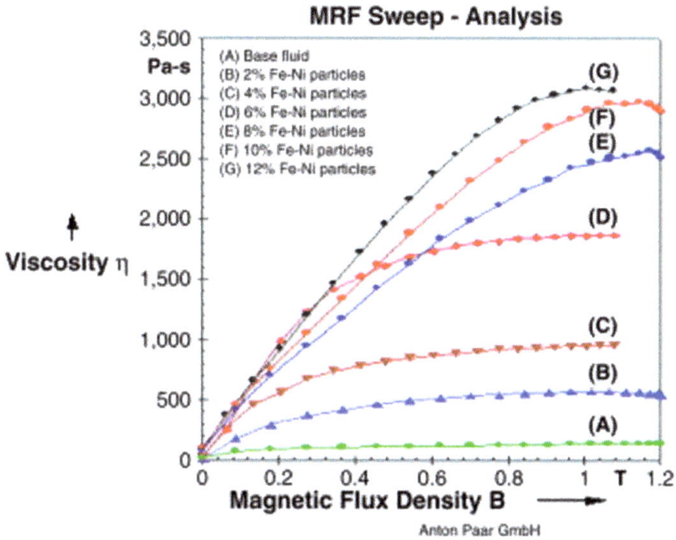

Figure 4.8 Viscosity evolution of Fe–Ni oil based nanofluids at 298 K as a function of nanoparticle content and magnetic flux. Reproduced from ref. 50 with permission from Elsevier, Copyright 2012.

to the thixotropic nature of these nanofluids and the authors concluded that the electrorheological behaviour of the nanofluids must be examined under different configurations of rheological measurements.

Finally, a comprehensive review on the influence of external magnetic field on the viscosity of magnetic nanofluids was published by Doganay *et al.*[52] As a main conclusion, it was reported that the alignment of the magnetic nanoparticles in the direction of the field lines under magnetic field led to the formation of a chain-like structure of nanoparticles due to their dipolar interaction. At the end, this leads to an enhancement in viscosity, as also reported in other previous studies. The readers are referred to this work for a detailed discussion on the impact of the type of nanoparticles, their concentration, impact of direction of the external magnetic field, and some inconsistencies in terms of experimental results.

4.6 Summary and Future Directions

This main aim of this chapter was to emphasize some fundamental and critical issues on the rheological behaviour of nanofluids. Some important key points were presented and discussed considering successively the shear flow behaviour of nanofluids and the evaluation of dynamic viscosity, the viscoelastic nature of nanofluids and possible time-dependence behaviour. Lastly, the impact of external stimuli such electric and magnetic fields on the rheological behaviour of nanofluids was also highlighted.

It should be noted here that adding even a low fraction of nanoparticles leads to a dramatic change in the rheological nature of well-known Newtonian HTFs. While most studies have focused on steady-shear rheometry that can be used to distinguish between Newtonian and non-Newtonian behaviour, nanofluids can also exhibit more complex properties, including both elastic and viscous ones, in particular for highly viscous HTFs and a high nanoparticle content. Consequently, one cannot reduce the investigation of rheological behaviour of nanofluids to a simple value of viscosity, in particular if Newtonian behaviour is not surely demonstrated or if the viscosity value is not associated with shear rate. Consequently, rheological measurements must be extended to complex properties to fully characterize nanofluids. In addition, all measurements must be carefully performed and well detailed to ensure that the results are relevant, reliable and reproducible. All these steps serve towards a better understanding of flow behaviour in relation to internal structure and nanoparticle organisation and arrangement. It was presented that the field-dependent behaviour of nanofluids requires a more advanced and methodical approach. A series of experimental studies on the complex electro- and magneto-rheological properties of nanofluids containing charged particles or particles endowed with a magnetic moment should be planned and performed. Along with this study, the theoretical work of physicists and fluid mechanic scientists is required to understand the phenomena observed therein.

In our opinion, the previous recommendations must also be accompanied in the near future by some new important developments. An interesting challenge is the development of rheological model(s) based on physical insights that can be used to predict dynamic viscosity enhancement with nanoparticle content and take into consideration all other parameters of composition. Such model(s) is/are lacking, as classical viscosity models do not accurately reflect the properties of nanofluids. This theoretical development could rely on molecular dynamic simulations[53] and flow visualization under shear.[54] A first step in this direction was initiated recently though the first special issue on this topic.[55]

In addition, to go further in the development of nanofluids for industrial applications and commercial usage, the characterization of rheological properties must be done as close as possible to real situations. This means that measurements need to be conducted under well-controlled pressure conditions, and higher than atmospheric pressure. Moreover, most of the experiments conducted thus far have been performed in the temperature range of 10–60 °C. Consequently, in view of the potential applications of nanofluids, this range must be greatly extended. For use of nanofluids in solar applications, few studies where the temperature reaches 400 °C have yet been performed to fill this gap.[56–59]

Abbreviations

EG	Ethylene glycol
DG	Diethylene glycol
Fe_2O_3	Ferric oxide
G'	Storage modulus
G''	Loss modulus
HFT	Heat transfer fluid
k	Consistency
n	Flow index
ND	Nanodiamonds
SnO_2	Tin dioxide
T	Temperature
TiN	Titanium nitride
TiO_2	Titanium dioxide
wt	Weight

Symbols

μ	Apparent dynamic viscosity
τ	Shear stress
τ_0	Yield stress
$\dot{\gamma}$	Shear rate
φ_m	Mass fraction in nanoparticles
γ	Deformation
ω or w	Frequency

Acknowledgements

This work has been developed by participants of the COST Action CA15119 "Overcoming Barriers to Nanofluids Market Uptake" (Nanouptake) and COST Innovators' Grant CIG15119 "Nanofluids for convective heat transfer devices" (Nanoconvex). The authors wish to thank D. Cabaleiro, from Universidad de Vigo, for useful comments and suggestions.

References

1. J. Traciak, J. Fal and G. Żyła, *Appl. Surf. Sci.*, 2021, **561**, 149878.
2. S. M. S. Murshed and P. Estellé, *Renewable Sustainable Energy Rev.*, 2017, **76**, 1134.
3. D. Cabaleiro, L. Colla, S. Barison, L. Lugo, L. Fedele and S. Bobbo, *Nanoscale Res. Lett.*, 2017, **12**, 53.
4. R. Gómez-Villarejo, P. Estellé and J. Navas, *Sol. Energy Mater. Sol. Cells*, 2020, **205**, 110266.
5. http://apps.webofknowledge.com (accessed on 09 September 2021).
6. E. J. Kim, A. Desforges, L. Speyer, J. Ghanbaja, J. Gleize, P. Estellé and B. Vigolo, *J. Nanofluids*, 2017, **6**(3), 603.
7. S. Hamze, D. Cabaleiro, T. Maré, B. Vigolo and P. Estellé, *J. Mol. Liq.*, 2020, **316**, 113875.
8. B. Jóźwiak and S. Boncel, *J. Mol. Liq.*, 2020, **302**, 112568.
9. P. Estellé, S. Halelfadl, N. Doner and T. Maré, *Curr. Nanosci.*, 2013, **9**(2), 225.
10. P. C. Mishra, S. Mukherjee, S. K. Nayak and A. Panda, *Int. Nano Lett.*, 2014, **4**, 109.
11. A. K. Sharma, A. K. Tiwari and A. R. Dixit, *Renewable Sustainable Energy Rev.*, 2016, **53**, 779.
12. K. Bashirnezhad, S. Bazri, M. Reza Safaei, M. Goodarzi, M. Dahari, O. Mahian, D. Ahmet Selim and S. Wongwises, *Int. Commun. Heat Mass Transfer*, 2016, **73**, 114.
13. A. A. Nadooshan, H. Eshgarf and M. Afrand, *Powder Technol.*, 2018, **338**, 342.
14. A. Asadi, S. Aberoumand, A. Moradikazerouni, F. Pourfattah, G. Żyła, P. Estellé, O. Mahian, S. Wongwises, H. M. Nguyen and A. Arabkoohsar, *Powder Technol.*, 2019, **352**, 209.
15. S. Hamze, D. Cabaleiro and P. Estellé, *J. Mol. Liq.*, 2021, **325**, 115207.
16. A. Einstein, *Ann. Phys.*, 1906, **19**, 289.
17. H. C. Brinkman, *J. Chem. Phys.*, 1952, **20**(4), 571.
18. G. K. Batchelor, *J. Fluid Mech.*, 1977, **83**(1), 97.
19. S. H. Maron and P. E. Pierce, *J. Colloid Sci.*, 1956, **11**, 80.
20. I. M. Krieger and T. J. Dougherty, *Trans. Soc. Rheol.*, 1959, **3**, 137.
21. H. Chen, Y. Ding and C. Tan, *New J. Phys.*, 2007, **9**, 367.
22. S. Halelfadl, P. Estellé, B. Aladag, N. Doner and T. Maré, *Int. J. Therm. Sci.*, 2013, **71**, 111.

23. M. A. Marcos, N. E. Podolsky, D. Cabaleiro, L. Lugo, A. O. Zakharov, V. N. Postnov, N. A. Charykov, S. V. Ageev and K. N. Semenov, *J. Mol. Liq.*, 2019, **294**, 111616.
24. A. Banisharif, M. Aghajani, S. Van Vaerenbergh, P. Estellé and A. Rashidi, *J. Mol. Liq.*, 2020, **302**, 112606.
25. A. Bakak, M. Lotfi, R. Heyd, A. Ammar and A. Koumina, *Entropy*, 2021, **23**, 979.
26. R. Pal, *Fluids*, 2016, **1**(2), 11.
27. D. S. Udawattha, M. Narayana and U. P. L. Wijayarathne, *J. King Saud Univ., Sci.*, 2019, **31**(3), 412.
28. J. P. Vallejo, S. Gómez-Barreiro, D. Cabaleiro, C. Gracia-Fernández, J. Fernández-Seara and L. Lugo, *Int. Commun. Heat Mass Transfer*, 2018, **91**, 150.
29. G. S. Fulcher, *J. Am. Ceram. Soc.*, 1925, **8**, 339–355.
30. H. Vogel, *Phys. Z.*, 1921, **22**, 645.
31. J. D. Ferry, *Viscoelastic Properties of Polymers*, Wiley, New York, 3rd edn, 1980.
32. G. Żyła, *J. Mol. Liq.*, 2020, **297**, 111732.
33. M. J. Pastoriza-Gallego, M. Perez-Rodriguez, C. Gracia-Fernandez and M. M. Piñeiro, *Soft Matter*, 2016, **12**(8), 2264.
34. M. J. Pastoriza-Gallego, L. Lugo, J. L. Legido and M. M. Piñeiro, *Nanoscale Res. Lett.*, 2011, **6**, 560.
35. J. Sobczak, J. P. Vallejo, J. Traciak, S. Hamze, J. Fal, P. Estellé, L. Lugo and G. Żyła, *J. Mol. Liq.*, 2021, **326**, 115255.
36. A. Mariano, M. J. Pastoriza-Gallego, L. Lugo, A. Camacho, S. Canzonieri and M. M. Piñeiro, *Fluid Phase Equilib.*, 2013, **337**, 119.
37. G. Żyła, J. Fal and P. Estellé, *Int. J. Heat Mass Transfer*, 2017, **113**, 1189.
38. G. Żyła, J. P. Vallejo, J. Fal and L. Lugo, *Int. J. Heat Mass Transfer*, 2018, **121**, 1201.
39. G. Żyła, J. Fal and P. Estellé, *Diamond Relat. Mater.*, 2017, **74**, 81.
40. P. Oswald, *Rheophysics*, Cambridge University Press, Cambridge, UK, 2009.
41. R. Shu, Y. Gan, H. Lv and D. Tan, *Colloids Surf., A*, 2016, **509**, 86.
42. B. Aladag, S. Halelfadl, N. Doner, T. Maré, S. Duret and P. Estellé, *Appl. Energy*, 2012, **97**, 876.
43. S. Scarassatti Freitas, V. Silveira Jr, J. M. Saìz Jabardo and A. Ceinos Arce, *J. Braz. Soc. Mech. Sci. Eng.*, 2020, **42**, 439.
44. A. P. Gast and C. F. Zukowski, *Adv. Colloid Interface Sci.*, 1989, **30**, 153.
45. G. Żyła, J. Grzywa, A. Witek and M. Cholewa, *Nanoscale Res. Lett.*, 2014, **9**, 170.
46. A. Pereira and C. McIntyre, *Appl. Rheol.*, 2014, **24**, 6347.
47. V. S. Raykar, S. K. Sahoo and A. K. Singh, *J. Appl. Phys.*, 2012, **108**, 034306.
48. P. D. Shima and J. Philip, *J. Phys. Chem. C*, 2011, **115**, 20097.
49. L. J. Felicia and J. Philip, *Langmuir*, 2013, **29**, 110.
50. A. Katiyar, A. N. Singh, P. Shukla and T. Nandi, *Powder Technol.*, 2012, **224**, 86.

51. C. Yang, Z. Liu, M. Yu and X. Bian, *J. Mol. Liq.*, 2020, **320(A)**, 114425.
52. S. Doganay, R. Alsangur and A. Turgut, *Mater. Res. Express*, 2019, **6**, 112003.
53. N. Razmara, H. Namarvari and J. Romano Meneghini, *J. Mol. Liq.*, 2019, **293**, 111438.
54. G. Żyła and M. Cholewa, *RSC Adv.*, 2014, **4**, 26057.
55. P. Estellé and G. Żyła, *J. Mol. Liq.*, 2022, **362**, 119669.
56. B. Muñoz-Sánchez, J. Nieto-Maestre, E. Veca, R. Liberatore, S. Sau, H. Navarro, Y. Ding, N. Navarrete, J. E. Juliá, Á. G. Fernández and A. García-Romero, *Sol. Energy Mater. Sol. Cells*, 2018, **176**, 357.
57. N. Navarrete, L. Hernández, A. Vela and R. Mondragón, *J. Mol. Liq.*, 2020, **302**, 112570.
58. Y. Grosu, A. Anagnostopoulos, B. Balakin, J. Krupanek, M. E. Navarro, L. González-Fernández, Y. Ding and A. Faik, *Sol. Energy Mater. Sol. Cells*, 2021, **220**, 110838.
59. M. Isaza-Ruiz, R. Mondragón, F. Bolívar Osorio, J. Ventura-Espinosa and L. Hernández, *Sol. Energy Mater. Sol. Cells*, 2021, **222**, 110923.

CHAPTER 5

Parametric Research and Theoretical Models for the Viscosity of Nanofluids

S. O. GIWA,[a] M. SHARIFPUR,*[b] S. M. S. MURSHED[c] AND J. P. MEYER[b,d]

[a] Department of Mechanical Engineering, Olabisi Onabanjo University, Ibogun 112104, Nigeria; [b] Department of Mechanical and Aeronautical Engineering, University of Pretoria, Pretoria 0002, South Africa; [c] IDMEC, Department of Mechanical Engineering, Instituto Superior Técnico, University of Lisbon, 1049-001 Lisbon, Portugal; [d] Department of Mechanical and Mechatronic Engineering, Stellenbosch University, Stellenbosch, Private Bag X1, Matieland 7602, South Africa

*Email: mohsen.sharifpur@up.ac.za

5.1 Introduction

Thermal transport in energy systems utilized for various applications, such as thermal plants, electronics, nuclear reactors, automobile engines, heat exchangers, aviation, space exploration, agriculture and so on, through the aid of working fluids is critical to the efficient and effective operation of the systems.[1] Because the existing working fluids have low thermal conductivity and technological advancement demands an increasing need to quickly and effectively remove a huge quantum of heat, the use of orthodox working fluids such as water, ethylene glycol, transformer oil, thermal oil, *etc.*, is proving to be impracticable.[2] To address this challenge, passive and active heat enhancing

Soft Matter Series No. 16
Fundamentals and Transport Properties of Nanofluids
Edited by S. M. Sohel Murshed

Published by the Royal Society of Chemistry, www.rsc.org

techniques (miniaturization, surface modification, and surface extension) have been deployed, which over time have reached their thresholds.[1,2] Nanotechnology has afforded the re-engineering of existing orthodox working fluids to high-performing and energy-efficient fluids through the suspension of nano-sized materials (metals, metal oxides, ceramics, polymers, non-metals, *etc.*). Since solid metals have a higher thermal conductivity than orthodox working fluids, the suspension of metal particles in working fluids was conceptualized by Maxwell in 1873, then Ahuja in 1975, Akoh in 1978, and finally by Masuda in 1993.[3,4] Prior to the work of Masuda *et al.*,[3] milli-scale particles were used that led to erroneous results due to the sedimentation and clogging of particles. Thus, nanoscale particles were used by Masuda *et al.*, which birthed nanofluids (NFs), a term coined by Choi.

Subsequent to these pioneering efforts, numerous researchers have studied the thermophysical (thermal conductivity, density, viscosity, *etc.*) and thermal-flow (coefficient of heat transfer, friction factor, Nusselt number, *etc.*) properties of different NFs made up of diverse nanoparticles (NPs) and base fluids. These studies revealed outstanding enhancements of these properties above those of base fluids in all cases in terms of the thermophysical properties (at all concentrations) but only up to certain concentrations for the thermal–fluid properties.[5,6] The improved performance of different NFs over those of base fluids in various thermal devices has been widely reported to be strongly linked to their enhanced thermal properties, of which viscosity and thermal conductivity are crucial.[1,6] By increasing the concentration of dispersed NPs in the base fluid, both the viscosity and thermal conductivity of the resulting NF are increased. However, an increase in the viscosity consequently leads to a denser and more viscous NF, leading to increased pumping power, frictional loss and pressure drop.[7,8] The heat transfer and thermal efficiency gain *via* the enhanced coefficient of heat transfer through the improved thermal conductivity could be lost through the penalties resulting from increased viscosity.[8]

The above shows the viscosity of NFs to be a property of huge importance in the application, design, and development of thermal systems utilizing advanced thermal fluids. Experimentally, the viscosity of NFs has been reported to be affected by parameters such as dispersion characteristics (sonication time, mode, frequency, amplitude, and stirring duration and speed); base fluid density, rheology, and type; NP weight/volume concentration, type (density), and nano-size; surfactant type and quantity; temperature, agglomeration, and nano-layer thickness.[9–12] However, the theoretical (classical and contemporary) development of models for predicting the viscosity of colloidal suspension (micro-, milli-, and nanoscale) link factors such as nano-layer, spacing distance, volume fraction, temperature, diameter of NPs and shape, agglomeration, Brownian motion, interparticle spacing, *etc.,* to viscosity.

The work of Einstein, who proposed a theoretical model for the prediction of the viscosity of colloidal suspensions (rigid, spherical, and milli-scale particles of very low concentration), pioneered the development of models for the prediction of colloidal suspension viscosity.[13,14] Thereafter, several attempts have been made to propose generalized models that can be employed to estimate the viscosity of colloidal suspensions using different theoretical backgrounds and mechanisms related to various parameters that affect viscosity. However, the existing classical models mainly under-predict the viscosity of NFs, with a few studies reporting good prediction accuracy.[15] A couple of new theoretical models for predicting the viscosity of NFs have been published in the open literature, which give better prediction accuracy than classical models. The development of empirical models from experimental viscosity data using various techniques is not generally practical for the prediction of different NFs[13,16] and in some cases the same NFs researched by different authors.

This chapter succinctly presents and discusses the key parameters affecting the viscosity of NFs. In addition, a compendium of models developed for the prediction of NF viscosity is highlighted herein, comprising classical (prior to NF development) and contemporary (during NF development) models. It is pertinent to know that single and hybrid NFs are also focused on concerning the viscosity parameters and predictive models.

5.2 Mechanisms Impacting Viscosity of Nanofluids

Viscosity is a fluid property that indicates its resistance to gradual deformation under increasing shear stress. It is also one of the key properties of NFs in terms of determining their hydrodynamic and heat transfer performance in thermal devices and systems. Pak and Choi[17] pioneered studies on the experimental measurement of the viscosity of NFs. They reported an enhancement in the viscosity of NFs compared with the base fluid, which has generally been established in subsequent studies to date.[18–21] Thus, the suspension of NPs in base fluids to synthesize NFs has been found to lead to unnatural and various degrees of viscosity enhancement. NFs are generally considered to be homogeneous liquid–particle interactions. The phenomenon responsible for the upsurge in the viscosity of NFs has been ascribed as Brownian motion – the motion of NPs in the base fluid –, NP clustering, ballistic phenomena in NPs, and liquid layering (molecular level) at the particle–liquid interface.[7,14]

The Brownian motion is a pervasive phenomenon that comes into play when particles of different sizes, concentration, and shapes are suspended in different aqueous solutions. Brownian motion augments the drag effect on particles owing to the velocity of the suspended particles in the suspension, which is linked to the thermophysical properties of the suspension. Particle size, concentration, shape, temperature and velocity gradient of flows are principal factors that influence the prominence of the rotatory Brownian motion (micro-convection) of nanoparticles. The mechanism of Brownian motion is well pronounced in particle suspension as it contributes immensely to thermal properties, especially the viscosity of micro- and nano-suspension.

Particle interaction due to liquid–particle and particle–particle interfaces in a suspension is a mechanism that impacts on viscosity enhancement.

This interaction leads to the generation of van der Waals, electrical, thermal and hydrodynamic forces within the suspension. At very low concentration, particle interactions are near negligible but more pronounced at high concentration of particles in the suspension as forces around the particles (electro-viscous effects) are altered, leading to the high viscosity of the suspension. Electro-viscous effects are strongly linked to agglomeration size and level of Brownian motion. As the size of the particles in a suspension is reduced and the concentration is increased, the particle–liquid and particle–particle interaction surface area increase lead to the augmentation of the electro-viscous effects and Brownian motion. The clustering of particles in suspension is regarded as one of the most important mechanisms that affects viscosity, leading to viscosity augmentation, as reported in experimental studies. The formation of clusters increases the effective concentration of nanoparticles, causing a surge in the drag and inertial forces and successive dissipation of energy, which consequently leads to viscosity increase of the suspension.

5.3 Factors Influencing the Viscosity of Nanofluids

5.3.1 Concentration and Temperature

The formation of NFs starts with the suspension of NPs in base fluids. Different quantities of NPs are involved, leading to the volume/weight concentration/fraction of a specific NF. Thus, the base fluid and NPs have their density, and the even suspension of both materials – NFs – has an effective viscosity. Therefore, NF viscosity is dependent on volume/weight fraction or concentration. As the NF exists at a particular temperature, the viscosity of the NF is related to both the volume fraction (φ) and temperature. The NF viscosity is dependent on several factors, as presented in Figure 5.1, but the most studied factors concerning the viscosity of NFs (either mono or hybrid NFs) are the temperature and φ.

Numerous studies have investigated the influence of φ and temperature on the viscosity of different types of NFs formulated from diverse NPs and base fluids.[11,12,21–27] These studies have involved the use of different ranges of temperature and φ to examine their effect on NF viscosity.[3,13] The suspension of different NPs with notably higher density than the base fluids augment the density of the base fluids, which consequently enhance the viscosity of the mixture – NF. An increase in φ leads to an increase in the amounts of NPs suspended and slows the motion of NPs and base fluid molecules, thereby improving the NF viscosity. The improvement in viscosity can be attributed to Brownian motion, liquid–particle and particle–particle interactions, nanolayering, base fluid, *etc.*[7,14,28] A rise in temperature *via* the introduction of thermal energy weakens the intermolecular forces and increases the liquid–particle and particle–particle velocity, thereby reducing the viscosity of the NF. Temperature is inversely proportional to viscosity, thus, increasing the temperature of the NF correspondingly contributes to a reduction in viscosity. Generally, it has been established (for both mono and hybrid NFs) that the

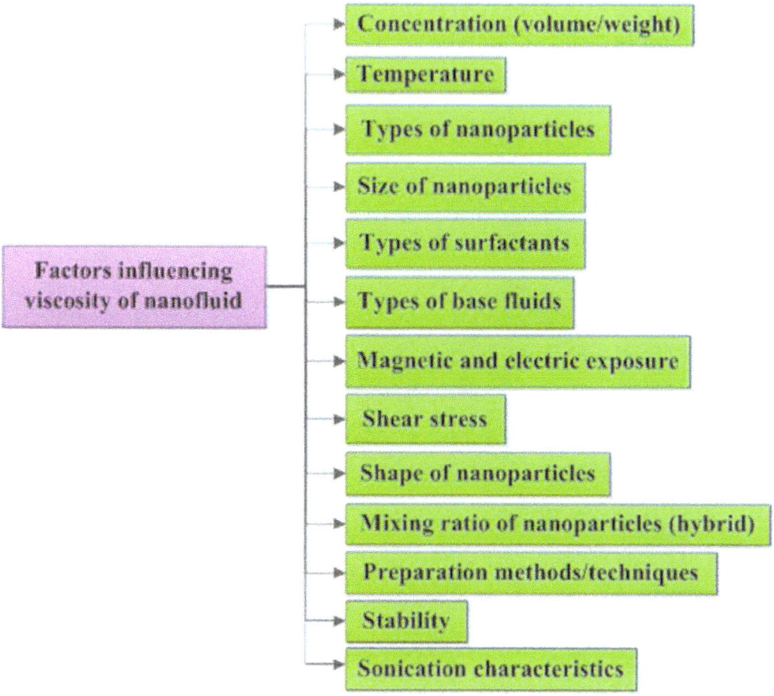

Figure 5.1 Factors affecting the viscosity of nanofluids.

viscosity of a NF is enhanced with a rise in φ and depreciates with a surge in temperature.[20,28–32]

Under various experimental conditions involving φ and temperature, different values of NF viscosity and improvement have been published in the literature. Yiamsawas *et al.*[24] studied the viscosity of ethylene glycol (EG)–water (20 : 80) based Al_2O_3 and TiO_2 NFs for increasing φ (0–4%) and temperature (15–60 °C) and found that the viscosity of Al_2O_3 NFs was higher than that of the other NFs and more pronounced at high concentrations. The viscosity of a SiO_2/EG–water NF covering a temperature of −35 to 50 °C and φ of 2 to 10% shows slight reduction with an increase in temperature and considerable increase with φ.[18] Oster *et al.*[33] measured the viscosity of 20 different types of ionic liquid-based NFs formulated using four different NPs and five ionic liquids at 25–90 °C and 0.5–3 wt%. They revealed the augmentation of viscosity with φ and reduction with temperature to be functions of the base fluid and nanoparticle type. The viscosities of bio-based NFs of coconut fiber carbon,[34] fruit bunch–GO,[35] and banana fiber[36] NPs were found to be augmented by 50% (1 wt% and 60 °C), 4.16% (0.06 wt% and 40 °C), and 22% (60 °C and 1.5 vol%), compared with EG–deionised water (DIW), EG, and DIW, respectively. In addition, studies concerning hybrid NFs show that viscosity is influenced by φ, temperature, and mixing ratio.[31,37–40] Other notable studies corroborated the research trend that the

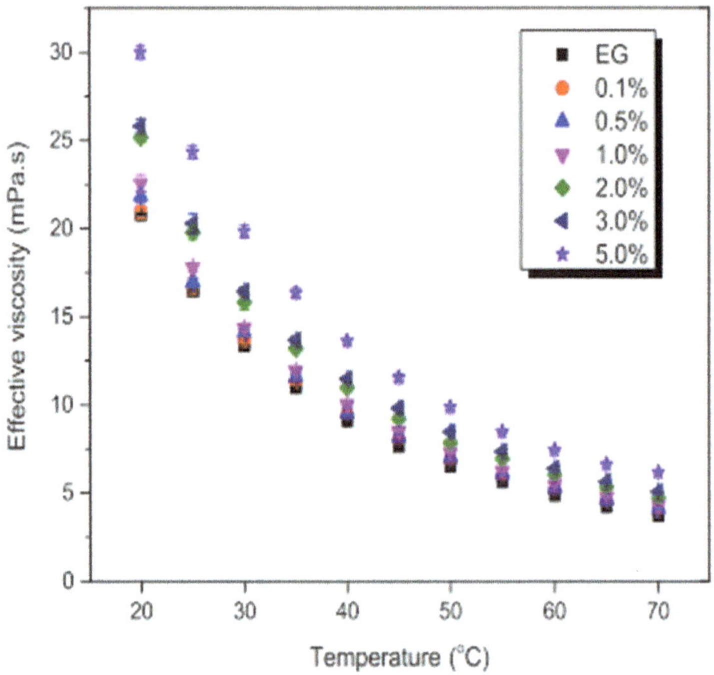

Figure 5.2 Influence of temperature on the viscosity of a 21 nm-MgO–EG nanofluid with varying volume fraction. Reproduced from ref. 50 with permission from Elsevier, Copyright 2016.

contribution of φ to NF viscosity is lower than temperature with an increase in the former improving viscosity and a rise in the latter reducing viscosity.[11,12,19,31,40–42] Figure 5.2 shows the relationship between the viscosity of NFs and φ and temperature.

5.3.2 Base Fluid

The thermophysical properties of base fluids such as viscosity, thermal conductivity, density, electrical properties, and pH contribute immensely to the overall properties of NFs formulated from them, such as stability, viscosity, electrical and thermal conductivity. The viscosity of base fluids used in the formulation of NFs has a strong impact on the viscosity of the resulting NFs. As the name suggests, "base fluid" is the basal fluid in which NPs are suspended to formulate NFs. The base fluids used mainly in the NF formulation are water (distilled water (DW) or deionised water (DIW)), glycerol, ionic fluids, engine oil, propylene glycol (PG), EG, water–EG, transformer oil, *etc.* It is pertinent to note that different base fluids have different viscosity values owing to their chemical compositions. The extent to which the viscosity of base fluids used in NF formulation affects the overall or effective viscosity of the NF has been scarcely studied compared to the number of publications on NFs. However, studies have shown that the suspension of NPs in base fluids augments the viscosity of the base fluids.[12,21,22,31,33,43,44]

Sundar *et al.*[45] examined the viscosity of Fe_3O_4 NFs formulated using different ratios (by weight) of EG–water (20 : 80, 40 : 60, and 60 : 40). In comparison with the corresponding base fluid, the NF viscosity was augmented by 1.42-, 1.61-, and 2.94-fold for 20 : 80, 40 : 60, and 60 : 40 EG–water, respectively, at 50 °C and 1 vol%. The measurement of the viscosity of EG and water-based Al_2O_3 NFs by Meyer *et al.*,[13] showed that at a volume fraction of 3%, the viscosity of an Al_2O_3–water NF was improved by 20–30%, whereas that of an Al_2O_3–EG NF was augmented by 40% at a volume fraction of 3.5%. Al-Waeli *et al.*[46] showed that the viscosity of SiC NFs was enhanced in comparison to water by 16.66%, 12.66%, and 0.063% using PG–water (35 : 65), EG–water (35 : 65), and water as base fluids, respectively. Hamid *et al.*[47] used three different base fluids (water–EG mixtures of 40 : 60, 50 : 50, and 60 : 40) to formulate Al_2O_3 NFs and observed enhancement of viscosity as the quantity of EG in the base fluid increased due to EG having a higher viscosity than water. Similarly, Timofeeva *et al.*[48] showed that the viscosity of a SiC/EG–water NF was higher than that of a SiC/water NF.

The influence of different base fluids on the viscosity of hybrid NFs has been examined in some studies. Giwa *et al.*[40] determined the viscosity of Al_2O_3–Fe_2O_3 NPs suspended in DIW and EG–DIW (50 : 50 vol) and found that the EG–DIW-based hybrid NFs had a slightly higher viscosity than DIW-based hybrid NFs at the studied temperature and concentration. In addition, Sundar *et al.*[26] showed that the use of water and EG–water mixtures as base fluids to formulate Nanodiamond (ND)–Fe_3O_4 NFs led to variation in the values of the viscosity of the hybrid NFs. They reported enhancement of 1.79-fold (60 : 40 EG–water), 1.50-fold (40 : 60 EG–W), 2.19-fold (20 : 80 EG–W), and 1.72-fold (water). Kumar *et al.*[49] formulated Al_2O_3–CuO NFs using EG–water and PG–water mixtures (20–55 : water by mass fraction) and found that the viscosity of the hybrid NFs enhanced as the EG and PG contents in the base fluids increased. It can be deduced from the above studies that the dilution of EG and PG in water led to an increase in the viscosity of water, as both EG and PG have viscosity values higher than water. The relationship between mono and hybrid NF viscosity and different base fluids at various temperatures and φ is shown in Figures 5.3 and 5.4.

5.3.3 Nanoparticle Size and Shape

Studies on the impact of nanoparticle size and shape have been conducted, with it being established that these two parameters affect NF viscosity.[11,12,38,48,50] For the nano-size effect on the viscosity of NFs, two contradictory results have been reported in the open literature. First, increasing the size of NPs has been observed to reduce NF viscosity. The formulation of Al_2O–glycerol NFs by the suspension of 20–30, 80, and 100 nm Al_2O_3 NPs revealed that the NF with 20–30 nm Al_2O_3 exhibits the highest viscosity.[11] With the use of 19-, 139-, and 160 nm Al_2O_3 to formulate Al_2O_3–glycerol NFs, the NF with the

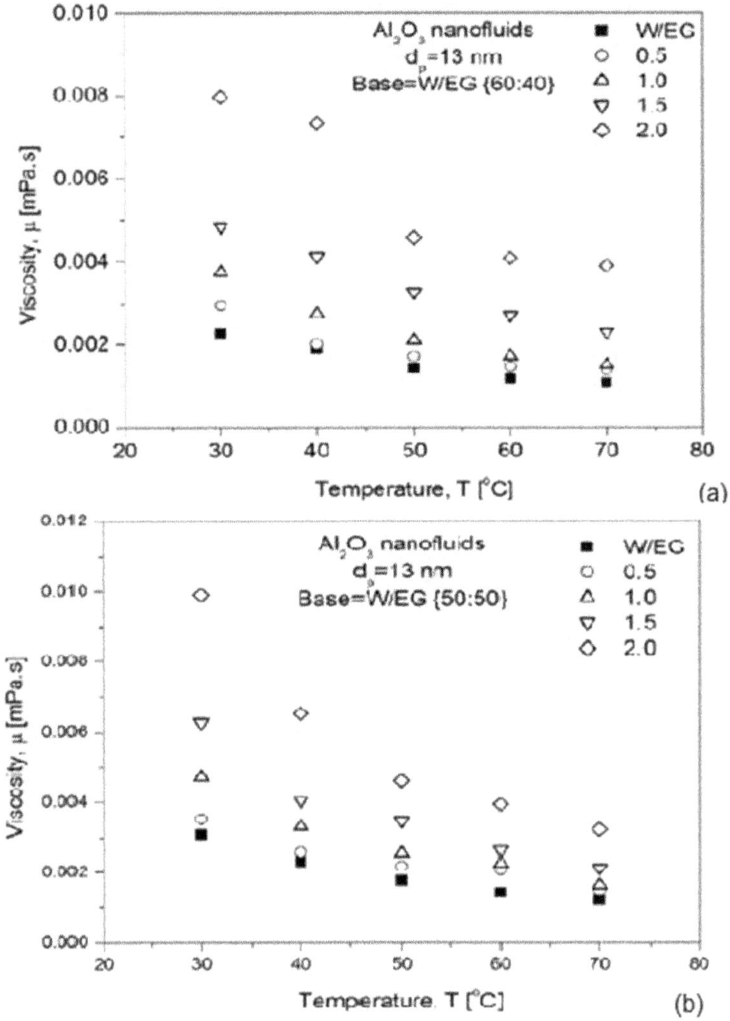

Figure 5.3 Influence of temperature on the viscosity of Al_2O_3 nanofluids with varying volume fractions in (a) EG–water (40 : 60) and (b) EG–water (50 : 50) Reproduced from ref. 47 with permission from Elsevier, Copyright 2015.

lowest nano-size exhibited the highest viscosity.[12] Under constant φ and temperature conditions, Adio *et al.*[50] demonstrated that the 21 nm MgO–EG NF exhibited the highest viscosity in comparison to the 105 nm and 125 nm MgO–EG NFs. The same outcome was reported for the viscosity of 2% water-based Al_2O_3, TiO_2, and SiO_2 NFs with a nano-size ranging from 10 to 150 nm.[28] Also, Timofeeva *et al.*[48] formulated and determined the viscosity of SiC/EG–water NFs with particle sizes of 16, 29, 66, and 90 nm. They found that particle size increase led to a reduction in viscosity.

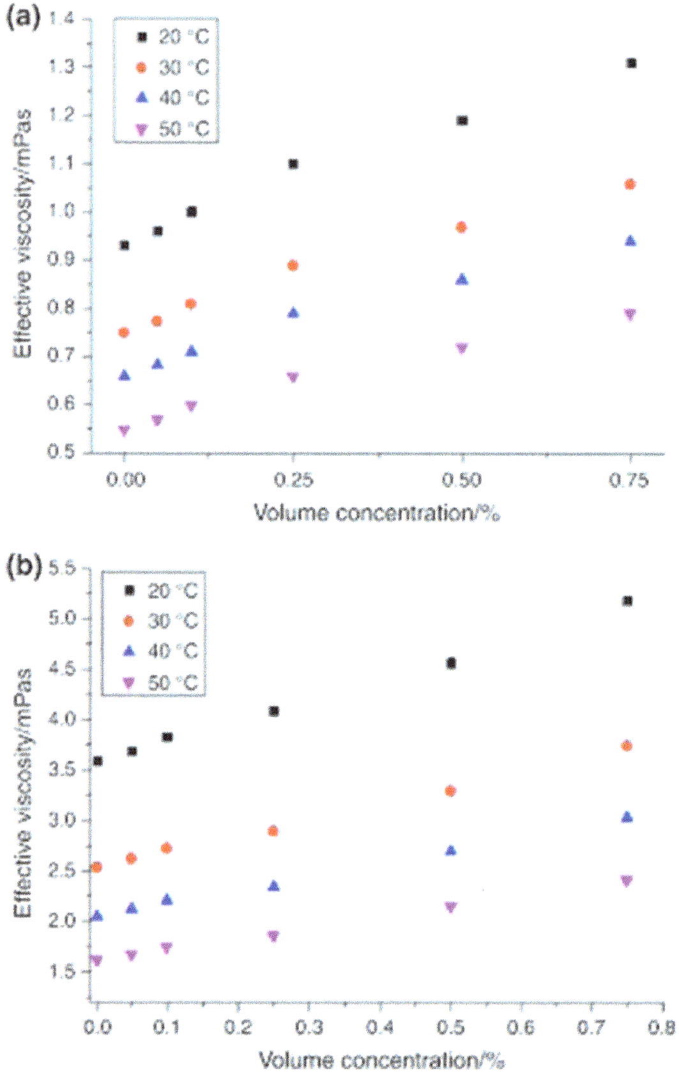

Figure 5.4 Influence of volume concentration on the viscosity of Fe_2O_3–Al_2O_3 hybrid nanofluids upon varying the temperature in (a) water and (b) EG–water (50 : 50). Reproduced from ref. 40 with permission from Springer Nature, Copyright 2020.

These studies reasoned that reducing the size of NPs increased the number of NPs present in the base fluids and surface area of NPs, thus leading to an augmentation in effective φ and consequently leading to a surge in the overall viscosity of the NFs. This thought is supported by the Brownian motion theory, which establishes an increase in particle–particle interactions and Brownian velocity for small-sized NPs. In addition, small NPs tend to form clusters compared to large NPs, which leads to an increase in NF viscosity.

Contrary to the above-published works, He *et al.*[51] studied the viscosity of TiO_2–water NFs formulated using three different particle sizes (≥95 nm) and revealed that viscosity increased in line with the size of the NPs. Using 36 nm and 47 nm Al_2O_3 NPs to formulate Al_2O_3–water NFs, Nguyen *et al.*[52] observed that 36 nm NFs exhibit a slightly higher viscosity than 47 nm NFs at elevated volume fraction (>5%). Abdelhalim *et al.*[53] showed that of the 10-, 20-, and 50 nm sizes of NPs used to formulate GNP NFs, the 50 nm-based sample yielded the highest viscosity. Recently, Giwa *et al.*[38] determined the viscosity of MgO–ZnO/water NFs with nanoparticle sizes of 20 nm and 100 nm for MgO and 20 nm for ZnO. They reported that the 100 nm MgO-based hybrid NFs exhibited higher viscosity compared to the 20 nm MgO-based hybrid NFs. It was observed that these studies did not give reasons for the observation that the viscosity of the NFs enhanced with an increase in the nanosize of the particles. There is thus a need to further and deeply explore the relationship between viscosity and the size of NPs of NFs to establish the right trend and scientific rationale behind it.

Very limited research studies have been performed to determine the influence of NP shapes on NF viscosity. Timofeeva *et al.*[32] examined the impact of different NP shapes (platelets, bricks, cylinders, and blades) on the viscosity of Al_2O_3/EG–water NFs. The highest and lowest viscosity was observed for NF samples formulated using platelets and bricks NPs, respectively. Ferrouillat *et al.*[54] studied the viscosity of SiO_2/water NFs (banana and spherical) and ZnO/water NFs (polygons and rod) and noted that spherical-SiO_2/water and rod-ZnO/water NFs showed the lowest viscosity. Recently, Kumar and Sahoo[49] investigated the viscosity of PG–water and EG–water mixtures based Al_2O_3–CuO NFs with various shapes of NPs (spheres, platelets, blades, bricks, and cylinders). The platelets and spherical NP-based hybrid NFs exhibited the highest and lowest viscosity, respectively. In addition, some studies have shown that different shapes of NFs have no apparent effect on viscosity values.[28] Obviously, studies in this regard are very scarce, and more investigation on the effect of particle shapes on NF viscosity and the physics contributing to it is of great importance.[28] The influence of NP shapes on NFs is presented in Figure 5.5.

5.3.4 Magnetic Excitation

The influence of magnetic field as an active technique of improving the viscosity of NFs (mono and hybrid), especially magnetic types, has been studied by a number of researchers.[55–60] Some representative materials are briefly dicussed here. Amani *et al.*[55] examined the influence of magnetic field strength on the viscosity of $MnFe_2O_4$–DIW NFs and demonstrated that the viscosity improved as the magnetic field strength increased, with a peak improvement of 1.38 times compared to in the absence of a magnetic field. Malekzadeh *et al.*[59] examined the viscosity of an Fe_3O_4–DIW NF exposed to magnetic field strength and they revealed 20% improvement at the studied φ and temperature. Shahsavar *et al.*[60] studied the effect of magnetic

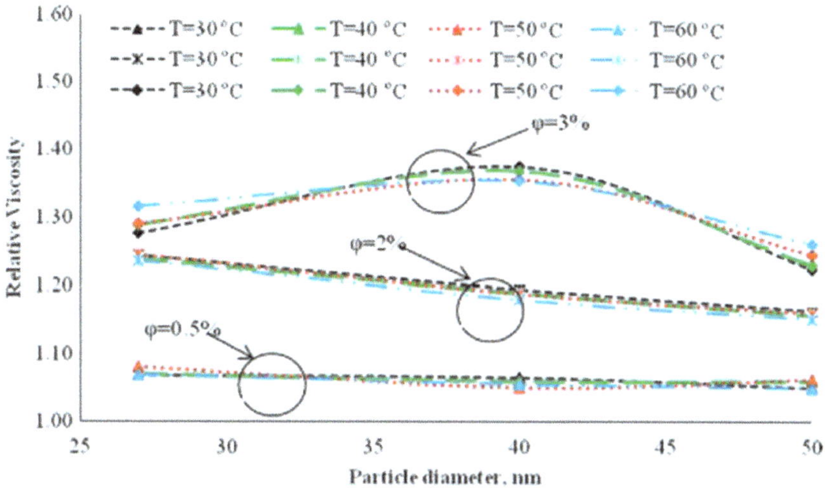

Figure 5.5 Influence of nano-size on the viscosity of Al_2O_3–PG nanofluids. Reproduced from ref. 5 with permission from Elsevier, Copyright 2012.

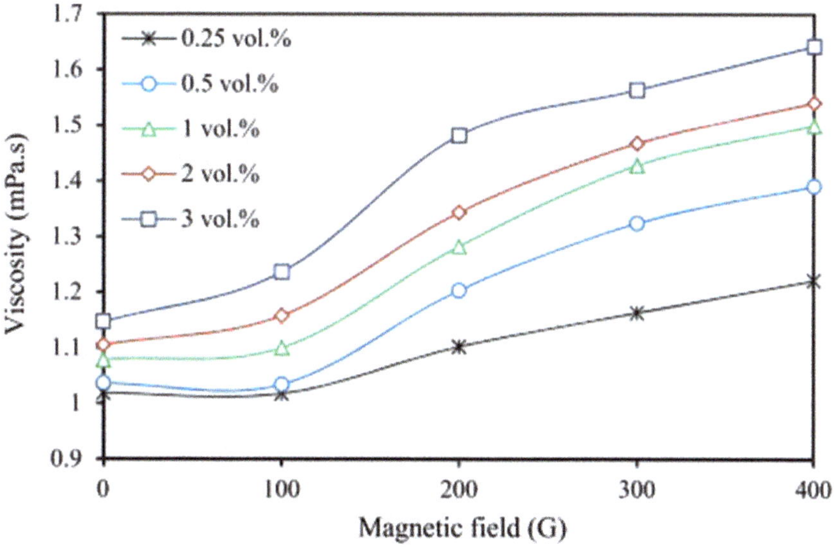

Figure 5.6 Influence of magnetic field on the viscosity of an $MnFe_2O_4$–water nanofluid with varying concentration. Reproduced from ref. 55 with permission from Elsevier, Copyright 2017.

field strength on the viscosity of Fe_3O_4–CNT/water NFs. They showed that the viscosity was improved as the magnetic field strength increased and reduced as the temperature increased. The impact of magnetic field strength on viscosity is shown in Figure 5.6.

5.3.5 Influence of Dispersion Characteristics

The stability of mono and hybrid NFs is significantly related to their preparation characteristics, which include stirring duration and speed; volume, density, and base fluid type; weight/volume concentration; density, nano-size, and type of mono or hybrid NPs used; type, and quantity of surfactant used; sonication time, temperature, mode, frequency, and amplitude. It is crucial to know that stability also has a direct effect on the viscosity of mono and hybrid NFs as the preparation characteristics strongly influence this property. Studies have shown how the preparation characteristics of NFs (mono and hybrid) affect viscosity, which is related to the dispersion characteristics.[11,25,50,61–63]

Asadi and Alarifi[61] showed that 60 min of sonication yielded the most stable and lowest viscosity of PVP + MWCNT–W NFs. Michael *et al.*[64] reported that the optimal viscosity of Al_2O_3–DW NFs was attained at different sonication times (60 min for 0.5 and 1 vol% and 90 min for 1.5 vol%), with a reduction in the average particle size. Mahbubul *et al.*[65] reported a stable 0.5 vol% Al_2O_3/DW NF at a sonication time of 5 h with a sonication energy of 42.3 kJ, cluster size of 105 nm, and viscosity of 1.63 mPa s. Ghadimi and Metselaar[66] reported that the use of sodium dodecyl sulfate (SDS) and a bath ultrasonicator yielded the most stable 0.1 wt% TiO_2/DW NF with a nano-size of 237.7 nm and maximum viscosity. Yang *et al.*[67] proved that optimal viscosity was attained with 30 min of sonication, an energy density of 1.4×10^6 kJ m^{-3}, and 3 wt% surfactant concentration for a 0.12 vol% MWCNT/W NF. Ruan and Jacobi[68] demonstrated that with 1355 min of sonication, minimum viscosity, CNT length, and average cluster size are attained under optimal stability conditions for a 0.5 wt% MWCNT/EG + 0.25 wt% GA NF.

In addition, Li *et al.*[69] revealed that for a Cu/EG NF, the lowest viscosity was reached at a sonication time of 60 min (low φ) and 45 min (high φ) coupled with aggregate size reduction. Adio *et al.*[11] stated that stable Al_2O_3/GL NFs were prepared for 20–30 nm-sized and 80 nm and 100 nm-sized NFs using a sonication time of 6 h and 3 h, corresponding to energy densities of 3×10^7 kJ m^{-3} and 1.5×10^7 kJ m^{-3}, respectively. Adio *et al.*[50] demonstrated that under stable conditions for MgO/EG NFs, sonication durations of 30, 60, and 180 min corresponding to sonication energy densities of 2.2×10^6, 4.4×10^6, and 13×10^6 kJ m^{-3} for 21, 105, and 125 nm-sized MgO/EG NFs, respectively, were achieved. This observation was found to be in agreement with the work of Yang *et al.*[67] but contradicted that of Adio *et al.*,[11] probably due to the use of different NPs and sizes, and base fluids. The influence of sonication time on the viscosity of NFs is presented in Figure 5.7.

5.4 Classical and Contemporary Viscosity Models

To aid the design and operation of thermal devices and systems, the idea of using and proposing a universal model to predict the viscosity of NFs is very much in order based on the particle–liquid nature of these superior thermal media. Owing to the importance of the viscosity of NFs as advanced thermal fluids and the complexity of the phenomenon associated with this property, researchers have made several attempts towards developing theoretical models

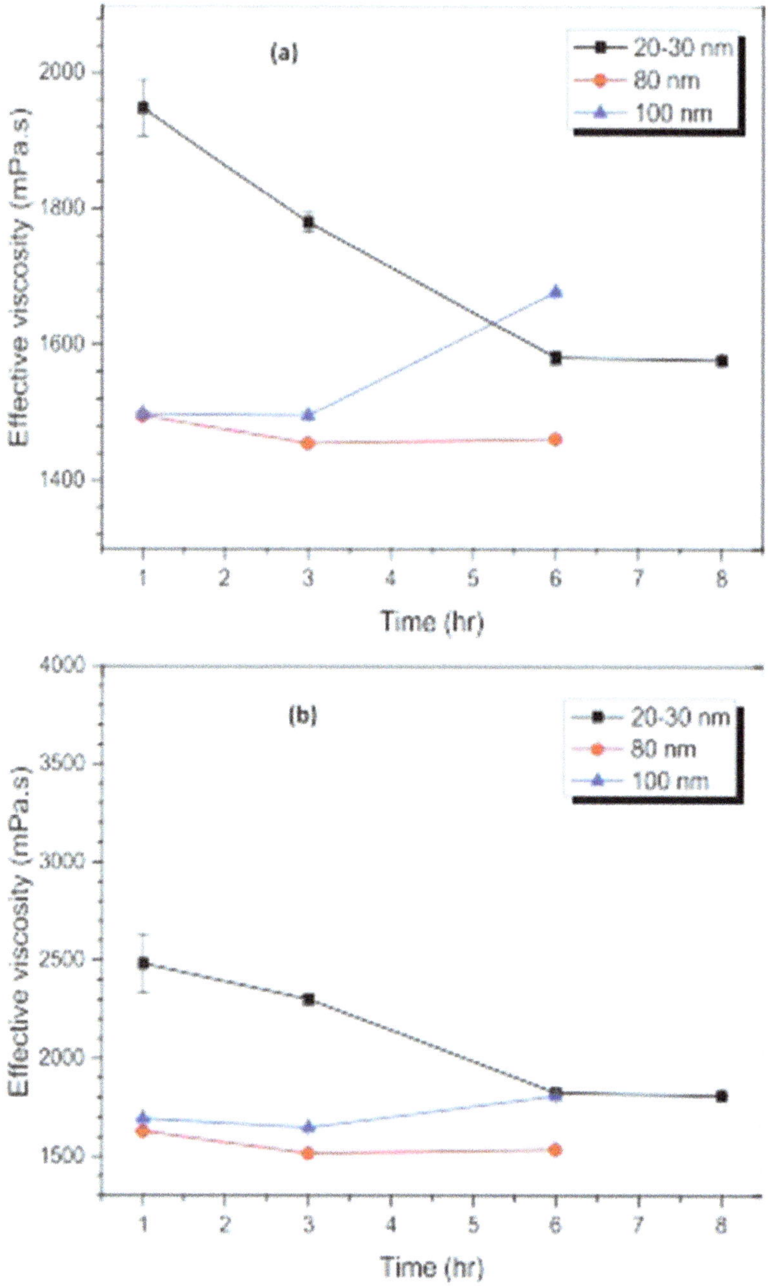

Figure 5.7 Influence of sonication time on the viscosity of an Al_2O_3/glycerol nanofluid with different nano-sizes at (a) 2% and (b) 3% concentration. Reproduced from ref. 11 with permission from Taylor & Francis Ltd., Copyright 2015.

for the effective prediction of viscosity of NFs. In this chapter, theoretical models proposed prior to the development of NFs for colloidal suspensions are coined classical models while those models developed since the advent of NFs are refer to as contemporary models. The models, different concepts behind their development, viscosity-dependent parameters, mechanisms, merits and demerits, limitations, application of experimental data, *etc.*, are presented.

5.4.1 Classical Viscosity Models

The work of Einstein commenced the development of models to predict NF viscosity.[3] Einstein developed a model on the vorticity of the shear field, energy dissipation, and non-interaction of suspension of rigid spherical particles (macro-scale) in a linearly viscous fluid, as expressed in eqn (5.1).[7,14]

$$\mu_s = \mu_{bf}\,(1 + [\mu]\varphi) = \mu_{bf}\,(1 + 2.5\varphi) \tag{5.1}$$

where $[\mu]$ is the intrinsic viscosity of the suspension for rigid spheres (2.5 for Einstein's model), φ, μ, s and bf represent the particle volume fraction, viscosity, suspension, and base fluid, respectively. For this model, φ is applied for $\leq$0.02, which is one of the shortcomings of this model as a typical φ value is mostly above the specified range. Brinkman[5] theoretically analysed the suspension of rigid spherical particles based on very slow flow in the absence of inertia as an extension of Einstein's model for φ up to 4%. Brinkman's model is expressed in eqn (5.2):

$$\mu_s = \mu_{bf}\,(1 - \varphi)^{-2.5} \tag{5.2}$$

In addition, Maron–Pierce considered the Ree–Eyring flow equations[30] to develop a model (see eqn (5.3)) to estimate the fibre suspension relative viscosity based on minimum principle incorporated to energy dissipation due to viscous effects.[70] φ_m is dependent on the aspect ratio of the fibers, which is inversely proportional to relative viscosity:

$$\mu_s = \mu_{bf}\left(1 - \frac{\varphi}{\varphi_m}\right)^{-2} \tag{5.3}$$

Owing to the concept of structure creation (mono-dispersed particle packing fraction) in dispersed systems and considering no agglomeration, Krieger and Dougherty[30] developed a semi-empirical model for the viscosity of particle suspension. Their model assumed that hydrodynamic interaction of particles is more pronounced at high concentration (φ >1%) and it applies to a wide range of concentrations, as expressed by eqn (5.4).

$$\mu_s = \mu_{bf}\left(1 - \frac{\varphi}{\varphi_m}\right)^{-[\mu]\varphi_m} \tag{5.4}$$

where φ_m is the peak concentration (0.495–0.54 under quiescent conditions and $\approx$0.605 at high shear rates). For a rigid sphere mono-dispersed suspension, $[\mu]$ is 2.5. Conversely, practical reality shows that particles are polydispersed. By considering the effect of maximum volume concentration and extending knowledge on Einstein's model, Frankel and Acrivos[16] proposed eqn (5.5) to predict the viscosity of a suspension of particles.

$$\mu_s = \mu_{bf}\left(1.125\left(\frac{\left(\frac{\varphi}{\varphi_m}\right)^{\frac{1}{3}}}{\frac{(\varphi_m - \varphi)^{\frac{1}{3}}}{\varphi_m^{\frac{1}{3}}}}\right)\right) \tag{5.5}$$

Nielsen[16] proposed a power law-based model for predicting the viscosity of a composite material suspension, as given by eqn (5.6), for a particle concentration of >0.02.

$$\mu_s = \mu_{bf}(1 + 1.5\varphi)e^{\left(\frac{\varphi}{1-\varphi_m}\right)} \tag{5.6}$$

Lundgren[5] considered Brownian motion and bulk stress in the development of a Taylor's series-based model for predicting the viscosity of suspension (isotropic) of rigid spherical particles. At very low concentration, this model (eqn (5.7)) reduces to Einstein's model.

$$\mu_s = \mu_{bf}\,(1 + 2.5\varphi + 6.25\varphi^2 + f(\varphi^3)) \tag{5.7}$$

Using the concept of Brownian motion and mutual interactions of particles (rigid and spherical), Batchelor[30] developed the model described in eqn (5.8) for evaluating the viscosity of a monodispersed suspension. The model is a modification of Einstein's model related to higher particle concentration ($\varphi \leq 4\%$). At low concentration, the non-interaction assumption of Einstein's model holds for Batchelor's model.

$$\mu_s = \mu_{bf}\,(1 + 2.5\varphi + 6.5\varphi^2) \tag{5.8}$$

Using cell-based theory and considering the influence of minimum and uniform diameter of equidistance spheres, zero inertial, electroviscous forces, van der Waals forces, and Brownian motion, Graham proposed a viscosity model for the suspension of a solid sphere, as described by eqn (5.9).[30,71]

$$\mu_s = \mu_{bf}\left\{1 + 2.5\varphi + \frac{4.5}{\left(\left(\frac{l}{d}\right)\left(2 + \frac{l}{d}\right)\left(1 + \frac{l}{d}\right)^2\right)}\right\} \tag{5.9}$$

where d is the particle radius and l is the inter-particle spacing. The Graham's model appears to be a generalized form of Lundgren's model, with both models reducing to Einstein's model at low concentration values.

It can be deduced from the above literature that the various theories coupled with the different established mechanisms used along the diverse parameters which influence viscosity are only used to develop the models for

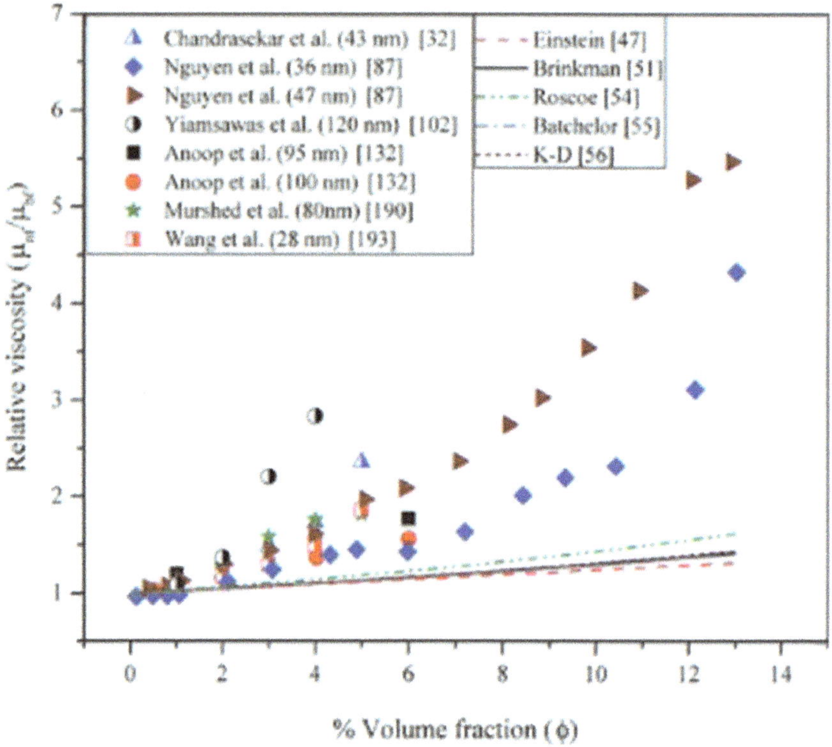

Figure 5.8 Viscosity prediction performance of classical models as a function of volume fraction for Al_2O_3–deionized water nanofluids. Reproduced from ref. 13 with permission from Taylor & Francis Ltd., Copyright 2015.

predicting the viscosity of milli- and micro-size suspensions of particles. The validation of these models primarily under-fits the experimental viscosity data, showing the inability of the models to predict the viscosity of NFs.[12,21,26,42,52] Also, these models predate NF emergence and some of the theories do not apply to nanosized particles. Figure 5.8 presents a comparison between viscosity experimental data and classical models.

5.4.2 Contemporary Viscosity Models

There are a number of models for the prediction of the viscosity of NFs that have been proposed by researchers. Choi *et al.*[72] proposed a model for the prediction of the viscosity of NFs (Ba–ferrite, CrO_2, and Fe_2O_3) with different shapes of particles (rod and plate) using mean-field theory and modifying Mooney's model. The substitution of $[\mu]$ (26 (CrO_2 and Fe_2O_3) and 12 (Ba–ferrite)) and other parameters into eqn (5.10) gives a good prediction of the experimental data for each NF. The need to estimate the $[\mu]$ (related to the NP shape) of different NFs seems to be a drawback to the use of this model. However, the authors failed to compare the model with experimental data and classical models to validate its predictive performance.

$$\mu_{\text{nf}} = \mu_{\text{bf}} e^{\left(\frac{[\mu]\varphi}{1-\frac{\varphi}{\varphi_{\text{m}}}}\right)} \tag{5.10}$$

The Pozhar–Gubbins expression based on the nonequilibrium statistical mechanical theory for inhomogeneous fluids was simplified for the prediction of NF viscosity by Pozhar *et al.*[73] The developed model presented in eqn (5.11) is related to the NF density (n), viscosity, pore width (H), and pore-average velocity (u) for applications in engineering, where F is the absolute force value. However, the model is not applicable for dense NFs. Specifically, the authors validated this model using simulation method and not with experimental data.

$$\mu_{\text{nf}} = \mu_{\text{bf}} \left(\frac{\langle n\rangle FH^2}{12\langle u\rangle}\right) \tag{5.11}$$

Avsec and Oblak[74] deployed statistical nanomechanics (with the possibility to model the effect of nanolayer interaction) in addition to the modification of the Cheng and Law, and Ward models[3] to propose a model for the prediction of NF viscosity. They revealed the inability of the Cheng and Law, and Ward models to predict the viscosity of NFs while predicting that of a suspension with particles of >100 nm in size. The Avsec and Oblak model is expressed in eqn (5.12). The model's predictive performance was verified using Al_2O_3 and TiO_2 NFs with excellent results.

$$\mu_{\text{nf}} = \mu_{\text{bf}} \left(1 + 2.5\varphi_{\text{eff}} + (2.5)^2\varphi_{\text{eff}}^{\ 2} + (2.5)^3\varphi_{\text{eff}}^{\ 3} + (2.5)^4\varphi_{\text{eff}}^{\ 4} + \cdots\right) \tag{5.12}$$

$$\text{where } \varphi_{\text{eff}} = \varphi\left(1 + \frac{h}{d}\right)^3 \tag{5.13}$$

φ_{eff} is the effective volume fraction, h is the nanolayer thickness, and d is the particle radius.

Chen *et al.*[75] modified the Krieger and Dougherty model expressed in eqn (5.4) based on the knowledge of the influence of agglomeration of particles on viscosity and the incorporation of the fractal index and maximum packing fraction. The proposed model is given by eqn (5.14).

$$\mu_{\text{nf}} = \mu_{\text{bf}} \left(1 - \frac{\varphi_{\text{a}}}{\varphi_{\text{m}}}\right)^{-[\mu]\varphi_{\text{m}}} \tag{5.14}$$

$$\text{where, } \varphi_{\text{a}} = \varphi\left(\frac{a_{\text{a}}}{a}\right)^{3-D} \tag{5.15}$$

φ_{a} is the effective aggregate volume concentration, D is the fractal index, and a_{a}/a is the ratio of effective aggregate and NPs. The substitution of some empirical data ($D = 1.8$, $[\mu] = 2.5$ and $\varphi_m = 0.605$) describing the case of extreme agglomeration of NPs is used to modify eqn (5.14) to eqn (5.16).

$$\mu_{nf} = \mu_{bf}\left(1 - \frac{\varphi}{0.605}\left(\frac{a_a}{a}\right)^{1.2}\right)^{-1.5125} \tag{5.16}$$

Using $a_a = 3.34a$ for spherical NPs, Chen *et al.*[75] showed good agreement between the viscosity increment data predicted using the modified Krieger and Dougherty (*i.e.*, eqn (5.4)) model and the experimental data of TiO_2/EG NFs and other NFs from other studies used for comparison. In another study by Chen *et al.*,[76] the modified Krieger and Dougherty model was also shown to be well fitted to the viscosity increment data of a TNG/glycol NF using $a_a = 9.46a$, $D = 2.1$, and $\varphi_m = 0.30$. The shortcoming of this model involves determining the values of a_a and D and φ is expected not to exceed 0.605, which is not applicable to all cases. By accounting for aggregate and NP size, Chevalier *et al.*[77] modified the Maron–Pierce model (eqn (5.3)) and used the resulting model (eqn (5.17)) to estimate the relative viscosity of SiO_2/ethanol NFs with 35, 94, and 195 nm to a good degree. For $\varphi > 0.6$, this model is may not be applicable.

$$\mu_{nf} = \mu_{bf}\left(1 - \frac{\varphi_a}{\varphi_m}\right)^{-2} \tag{5.17}$$

where $\varphi_a = \varphi\left(\frac{a_a}{a}\right)^{1.2.}$

Considering factors such as temperature, base fluid, NP volume fraction and diameter, and capping layer thickness, Hosseini *et al.*[78] developed a model to predict the viscosity of NF using dimensional analysis. The model is given as eqn (5.18):

$$\mu_{nf} = \mu_{bf} e^{\left[m + \alpha\left(\frac{T}{T_0}\right) + \beta(\varphi) + \gamma\left(\frac{d}{1+r}\right)\right]} \tag{5.18}$$

where m is a parameter depending on the above factors and α, γ, and β are empirical constants calculated using experimental data. The developed model accurately fits the experimental viscosity data of 36 and 47 nm Al_2O_3/water NFs with φ of 1–9.4% at 20–60 °C, and it performs better than the classical models.

Similarly, Masoumi *et al.*[79] considered Brownian motion and the relative velocity between a liquid (base fluid) and solid (NPs) interface in a NF to develop a model for the prediction of NF viscosity. The model as expressed in eqn (5.19) is dependent on mean diameter, temperature, φ, density of NPs, and base fluid properties.

$$\mu_{nf} = \mu_{bf}\left(1 + \frac{V_B \rho_p d_p^2}{72 C \delta \mu_{bf}}\right) \tag{5.19}$$

$$\text{where } V_B = \frac{1}{d_p}\sqrt[3]{\frac{18 k_b T}{\pi d_p \rho_p}} \tag{5.20}$$

$$\delta = \sqrt[3]{\frac{\pi}{6\varphi}} d_{\rm p} \tag{5.21}$$

$$C = \frac{\left[\left(C_1 d_{\rm p} + C_2\right)\varphi + \left(C_3 d_{\rm p} + C_3\right)\right]}{\mu_{\rm bf}} \tag{5.22}$$

where $V_{\rm B}$, T, $k_{\rm b}$, $\rho_{\rm p}$, δ, $d_{\rm p}$, C are the Brownian velocity, temperature, Boltzmann constant, particle density, distance between particle centres, particle diameter, and correction factor, respectively. The model was tested using the experimental data of the viscosity of Al_2O_3/water NFs with 13 nm and 28 nm NPs and a high degree of accuracy was reported. Also, using experimental data of the viscosity of different NFs with various types of base fluids and nano sizes, the developed model was found to fit the data best in comparison with existing classical models and experimental data-fitted models. The authors incorporated temperature as a parameter in their model, which was significant to the viscosity of NFs and their applications.

Hosseini and Ghader[80] proposed a model for the prediction of NF viscosity based on the non-random two liquid and incorporation of Eyring's viscosity model. The model as expressed in eqn (5.23) is dependent on φ and temperature. This model fairly predicts the viscosity of water-based Al_2O_3 and CuO NFs at varying temperatures (0–50 °C) and φ (1–9.4%) better than classic models. However, it appears complex to estimate NF viscosity as it requires numerous parameters in the computation, especially for a three-component NF. The effect of agglomeration is not accounted for and there is a need to further understand the influence of Gibb's free energy on NF viscosity.

$$\mu_{\rm nf} = \frac{\mu_{\rm bf} {\rm e}^{\xi}}{V} \tag{5.23}$$

$$\xi = \varphi_1\varphi_2\left(\frac{A_{21}G_{21}}{\varphi_1 + \varphi_2 G_{21}} + \frac{A_{21}G_{12}}{\varphi_2 + \varphi_1 G_{12}}\right) + \varphi_1\xi_1 + \varphi_2\xi_2 \; (2\text{–component NF}) \tag{5.24}$$

$$\begin{aligned} \xi = {} & \varphi_1\left(\frac{\varphi_2 G_{21}}{\varphi_1 + \varphi_2 G_{21} + \varphi_3 G_{31}} A_{21} + \frac{\varphi_2 G_{31}}{\varphi_1 + \varphi_2 G_{21} + \varphi_3 G_{31}} A_{31}\right) \\ & + \varphi_2\left(\frac{\varphi_1 G_{12}}{\varphi_2 + \varphi_1 G_{12} + \varphi_3 G_{32}} A_{12} + \frac{\varphi_3 G_{32}}{\varphi_2 + \varphi_1 G_{12} + \varphi_3 G_{32}} A_{32}\right) \\ & + \varphi_3\left(\frac{\varphi_1 G_{13}}{\varphi_3 + G_{13} + \varphi_2 G_{23}} A_{13} + \frac{\varphi_2 G_{23}}{\varphi_2 + \varphi_1 G_{13} + \varphi_2 G_{23}} A_{23}\right) \\ & + \varphi_1\xi_1 + \varphi_2\xi_2 + \varphi_3\xi_3 \; (\text{three} - \text{component NF}) \end{aligned} \tag{5.25}$$

where $A_{\rm ij} = a_{\rm ij} + b_{\rm ij}T$ and $G_{\rm ij} = {\rm e}\left(\frac{-\alpha A_{\rm ij}}{RT}\right)$ are characteristic interactions, α is an empirical constant, V is the molar volume, R is the gas constant, and ξ is a pure component parameter.

Yang *et al.*[71] theoretically proposed models to estimate the viscosity of NFs subject to the contribution of volume fraction of the surfactant (SDBS and

PO-10) and NPs (carbon black, Fe_2O_3, and $ZnFe_2O_4$) and types of dispersion (monolayer and electrical double layer (EDL) adsorption). The surfactant–base fluid, surfactant–NPs, and base fluid–NPs interactions and their contribution to viscosity were deployed in proposing the models along with the use of Einstein's model:

$$\mu_{\rm nf} = \mu_{\rm bf}\left(1 + 2.5\left(\frac{R+t+h}{R}\right)^3 \varphi\right) \text{ (monolayer; PO} - 10) \tag{5.26}$$

$$\mu_{\rm nf} = \mu_{\rm bf}(0)\left(1 + 2.5K\frac{w_{\rm i}}{J\rho}\right)\left(1 + 2.5\left(\varphi - \frac{w_{\rm i}}{J\rho}\right)\right) \text{ (EDL; SDBS)} \tag{5.27}$$

where $K = \frac{\varphi_{\rm new}}{\varphi}$, $\varphi > \frac{w_{\rm i}}{J\rho}$, J is the volume ratio of surfactant to NPs, R is the radius of the NPs, w is the mass fraction of surfactants, t is the thickness of the monolayer adsorption layer, h is the thickness of the salvation shell, $w_{\rm i}$ is the initial mass fraction of the surfactant, ρ is the surfactant density. Also, $\mu_{\rm bf}(0)$ characterizes the ammonia water viscosity, $1 + 2.5K\frac{w_{\rm i}}{J\rho}$ represents NP viscosity with EDL and $1 + 2.5\left(\varphi - \frac{w_{\rm i}}{J\rho}\right)$ signifies the contribution of remaining surfactant-free NPs, which aligns with Einstein's model (for eqn (27)). Eqn (5.26) and (5.27) are applicable for predicting the viscosity of uncharged and charged NFs, respectively, to a very high degree. The proposed models predict the viscosity of carbon black, Fe_2O_3 and $ZnFe_2O_4$ NFs better than classical models and under-fit the viscosity data compared to Masoumi's model (eqn (5.19)) as the Brownian motion and relative motion between the base fluid and NPs are not accounted for.

To propose a model to predict the viscosity of NFs, Song *et al.*[81] deployed the modified population balance model to account for the aggregation of NPs and incorporated particle distribution, Brownian motion, shear deformation, and magnetic field effects. The model in eqn (5.28) reveals that accounting for Brownian motion results in the better fitting of experimental data. Also, Brownian motion is reported to influence particle size distribution and agglomeration. The predictive performance of this model using viscosity experimental data of NFs other than the one used to fit the model seems questionable as R and $D_{\rm f}$ must be estimated for different NFs and most especially when $\varphi > 0.6$.

$$\mu_{\rm nf} = \mu_{\rm bf}\left(1 - \frac{\varphi}{\varphi_{\rm m}}\right)^{-2.5\left(\frac{R}{r}\right)^{\frac{3}{D_{\rm f}}}\varphi_{\rm m}} \tag{5.28}$$

where R is the average gyration radius, $D_{\rm f}$ (= 1.9) is the fractal dimension, $\varphi_{\rm m}$ (= 0.6) is the volume fraction of the aggregate, and r is the primary particle radius.

As an improvement of the work of Masoumi *et al.*,[79] Shahriari *et al.*[82] proposed a model incorporating Brownian motion, Stokes' law, and the homogeneous suspension of NPs for predicting the viscosity of NFs. The influence of temperature, NP size and φ, and the thermal properties of the base fluid and NPs on viscosity are accounted for. The model described in eqn (5.29) to (5.31) is valid for different NFs with φ, nanosize, and temperature ranges of 0–11%, 13–95 nm, and 290–350 K, respectively. Shahriari's model was found to fit the experimental data of different (Al_2O_3 and CuO) NFs better than classical models and the Masoumi's model. The potency of the model in predicting the viscosity of different NFs needs to be further investigated as the model appears to exhibit excellent performance.

$$\mu_{nf}=\mu_{bf}\left(1+\frac{A}{B-\varphi}\left(\frac{V_B\rho_p d_p^2}{36\delta\sqrt{T}}\right)\right) \tag{5.29}$$

$$\text{where}\,A=\frac{36\delta\sqrt{T\varphi^{\frac{1}{3}}}}{V_B\rho_p d_p^2}\times\left(9.193\times10^{28}d_p^4-2.404\times10^{22}d_p^3+2.025\times10^{15}d_p^2\right.$$
$$\left.-5.886\times10^7 d_p+\frac{1}{488774}\rho_p+\frac{2507}{4446}\right) \tag{5.30}$$

$$B=-8.344\times10^{36}d_p^5+2.212\times10^{30}d_p^4-2.16\times10^{23}d_p^3+9.492\times10^{15}d_p^2$$
$$-1.804\times10^8 d_p-\frac{2}{327327}\rho_p+\frac{3575}{291360} \tag{5.31}$$

Sun *et al.*[83] developed a model to estimate the viscosity of NFs using the two-body DLVO potential. The model is based on the influence of interparticle potentials on the viscosity of NFs. The repulsion between NPs as well as the effective charge number is noted to influence the viscosity of NFs. For different φ, a change in particle agglomeration structure is observed. The model as expressed in eqn (5.32) fairly predicts the experimental values of SiO_2/ethanol NF viscosity for different ranges of nano-sized particles (35–190 nm) and φ (1–7%) better than the Batchelor model. The need to estimate some uncommon parameters in the process of predicting the viscosity of NFs makes the model cumbersome. The ability of the model to predict the viscosity of other NFs outside the SiO_2/ethanol NF is doubtful.

$$\mu_{nf}=\mu_{bf}\left(0.97\varphi^2\right)\left\{\left(\frac{\pi^2\varepsilon k_B TZ}{e^2 n}\right)^{\frac{1}{2}}\left(\frac{1-\varphi}{1-\frac{\rho_i}{nZ}}\right)^{\frac{3}{2}}\right.$$
$$\times\left(\frac{0.3238(rk_D)^3+3.0667(rk_D)^2+2.6667(rk_D)-0.9241}{(1+rk_D)^2}\right)$$
$$+\pi Z\left(\frac{1-\varphi}{1-\frac{\rho_i}{nZ}}\right)\frac{0.435(rk_D)^2}{(1+ak_D)^2}-\left(\frac{\pi^2 e^2 TZ^5}{16\varepsilon k_B n}\right)^{\frac{1}{2}}\left(\frac{1-\varphi}{1-\frac{\rho_i}{nZ}}\right)^{\frac{3}{2}}$$
$$\left.\times\left(\frac{0.1584(rk_D)^4+0.6931(rk_D)^3+0.7686(rk_D)^2}{(1+rk_D)^4}\right)\right\} \tag{5.32}$$

where Z is the effective charge of the particle, ε is the permittivity of the base fluid, k_D is the inverse Debye length, ρ_i is the liquid ionic charge density, and e is the electron charge.

Udawattha *et al.*[84] theoretically developed a model for the prediction of mono- and hybrid NFs taking into account the interfacial layer and Brownian motion. Eqn (5.33) and (5.34) expressed the model for mono and hybrid NFs, respectively. Testing of eqn (5.34) with 501 viscosity data from 22 different NFs with varying φ, nanosize, temperature, and base fluid showed good performance. However, eqn (5.34), which is a modification of eqn (5.33), was not compared with experimental data. Additionally, both models are only valid for Newtonian NFs. At a particle size of >1 μm, the model reduces to the Einstein model.

$$\mu_{nf} = \mu_{bf}\left(1 + 2.5\varphi + \frac{V_B \rho_p d_p^2}{72\delta(10^{-10}\varphi^{-0.002T-0.284}T)}\right) \tag{5.33}$$

$$\mu_{nf} = \mu_{bf}\left(1 + 2.5\varphi_1 + \frac{V_{B1}\rho_{p1}d_{p1}^2}{72\delta_1(10^{-10}\varphi_1^{-0.002T-0.284}T)} + 2.5\varphi_2 + \frac{V_{B2}\rho_{p2}d_{p2}^2}{72\delta_2(10^{-10}\varphi_2^{-0.002T-0.284}T)}\right) \tag{5.34}$$

Assuming the NF to be a Newtonian and two-phase fluid and extending Einstein's model, Lebon and Machrafi[85] proposed an NF viscosity prediction model using an extended irreversible thermodynamics approach. The proposed model (expressed in eqn (5.35)) accounted for the effect of nanolayer and nanosize on viscosity and was used to predict experimental data reasonably. However, the model was only compared with Al_2O_3/water and Li/Ar NFs and the effect of temperature was not considered in its development.

$$\mu_{nf} = \mu_{bf}\left(\frac{1 + 2.5\varphi\left(1 + \frac{h}{r}\right)^3}{1 + 4\pi^2\varphi^2\left(1 + \frac{h}{r}\right)^4\left(\frac{l^2}{r^2}\right)}\right) \tag{5.35}$$

Machrafi[86] proposed models for predicting the viscosity of nanodispersions based on extended knowledge of fluid mechanics and thermodynamics. These models accounted for the influence of mass density, interfacial nanolayer, particle kinetics, nanosize, and φ on viscosity. Eqn (5.36) and (5.37) are models for the spherical and cylindrical viscosity of nano-dispersions, respectively. The models showed a good prediction of the experimental data of NF viscosity (from eight NFs with four base fluids at different nanosizes) and gave insight into the contribution of NP density to viscosity. The model did not take cognizance of temperature as a parameter that influences the viscosity of NFs.

$$\mu_{nf}=\mu_{bf}\left(1+\left(\frac{5}{2}+\frac{1}{4}\frac{\rho_p}{\rho}\left(1+\xi_{p,s}\frac{\rho_f}{\rho}\right)\right)\left(c+\xi_{p,s}c(1-c)\right)\right) \tag{5.36}$$

$$\mu_{nf}=\mu_{bf}\left(1+\left(\frac{5}{2}+\frac{1}{4\frac{\pi}{2+\pi}}\frac{\rho_p}{\rho}\left(1+\xi_{p,s}\frac{\rho_f}{\rho}\right)\right)\left(c+\xi_{p,s}c(1-c)\right)\right) \tag{5.37}$$

where $\xi_{p,s}=2\left(\left(1+\frac{l_1}{a_{p,s}}\right)^2-1\right)$, $\xi_{p,s}=2\frac{l_1}{a_{p,s}}$, and $c=\frac{\varphi\rho_p}{\rho}$

ρ_f is the density of the base fluid, c is the mass fraction of the fluid, l_i is the mass flux mean free path, $\xi_{p,s}$ is the interaction parameter, and $a_{p,}$ is the cylindrical radius. Figure 5.9 provides a comparison of viscosity experimental data and contemporary models.

All contemporary models developed based on the knowledge and existence of NFs show significant improvement in the prediction of NF viscosity over classical models, which were proposed considering micro- and milli-sizes particle suspensions. This observation can be attributed to the deployment of some relatively new theories, the involvement of more mechanisms, and the introduction of several factors contributing to the viscosity of NFs. Temperature, agglomeration, base fluid, and NP characteristics are among the key

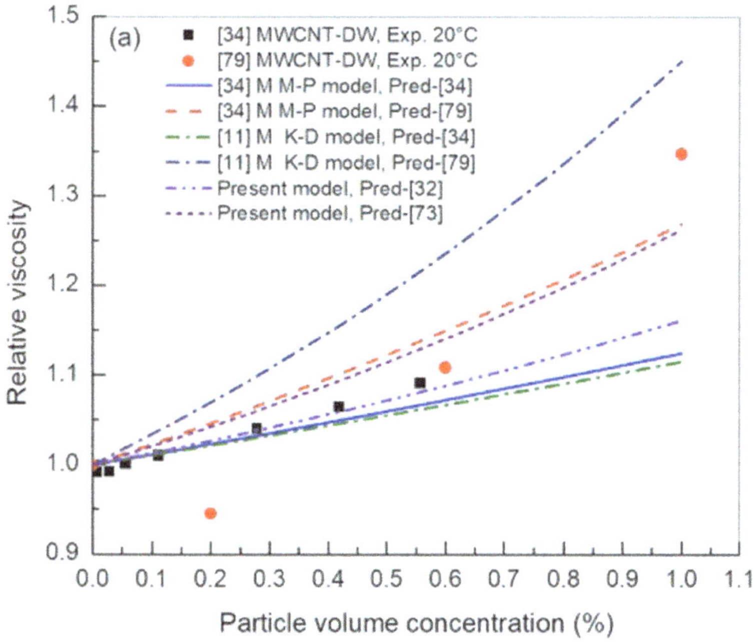

Figure 5.9 Comparison of experimental viscosity data of contemporary models (M M–P (modified Maron–Pierce), M K–D (modified Krieger and Dougherty), and present (Ye *et al.*) model). Reproduced from ref. [14] with permission from Springer Nature, Copyright 2019.

parameters behind the difference in the predictive performance between classical and contemporary models. Nonetheless, the main drawback of most of these models is the limited validation of their prediction capacity using very few experimental data of NF viscosity. In addition, the effect of temperature on NF viscosity is not accounted for in most of the models. However, the Shahriari, Masoumi and Yang models appear to be the best.

5.5 Concluding Remarks

This chapter has explored the importance of viscosity to thermal management and the performance of energy systems using NFs. The effect of some factors such as φ, nanosize and shape, temperature, dispersion characteristics, and base fluids on NF viscosity has been discussed. Also, various theories, mechanisms, assumptions, and parameters have been engaged to propose different models to predict the viscosity of NFs with varying degrees of accuracy and acceptance coupled with their merits and demerits. It can be noted that the proposed classical models that consider the suspension of larger particles poorly predict the viscosity of NFs while contemporary models perform better with relatively high accuracy owing to the physics and understanding of NFs. The key shortcoming of the contemporary models is that most of them are proposed without accounting for temperature and validated using very few viscosity experimental data. Future research involves the need for more studies on the influence of shapes, nanosize, mixing ratio, and preparation characteristics on the viscosity of NFs, and the development of improved models, bearing in mind the emergence of novel base fluids (*e.g.*, green base fluids), green NPs, and magnetic and electrical fields. Temperature, base fluid, NP preparation and characteristics and thermophysical properties, nanosize physics-based theories coupled with existing viscosity mechanisms are suggested to develop improved models.

Acknowledgements

The support provided by Dr S. A. Aasa during the preparation of this chapter is highly acknowledged.

References

1. S. O. Giwa, M. Sharifpur, M. H. Ahmadi and J. P. Meyer, *J. Therm. Anal. Calorim.*, 2021, **145**, 2581–2623.
2. S. O. Giwa, M. Sharifpur, M. H. Ahmadi and J. P. Meyer, *Math. Models Methods Appl. Sci.*, 2020, **2020**, 1–59.
3. L. S. Sundar, K. V. Sharma, M. T. Naik and M. K. Singh, *Renewable Sustainable Energy Rev.*, 2013, **25**, 670–686.
4. M. Mahdavi, M. Sharifpur and J. P. Meyer, *Int. J. Therm. Sci.*, 2016, **110**, 36–51.
5. I. M. Mahbubul, R. Saidur and M. A. Amalina, *Int. J. Heat Mass Transfer*, 2012, **55**, 874–885.

6. S. M. S. Murshed, M. Sharifpur, S. Giwa and J. P. Meyer, *Nanomaterials*, 2020, **10**, 1–35.
7. P. N. Nwosu, J. Meyer and M. Sharifpur, *J. Nanotechnol. Eng. Med.*, 2014, **5**, 1–11.
8. R. B. Ganvir, P. V. Walke and V. M. Kriplani, *Renewable Sustainable Energy Rev.*, 2016, **75**, 451–460.
9. A. Afzal, I. Nawfal, I. M. Mahbubul and S. S. Kumbar, *J. Therm. Anal. Calorim.*, 2019, **135**, 393–418.
10. I. M. Mahbubul, R. Saidur, M. A. Amalina and M. E. Niza, *Int. Commun. Heat Mass Transfer*, 2016, **76**, 33–40.
11. S. A. Adio, M. Sharifpur and J. P. Meyer, *J. Exp. Nanosci.*, 2016, **11**, 630–649.
12. M. Sharifpur, S. A. Adio and J. P. Meyer, *Int. Commun. Heat Mass Transfer*, 2015, **68**, 208–219.
13. J. P. Meyer, S. A. Adio, M. Sharifpur and P. N. Nwosu, *Heat Transfer Eng.*, 2016, **37**, 387–421.
14. X. Ye, S. G. Kandlikar and C. Li, *Eur. Phys. J. E: Soft Matter Biol. Phys.*, 2019, **42**, 159.
15. K. B. Anoop, T. Sundararajan and S. K. Das, *Int. J. Heat Mass Transfer*, 2009, **52**, 2189–2195.
16. K. Bashirnezhad, S. Bazri, M. R. Safaei, M. Goodarzi, M. Dahari, O. Mahian, A. S. Dalkilica and S. Wongwises, *Int. Commun. Heat Mass Transfer*, 2016, **73**, 114–123.
17. B. C. Pak and Y. I. Cho, *Exp. Heat Transfer*, 1998, **11**, 151–170.
18. P. K. Namburu, D. P. Kulkarni, A. Dandekar and D. K. Das, *Micro Nano Lett.*, 2007, **2**, 67–71.
19. Q. Li, Y. Xuan and J. Wang, *Exp. Therm. Fluid Sci.*, 2005, **30**, 109–116.
20. P. K. Namburu, D. P. Kulkarni, D. Misra and D. K. Das, *Exp. Therm. Fluid Sci.*, 2007, **32**, 397–402.
21. S. M. S. Murshed, K. C. Leong and C. Yang, *Int. J. Therm. Sci.*, 2008, **47**, 560–568.
22. S. M. S. Murshed, K. C. Leong and C. Yang, *Appl. Therm. Eng.*, 2008, **28**, 2109–2125.
23. S. M. S. Murshed, C. A. N. de Castro, M. J. V Lourenço, M. L. M. Lopes and F. J. V Santos, *J. Phys.: Conf. Ser.*, 2012, **395**, 012117.
24. T. Yiamsawas, O. Mahian, A. S. Dalkilic, S. Kaewnai and S. Wongwises, *Appl. Energy*, 2013, **111**, 40–45.
25. S. O. Giwa, M. Sharifpur, J. P. Meyer, S. Wongwises and O. Mahian, *J. Therm. Anal. Calorim.*, 2021, **143**, 1037.
26. L. S. Sundar, E. Venkata Ramana, M. P. F. Graça, M. K. Singh and A. C. M. Sousa, *Int. Commun. Heat Mass Transfer*, 2016, **73**, 62–74.
27. L. S. Sundar, M. K. Singh, M. C. Ferro and A. C. M. Sousa, *Int. Commun. Heat Mass Transfer*, 2017, **84**, 1–10.
28. K. Apmann, R. Fulmer, A. Soto and S. Vafaei, *Materials*, 2021, **14**, 1–75.
29. L. S. Sundar, M. K. Singh and A. C. M. Sousa, *Int. Commun. Heat Mass Transfer*, 2013, **44**, 7–14.

30. S. M. S. Murshed and P. Estellé, *Renewable Sustainable Energy Rev.*, 2017, **76**, 1134–1152.
31. S. O. Giwa, M. Sharifpur, M. H. Ahmadi, S. M. S. Murshed and J. P. Meyer, *Nanomaterials*, 2021, **11**, 136.
32. E. V. Timofeeva, J. L. Routbort and D. Singh, *J. Appl. Phys.*, 2009, **106**, 014304.
33. K. Oster, C. Hardacre, J. Jacquemin, A. P. C. Ribeiro and A. Elsinawi, *Pure Appl. Chem.*, 2019, **91**, 1309–1340.
34. G. A. Adewumi, F. Inambao, M. Sharifpur and J. P. Meyer, *Int. J. Appl. Eng. Res.*, 2018, **13**, 8336–8342.
35. H. Yarmand, S. Gharehkhani, S. F. S. Shirazi, A. Amiri, E. Montazer, H. K. Arzani, R. Sadri, M. Dhari and S. N. Kazi, *Int. Commun. Heat Mass Transfer*, 2016, **72**, 10–15.
36. U. M. Kallamu, J. S. Ibrahim, M. Sharifpur and J. P. Meyer, *12th International Conference on Heat Transfer, Fluid Mechanics and Thermodynamics*, 2016, pp. 1713–1718.
37. K. A. Hamid, W. H. Azmi, M. F. Nabil, R. Mamat and K. V Sharma, *Int. J. Heat Mass Transfer*, 2018, **116**, 1143–1152.
38. S. O. Giwa, M. Momin, C. N. Nwaokocha, M. Sharifpur and J. P. Meyer, *J. Therm. Anal. Calorim.*, 2021, **143**, 1063–1079.
39. F. R. Siddiqui, C. Y. Tso, K. C. Chan, S. C. Fu and C. Y. H. Chao, *Int. J. Heat Mass Transfer*, 2019, **132**, 1200–1216.
40. S. O. Giwa, M. Sharifpur, M. Goodarzi, H. Alsulami and J. P. Meyer, *J. Therm. Anal. Calorim.*, 2021, **143**, 4149–4167.
41. M. Kole and T. K. Dey, *Int. J. Therm. Sci.*, 2011, **50**, 1741–1747.
42. M. Sharifpur, S. O. Giwa, K. Y. Lee, H. Ghodsinezhad and J. P. Meyer, *Heat Transfer Eng.*, 2020, **42**, 1675–1687.
43. M. Hemmat Esfe, H. Rahimi Raki, M. R. Sarmasti Emami and M. Afrand, *Powder Technol.*, 2019, **342**, 808–816.
44. L. S. Sundar, G. O. Irurueta, E. Venkata Ramana, M. K. Singh and A. C. M. Sousa, *Case Stud. Therm. Eng.*, 2016, **7**, 66–77.
45. L. S. Sundar, E. Venkata Ramana, M. K. Singh and A. C. M. De Sousa, *Chem. Phys. Lett.*, 2012, **554**, 236–242.
46. A. H. A. Al-Waeli, M. T. Chaichan, K. Sopian and H. A. Kazem, *Case Stud. Therm. Eng*, 2019, **13**, 1–8.
47. K. A. Hamid, W. H. Azmi, R. Mamat, N. A. Usri and G. Najafi, *Energy Procedia*, 2015, **79**, 354–359.
48. E. V. Timofeeva, W. Yu, D. M. France, D. Singh and J. L. Routbort, *J. Appl. Phys.*, 2011, **109**, 014914.
49. V. Kumar and R. R. Sahoo, *Adv. Heat Transfer – Asian Res.*, 2019, **48**, 3144–3161.
50. S. A. Adio, M. Mehrabi, M. Sharifpur and J. P. Meyer, *Int. Commun. Heat Mass Transfer*, 2016, **72**, 71–83.
51. Y. He, Y. Jin, H. Chen, Y. Ding, D. Cang and H. Lu, *Int. J. Heat Mass Transfer*, 2007, **50**, 2272–2281.
52. C. T. Nguyen, F. Desgranges, N. Galanis, G. Roy, T. Maré, S. Boucher and H. Angue Mintsa, *Int. J. Therm. Sci.*, 2008, **47**, 103–111.

53. A. Boualit, N. Zeraibi, T. Chergui, M. Lebbi, L. Boutina and S. Laouar, *Int. J. Hydrogen Energy*, 2017, **42**, 8611–8623.
54. S. Ferrouillat, A. Bontemps, O. Poncelet, O. Soriano and J. A. Gruss, *Appl. Therm. Eng.*, 2013, **51**, 839–851.
55. M. Amani, P. Amani, A. Kasaeian, O. Mahian, F. Kasaeian and S. Wongwises, *J. Magn. Magn. Mater.*, 2017, **428**, 457–463.
56. P. D. Shima, J. Philip and B. Raj, *Appl. Phys. Lett.*, 2009, **95**, 1–4.
57. L. Wang, Y. Wang, X. Yan, X. Wang and B. Feng, *Int. Commun. Heat Mass Transfer*, 2016, **72**, 23–28.
58. K. Paulovičová, J. Tóthová, M. Rajňák, Z. Wu, B. Sundén, L. Wadsö, T. Tobiáš, P. Kopčansky, M. Timko and V. Lisý, *Acta Phys. Pol., A*, 2018, **133**, 564–566.
59. A. Malekzadeh, A. R. Pouranfard, N. Hatami, A. Kazemnejad Banari and M. R. Rahimi, *J. Appl. Fluid Mech.*, 2016, **9**, 693–697.
60. A. Shahsavar, M. R. Salimpour, M. Saghafian and M. B. Shafii, *J. Mech. Sci. Technol.*, 2016, **30**, 809–815.
61. A. Asadi and I. M. Alarifi, *Sci. Rep.*, 2020, **10**, 15182.
62. S. A. Adio, M. Sharifpur and J. P. Meyer, *Heat Transfer Eng.*, 2015, **36**, 1241–1251.
63. S. O. Giwa, M. Sharifpur and J. P. Meyer, *Appl. Therm. Eng.*, 2020, **170**, 115004.
64. M. Michael, A. Zagabathuni, S. Sikdar, S. K. Pabi and S. Ghosh, *Int. Nano Lett.*, 2020, **10**, 207–217.
65. I. M. Mahbubul, I. M. Shahrul, S. S. Khaleduzzaman, R. Saidur, M. A. Amalina and A. Turgut, *Int. J. Heat Mass Transfer*, 2015, **88**, 73–81.
66. A. Ghadimi and I. H. Metselaar, *Exp. Therm. Fluid Sci.*, 2013, **51**, 1–9.
67. Y. Yang, E. A. Grulke, Z. G. Zhang and G. Wu, *J. Appl. Phys.*, 2006, **99**, 114307.
68. B. Ruan and A. M. Jacobi, *Nanoscale Res. Lett.*, 2012, **7**, 127.
69. F. Li, L. Li, G. Zhong, Y. Zhai and Z. Li, *Int. J. Heat Mass Transfer*, 2019, **129**, 278–286.
70. S. Halelfadl, P. Estellé, B. Aladag, N. Doner and T. Maré, *Int. J. Therm. Sci.*, 2013, **71**, 111–117.
71. L. Yang, K. Du, Y. H. Ding, B. Cheng and Y. J. Li, *Powder Technol.*, 2012, **215–216**, 210–218.
72. H. J. Choi, C. A. Kim, T. M. Kwon and M. S. Jhon, *J. Magn. Magn. Mater.*, 2000, **209**, 228–230.
73. L. A. Pozhar, E. P. Kontar and M. Z. C. Hu, *J. Nanosci. Nanotechnol.*, 2002, **2**, 209–227.
74. J. Avsec and M. Oblak, *Int. J. Heat Mass Transfer*, 2007, **50**, 4331–4341.
75. H. Chen, Y. Ding and C. Tan, *New J. Phys.*, 2007, **9**, 367.
76. H. Chen, Y. Ding, A. Lapkin and X. Fan, *J. Nanopart. Res.*, 2009, **11**, 1513–1520.
77. J. Chevalier, O. Tillement and F. Ayela, *Appl. Phys. Lett.*, 2007, **91**, 1–4.
78. S. M. Hosseini, A. R. Moghadassi and D. E. Henneke, *J. Therm. Anal. Calorim.*, 2010, **100**, 873–877.

79. N. Masoumi, N. Sohrabi and A. Behzadmehr, *J. Phys. D: Appl. Phys.*, 2009, **42**, 055501.
80. M. Hosseini and S. Ghader, *J. Mol. Liq.*, 2010, **153**, 139–145.
81. D. Song, D. Jing, J. Geng and Y. Ren, *Powder Technol.*, 2015, **283**, 561–569.
82. A. Shahriari, N. Jahantigh and F. Rakani, *Int. J. Eng., Trans. B*, 2017, **30**, 1919–1924.
83. M. H. Sun and X. R. Zhang, *Int. J. Heat Mass Transfer*, 2018, **122**, 150–160.
84. D. S. Udawattha, M. Narayana and U. P. L. Wijayarathne, *J. King Saud Univ., Sci.*, 2019, **31**, 412–426.
85. G. Lebon and H. Machrafi, *J. Nonnewton. Fluid Mech.*, 2018, **253**, 1–6.
86. H. Machrafi, *Nanoscale*, 2020, **12**, 15081–15101.

CHAPTER 6

Thermal Conductivity of Nanofluids – Experiments, Models, and their Advancements

T. M. KOLLER,* F. E. BERGER BIOUCAS AND A. P. FRÖBA

Institute of Advanced Optical Technologies – Thermophysical Properties (AOT-TP), Department of Chemical and Biological Engineering (CBI) and Erlangen Graduate School in Advanced Optical Technologies (SAOT), Friedrich-Alexander-Universität Erlangen-Nürnberg (FAU), Paul-Gordan-Straße 8, 91052 Erlangen, Germany
*Email: thomas.m.koller@fau.de

6.1 Introduction

Nanofluids represent a class of heterogeneous systems or dispersions, which consist of a continuous liquid dispersion medium and a dispersed phase of solid nanoparticles in the nanometer range.[1] For the nanoparticles, typically employed materials are metals, metal oxides, and carbon materials, while water and ethylene glycol represent commonly used liquids. A key thermophysical property and, in specific, transport property, characterizing nanofluids and their performance in applications, is the effective thermal conductivity λ_{eff} as a measure of the ability of a material to conduct heat. Since the thermal conductivity λ in units of $W \cdot m^{-1} \cdot K^{-1}$, is a single-phase property, the additional term "effective" is required for nanofluids, considering that they refer to multiphase systems with at least two phases.[2] Yet, the term is omitted throughout this chapter due to simplicity reasons.

Soft Matter Series No. 16
Fundamentals and Transport Properties of Nanofluids
Edited by S. M. Sohel Murshed

Published by the Royal Society of Chemistry, www.rsc.org

The research field of nanofluids, with a strong focus on their thermal conductivity, as shall be shown later on, originated from the "discovery" of anomalous behavior by Choi *et al.*[3] in 1995. Here, experimental investigations indicated that nanofluids containing copper particles with a size of around 10 nm showed thermal conductivities which are about three times higher than the thermal conductivity of the corresponding base fluid. As a result, "dramatic reductions in heat exchanger pumping power" were expected.[3] An extension of these anomalous observations led to the first publication by the same group in 2001,[4] which reported measurement results for λ_{eff} that were significantly larger than the values predicted by classical theories at that time. To date, the work of Choi *et al.*[4] has been cited more than 2000 times,[5] and marks the starting point for the development of a new, rapidly growing research field. This field is connected to the investigation of the properties of nanofluids *via* experimental and modeling approaches, and, arising from the anticipated favorable features of nanofluids, their application in various fields of engineering.

The increasing interest in nanofluids has been accompanied by an increase in the number of publications. This applies for various research disciplines and is also strongly reflected on the thermal conductivity of nanofluids, which is the subject of this chapter. Based on a search in the ISI Web of Science[6] using the search criteria "nanofluid* thermal conductivity", Figure 6.1 shows the number of articles published between 2000 and 2020 connected to the determination of the thermal conductivity of nanofluids using either experimental methods or modeling in the form of predictions, correlations, and/or simulations. The rapid increase in the number of publications related to the measurement of the thermal conductivity of

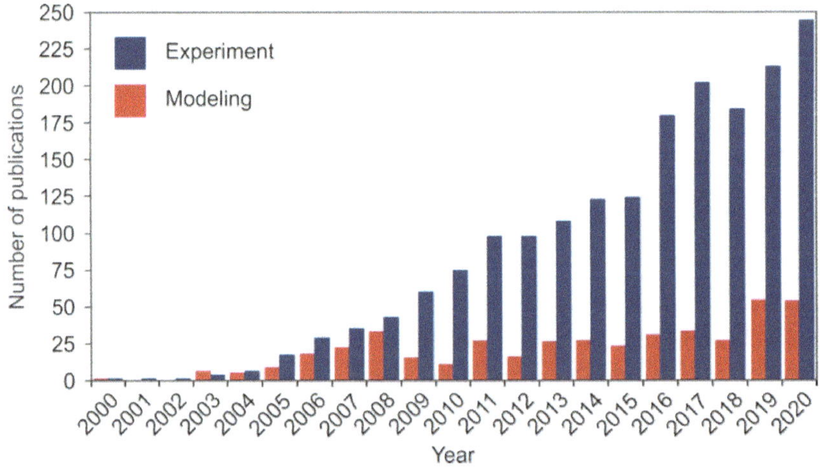

Figure 6.1 Number of publications related to the topic "nanofluid* thermal conductivity" between 2000 and 2020 accessed from webofscience.com.[6] For articles addressing both experimental investigation and modeling, the publication is counted as half for each topic.

nanofluids has also been encouraged by the fact that different research works have reported a large improvement of this property with respect to the base fluid of nanofluids. The strong increase in the number of experimental studies was also promoted by the easier access to commercial equipment for thermal conductivity measurements starting from 2004 to 2007, as is the case for the commonly used transient hot-wire[7–12] and transient hot-disk instruments.[13–17]

The seemingly pronounced increase in the λ_{eff} values of nanofluids relative to the thermal conductivity of the liquid triggered interest in different application areas, where the thermal conductivity of nanofluids and, thus, heat transfer in general are relevant.[18] By serving as potential heat transfer fluids for heating and cooling purposes, nanofluids can be processed in a continuous flow in, *e.g.*, heat exchangers,[19,20] heat pipes,[21,22] solar panels,[23,24] material processing units,[25,26] or in a quiescent state in, *e.g.*, electrical transformers,[27,28] electrical devices,[29,30] and pharmaceuticals.[31,32] Two of the mentioned applications, where the thermal conductivity of nanofluids is essential, shall be detailed representatively below.

The classical field of application for nanofluids refers to their use as heat transfer fluids in heat exchangers. Here, nanofluids have been investigated in flow pipes,[21] annular counter flow heat exchangers,[33] and plate heat exchangers,[19] but also in refrigerators[34] and electronic devices.[29,30] Since the thermal conductivity of nanofluids is usually higher than that of the base fluids, the overall heat transfer in the aforementioned devices can be increased. Based on experimental studies, Buschmann *et al.*[35] showed that correlations for the Nusselt number describing the convective heat transfer between a conventional fluid and a solid surface can also be applied to nanofluids if their correct thermophysical properties, including λ_{eff}, are used. The authors also discovered that the increase in the convective heat transfer coefficient achieved using nanofluids is similar to the increase in the λ_{eff} value of the nanofluid compared to the base fluid.[35] In a review published by Tertsinidou *et al.*[2] on the heat transfer capabilities of nanofluids in a tube bundle heat exchanger, it was concluded that the benefits in the overall heat transfer were rather modest when replacing a conventional fluid with a nanofluid. As the addition of nanoparticles to liquids also increases the effective viscosity compared to the viscosity of the base fluid, the resulting pressure loss needs to be compensated for by an increase in the pumping power. This effect outweighs the weak gain in the convective heat transfer coefficient by employing nanofluids with increased λ_{eff} values.

Another application of nanofluids is in biomedical engineering, where thermal medicine is often used for the treatment of cancers and tumors. Here, examples are thermal ablation[36,37] or radiofrequency ablation (RFA)[32,38] techniques. In the latter, saline solutions[39] or ethanol[40] are infused into the target cancer or tumor tissue, before an oscillating current is applied between an electrode inserted in the target tissue and a grounding source. By applying radiofrequency pulses, the electrical energy dissipates to thermal energy used to destroy the target tissue. To further optimize the technique, the thermal conductivity of the injected fluid should be increased compared to that of the target tissue. This serves to enhance the heat dissipation and to decrease the ablation temperature, which prevents tissue charring above around 373 K. For example, Wu *et al.*[32] compared the performance of RFA when using a saline solution or a solution containing dispersed carbon-coated iron nanoparticles. By adding 0.11% nanoparticles to the base fluid, a 41% increase in the λ_{eff} value of the nanofluid relative to the thermal conductivity of the base fluid was reported. Mainly due to this, the nanofluid injection improved heat dissipation during ablation and suppressed tissue charring.

Considering the aforementioned and other applications, detailed knowledge about the thermal conductivity of nanofluids is of fundamental importance for the design and performance of related processes and equipment. The accurate determination of the λ_{eff} values of nanofluids is thus an important, yet challenging task. This challenge holds for both the measurement of this transport property, including the associated sample preparation and characterization, but also for the modeling of λ_{eff} based on physically-sound theories. Even today, the experimental and theoretical investigations of different research groups have revealed no consensus regarding the value of the thermal conductivity of nanofluids, as reviewed by Tertsinidou *et al.*[2] Similar to the discoveries of Choi *et al.*,[4] most of the studies have reported a distinct increase in the λ_{eff} relative to the thermal conductivity of the base fluid upon the addition of a small amount of nanoparticles. In other studies,[2,41–43] only a moderate enhancement was observed upon increasing nanoparticle concentration. This current state-of-the-art is exemplarily reflected in Figure 6.2, which shows the ratio of the thermal conductivity of the nanofluid λ_{eff} to the thermal conductivity of the base fluid λ_{bf} as a function of the volume fraction of the nanoparticles φ_p. This was conducted for a commonly probed nanofluid consisting of water and dispersed titanium dioxide (TiO_2) nanoparticles of nearly spherical shape with a maximum diameter of 155 nm. For all accessible 21 experimental studies reporting in total 157 measurement points for λ_{eff}, the large disparity between the results of studies becomes obvious. The line in Figure 6.2 represents the predicted "normal" behavior for the thermal conductivity of two-phase systems according to the effective medium theory of Hamilton and Crosser,[44] which the minority of experimental results follows. The inset in Figure 6.2 focusing on extremely small φ_p values of up to 0.75×10^{-3} highlights the abnormally large $\lambda_{eff}/\lambda_{bf}$ ratios.

The ongoing controversy on the thermal conductivity of nanofluids is connected to the reliability of the underlying experimental and modeling approaches. Regarding the latter, a fundamental understanding of the heat transport mechanisms in nanofluids can only be achieved using theoretical models. Here, various mechanisms including Brownian motion,[67–69] phonon energy transport,[70] aggregation,[12,71] and/or interfacial layer[72,73] have been controversially discussed. In this context, the Brownian motion of the particles has attracted the most attention for explaining the anomalous increases in λ_{eff}

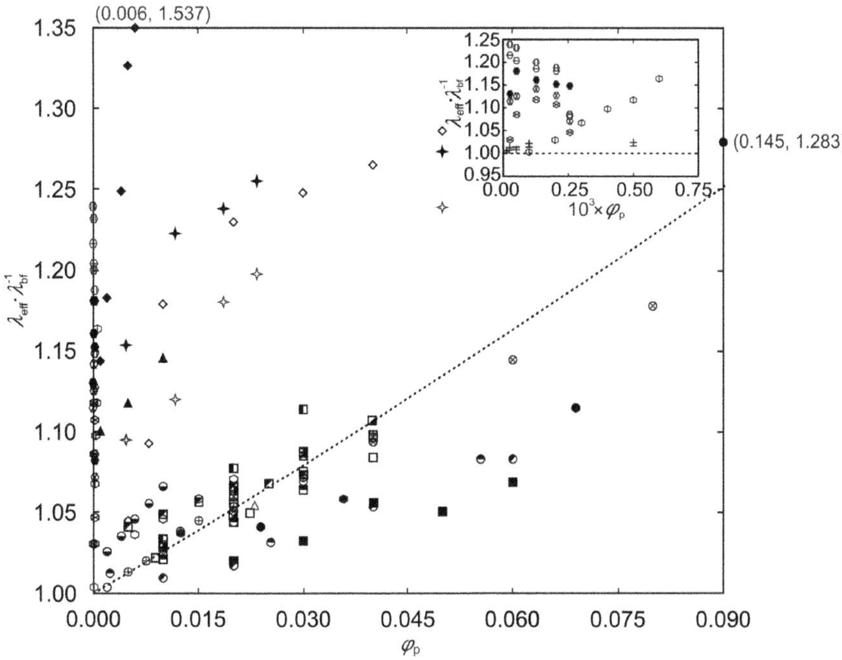

Figure 6.2 Thermal conductivity ratio for water-based nanofluids containing nearly spherical TiO_2 particles as a function of the particle volume fraction. Experimental results at temperatures between (293.15 to 303.15) K: ⬢, 40 nm, 303.15 K, Zhang *et al.*;[45] ◇, 15 nm, room temperature (RT), Murshed *et al.*;[46] ▲, 25 nm, Yoo *et al.*;[47] ◧, 10 nm, stabilized by sodium dodecyl sulfate stabilized (SDS), Kim *et al.*;[48] ◨, 34 nm, SDS, Kim *et al.*;[48] ⬒, 70 nm, SDS, Kim *et al.*;[48] ○, 21 nm, 296.15 K, Turgut *et al.*;[49] ⬡, 21 nm, 298.15 K, Duangthongsuk *et al.*;[50] ◐, 155 nm, 293.15 K, Longo *et al.*;[51] ◓, 74 nm, 294.2 K, stabilized by acetic acid, Fedele *et al.*;[52] ⊗, 21 nm, 293.15 K, stabilized by surfactant, Yiamsawasd *et al.*;[53] ◒, 21 nm, 303.15 K, stabilized by cetyl trimethyl ammonium bromide, Reddy *et al.*;[54] △, 20 nm, 293.15 K, stabilized by ammonium polyacrylate, Haghighi *et al.*;[55] ◩, 30 nm, 303.15 K, Azmi *et al.*;[56] ⦶, 9 nm, 303 K, Nishant *et al.*;[57] □, 30 to 50 nm, 298.15 K, stabilized by polyvinylpyrrolidone (PVP), Tertsinidou *et al.*;[58,59] ⊕, 298.15 K, 5 to 15 nm, PVP, Tertsinidou *et al.*;[58,59] ⊗, 5 to 15 nm, 298.15 K, PVP, Tertsinidou *et al.*;[58,59] ■, 5 nm, 298.15 K, PVP, Tertsinidou *et al.*;[58,59] ◆, 31 nm, 303 K, Maheshwary *et al.*;[60] +, 11 nm, 293.15 K, PVP, Oliveira *et al.*;[61] –, 11 nm, 293.15 K, PVP, Oliveira *et al.*;[61] ⊕, 40 to 60 nm, 298.15 K, Singh *et al.*;[62] ⦶, 21 nm, 300 K, Mitra *et al.*;[63] ⊖, 18 nm, 300 K, Mitra *et al.*;[63] ⬡, 16 nm, 300 K, Mitra *et al.*;[63] ⊗, 14 nm, 300 K, Mitra *et al.*;[63] ⊗, 13 nm, 300 K, Mitra *et al.*;[63] ◪, 23 nm, 291 K, Wang *et al.*;[64] ✦, 18 nm, 298.15 K, Kazemi-Beydokhti *et al.*;[65] ✧, 35 nm, 298.15 K, Kazemi-Beydokhti *et al.*;[65] ●, 66 nm, 298.15 K, Bioucas *et al.*;[43] Modeling: ---, Hamilton–Crosser model[44] at $T = 298.15$ K using data for the thermal conductivity of water[66] and TiO_2.[2]

values. Recent models[43,74,75] contradict this finding and argue instead that Brownian motion introduces an additional heat transfer resistance, which decreases with an increase in particle velocity. Corresponding predictions are in agreement with the Hamilton–Crosser (HC) model,[44] suggesting that there are no extraordinary changes in λ_{eff} upon adding particles. For comparison with theory, a reliable experimental database for λ_{eff} is required, yet not completed so far. In many cases, routine laboratory techniques are used, for which a critical assessment of their application to nanofluids and/or of their experimental uncertainty is often missing.[2] Furthermore, the results may be affected by different sample preparation methods, which further impedes a trustable data comparison. In a benchmark study by Buongiorno *et al.*[41] involving 30 research groups worldwide, the experimental results showed no anomalous increase in λ_{eff} with increasing particle volume fraction and were in accordance with the HC model.[44] In 2015, Tertsinidou *et al.*[2] reviewed the existing literature data on the λ_{eff} values of commonly probed nanofluids. Anomalously high thermal conductivities observed for small particle fractions, such as the ones shown in Figure 6.2, were mainly attributed to the use of inadequate experimental techniques of unproven validity. The authors concluded that, if the thermal conductivity is determined from reliable techniques, agreement within $\pm 10\%$ or better between the experimental data themselves and the HC model[44] is observed.[2]

In the following sections of this book chapter, a summary of the scientific activities and findings in connection with the experimental investigation and modeling of the thermal conductivity of nanofluids and their advancements is presented. In Section 6.2, the experimental determination of the thermal conductivity of nanofluids is discussed, which includes, besides the preparation and characterization of the nanofluid samples, the measurement of the thermal conductivity. Here, the focus lies on methods that are favorable for their application to nanofluids. Section 6.3 then highlights the main developments in the modeling of the thermal conductivity, with the main emphasis on theoretical approaches for the prediction of this property. In Section 6.4, results on the thermal conductivity of selected types of nanofluids are analyzed as a function of influencing parameters and mechanisms, and are discussed in the context of new advancements in the field. Concluding remarks are given in Section 6.5.

6.2 Experimental Determination of the Thermal Conductivity of Nanofluids

An accurate and reliable determination of the thermal conductivity of nanofluids is essential to obtain a fundamental understanding of heat transfer mechanisms in such dispersed systems. In this section, the experimental methodology ranging from the preparation of nanofluids over their characterization to the measurement of the thermal conductivity, with the latter representing the focus of this section, are addressed and suggestions for its proper application are given. For more detailed information on the preparation and characterization of nanofluids, the reader is referred to references cited in this section.

6.2.1 Sample Preparation

Ideally, nanofluids represent colloidal suspensions, where all nanoparticles are dispersed within a liquid continuous phase of the dispersing medium, *i.e.* the base fluid. In nanofluids used for experimental investigations, including thermal conductivity measurements, the particles should be well dispersed and separated from each other. In practice, however, complete individualization of all primary particles is not achievable, so real nanofluids also contain particle aggregates to a certain extent. There are two common methods for the preparation of nanofluids: the one-step method[4,76,77] and the two-step method.[78–80] The former method is based on the synthesis of the particles in the dispersing fluid, *i.e. in situ*, while in the latter, a dry powder is produced *via* chemical or physical methods and dispersed in the fluid afterwards. In general, the two-step method is faster, cheaper, and allows for the production of larger nanofluid batches with high particle concentrations than the one-step method. However, the latter yields stable nanofluids with a high degree of monodispersity, yet is highly susceptible to contamination during particle synthesis.

Focusing on the traditionally used two-step method, the mechanical dispersion of the particles synthesized in the first step can be accomplished in the second step using different instruments, such as ball milling processing units,[81,82] high-pressure homogenizers,[81,82] ultrasonic baths,[81,82] and sonicators.[58,82] The latter disperse the particles by applying sound waves to the sample *via* a probe, which is typically more efficient in terms of dispersion quality than ultrasonic baths.[82,83] According to Arrigo *et al.*[84] and Pal *et al.*,[85] when using sonicators as the most commonly employed devices, care has to be taken to not break the particles by applying too much energy and to not overheat the nanofluid sample. Such effects could be harmful to the dispersing medium and/or added stabilizing agents.[86–88]

Although nanofluids have been subjected to mechanical treatment, in most particle–fluid systems, dispersion stability without the use of further additives can only be observed a certain time, *i.e.* stability is a kinetic issue. In general, particles have a larger density than the dispersing fluid, leading to sedimentation of the particles over time due to the influence of gravity. Furthermore, attractive particle–particle interactions can be strong enough to enable the formation of aggregates. Such systems where sedimentation and/or a certain degree of aggregates is present do not correspond to a fully dispersed system, as often anticipated, and are likely to be associated with different λ_{eff} values than for dispersed nanofluids.

To achieve the long-time stability of nanofluids over at least a few days, different stabilization techniques are often applied for nanofluids in addition to mechanical stabilization. The two main stabilization mechanisms are electrostatic and steric stabilization, as illustrated in Figure 6.3. Electrostatic stabilization is a kinetic stabilization method and basically restricted to aqueous nanofluids. The stabilization results from inter-particle repulsion between similarly charged electrical double layers around the

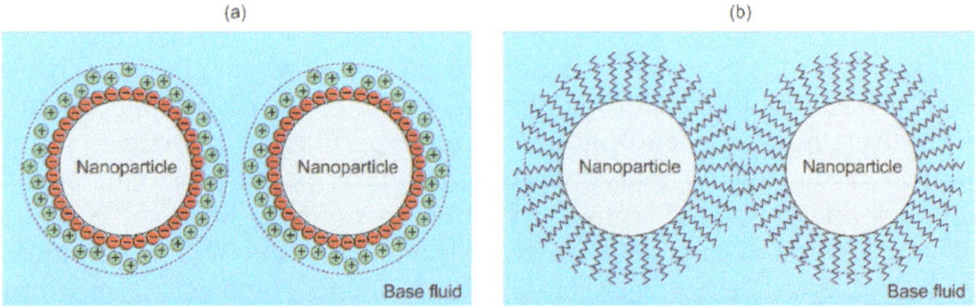

Figure 6.3 Two-dimensional representation of the (a) electrostatic stabilization and (b) steric stabilization used for the dispersion stability of nanoparticles in nanofluids.

surface of a charged particle, as shown schematically in Figure 6.3(a). For charging the dispersed particles, two different processes can be used.[89] One is related to the adjustment of the pH *via* the dissociation of strong electrolytes, *e.g.*, hydrochloride acid or potassium hydroxide, that will be attached to the particle surface.[89] The second method relies on the adsorption of sufficiently soluble ionic species in the form of cationic, anionic, or amphoteric surfactants on the surface of the particles.[90] For example, in water-based nanofluids, the hydrophobic tail of the surfactant sticks to the particle, while the hydrophilic head group points towards the dispersing fluid. Although the use of such surfactants such as sodium dodecyl sulfate (SDS) is economic, they are sensitive to high temperature, temperature cycles, and high ultrasonic input power, leading to degradation of the surfactant[86–88] or foam formation in the nanofluid.

In the steric stabilization shown in Figure 6.3(b), the repulsion between particles results from stabilizing agents in the form of absorbed or grafted nonionic polymeric surfactants consisting of hydrophilic and hydrophobic functional repeating units, *i.e.*, these agents are amphiphilic in nature. These surfactants, among which polyvinylpyrrolidone (PVP) is commonly used,[58,91] cover the surface of the particles with a thicknesses on the order of several nanometers. When the particles become close to each other, the polymer chains may overlap, leading to increased local chain density. This creates a repulsion between the particles due to an increase in osmotic pressure and the energetically unfavorable reduction of the configurational entropy of the chains.[89] Thus, steric stabilization is a thermodynamic stabilization method that can be used for a wide spectrum of particles up to relatively high concentrations. Furthermore, this method allows a redispersion of the aggregated particle clusters, is flexible regarding the polarity of the dispersing medium, and is not sensitive to electrolytes or pH changes.[92]

6.2.2 Sample Characterization

After the preparation of the nanofluids, it is required to perform a characterization of the dispersion stability as well as of the morphology of the involved particles in terms of size and shape. These parameters may have a strong influence on the thermal conductivity of nanofluids, as will be explained later on in this section. In the upcoming section, different methods used for the sample characterization of nanofluids will be addressed and reviewed.

The simplest method to evaluate the nanofluid stability in a qualitative way is visual inspection. For this, a nanofluid is placed in a sufficiently long tube with a small diameter over a period of time.[43,93] From this, it is possible to see if sedimentation occurs over time.[43] For a quantitative analysis of stability, ultraviolet–visible (UV–vis) spectroscopy,[94–96] which resolves changes in the transmitted light due to the light scattered or absorbed by the nanoparticles in a liquid, is often applied. A further common tool for the stability analysis of nanofluids[82,97] is the measurement of the zeta potential ζ. This represents the electrical potential in the double layer at the interface between the fluid loosely attached to the particle and the mobile fluid phase. The ζ potential is influenced by the pH, the presence of ionic species, and surfactants and polymeric additives attached to the particle surface.[98] It is commonly accepted in the scientific literature that a nanofluid is highly stable for ζ above +30 mV or below −30 mV, while for ζ between −10 mV and +10 mV the nanofluid becomes unstable.[99] Here, the net electrokinetic charge density at the particle surface becomes too low and, in the case of the isoelectric point (IEP), even zero. Thus, the stabilizing electrostatic repulsion forces between particles become too weak compared to their attractive van der Waals forces, resulting in particle aggregation.

Prevalent techniques that enable a quantitative characterization of the morphology of the particles in nanofluids are differential centrifugal sedimentation (DCS)[43,100] and scanning electron microscopy (SEM).[41,43,58] For the latter technique, discussed here as an example, an electron beam with energies of between 1 and 30 keV is used.[101] While scanning the sample in a vacuum chamber with an electron beam, primary and secondary electrons emitted from the sample are collected by detectors. From the analysis of the electron signals, not only the size and shape of the particles, but also their composition and surface topography can be obtained.[101] One main disadvantage of the SEM method is the necessity of drying the particles dispersed in a base fluid under a vacuum. This required step may result in a change in the morphology of the particles in comparison to that present during thermal conductivity measurements. In addition, if the particles are electrically nonconductive or only weakly conductive, a conductive coating has to be added to the particles to remove or reduce the electric charge that builds up in a nonconducting sample when it is scanned by the electron beam.[102]

To characterize nanofluids in the same fluid state as in thermal conductivity measurements, dynamic light scattering (DLS) is a suitable technique that can be used to study the size, size distribution, and also, in some cases, the shape of the particles in dispersed systems.[43,103–106] By studying statistical fluctuations in particle concentration that originate from the random thermal movement of molecules and particles, and whose dynamics are reflected in the time-dependent correlation function of the scattered light intensity, information about the particle diffusion process and, thus, about the particle size and shape is accessible. For the investigation of opaque or non-transparent samples, as is often the case for nanofluids, it is required to study the scattered light in the reflection direction at scattering angles close to 180°. The latter configuration is of advantage since the scattering volume can be positioned close to the boundary of the sample, which enables the application of low laser powers and reduces multiple scattering effects.[104] Due to external influences such as light scattered from the windows of the sample cell, a pure homodyne detection scheme analyzing solely light scattered from the fluid sample itself can hardly be realized,[104] yet is often assumed in the literature.[12,52,58] As a result, data evaluated for the particle diffusion coefficient and, in turn, for the particle size may be erroneous, as shown by Berger Bioucas *et al.*[104] In a heterodyne detection scheme, where the light scattered by the particles is superimposed coherently with much more intense reference light, unambiguous data evaluation is ensured.[43,104]

Assuming monodisperse spherical particles, close to infinite dilution, *i.e.*, at particle volume fractions below around 10^{-5}, the diffusion coefficients determined by DLS correspond to translational diffusion coefficients. Employing the translational diffusion coefficient *via* the Stokes–Einstein equation, the hydrodynamic diameter of the particles can be obtained. In the case of multimodal particle dispersions, by corresponding evaluation of the correlation functions *via* CONTIN[107] or cumulant[108] analysis, information about the particle size distribution can also be accessed. For different water-based nanofluids, the results derived from DLS have been shown to be in agreement with the particle sizes measured by SEM and DCS.[43] For non-spherical particles such as rods or tubes, an additional rotational diffusion is measurable by DLS. From both the translation and rotational diffusion coefficients, information about the morphology in terms of diameter and length of nanorods can be obtained.[104] At higher particle concentrations beyond infinite dilution, direct particle interactions and indirect hydrodynamic interactions affect the diffusion coefficients. Although at high concentrations a quantitative determination of particle size and shape is impeded, yet still often performed,[52,55,58] qualitative conclusions on the dispersion stability can be drawn by evaluating the diffusion coefficient as a function of particle concentration and/or time.[43]

6.2.3 Measurement of the Thermal Conductivity

The accurate measurement of the thermal conductivity of fluids, including nanofluids, is a challenging task. Although thermal conduction through nanofluids is of primary interest, other mechanisms of heat transfer in the form of convection and/or radiation and parasitic heat fluxes can also be present at the same time. If not considered in the experimental realization or data evaluation, these effects can falsify the measurements. Furthermore, in contrast to single-phase fluids, the measurement results for multi-phase nanofluids depend not only on thermodynamic state variables, but may also be affected by the measurement technique and procedure itself.[2]

The thermal conductivity of a material describes the heat flux density resulting from a temperature gradient. The underlying mathematical description is known as Fourier's law of thermal conduction and forms the theoretical basis of thermal conductivity measurements. By subjecting the sample itself or a body in contact with the sample to a macroscopic temperature gradient, the resulting heat flux can be probed. Here, the gradient has to be sufficiently large to obtain a measurable signal, but small enough to suppress disturbances of the equilibrium. Methods for the measurement of thermal conductivity can be divided into steady-state and transient techniques.[109] While in steady-state methods the temperature gradient is constant during the measurements, transient ones are characterized by a gradient that changes as a function of time.

In the following, an overview of the principles and the experimental realization of methods for the measurement of the thermal conductivity of nanofluids is given. Here, the focus lies on techniques, which are backed by valid theory and are thus suitable for the absolute determination of the thermal conductivity of nanofluids. A detailed description on measurement techniques for thermal conductivity is given in the textbook by Wakeham *et al.*[109]

6.2.3.1 *Transient Methods*

In transient methods, the fluid originally kept at a constant temperature is subjected to a small temperature perturbation as a result of an applied heat flux. The evaluation of this perturbation as a function of time serves for the determination of the thermal conductivity. For its accurate determination, it has to be ensured that the propagation time of the resulting temperature wave through the fluid caused by the localized temperature gradient is much shorter than the characteristic time for the acceleration of the fluid by buoyancy forces associated with convection.[109] In connection with nanofluids, there are two suitable transient methods given by the transient hot-wire technique and the transient hot-disk method, which are of interest in this section. Other transient methods, such as the needle-probe instrument,[56,57,62,63] the temperature-oscillation method,[110–112] and 3ω methods[49,64,113] have a limited theoretical basis, often require calibration, and are likely to be affected by the presence of convection during the measurements.[2] For an accurate determination of the thermal conductivity of nanofluids, these methods will not be addressed here, although they are commonly employed in the literature.

6.2.3.1.1 Transient Hot-wire Technique. The transient hot-wire (THW) technique is the most commonly used method for nanofluids; see, *e.g.*, ref. 46, 48, 50, 58 and 60. When employed in an appropriate manner, the technique enables absolute, fast, and accurate measurements of the λ_{eff} values of nanofluids. For this, several experimental and theoretical requirements briefly discussed below have to be met, which are typically not given in the literature, as reviewed by Antoniadis *et al.*[114] A detailed description of the theoretical framework of the THW technique and its experimental realization can be found in ref. 109 and 114.

The measuring principle is based on applying a heat flux generated *via* an electrical current over a cylindrical metallic wire of sufficiently long length and small diameter to the fluid, which is initially in thermodynamic equilibrium at temperature T_0. The time-dependent rise in the temperature of the wire, $T(t)$, accessible by its resistance, corresponds to that of the surrounding fluid and is thus connected to the thermal conductivity of the fluid. By solving Fourier's partial differential equation of heat conduction for an ideal configuration, the simplest form of the working equation is given by:[109]

$$\Delta T(r_0, t) = \frac{\dot{q}}{4\pi\lambda}\ln\left(\frac{4at}{r_0^2 e^{\gamma}}\right). \tag{6.1}$$

In eqn (6.1), ΔT $(= T - T_0)$ is the temperature rise of the wire, $\dot{q}$ is the heat flux per unit length, a is the thermal diffusivity of the fluid, t is the time, and γ is the Euler constant. From the slope of the linear fit of ΔT as a function of $\ln(t)$, the thermal conductivity of the fluid λ is accessible.[114]

The aim is to design a THW instrument that matches very closely the working equation in eqn (6.1) by eliminating deviations from it through proper design and operation as well as by considering remaining deviations *via* correction terms, which only account for as much as around 1% of the measured temperature rise.[114] The main precautions required for accurate λ measurements are briefly highlighted as follows. Firstly, it is essential to use a vertically arranged wire with a length of at least 50 mm and a small diameter of about 25 μm, which is immersed in a sufficiently large vessel containing the fluid and is surrounded by a thermostatted bath to control the temperature, $T_{control}$; see Figure 6.4(a). As the wire cannot be infinitely long in practice, end effects on the edges of the finite wires at the point of connection with the respective support may arise, and disturbances in the T-field at the wire edges are inevitable.[115] To eliminate this problem, a

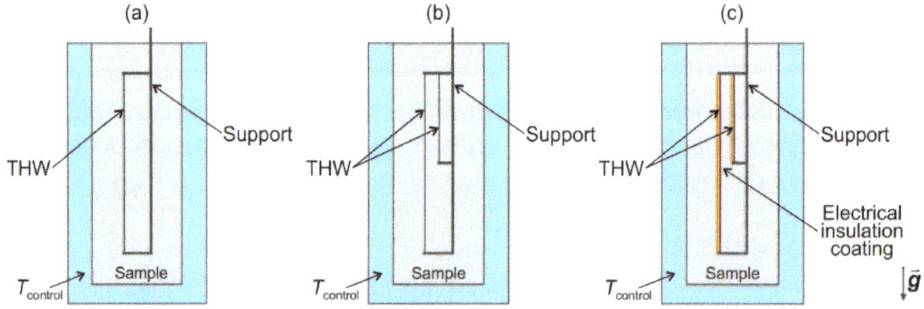

Figure 6.4 Schematic drawing of a THW instrument for the measurement of the thermal conductivity of nanofluids: (a) single-wire configuration; (b) two-wire configuration; (c) two-wire configuration including insulating coating (optimized configuration).

two-wire configuration with different lengths of about 50 mm and 20 mm is required,[114] as shown in Figure 6.4(b). Furthermore, if the wires are not electrically insulated, current leakages and/or electrical distortions will occur, falsifying the thermal conductivity analysis for polar or electrically conducting fluids, as is often the case for nanofluids. Thus, it is essential to apply a thin insulation layer on the surface of the wires; see Figure 6.4(c). Antoniadis *et al.*[114] and Tertsinidou *et al.*,[58] for example, used two metallic wires made of tantalum with an oxidized layer of tantalum oxide.

Based on proper design of the THW instrument according to Figure 6.4(c), further operational considerations have to be taken into account for an accurate thermal conductivity measurement. While heat transfer due to radiation is negligible from the choice of small wires, the influence of convection can be eliminated if the measurement time is below around 1 s, which corresponds to a temperature rise for water and related nanofluids of below around 4 K.[114] Validation of the THW technique can be achieved by comparing the experimental signal for ΔT with the theoretical value, which should match within the experimental uncertainty over the entire period of observation; see in Antoniadis *et al.*[114] Considering the aforementioned criteria, the THW technique allows for the measurement of the thermal conductivity of nanofluids with an expanded uncertainty (coverage factor $k=2$) of 2%,[58] while values down to 0.5% are possible for pure fluids such as water[114] and alcohols.[116]

For nanofluids of fairly spherical, metallic oxide particles in water or ethylene glycol up to volume fractions of 0.07, the reliable experimental λ_{eff} data published by Tertsinidou *et al.*[58] with an uncertainty of <2% ($k=2$) show only a moderate increase with increasing particle volume fraction and are not distinctly affected by the addition of agents to improve dispersion stability. Similar statements hold for investigations of the same research group on nanofluids featuring TiO_2 or multi-wall carbon nanotubes (MWCNTs).[114] In other THW experiments, inadequate setups using, *e.g.*, very short[45,65] or thick[46,53,65,117–119] wires, partially in combination without electrical insulation,[117,119,120] have been applied. In all these cases, validation of the experiment was not presented, which makes specified uncertainties of (1 or 1.5)%[70,120] questionable, if they are specified at all. Considering corresponding studies,[46,65,120] the measured results for λ_{eff} are often distinctly higher than predictions using the HC model.[44] Similar observations have also been reported in ref. 2 and 114.

In conclusion, the THW technique is a very suitable method for the determination of the thermal conductivity of nanofluids, as it represents an absolute technique and is based on a solid theoretical framework. By fulfilling an appropriate design and validating the method on the basis of the theoretical description, reliable results for λ_{eff} of nanofluids with an uncertainty of around 2% are possible; see, *e.g.*, ref. 58 and 114. Even today, in most cases, the required criteria for proper application of the THW technique are not met. This may partly explain why there is a large discrepancy in the published values for the thermal conductivity of nanofluids.

6.2.3.1.2 Transient Hot-disk Technique. The transient hot-disk (THD) technique, also known as the transient plane-source method, is considered as an absolute method[121] for the measurement of the thermal conductivity and can be used for the study of solids and liquids, including nanofluids.[51,52,55,58,61] Besides the closely related THW technique, the THD method can be considered as a second source for attaining reliable λ_{eff} values of nanofluids if the underlying theory is realized as accurately as possible in the experiment. In the following, a brief description of the working principle of the THD technique is provided, while a detailed discussion of the method can be found in ref. 122–124.

As illustrated in Figure 6.5(a), the core element of the THD technique is a hot-disk sensor that acts as a heating source and, at the same time, as a temperature sensor, *i.e.*, in the same way as the wire in the THW technique. The THD sensor consists of a double spiral of a thin, around 10 μm thick metallic wire, which is typically made of nickel and encapsulated by an electrically insulating polymeric film.[121,124] The sensor is immersed in a vertical configuration into the fluid of investigation contained inside a cylindrical cell and surrounded by a temperature-controlled reservoir set at T_{control}; see Figure 6.5(b). From time $t=0$, where at the sensor operated with a constant power electric energy dissipates and, as a result, a heat flux from the sensor to the fluid can be observed, the temperature increase in the sensor and, accordingly, in the fluid, is recorded as a function of time *via* resistance measurements. The underlying mathematical description of the transient heat conduction problem is expressed *via:*[124]

$$\frac{\partial T}{\partial t} = a\nabla^2 T + \frac{\dot{Q}}{\rho c_{\text{p}}}. \tag{6.2}$$

Here, $\dot{Q}$ is the heat flux dissipated by the hot-disk sensor per unit volume and t is the time, while a, ρ, and, c_{p} are the thermal diffusivity, density, and isobaric heat capacity of the fluid sample, respectively, which are related to the thermal conductivity by $\lambda = a\rho c_{\text{p}}$. A solution of eqn (6.2) by integration over time and volume was reported by He *et al.*[124] Here, the theoretical form

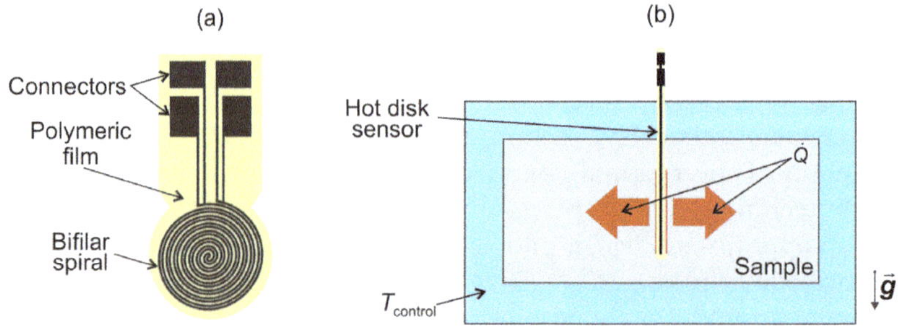

Figure 6.5 Schematic drawing of a THD instrument for the measurement of the thermal conductivity of nanofluids: (a) THD sensor; (b) setup configuration for the experimental realization.

of the measurement signal in the form of the temperature rise, $\Delta\bar{T}(\tau)$, at the surface of the hot-disk sensor averaged over the length of its concentric rings is displayed as a function of a relatively complicated function, $D(\tau)$, with the dimensionless time $\tau=\sqrt{at}/r_{\text{disk}}$ and the hot-disk radius r_{disk}. By plotting $\Delta\bar{T}$ as a function of $D(\tau)$, the λ value of the sample can be obtained, with or without knowledge of ρ and c_{p}, from the slope of a straight line in an iterative optimization procedure, as detailed by Zheng *et al.*[125]

In an analogous way to that shown for the THW technique, accurate use of the THD technique requires different experimental considerations and corrections to the working equation; see ref. 123 and 124. As the theory of the THD method assumes an infinitely large sample, it has to be ensured that the sample thickness between the surface of the hot-disk sensor and the sample boundary is larger than the penetration depth of the heat flux, $\Delta p=2\sqrt{a\cdot t_{\text{tot}}}$, where t_{tot} corresponds to the total measurement time.[121,123,124] Another crucial issue is to avoid convection, which can arise from the use of overly high heat fluxes subjected to the sample and measurement times that are too long. If the onset of natural convection is not considered in the representation of the measurement signal, artificially high thermal conductivities are obtained, as demonstrated experimentally by Warhoza and Fleischer[123] and Ma *et al.*[126] Depending on the thermophysical properties of the fluid, including its thermal conductivity and viscosity, heat fluxes of <100 mW applied over measurement times of below 5 s are typically required to obtain reliable λ results, which are in agreement within their expanded uncertainties ($k=2$) of around 5% with reference data.[123]

With respect to nanofluids, it appears that only commercial THD devices, mostly of the type TPS2500S,[51,55,58,82] have been applied. Yet, some reports[55,127] neither prove the validity of their instruments nor perform a critical assessment of the measurement uncertainty and/or accuracy. Exceptions to this are studies by Tertsinidou *et al.*[58] and Fedele *et al.*,[52] who tested their instruments against reference fluids and found deviations from reference values of $<1.5\%$, which is below the specified measurement accuracy of 5%. It could be demonstrated that if the THD method is applied properly, the measured λ_{eff} values of the nanofluids are in agreement with those obtained using the THW technique.[58] This shows that although both transient methods probe significantly different temporal and spatial regions in the nanofluid sample, the results from both techniques can be used for mutual validation.

Overall, the THD method may serve as a further convenient method for the determination of the thermal conductivity of nanofluids. If the application of the theoretical working equation can be justified for the experimental situation, it is anticipated that the λ_{eff} experimental data of nanofluids are reliable with an uncertainty of around 5%,[58,82] which is higher than in the case of the THW method. Since mostly commercial THD equipment is used, whose validity check against the exclusion of convection during the measurement is often not proven, it is likely that corresponding results for the thermal conductivity of nanofluids will be inaccurate to some extent.

6.2.3.2 Steady-state Methods

The working principle of steady-state methods for the measurement of thermal conductivity is based on the determination of the heat flux required to maintain a time-invariant temperature difference between two surfaces of the fluid. Depending on whether the surfaces are provided by concentric cylinders or two parallel plates, steady-state methods can be divided into concentric-cylinder (CC) or the parallel-plate (PP) methods.[109] For an accurate measurement of λ on the grounds of relatively simple working equations, other heat transfer mechanisms, *e.g.*, due to convection or radiation, and/or parasitic heat fluxes to the ambient environment, have to be suppressed or considered in the data evaluation. In contrast to transient methods, there exists only a limited number of investigations on nanofluids applying CC[42,54,93,128–133] and PP[43,134–138] methods. In most of these studies, a calibration is performed, except for ref. 43, where the PP version is employed. As the CC method is closely related to the PP method, yet more susceptible to the need for corrections and the occurrence of convection effects,[109] only the PP technique is discussed below.

6.2.3.2.1 Parallel-plate Method. The steady-state PP method is a well-established technique for the measurement of the thermal conductivity of solids and fluids in the form of gases and liquids.[109,139,140] Recently, it has also been applied for the study of nanofluids.[43] For absolute measurement of thermal conductivity in a reliable way, suitable design of the apparatus and appropriate measurement conditions are necessary, as is outlined below. A detailed presentation of the theoretical background of the PP method and its experimental realization is given in ref. 109 and 140.

The working principle of the PP method is to fulfill the ideal one-dimensional form of the Fourier law of heat conduction for a planar sample,

$$\dot{Q}_{\text{cond}} = A\frac{\lambda}{s}(T_{\text{hot}} - T_{\text{cold}}), \tag{6.3}$$

as close as possible. From an accurate determination of the heat flux $\dot{Q}_{\text{cond}}$ that arises purely by conduction through the sample, the heat transfer area A, the sample thickness s, and the temperature difference $\Delta T\,(= T_{\text{hot}} - T_{\text{cold}})$ between the two outer surfaces of the sample, the thermal conductivity of the sample λ is directly accessible.[140] In this ideal case, the experimentally determined heat flux $\dot{Q}_{\text{exp}}$ is attributed to pure thermal conduction through the fluid, which is challenging to realize experimentally. Thus, to access the thermal conductivity in a reliable way, any additional contributions to the measured heat flux have to be minimized or considered reliably, which is discussed in the following.

As shown in a schematic drawing of a PP instrument in Figure 6.6, the sample is confined in a thin horizontal layer between two parallel plates made of a material with a high thermal conductivity, such as copper, and oriented perpendicular to the gravitational field. By this arrangement requiring a precise adjustment of the sample thickness, and by guiding the heat flux from the upper hot plate, used for balancing $\dot{Q}_{exp}$, to the lower cold plate, convection effects inside the sample layer are minimized. Yet, with this design, as illustrated in Figure 6.6(a), the one-dimensional Fourier law is not fulfilled because a part of $\dot{Q}_{exp}$ is partially transferred as parasitic heat leakage from the heating plate to the surrounding. To avoid this effect, which would lead to λ values that are systematically too large, the heating plate is immersed in a guard system;[140] see Figure 6.6(b). It has to be ensured that this guard comprising a multi-part system maintains the same temperature as the hot plate in order to guarantee, as accurately as possible, a one-dimensional heat flux from the heating to the cooling plate. In the configuration as per Figure 6.6(b), convective flows at the boundary of the fluid volume may still affect the heat flux through the sample in the balanced region and, thus, need to be suppressed. For this purpose, the cooling plate can be designed to have a larger surface area than the hot plate. With this optimized design given in Figure 6.6(c), any possible convective effects are shifted to the outer edges of the sample and do not disturb the horizontal isometric lines between the hot and the cold plate.

To evaluate the occurrence of two types of convection, *i.e.*, Bénard and laminar convection, the thermal Rayleigh number Ra_T is a governing parameter. For an infinite horizontal fluid layer heated from below, Mostert *et al.*[141] concluded that Bénard convection cells start to become relevant for $Ra_T > 1708$. A similar threshold of 1000 was stated by Naziev *et al.*[142] for vertically arranged coaxial cylinders. Since in PP instruments the fluid layer is typically heated from above, Bénard convection is unlikely to occur.[143] Regarding laminar convection arising from the deviation of the sample layer from a horizontal plane, the corresponding power dissipated by convection can be estimated[144] and scales

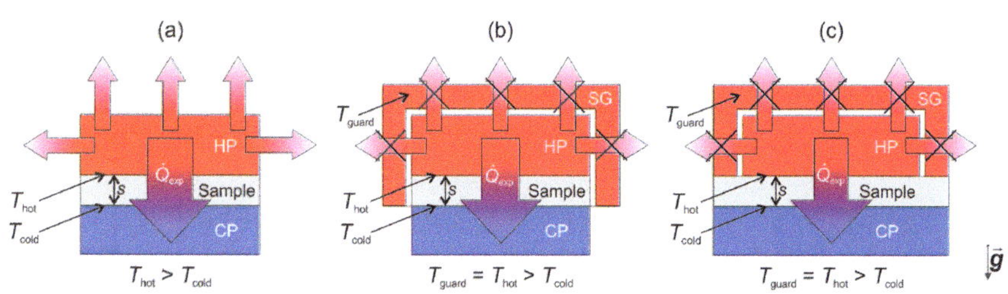

Figure 6.6 Schematic drawing of the PP instrument for the measurement of the thermal conductivity of nanofluids: (a) simple two-plate configuration with upper hot heating plate (HP) and lower cold cooling plate (CP); (b) configuration as in (a) including a side-guard (SG) system; (c) guarded PP instrument using a larger diameter for CP than for HP, as realized in ref. 43 and 140 (optimized configuration).

linearly with Ra_T. Thus, special care has to be taken to ensure full planarity of the fluid layer, which can be realized by the use of, *e.g.*, micrometer screws.[43,140] The insignificance of convection can be corroborated by experiments, where no dependence of the resulting λ values on the T-difference applied between the plates has been observed for different fluids.[140]

In the case of nanofluids, not only conductive energy transport caused by a temperature gradient, but also energy transport caused by a concentration gradient, the so-called Dufour effect, may arise. This effect originates from the mass transfer of the particles due to the applied temperature gradient. According to the analytical expression proposed by Ali Pakravan and Yaghoubi[145] for determining energy transport due to the Dufour effect, calculations showed that the contribution to the λ_{eff} value of common nanofluids caused by the Dufour effect was distinctly <1% and, thus, negligible. To reduce the contribution of the radiative heat transfer between the plates to the total heat flux, their polished surfaces can be covered with a thin layer of a low emission coefficient such as silver[93] or chrome.[43,140] If the sample is a strongly absorbing fluid such as water, radiation contributions can be neglected as they make up <0.1% of the measured thermal conductivities.[145] In the case of weakly or non-absorbing fluids, the contribution of the radiation to the measured thermal conductivity can be estimated, *e.g.*, using an approximate calculation method proposed by Braun *et al.*[146] For nanofluids, which are often opaque or non-transparent, and thus strongly absorb radiation, radiation effects on the measurement results for λ_{eff} can be neglected.

A guarded PP instrument was developed by the Fröba group based on the design shown in Figure 6.6(c) and that was used for the absolute measurement of the thermal conductivity of various fluid types under ambient pressure at temperatures between 263 and 373 K.[43,140,147,148] By applying small T-gradients of around 1 K mm^{-1} and using a multi-guard system, an expanded uncertainty ($k=2$) of the measured λ values between 3 and 5% was achieved. Their deviations from reference fluids such as water or toluene were typically smaller than the specified uncertainty.[140] In 2020, the technique was applied for the measurement of λ_{eff} of nanofluids with particle volume fractions up to 0.31 between 298 and 323 K.[43] For dispersions of nearly-spherical TiO_2, silicon dioxide (SiO_2), and polystyrene (PS) particles in water, the measurement results show only moderate changes over the entire concentration range and are in agreement with reliable THW-based results[2] for TiO_2-based nanofluids. The only other five studies on nanofluids employing the PP method have been reported by Shalkevich *et al.*,[134] Wang *et al.*,[135] and Li *et al.*[136–138] Their instruments feature no real guard system around the HP[134–138] and a CP of the same size as the HP.[134,136–138] Furthermore, these studies rely on a relative measurement in which the thickness of the sample layer or the heat transfer area is calibrated using reference fluids.[134,136–138] The latter procedure is also often performed in studies based on CC devices.[42,54,93,128–133] With specified uncertainties of between 0.8 and 2.5%,[42,93,130] the λ_{eff} data for nanofluids obtained by these PP methods[54,129,131] are by trend larger than the predictions of the HC model.[44]

In summary, steady-state methods and, in particular, their PP designs are useful alternatives to transient methods for the measurement of the thermal conductivity of nanofluids as they are based on simple working equations and are unaffected by electrical polarization effects. However, besides possibly occurring convective effects that shall be suppressed as much as possible, steady-state methods require sophisticated instrumentation in the form of a guard system to prevent heat leakages, and precise horizontal alignment of the two plate surfaces. Once such boundary conditions are fulfilled, which is mostly not the case in the available sparse number of nanofluid studies, reliable measurements of λ_{eff} with expanded uncertainties of 5% and below are possible.

6.3 Modeling of the Thermal Conductivity of Nanofluids

The accurate determination of the thermal conductivity of nanofluids is a challenging task, as has previously been reviewed. Such measurements are also highly demanding with respect to time, resources, and costs. It is thus of interest, both from engineering and scientific perspectives, to develop reliable methods for the prediction of the thermal conductivity of nanofluids. For this, available approaches can be separated into theoretical and empirical models. To obtain a fundamental understanding of the heat transport mechanisms in dispersions, theoretical models are necessary and, thus, will be the focus of the following section.

6.3.1 Empirical Models

Most prediction approaches for the thermal conductivity of nanofluids are empirical in nature. Here, available experimental λ_{eff} data are correlated as a function of relevant parameters that characterize the system, such as the particle volume fraction, temperature, particle geometry, and/or features of a fluid layer around the particles; see, *e.g.*, ref. 120 and 149–151. However, there is little benefit of empirical models if the experimental data used for the model development are not critically reviewed regarding the methods applied, measurement conditions, and uncertainties. Furthermore, empirical models are often system-specific, with parameters optimized for a certain combination of particle and base fluid, which limits their transferability to other systems.

An example of an empirical approach is given by Corcione *et al.*[150] Here, for systems containing relatively highly conductive TiO_2, CuO, Al_2O_3, or Cu particles dispersed in water or EG, selected experimental λ_{eff} data were considered and correlated as a function of temperature T, volume fraction φ_p, thermal conductivities of particle (λ_p), and base fluid (λ_{bf}) with a standard deviation of around 1.9%. This was achieved by introducing an abnormally high temperature dependency in the thermal conductivity ratio according to $\lambda_{eff}/\lambda_{bf} \sim T^{10}$. However, as will be discussed later, the $\lambda_{eff}/\lambda_{bf}$

ratio of nanofluids is not distinctly affected by temperature, which casts doubt on the reliability of the model of Corcione *et al.*[150] and its underlying database.

6.3.2 Theoretical Models

Theoretical models for the thermal conductivity aim to represent the heat transport mechanisms in nanofluids based on physical concepts. Here, two classes can be differentiated: static and dynamic models. While static models do not consider the movement of nanoparticles in liquids, dynamic models include this phenomenon in addition to other mechanisms. As it is impossible to discuss all available theoretical models here, an overview of common theoretical approaches and considerations, often with controversial conclusions, will be given in the following.

In connection with static models, the series and parallel models of thermal conduction represent the simplest predictions for the λ_{eff} values of nanofluids. In the series model, the heat flux is conducted through the bulk fluid and the particles in series, as given in Figure 6.7(a), whereas parallel paths are assumed in the parallel model; see Figure 6.7(d). Considering the thermal conductivities of both phases, *i.e.*, particles (λ_p) and base fluid (λ_{bf}), as well as their corresponding volume fractions φ_p and $(1-\varphi_p)$, the thermal conductivity of the nanofluid is given by

$$\frac{1}{\lambda_{eff}} = \frac{1-\varphi_p}{\lambda_{bf}} + \frac{\varphi_p}{\lambda_p} \tag{6.4}$$

for the serial model and by

$$\lambda_{eff} = \left(1-\varphi_p\right) \cdot \lambda_{bf} + \varphi_p \cdot \lambda_p \tag{6.5}$$

for the parallel model. These two extreme cases can be considered as the lower (serial model) and upper (parallel model) boundaries of the thermal conductivity of nanofluids, in the case that the entire particle ensemble is arranged either perpendicular or parallel to the heat flux, respectively.

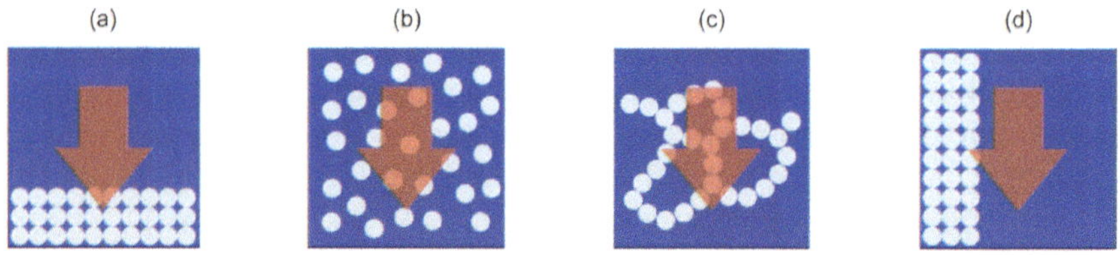

Figure 6.7 Two-dimensional representation of a (a) serial, (b) dispersed, (c) fractal, or (d) parallel configuration of particles (gray color) dispersed in a liquid continuous phase (blue color) for modeling the thermal conductivity of nanofluids associated with a corresponding heat flux (red arrows). Values of λ_{eff} increase from left to right.

For nanofluids, a fully dispersed state of the particles in the liquid is desired, as is shown in Figure 6.7(b). For such dispersions, Maxwell[152] was one of the first to use the effective medium theory to describe the electrical conductivity of two-component solid systems consisting of one material distributed as spherical inclusions within a continuous material. Given the analogy of the transport phenomena for electrical and thermal energy, Maxwell's expression for the electrical conductivity can be transferred accordingly to the thermal conductivity of heterogeneous two-phase systems including nanofluids.[152] In the derivation, Laplace's equations for the description of pure thermal conduction can be solved for both the dispersed and continuous phase in the steady state. Here, it is assumed that the particles do not move relative to the fluid and that no thermal resistance exists at the interface between the two phases. For spherical particles dispersed in the base fluid, the resulting equation for the thermal conductivity according to Maxwell[152] is given by

$$\lambda_{\mathrm{eff}} = \lambda_{\mathrm{bf}} \cdot \left(\frac{\lambda_{\mathrm{p}} + 2\lambda_{\mathrm{bf}} - 2\varphi_{\mathrm{p}}\left(\lambda_{\mathrm{bf}} - \lambda_{\mathrm{p}}\right)}{\lambda_{\mathrm{p}} + 2\lambda_{\mathrm{bf}} + \varphi_{\mathrm{p}}\left(\lambda_{\mathrm{bf}} - \lambda_{\mathrm{p}}\right)} \right). \tag{6.6}$$

Based on their own measurements of two-phase solid systems, Hamilton and Crosser[44] further extended the Maxwell model to also include the influence of the shape of the dispersed phase. The modified equation

$$\lambda_{\mathrm{eff}} = \lambda_{\mathrm{bf}} \cdot \left(\frac{\lambda_{\mathrm{p}} + (n-1)\lambda_{\mathrm{bf}} - (n-1)\varphi_{\mathrm{p}}(\lambda_{\mathrm{bf}} - \lambda_{\mathrm{p}})}{\lambda_{\mathrm{p}} + (n-1)\lambda_{\mathrm{bf}} + \varphi_{\mathrm{p}}(\lambda_{\mathrm{bf}} - \lambda_{\mathrm{p}})} \right) \tag{6.7}$$

is often referred to as the Hamilton–Crosser (HC) model and incorporates the shape factor n. The latter is related *via* $n = 3/\psi$ to the sphericity of the particle ψ, which is defined by the ratio of the surface area of a sphere that has the same volume as the particle to the surface area of the particle. For spherical particles with $\psi = 1$ and $n = 3$, eqn (6.7) reduces to the original Maxwell model, *i.e.*, eqn (6.6). For particles with deviations from the spherical shape, such as rods or plates, smaller ψ and, thus, larger λ_{eff} values are obtained at a given particle volume fraction. For the common case where $\lambda_{\mathrm{p}} > \lambda_{\mathrm{bf}}$, the thermal conductivity ratios $\lambda_{\mathrm{eff}}/\lambda_{\mathrm{bf}}$ predicted by the HC model show a moderate and almost linear increase with increasing φ_{p} values and are slightly larger than the values related to the serial model.

In the so-called fractal configuration, the particles form a continuous phase, while the base fluid constitutes the dispersed phase; see Figure 6.7(c). Here, the particles are connected *via* a chain-like particle network that separates large non-interacting fluid regions. In the related fractal model where the thermal conductivities and volume fractions of the two phases are exchanged in eqn (6.6), the λ_{eff} values are between those calculated using the HC and parallel models.[75] For typical nanofluids, where besides fully dispersed particles also aggregated ones can be found to some extent, the thermal conductivity is considered to be between the predictions of the HC and fractal models.[12,75] Under such conditions, the Bruggeman model[153] is often used, which includes the same parameters as the HC model.[44] Other models[69,75] also conclude that particle aggregation in the liquids leads to an increase of λ_{eff} compared to fully dispersed systems, as energy can be transferred over longer distances in the solid material along the temperature gradient. Further details on the influence of aggregation are given in Section 6.4, as the main focus in the present section lies on fully dispersed nanofluids.

In other static models,[72,73,154–156] modifications of the HC model have been proposed that account for the influence of an interfacial layer around the dispersed particles on λ_{eff}. This layer is formed as a result of the affinity of liquid molecules to the particle surface. Some approaches[72,73,154,155] attribute the distinct increase in measured thermal conductivity ratios to highly ordered layers, for which thicknesses up to 10 nm and thermal conductivities several times larger than the ones of the base fluid are assumed. While the latter assumption of a highly conductive thermal bridge is questionable and conflicts with recent molecular dynamics (MD) simulations,[157,158] the former is in contrast to findings from experiments[159] and simulations,[160] which reveal layer thicknesses in the order of 1 nm. Other models[12,156] disproving the layer concept claim that the particle–fluid interface provides interfacial thermal resistance. The latter reduces the thermal conductivity ratio compared to a system with no contact resistance, as assumed in the HC model. According to the model proposed by Nan *et al.*[156] shown later on, the interfacial resistance can be accounted for in the solution of Laplace's equations by considering a temperature jump between the two phases that depends on the thermal conductivity and size of the particles as well as on an interfacial resistance coefficient. This jump is reflected by the Biot number associated with the nanoparticles, $\alpha = (2 \cdot R_b \cdot \lambda_{bf})/d_p$, and depends on the thermal conductivity of the base fluid λ_{bf}, the diameter of the particles d_p, and the interfacial resistance coefficient R_b. This system-specific coefficient is usually determined by fitting the predictions of the model to the measured λ_{eff} data,[69,154,161] which limits the transferability to arbitrary nanofluids.

In dynamic models of thermal conductivity, diffusive processes in colloidal suspensions can be considered by means of which the random Brownian motion of the dispersed particles is often addressed.[75] The widespread opinion in the literature, *e.g.*, in ref. 67–69 and 161, is that Brownian motion is the main reason for the abnormal increase in the λ_{eff} values of nanofluids. It is hypothesized that this effect is caused by the diffusing nanoparticles carrying energy from hot to cold regions along the temperature gradient.[68] For the calculation of λ_{eff}, the dynamic contribution related to Brownian motion is considered to be additive to the static contribution given by, *e.g.*, the HC model[44] or its modified versions considering interfacial resistance[69,161] or a layer.[162,163] This dynamic contribution is expressed explicitly as a function of the Brownian velocity or the self-diffusion coefficient of the particles[67,68,161] or implicitly in correlations for the description of the micro-convective heat transfer between the base fluid and particles.[67,69] One such dynamic model that is often considered in the literature is that of Prasher *et al.*,[69] who expressed the thermal conductivity ratio in the form of

$$\frac{\lambda_{\mathrm{eff}}}{\lambda_{\mathrm{bf}}} = \left(1 + A{\cdot}Re^{m}Pr^{0.333}\varphi_{\mathrm{p}}\right)\cdot\left(\frac{\left[\lambda_{\mathrm{p}}(1+2\alpha)+2\lambda_{\mathrm{bf}}\right] - 2\varphi_{\mathrm{p}}\left[\lambda_{\mathrm{bf}} - \lambda_{\mathrm{p}}(1+2\alpha)\right]}{\left[\lambda_{\mathrm{p}}(1+2\alpha)+2\lambda_{\mathrm{bf}}\right] + \varphi_{\mathrm{p}}\left[\lambda_{\mathrm{bf}} - \lambda_{\mathrm{p}}(1+2\alpha)\right]}\right). \tag{6.8}$$

Here, following the work by Nan *et al.*,[156] the HC model for spherical particles as a static contribution is modified by the integration of the aforementioned Biot number, α, reflecting an interfacial resistance. This contribution given as the second term in brackets on the right-hand side of eqn (6.8) represents the model proposed by Nan *et al.*[156] The dynamic contribution to λ_{eff} is given by weighing Nan's model with the term containing the Reynolds number *Re*, the Prandtl number *Pr*, and the fit parameters *A* and *m*. While *A* is considered to be independent of the base fluid, *m* is weakly fluid-dependent when optimizing the parameters against relatively high λ_{eff} values of selected nanofluids. In contrast, other studies have revealed that the effects of Brownian motion on λ_{eff} are negligible.[71,75,164–166] Based on kinetic theory and MD simulations, Evans *et al.*[165] showed that there is no significant increase in the thermal conductivity due to the Brownian motion and that their predictions are in agreement with the effective medium theory. This is supported by the work of Eapen *et al.*,[75] who reported that the dynamics of the particles or *Nu* numbers used in related models are highly overestimated, leading to overly large predictions of λ_{eff}. Since particle self-diffusion coefficients are several orders of magnitude smaller than the thermal diffusivity of the base fluid, the Brownian motion of the particles causes only very weak transportation of energy and, thus, cannot contribute distinctly to enhanced λ_{eff} values of nanofluids.[75,165]

In 2016, an analytical dynamic model for the thermal conductivity of nanofluids was developed at the Institute of Advanced Optical Technologies – Thermophysical Properties (AOT-TP).[74] The model accounts for the thermal resistances due to heat conduction through the particles and the base fluid as well as due to the micro-convective heat transfer through the hydrodynamic boundary layer related to the Brownian motion of the particles. The consideration of the latter resistance reduces the λ_{eff} values distinctly compared to a system where no such resistance is present. This shows that the Brownian motion is a limiting factor for the enhancement of λ_{eff} relative to λ_{bf}. Only a weak increase in λ_{eff} is expected *via* intensification of the particle Brownian motion due to, *e.g.*, an increase in temperature. By describing the influences of the thermal conductivities of the base fluid and particles, volume fraction, size, and shape of the particles as well as temperature on λ_{eff}, agreement with the HC model[44] was observed for nanofluids with spherical and non-spherical particles of relatively large thermal conductivities.[74] In 2020, the model was further improved to also allow for predictions in the case of particles with thermal conductivities similar or lower than that of the base fluid. In this so-called AOT-TP model[43,74] given by

$$\frac{\lambda_{\mathrm{eff}}}{\lambda_{\mathrm{bf}}} = (1 - \varphi_{\mathrm{p}}) + \varphi_{\mathrm{p}} \left[\frac{\lambda_{\mathrm{bf}}}{\lambda_{\mathrm{p}}} + \left(\frac{\pi}{6\varphi_{\mathrm{p}}} \right)^{1/3} \cdot \frac{\psi}{\pi \cdot Nu} \right]^{-1}, \tag{6.9}$$

the first, second, and third term on the right-hand side reflect the contributions by the heat conduction through the fluid, heat conduction through the particles, and convective heat transfer between particle and fluid, respectively. In eqn (6.9), ψ is the sphericity of the particles, which is 1 for spherical particles. Considering particles of the same volume, decreasing sphericity of the particles ($\psi < 1$) is connected to an increase in surface area and, thus, a decrease in the convective heat transfer resistance, which increases λ_{eff}. The Nusselt number *Nu* can be obtained from a correlation of the Péclet number (*Pe*) and, thus, from the Prandtl (*Pr*) and Reynolds (*Re*) numbers.[43] For macroscopically static nanofluids, $Pe < 1$ and $Re \ll 1$, which results in *Nu* numbers close to 2 in the case of spherical particles.

On the basis of the aforementioned review, a comparison of the predictions for the thermal conductivity ratio $\lambda_{\mathrm{eff}}/\lambda_{\mathrm{bf}}$ from the serial and parallel model, the HC model,[44] the model of Prasher *et al.*,[69] and the AOT-TP model[43,74] is exemplarily shown in Figure 6.8 as a function of particle volume fraction φ_{p} for water-based nanofluids containing spherical TiO_2 nanoparticles of 66 nm diameter at $T = 298.15$ K. To check these models, only those of the experimental data summarized in Figure 6.2 are considered in Figure 6.8, which are based on reliable and absolute measurements from the THW,[58,59] THD,[51,52,55,61] and PP[43] methods. All other data obtained from inappropriate THW[46–48,50,53,60,167] and PP[54] instruments as well as measurements by transient needle[45,56,57,62,63] and 3ω[49,64] techniques were disregarded.

As can be seen in Figure 6.8, the experimental data situation becomes very clear and consistent with matching results within expanded uncertainties. The remaining 38 data points are constrained within the upper and lower boundaries given by the parallel and serial model, respectively, and show only a moderate enhancement. The results are in good agreement with the HC[44] and AOT-TP[43,74] models, where the latter seems to represent the behavior more accurately at larger particle volume fractions above $\varphi_{\mathrm{p}} = 0.05$. In terms of the model of Prasher *et al.*[69] with an employed interfacial resistance value of $R_{\mathrm{b}} = 0.77 \times 10^{-8}$ K m^2 W^{-1}, sound predictions close to the HC model can be obtained using a value for $m = 2.5$ optimized in their study for water-based nanofluids. Yet, if a slightly different value of $m = 2.0$ is used, as suggested by Holman *et al.*,[168] systematically too high $\lambda_{\mathrm{eff}}/\lambda_{\mathrm{bf}}$ ratios are obtained, showing the limitations of the semi-empirical approach proposed by Prasher *et al.*[69] The trends observed for the TiO_2–water nanofluid system have also been observed for other types of nanofluids in a similar way; see, *e.g.*, ref. 2, 43 and 58. For the following discussion on factors and mechanisms that influence the thermal conductivity of nanofluids, it is thus advisable to only consider results from reliable measurements and physically solid theoretical approaches.

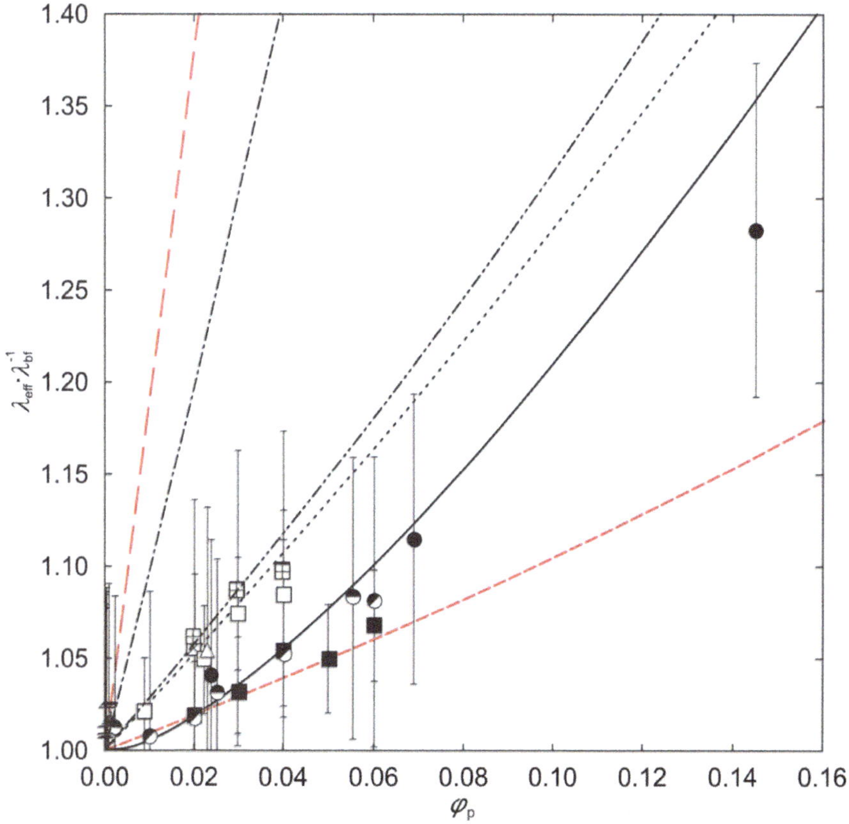

Figure 6.8 Thermal conductivity ratio for water-based nanofluids containing spherical TiO_2 particles as a function of the particle volume fraction. Experimental results at temperatures between 293.15 to 303.15 K excluding the data from less reliable measurements: ◐, 155 nm, 293.15 K, Longo *et al.*;[51] ◓, 74 nm, 294.2 K, Fedele *et al.*;[52] △, 20 nm, 293.15 K, Haghighi *et al.*;[55] □, 30 to 50 nm, 298.15 K, Tertsinidou *et al.*;[58,59] ⊞, 298.15 K, 5 to 15 nm, Tertsinidou *et al.*;[58,59] ⊠, 5 to 15 nm, 298.15 K, Tertsinidou *et al.*;[58,59] ■, 5 nm, 298.15 K, Tertsinidou *et al.*;[58,59] +, 11 nm, 293.15 K, Oliveira *et al.*;[61] –, 11 nm, 293.15 K, Oliveira *et al.*;[61] ●, 66 nm, 298.15 K, Bioucas *et al.*[43] Modeling results at $T = 298.15$ K for spherical particles of a diameter of 66 nm using data for the thermal conductivity of water[66] and TiO_2:[2] ----, serial model, eqn (6.4); — —, parallel model, eqn (6.5); ---, Hamilton-Crosser model,[44] eqn (6.7); ·-·, model of Prasher *et al.*[69] using $m = 2.0$, eqn (6.8); —·-—, model of Prasher *et al.*[69] using $m = 2.5$, eqn (6.8); –, AOT-TP model,[43,74] eqn (6.9).

6.4 Results and Discussion – Influences on the Thermal Conductivity of Nanofluids

As has been pointed out in the above experimental and theoretical sections, the controversy on the behavior of the thermal conductivity of nanofluids is still ongoing. Nevertheless, it is now broadly understood that there is no anomalous or distinct change in this property upon the addition of particles to liquids, as has been often believed over the last twenty years and even today. In the remainder of this section, the relevant influences on the thermal conductivity of nanofluids, which have been discussed controversially in the literature, are reviewed through a compilation of selected results from experimental and theoretical studies, which serves to give a representative and basic view on the behavior of nanofluids. The focus lies on properties of state in terms of particle concentration and temperature, the thermal conductivities of the particle and base fluid, and the morphology of the individual particles and their aggregates. In this context, light is also shed on the role of the Brownian motion of the particles and the interfacial layer around these.

6.4.1 Particle Concentration and Thermal Conductivity of Particles and Base Fluid

One key property that influences the thermal conductivity of nanofluids is the particle concentration, which is usually expressed by the volume fraction φ_p. This parameter is typically varied in experiments up to values of around 0.05, while larger values are barely investigated, mostly due to reduced dispersion stability. Regarding nanofluid systems, most literature studies have considered the base fluids water and EG in combination with particles of distinctly larger thermal conductivities, such as metal-based particles or MWCNTs. Nanofluids with weakly conductive particles, showing similar or even lower thermal conductivities than the base fluid, have been investigated to a much lesser extent.

In Figure 6.9, measurement results for the thermal conductvity ratio $\lambda_{eff}/\lambda_{bf}$ of four different water-based nanofluids obtained with the help of the THW technique and the PP method by Tertsinidou *et al.*[58,59] and Bioucas *et al.*,[43] respectively, are compared with the theoretical predictions for well-dispersed nanofluids according to the HC[44] and AOT-TP[43,74] models as a function of the particle volume fraction φ_p at 298.15 K. In detail, the experimentally considered particles of nearly spherical shape and average diameters d_p of between 30 and 610 nm are aluminum oxide (Al_2O_3, $\lambda_p \approx 40\ W\,m^{-1}\,K^{-1}$),[2] titanium oxide ($TiO_2$, $\lambda_p \approx 12\ W\,m^{-1}\,K^{-1}$),[2] silicon dioxide ($SiO_2$, $\lambda_p \approx 1.5\ W\,m^{-1}\,K^{-1}$),[169] and polystyrene (PS, $\lambda_p \approx 0.10\ W\,m^{-1}\,K^{-1}$)[170] and cover a wide range of φ_p values up to 0.31. For the selected nanofluids showing a broad range of λ_p/λ_{bf} ratios of between 0.2 and 66 down, the $\lambda_{eff}/\lambda_{bf}$ ratios are moderate and increase with increasing values for λ_p/λ_{bf}. For systems with Al_2O_3, TiO_2, and SiO_2 particles with higher thermal conductivities than water ($\lambda_{bf} \approx 0.6\ W\,m^{-1}\,K^{-1}$), the $\lambda_{eff}/\lambda_{bf}$ ratios are larger than one. For systems containing PS particles of lower thermal conductivity than water, the thermal conductivity of the fluid is reduced upon the addition of particles, which has been observed experimentally for the first time.

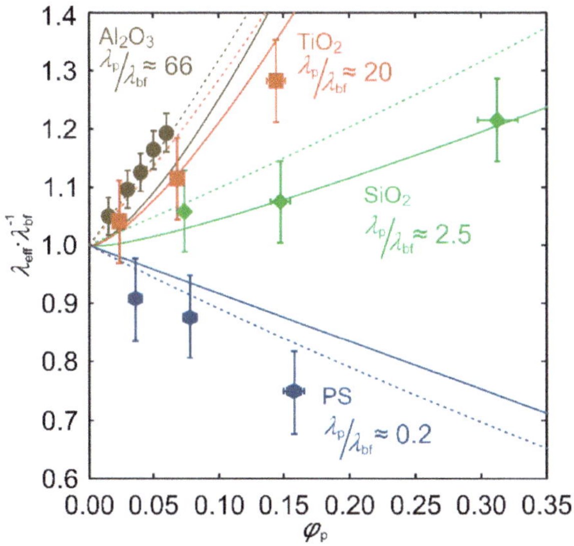

Figure 6.9 Comparison of measurement results and models for the thermal conductivity ratio of water-based nanofluids containing different spherical particles at $T = 298.15$ K as a function of the particle volume fraction. Experimental data: ●, Al_2O_3 ($d_p \approx 30$ nm), Tertsinidou *et al.*;[58,59] ■, TiO_2 ($d_p \approx 66$ nm), Bioucas *et al.*;[43] ◆, SiO_2 ($d_p \approx 36$ nm), Bioucas *et al.*;[43] ⬢, PS ($d_p \approx 610$ nm), Bioucas *et al.*[43] Modeling results for spherical particles of same diameters as in the respective experiments using data for the thermal conductivity of water[66] and particles:[2,169,170] ---, Hamilton–Crosser model,[44] eqn (6.7); –, AOT-TP model,[43,74] eqn (6.9).

In all cases, the concentration dependency of the experimental data is well represented by the HC[44] and AOT-TP[43] models. While the former model considers only thermal conduction, an additional thermal resistance due to convective heat transfer at the liquid–particle interface is taken into account in the latter model and reflected by the term associated with *Nu* in eqn (6.9). Thus, lower $\lambda_{eff}/\lambda_{bf}$ values are predicted by our model[43] than by the HC model[44] for nanofluids with Al_2O_3, TiO_2, and SiO_2 particles. Here, the convective heat transfer resistance $R_{p,conv}$ is distinctly higher than or comparable to the conductive resistance caused by the particle $R_{p,cond}$. When $R_{p,cond}$ is much greater than $R_{p,conv}$, as in PS-based systems where $\lambda_p/\lambda_{bf} < 1$, the AOT-TP model provides slightly higher $\lambda_{eff}/\lambda_{bf}$ ratios than the HC model.[44] In this case where the dispersed particles act as thermal insulators, the heat flux is mainly transferred through the liquid continuous phase, resulting in predictions close to those of a parallel model, as shown in eqn (6.5). The deviations in the values predicted by the HC and AOT-TP models from the measured data may be a result of the gradual sedimentation of the PS particles in the fluid layer within the PP instrument. This behavior shows similarities with the serial model shown in Figure 6.7(a), which gives lower λ_{eff} values than models considering fully dispersed particles, as is the case in the two aforementioned ones.

6.4.2 Temperature – Brownian Motion

Besides composition, a further thermodynamic property commonly addressed in studies on the thermal conductivity of nanofluids is temperature. In most cases, experimental investigations are limited to a narrow temperature range around an ambient temperature of 293 K. Measurements at elevated temperatures of up to 353.13 K[52] are scarce and are usually only possible for low particle concentrations due to weakened particle stability. As an increase in temperature is associated with an increase in the dynamics of the particles in the liquid, the effect of their enhanced Brownian motion on the values of λ_{eff} can indirectly be analyzed, if a dispersed state of the particles can be ensured.

Figure 6.10 shows the temperature dependency of the $\lambda_{eff}/\lambda_{bf}$ ratio for two water-based nanofluids containing spherical TiO_2 or SiO_2 particles of comparable size with medium or large particle volume fractions, respectively. For a system with TiO_2 at $\varphi_p = 0.05$ or 0.06, as shown in Figure 6.10(a), the $\lambda_{eff}/\lambda_{bf}$ ratio obtained from measurements by Bioucas *et al.*[43] between 298 and 323 K using the PP method is constant, which implies only a weak increase of λ_{eff} with increasing temperature in the same way as that of λ_{bf} of pure water. In this temperature range, the same behavior is observed in measurements by Fedele *et al.*[52] using a commercial THD device, whereas the thermal conductivity ratio begins to show a steep increase at T above 333 K up to 353 K. Here, the authors observed the formation of bubbles by evaporating water, which seemed to affect the measurement results. Longo *et al.*[51] who

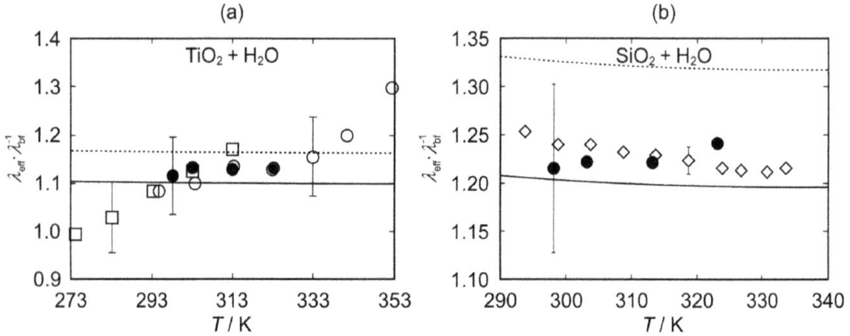

Figure 6.10 Comparison of measurement results and models for the thermal conductivity ratio of water-based nanofluids containing spherical (a) TiO_2 particles at $\varphi = 0.05$ or 0.06 or (b) SiO_2 particles at $\varphi = 0.31$ as a function of temperature. Experimental data: ●, TiO_2 ($d_p \approx 66$ nm), Bioucas *et al.*;[43] ○, TiO_2 ($d_p \approx 74$ nm), Fedele *et al.*;[52] □, TiO_2 ($d_p \approx 155$ nm), Longo *et al.*;[51] ●, SiO_2, $d_p \approx 36$ nm, Bioucas *et al.*,[43] ◇, SiO_2 ($d_p \approx 22$ nm), Ehle *et al.*[93] Modeling results for spherical particles of diameters of 66 nm for TiO_2 and 33 nm for SiO_2 using data for the thermal conductivity of water[66] and particles:[2,169] ---, Hamilton–Crosser model,[44] eqn (6.7); –, AOT-TP model,[43,74] eqn (6.9). Error bars incorporating the experimental uncertainties for λ_{eff} and λ_{bf} are exemplarily shown.

employed the same type of device as Fedele *et al.*,[52] reported a linear increase of the $\lambda_{eff}/\lambda_{bf}$ ratio with temperature starting from 274 K up to 313 K, for which the increasing contribution of the Brownian motion of the particles was proposed to be responsible. However, it appears to be likely that the behavior is caused by a gradual aggregation of the TiO_2 nanoparticles due to their lack of stabilization upon the use of surfactants, which was detected from corresponding DLS measurements. A negligible temperature effect on the λ_{eff} values of nanofluids and on the $\lambda_{eff}/\lambda_{bf}$ ratio can also be observed at large volume fractions of 0.31 for a SiO_2-based system. This is visible in Figure 6.10(b) from the matching measurements of Bioucas *et al.*[43] between 298 and 333 K using the PP method and from the measurements of Ehle *et al.*[93] between 294 and 334 K using the CC method, where both studies represent investigations of the same type of well-dispersed nanofluids.

For both nanofluids studied, the calculations from the classical effective medium theory according to the HC[44] and AOT-TP[43,74] models are shown in Figure 6.10 for comparison. Both models suggest a negligible effect of temperature on the thermal conductivity of the nanofluids. The HC model includes the temperature dependency implicitly *via* the thermal conductivities of the base fluid and nanoparticles, which is generally weak, in particular for the latter. Besides the T-dependent values of λ_{bf} and λ_p, the AOT-TP model also accommodates the Nu number, which depends on the Re and Pr number and, thus, on the further thermophysical properties of the base fluid. Calculations have shown[43,74] that the increase in the Nu number upon an increase in temperature is relatively small, because it is related to a weak decrease of the convective heat transfer resistance between particle and liquid associated with the particle Brownian motion. As a result, only a very minor increase in λ_{eff} with increasing temperature is predicted. Negligible influences of temperature and, thus, of Brownian motion on λ_{eff} suggested by the aforementioned models, including others,[75,156] and experiments, were also found by arguments from kinetic theory and MD simulations,[165] as has been discussed in Section 6.3.2. In conclusion, there are many indications, both experimental and theoretical, to believe that the Brownian motion has no significant influence on the thermal conductivity of nanofluids. In the plethora of literature studies that support the idea that Brownian motion is the main factor for a distinct increase in the thermal conductivity compared to that of the base fluid, it appears that inaccurate experiments were performed and used as a basis for theoretical and, in most cases, empirical modeling approaches.

6.4.3 Particle Morphology and Interfacial Layer

In addition to influences of the properties of state, the morphology of the dispersed nanoparticles in terms of their size and shape represents another important parameter that influences the thermal conductivity of nanofluids. So far, almost all experimental investigations have focused on more or less spherical particles as a result of their lower production costs and higher market accessibility compared to non-spherical particles. For the latter, very few studies on metal-based particles[171–173] have been performed, while MWCNTs have seen somewhat more attention; see, *e.g.*, ref. 58, 78, 128, 174 and 175. In the context of geometrical aspects, the role of an interfacial layer at the particle–fluid interface can also be addressed. As highlighted in Section 6.3.2, the presence of such a layer is considered to have either an enhancing or a weakly impeding effect on the λ_{eff} values.

In terms of fairly spherical particles, Figure 6.11 illustrates the behavior of the thermal conductivity ratio $\lambda_{eff}/\lambda_{bf}$ for a TiO_2–water nanofluid as a function of the diameter d_p of the dispersed particles. As no systematic analysis of the size dependency has been conducted within a single study, experimental results from different sources were considered here. In Figure 6.11(a) where the results for a low particle volume fraction $\varphi_p \approx 0.01$ are displayed, all the results from Figure 6.2 were included, with the exception of the abnormally large $\lambda_{eff}/\lambda_{bf}$ ratios larger than 1.1. In the case of the larger volume fraction $\varphi_p \approx 0.06$, only the carefully evaluated, reliable data according to Figure 6.8 are displayed in Figure 6.11(b). For the remaining experimental data points in Figure 6.11, the associated mean diameters of the fairly dispersed particles were obtained from corresponding characterization methods, which are discussed in Section 6.2.2.

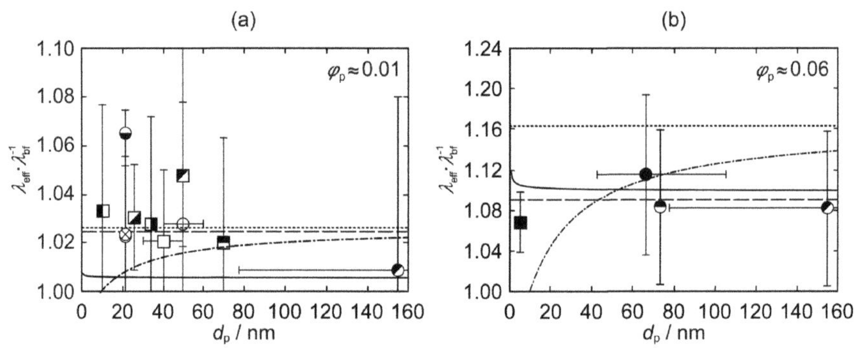

Figure 6.11 Comparison of measurement results and models for the thermal conductivity ratio of water-based nanofluids containing spherical TiO_2 particles at (a) $\varphi_p \approx 0.01$ or (b) $\varphi_p \approx 0.06$ as a function of the particle diameter for temperatures between 291 and 303 K. Experimental data: ◧, 10 nm, 293 K, Kim *et al.*;[48] ◨, 34 nm, 293 K, Kim *et al.*;[48] ⬒, 70 nm, 293 K, Kim *et al.*;[48] ○, 21 nm, 296.15 K, Turgut *et al.*;[49] ◐, 155 nm, 293.15 K, Longo *et al.*;[51] ⊗, 21 nm, 293.15 K, Yiamsawasd *et al.*;[53] ⬓, 21 nm, 303.15 K, Reddy *et al.*;[54] ◩, 30 nm, 303.15 K, Azmi *et al.*;[56] □, 30 to 50 nm, 298.15 K, Tertsinidou *et al.*;[58,59] ■, 5 nm, 298.15 K, Tertsinidou *et al.*;[58,59] ⊕, 40 to 60 nm, 298.15 K, Singh *et al.*;[62] ◪, 23 nm, 291 K, Wang *et al.*;[64] ●, 66 nm, 298.15 K, Bioucas *et al.*[43] Modeling results using data for the thermal conductivity of water[66] and the particles.[2] ---, HC model,[44] eqn (6.7); –, AOT-TP model,[43,74] eqn (6.9); -·-, model of Nan *et al.*[156] using $R_b = 0.77 \times 10^{-8}$ K m^2 W^{-1}, static term in eqn (6.8); —–, weighted average of experimental data.

As can be seen from Figure 6.11, no clear dependence of the thermal conductivity of TiO_2–water nanofluids on the size of the particles can be observed for the two different volume fractions studied. This is visualized by the matching of the individual experimental data points for $\lambda_{eff}/\lambda_{bf}$ with their uncertainty-weighted average value, where agreement mostly within the expanded uncertainties of the measurements is observed. The same conclusions on the size independence of λ_{eff} can also be drawn for other nanofluid systems, *e.g.*, by reviewing the experimental data analyzed by Tertsinidou *et al.*[2] It should be emphasized that also the aforementioned HC[44] and AOT-TP[43,74] models predict no variation in λ_{eff} values with changing particle size, as shown in Figure 6.11. While the HC model implies no size dependence, the AOT-TP model shows only a minor and insignificant increase in $\lambda_{eff}/\lambda_{bf}$ at diameters of below about 10 nm, arising from a slight increase in the *Nu* number, as shown in eqn (6.9).

To account for the effect of the interfacial layer around the particles, the model proposed by Prasher *et al.*[69,161] in eqn (6.8) was simplified to the original model of Nan *et al.*[156] as the dynamic contribution from the particle Brownian motion seems to be negligible according to the previous discussion. As is obvious from Figure 6.11, the resulting calculations from Nan's model based on an optimized interfacial resistance, $R_b = 0.77 \times 10^{-8}$ $K\,m^2\,W^{-1}$, reveal a decreasing trend of λ_{eff} with decreasing d_p, especially below around 40 nm. Furthermore, the model gives *per se* lower $\lambda_{eff}/\lambda_{bf}$ ratios than the HC model and is in good agreement with the AOT-TP model.[43,74] This comparison suggests that the empirical interfacial resistance introduced in the model proposed by Nan *et al.*[156] is already considered sufficiently by the convective heat transfer resistance in the analytical AOT-TP model. As a result, the interfacial layer at the particle–liquid interface of around 1 nm in thickness[158] reduces the λ_{eff}, which implies that this effect is often erroneously made responsible for anomalous enhancements in the thermal conductivity of nanofluids relative to that of the base fluid.

To investigate the behavior of nanofluids containing non-spherical particles, the measurement results of water-based nanofluids containing silicon carbide (SiC) nanoparticles obtained by Xie *et al.*[176] using the THW method at 277.15 K are shown in Figure 6.12 as an example. In addition to spherical particles with an average diameter of $d_p = 26$ nm and a sphericity $\psi = 1$, also cylindrical particles of $d_p = 600$ nm and $\psi = 0.5$ were studied. For given particle volume fractions up to about 0.04, the results show a larger $\lambda_{eff}/\lambda_{bf}$ ratio for the cylindrical particles than for the spherical particles. For comparison, predictions from the HC[44] and AOT-TP[43,74] models are also shown, in which the non-sphericity of the particles is reflected by n and ψ, respectively. Both models represent the influence of the particle shape on the λ_{eff} values of nanofluids, yet underestimate the experimental results, which is more pronounced in case of the AOT-TP model.

Due to the lack of systematic experimental studies on nanofluids featuring particles of different shape, the HC[44] and AOT-TP[43,74] models were applied to predict the $\lambda_{eff}/\lambda_{bf}$ ratios of water-based nanofluids with non-spherical

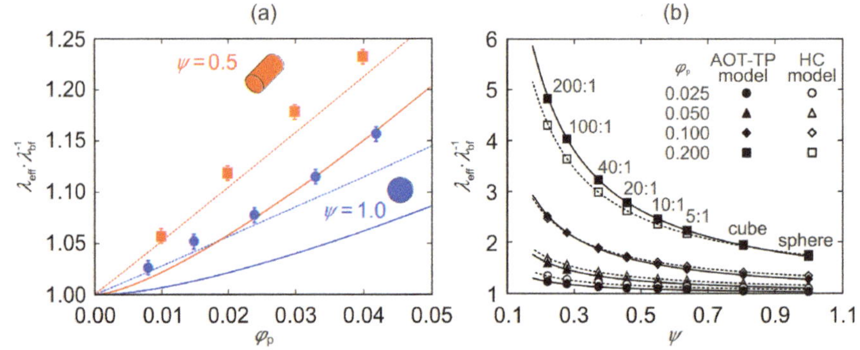

Figure 6.12 (a) Comparison of measurement results and models for the thermal conductivity ratio of water-based nanofluids containing SiC particles of varying shape at 277.15 K as a function of the particle volume fraction. Experimental data: ●, spheres ($d_p = 26$ nm, $\psi = 1$), Xie *et al.*;[176] ■, cylinders ($d_p = 600$ nm, $\psi = 0.5$), Xie *et al.*;[176] Modeling results: ---, HC model,[44] eqn (6.7); –, AOT-TP model,[43,74] eqn (6.9). (b) Predicted thermal conductivity ratio of water-based nanofluids containing spherical, cubic, or prolate TiO_2 particles as a function of the sphericity of the particles at different particle volume fractions at 298.15 K: ---, HC model,[44] eqn (6.7); –, AOT-TP model,[43,74] eqn (6.9).

TiO_2 nanoparticles at 298.15 K, as shown in Figure 6.12(b). Here, particles were considered that feature broadly varying values for the sphericity ψ and, thus, for the aspect ratio, *i.e.* the ratio of length L_p to the diameter d_p. Prolate cylindrical particles, which are shown in the figure only, exhibit aspect ratios larger than 1, while they are smaller for oblate particles. At varying φ_p values, both models are in good agreement in terms of their various particle geometries and show that a decreasing sphericity of the nanoparticles results in a distinct increase in the thermal conductivity of nanofluids, in particular at large particle volume fractions. It seems that the reduced convective heat transfer resistance caused by the higher specific surface area of non-spherical particles compared to spherical particles reasonably accounts for the increased λ_{eff} values in the AOT-TP model.

6.4.4 Particle Aggregation

So far, nanofluids containing fully dispersed particles have been addressed in cases, where a good dispersion stability is typically ensured by electrostatic and/or steric stabilization mechanisms. However, a fully dispersed state cannot be achieved experimentally as aggregation naturally occurs over time. Particle aggregation is often accompanied by sedimentation, especially for particles of larger density than that of the base fluid. For both cases where the particles sediment at the bottom of the fluid volume in the measurement instrument or in the measurement probe itself, the experimental results for the thermal conductivity will be inaccurate and, thus, do not correspond to the values for a dispersed nanofluid.[177] In this section, the focus is on the effect of particle aggregation on the thermal conductivity of nanofluids, which is supposed to have an enhancing effect on λ_{eff} values according to the prevalent opinion in the literature,[12,69,75,153] as addressed in Section 6.3.2.

Experimentally, the aggregation of particles can be induced by variations in the pH through the addition of electrolytes and/or in the surfactant concentration. As the volume fraction of these stabilizing agents is typically $<1\times10^{-2}$, their impact on the properties of nanofluids, including their thermal conductivity, can be neglected for both dispersed or aggregated states. Ismay *et al.*,[178] for example, studied relationships between pH and the thermal conductivity ratio $\lambda_{eff}/\lambda_{bf}$ for water-based nanofluids with rutile spherical TiO_2 particles of 21 nm in diameter at 298.15 K and $\varphi_p = 0.01$, as shown in Figure 6.13(a). Although the employed commercial needle instrument cannot be regarded as a fully reliable thermal conductivity measurement method, only relative trends for λ_{eff} are of interest here. For fairly stable nanofluid samples at sufficiently high ζ potentials of $|\zeta| > 20$ mV, the measured moderate $\lambda_{eff}/\lambda_{bf}$ ratios are in agreement with the HC[44] and AOT-TP[43,74] models, with values of 1.026 and 1.006, respectively. In the unstable region between pH = 5.5 and 7.5, where $|\zeta| < 20$ mV and the IEP of around pH ≈ 6.5 is included, a slight increase in λ_{eff} of around 2 to 3% relative to the values outside the pH region can be observed. This behavior correlates with the average size of the particle aggregates, which show a strong increase around the IEP with values up to about 4 μm.[178] From the experimental investigations conducted by Ismay *et al.*,[178] it was suggested that particle

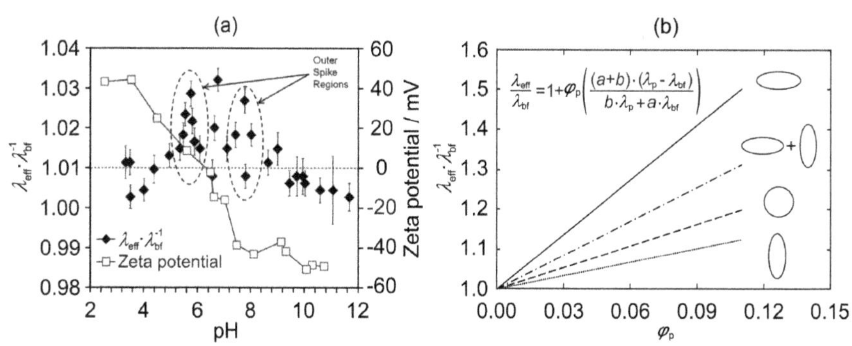

Figure 6.13 (a) Influence of pH on the ζ-potential and thermal conductivity ratio of water-based nanofluids with rutile TiO_2 particles at a particle volume fraction of 0.01 at 298.15 K according to the experiments of Ismay *et al.*[178] Figure adapted from Ismay *et al.*[178] with the permission from Elsevier. (b) Thermal conductivity ratio of water-based nanofluids with TiO_2 particles at 298.15 K as a function of particle volume fraction for aggregated spheres (aspect ratio 1 : 1) or ellipses (aspect ratio 5 : 1) of different orientations according to 2D model of Timofeeva *et al.*[12] (axes a along T-gradient and b normal to T-gradient): ---, sphere; –, ellipse parallel to T-gradient; ···, ellipse perpendicular to T-gradient; –·–, random distribution of ellipse orientations. Adapted from ref. 12 with permission from The American Physical Society, Copyright 2007.

aggregation leads to larger λ_{eff} values than without it, which is in agreement with the corresponding studies of Azizian *et al.*[179] and Wamkam *et al.*[180] The latter group attributed the effect caused by aggregation to particle chains with an irregular structure that favor thermal transport due to their extended network. Since the aggregation effect can hardly be quantified experimentally due to lacking information on the degree of agglomeration and the detailed structure of the individual agglomerates, modeling approaches can provide more robust explanations on this issue and shall be of interest in the following.

As the fractal configuration shown in Figure 6.7(c) represents an idealized aggregation state, nanofluids feature some degree of aggregation of single particles to clusters, which represents a mix between fully dispersed and fully interconnected fractal states. In this case, the prevalent opinion[12,75,181] is that particle aggregation in nanofluids leads to higher λ_{eff} values compared to those of fully dispersed systems. Among these models, the one proposed by Timofeeva *et al.*,[12] which is based on the classical effective medium theory of Maxwell,[152] is discussed here as an example of a case of two-dimensional (2D) heat transfer in water-based nanofluids containing TiO_2 particles in water at 298.15 K. Since the 2D case can be calculated easier but provides the same trends as the 3D case, it is employed here in Figure 6.13(b). In the equation for $\lambda_{eff}/\lambda_{bf}$ given in the figure, the particle aggregates are treated as ellipses with axes a and b along and normal to the temperature gradient ∇T, respectively. As can be seen, the increase in λ_{eff} is moderate for spherical particle aggregates ($a:b=1:1$), which corresponds to the HC model.[44] Compared to this situation, the λ_{eff} values are higher for extended elliptical aggregates ($a:b=5:1$) oriented parallel to ∇T compared to a perpendicular orientation, where the values for a and b are reversed. This is because in the parallel orientation, energy is transferred over longer distances in the solid particles with higher thermal conductivity than that of the base fluid. For randomly oriented ellipses that are statistically most likely in nanofluids, the λ_{eff} values as averages from the parallel and perpendicular orientation are higher than for spherical particle aggregates. As this random distribution of particles is related to enhancements in λ_{eff} that can be well fitted using the HC[44] or AOT-TP[43,74] models considering fully dispersed, non-spherical particles, it can be concluded that particle aggregates can be considered in modeling schemes as single non-spherical particles of larger dimension. Possible influences of an interfacial layer around such particle aggregates were not included in the present discussion, yet can be included accordingly,[12] as detailed in Section 6.4.3.

6.5 Summary

For more than two decades, after the extraordinary discoveries of Choi *et al.*,[3,4] research activities on nanofluids and, in particular, on their thermal conductivity, have undergone rapid development. To date, the various controversies on the behavior of the thermal conductivity of nanofluids have not been fully resolved. Probably the biggest advancement in this field, which seems to be propagating steadily throughout the scientific community and has been pointed out in this chapter, is the finding that there is no anomalous or significant change in the thermal conductivity of a fluid upon the addition of several volume percent of particles. At first glance, this may sound sobering, but it is notwithstanding highly relevant since accurate knowledge of the thermal conductivity of nanofluids is required for the design of processes and apparatus in various fields of applications, which range, *e.g.*, from classical heat engineering to biomedical technology.

Accurate measurements and predictions of the thermal conductivity of nanofluids are challenging tasks, yet indispensable to obtain a fundamental understanding of their properties. For reliable experiments, several requirements need to be considered with respect to the sample preparation and characterization and the measurement of the thermal conductivity itself. For the latter task, a detailed understanding of the experimental techniques applied is essential to obtain meaningful results, which is often lacking in available studies. Reliable results can be expected *via* proper application of the transient hot-wire and hot-disk techniques as well as of steady-state methods, including the PP method, all of which are based on physically sound working equations. In connection with modeling approaches, the origins were set by the classical effective medium theory developed by Maxwell,[152] which was further advanced and publicized by Hamilton and Crosser.[44] Mainly stimulated from often abnormally large enhancements in the thermal conductivity observed in experiments, numerous modeling approaches have been developed that repeatedly address effects due to Brownian motion, interfacial layering, aggregation, or an interplay of these. If reliable experimental data for the thermal conductivity of nanofluids are selected, it could be demonstrated that they are in agreement with experimental data to within around 10% using the classical effective medium theory of Hamilton and Crosser,[44] but also with the recently developed analytical AOT-TP model, which accounts for thermal resistance in nanofluids.[43,74]

As has been discussed on the basis of representative experimental and theoretical work, the thermal conductivity of nanofluids is mainly influenced by the thermal conductivities of the base fluid and the particles as well as the particle concentration and shape in terms of sphericity. The often-postulated distinct enhancement of the thermal conductivity of nanofluids in comparison to their base fluids due to Brownian motion could be debilitated. Moreover, a rigid consideration of the thermal resistance present in a nanofluid indicates further resistance that results from Brownian motion and is associated with convective heat transfer. This additional contribution reduces the thermal conductivity of the nanofluid in comparison to a configuration where no convective heat transfer is observed. Furthermore, the resistance decreases only slightly with an intensification in the Brownian motion caused by, *e.g.*, an increase in temperature. While the size of the dispersed particles does not affect the thermal conductivity distinctly, it has been rationalized that the presence of a thin interfacial layer at the particle surface may have a weakly reducing impact on the thermal conductivity. This effect seems to be already considered reasonably by the convective heat transfer resistance at the fluid–particle interface in the analytical resistance model. In the case of particle aggregation, an enhancement in the thermal conductivity compared to a completely dispersed nanofluid is observed from experiments and also anticipated from modeling approaches. The latter suggest that particle aggregates can be considered as single non-spherical particles of increased dimension, which are associated with a higher thermal conductivity of the nanofluid than the one featuring fully dispersed particles.

Acknowledgements

The authors acknowledge funding from the German Research Foundation (Deutsche Forschungsgemeinschaft, DFG) *via* the project grant FR 1709/20-1. This work was also funded by DFG – Project-ID 416229255 – SFB 1411 (project C03). Furthermore, the authors acknowledge funding from the Erlangen Graduate School in Advanced Optical Technologies (SAOT) by the Bavarian State Ministry for Science and Art. The authors also gratefully thank Annika Kraftzyk for proofreading this book chapter.

References

1. M. Buschmann, in *Topical Problems of Fluid Mechanics*, Prague, 2018, pp. 25–30.
2. G. Tertsinidou, M. J. Assael and W. A. Wakeham, *Int. J. Thermophys.*, 2015, **36**, 1367–1395.
3. S. U. S. Choi and J. A. Eastman, *Proc. ASME Int. Mech. Eng. Congr. Expo.*, 1995, **66**, 99–105.
4. S. U. S. Choi, Z. G. Zhang, W. Yu, F. E. Lockwood and E. A. Grulke, *Appl. Phys. Lett.*, 2001, **79**, 2252–2254.
5. Clarivate Web of Science, Anomalous thermal conductivity enhancement in nanotube suspensions, https://www.webofscience.com/wos/woscc/full-record/WOS:000171134900043 (accessed 15 September 2021).
6. Clarivate Web of Science, nanofluid* thermal conductivity, https://www.webofscience.com/wos/woscc/summary/1f650ce6-4de6-4d51-a0ae-42229daca487-0872ae2c/relevance/1 (accessed 1 March 2021).
7. KD2 Pro Thermal Properties Analyzer – Operator's Manual, Decagon Devices, Inc., 2016, 1–67.
8. D. Wen and Y. Ding, *Int. J. Heat Mass Transf.*, 2004, **47**, 5181–5188.
9. C. Jwo and T.-P. Teng, *Rev. Adv. Mater. Sci.*, 2005, **10**, 79–83.
10. Y. Ding, H. Alias, D. Wen and R. A. Williams, *Int. J. Heat Mass Transfer*, 2006, **49**, 240–250.
11. T. X. Phuoc, Y. Soong and M. K. Chyu, *Opt. Lasers Eng.*, 2007, **45**, 1099–1106.
12. E. V. Timofeeva, A. N. Gavrilov, J. M. McCloskey, Y. V. Tolmachev, S. Sprunt, L. M. Lopatina and J. V. Selinger, *Phys. Rev. E*, 2007, **76**, 061203.

13. Hot Disk|Technology, https://www.hotdiskinstruments.com/technology/ (accessed 15 September 2021).
14. Z. H. Liu, X. F. Yang and G. L. Guo, *J. Appl. Phys.*, 2007, **102**, 013526.
15. H. Hong, B. Wright, J. Wensel, S. Jin, X. R. Ye and W. Roy, *Synth. Met.*, 2007, **157**, 437–440.
16. H. Hong, Y. Zheng and W. Roy, *J. Nanosci. Nanotechnol.*, 2007, **7**, 3180–3184.
17. B. Wright, D. Thomas, H. Hong, L. Groven, J. Puszynski, E. Duke, X. Ye and S. Jin, *Appl. Phys. Lett.*, 2007, **91**, 173116.
18. *Handbook on Industrial Applications of Nanofluids in Energy Sector*, 2020, ed. M. Buschmann, L. Hernández López and L. Buj.
19. D. Huang, Z. Wu and B. Sunden, *Int. J. Heat Mass Transf.*, 2015, **89**, 620–626.
20. R. Barzegarian, M. K. Moraveji and A. Aloueyan, *Exp. Therm. Fluid Sci.*, 2016, **74**, 11–18.
21. R. Faridi Khouzestani and A. Ghafouri, *SN Appl. Sci.*, 2020, **2**, 298.
22. R. Sureshkumar, S. T. Mohideen and N. Nethaji, *Renewable Sustainable Energy Rev.*, 2013, **20**, 397–410.
23. M. Amjad, H. Jin, X. Du and D. Wen, *Sol. Energy Mater. Sol. Cells*, 2018, **182**, 255–262.
24. F. E. B. Bioucas, S. I. C. Vieira, M. J. V. Lourenço, F. J. V. Santos and C. A. Nieto de Castro, *Sol. Energy*, 2018, **172**, 171–176.
25. P. Krajnik, F. Pusavec and A. Rashid, *Nanofluids: Properties, Applications and Sustainability Aspects in Materials Processing Technologies*, Springer, Berlin, Heidelberg, 2011.
26. Y. Shokoohi and E. Shekarian, *J. Nanosci. Technol*, 2016, **2**, 59–63.
27. D. H. Fontes, G. Ribatski and E. P. Bandarra Filho, *Diam. Relat. Mater.*, 2015, **58**, 115–121.
28. M. Rafiq, M. Shafique, A. Azam and M. Ateeq, *Ain Shams Eng. J.*, 2021, **12**, 555–576.
29. M. H. Al-Rashed, G. Dzido, M. Korpyś, J. Smołka and J. Wójcik, *Microelectron. Reliab.*, 2016, **63**, 159–165.
30. G. Colangelo, E. Favale, M. Milanese, A. de Risi and D. Laforgia, *Appl. Therm. Eng.*, 2017, **127**, 421–435.
31. D. L. C. López, R. R. Méndez and E. Berjano, in *Memorias Del XXIV Congreso Internacional Anual de la Somim*, México, 2018.
32. Q. Wu, H. Zhang, M. Chen, Y. Zhang, J. Huang, Z. Xu and W. Wang, *J. Biomed. Mater. Res., Part B*, 2015, **103**, 908–914.
33. V. Mikkola, S. Puupponen, H. Granbohm, K. Saari, T. Ala-Nissila and A. Seppala, in *12th International Conference on Heat Transfer, Fluid Mechanics and Thermodynamics*, Spain, 2016.
34. A. Bhattad, J. Sarkar and P. Ghosh, *Renewable Sustainable Energy Rev.*, 2018, **82**, 3656–3669.
35. M. H. Buschmann, R. Azizian, T. Kempe, J. E. Juliá, R. Martínez-Cuenca, B. Sundén, Z. Wu, A. Seppälä and T. Ala-Nissila, *Int. J. Therm. Sci.*, 2018, **129**, 504–531.

36. Z. Ashikbayeva, D. Tosi, D. Balmassov, E. Schena, P. Saccomandi and V. Inglezakis, *J. Nanomater.*, 2019, **9**, 1195.
37. R. L. Manthe, S. P. Foy, N. Krishnamurthy, B. Sharma and V. Labhasetwar, *Mol. Pharmaceutics*, 2010, **7**, 1880–1898.
38. G. S. Gazelle, S. N. Goldberg, L. Solbiati and T. Livraghi, *J. Radiol.*, 2000, **217**, 633–646.
39. T. Livraghi, S. N. Goldberg, F. Monti, A. Bizzini, S. Lazzaroni, F. Meloni, S. Pellicanò, L. Solbiati and G. S. Gazelle, *J. Radiol.*, 1997, **202**, 205–210.
40. S. N. Goldberg, J. B. Kruskal, B. S. Oliver, M. E. Clouse and G. S. Gazelle, *J. Radiol.*, 2000, **217**, 827–831.
41. J. Buongiorno, *et al.*, *J. Appl. Phys.*, 2009, **106**, 094312.
42. M. H. Buschmann, *Int. J. Therm. Sci.*, 2012, **62**, 19–28.
43. F. E. Berger Bioucas, M. H. Rausch, J. Schmidt, A. Bück, T. M. Koller and A. P. Fröba, *Int. J. Thermophys.*, 2020, **41**, 55.
44. R. L. Hamilton and O. K. Crosser, *Ind. Eng. Chem. Fundam.*, 1962, **1**, 187–191.
45. X. Zhang, H. Gu and M. Fujii, *Int. J. Thermophys.*, 2006, **27**, 569–580.
46. S. M. S. Murshed, K. C. Leong and C. Yang, *Int. J. Therm. Sci.*, 2005, **44**, 367–373.
47. D.-H. Yoo, K. S. Hong and H.-S. Yang, *Thermochim. Acta*, 2007, **455**, 66–69.
48. S. H. Kim, S. R. Choi and D. Kim, *J. Heat Transfer*, 2007, **129**, 298–307.
49. A. Turgut, I. Tavman, M. Chirtoc, H. P. Schuchmann, C. Sauter and S. Tavman, *Int. J. Thermophys.*, 2009, **30**, 1213–1226.
50. W. Duangthongsuk and S. Wongwises, *Exp. Therm. Fluid Sci.*, 2009, **33**, 706–714.
51. G. A. Longo and C. Zilio, *Exp. Therm. Fluid Sci.*, 2011, **35**, 1313–1324.
52. L. Fedele, L. Colla and S. Bobbo, *Int. J. Refrig.*, 2012, **35**, 1359–1366.
53. T. Yiamsawasd, A. S. Dalkilic and S. Wongwises, *Thermochim. Acta*, 2012, **545**, 48–56.
54. M. C. S. Reddy and V. V. Rao, *Int. Commun. Heat Mass Transf.*, 2013, **46**, 31–36.
55. E. B. Haghighi, M. Saleemi, N. Nikkam, R. Khodabandeh, M. S. Toprak, M. Muhammed and B. Palm, *Int. Commun. Heat Mass Transf.*, 2014, **52**, 1–7.
56. W. H. Azmi, K. V. Sharma, P. K. Sarma, R. Mamat and S. Anuar, *Int. J. Therm. Sci.*, 2014, **81**, 84–93.
57. N. Kumar and S. S. Sonawane, *Int. Commun. Heat Mass Transf.*, 2016, **76**, 98–107.
58. G. J. Tertsinidou, C. M. Tsolakidou, M. Pantzali, M. J. Assael, L. Colla, L. Fedele, S. Bobbo and W. A. Wakeham, *J. Chem. Eng. Data*, 2017, **62**, 491–507.
59. G. J. Tertsinidou, C. M. Tsolakidou, M. Pantzali, M. J. Assael, L. Colla, L. Fedele, S. Bobbo and W. A. Wakeham, *J. Chem. Eng. Data*, 2018, **63**, 4277–4279.
60. P. B. Maheshwary, C. C. Handa and K. R. Nemade, *Appl. Therm. Eng.*, 2017, **119**, 79–88.

61. L. R. Oliveira, A. C. A. Silva, N. O. Dantas and E. P. Bandarra Filho, *Int. J. Heat Mass Transf.*, 2017, **115**, 795–808.
62. R. P. Singh, K. Sharma and A. K. Tiwari, *J. Sci. Ind. Res.*, 2019, **78**, 620–623.
63. D. Mitra, P. Howli, B. K. Das, N. S. Das, P. Chattopadhyay and K. K. Chattopadhyay, *J. Mol. Liq.*, 2020, **302**, 112499.
64. Z. L. Wang, D. W. Tang, S. Liu, X. H. Zheng and N. Araki, *Int. J. Thermophys.*, 2007, **28**, 1255–1268.
65. A. Kazemi-Beydokhti, S. Z. Heris, N. Moghadam, M. Shariati-Niasar and A. A. Hamidi, *Chem. Eng. Commun.*, 2014, **201**, 593–611.
66. M. L. Huber, R. A. Perkins, D. G. Friend, J. V. Sengers, M. J. Assael, I. N. Metaxa, K. Miyagawa, R. Hellmann and E. Vogel, *J. Phys. Chem. Ref. Data*, 2012, **41**, 033102.
67. S. P. Jang and S. U. S. Choi, *Appl. Phys. Lett.*, 2004, **84**, 4316–4318.
68. J. Koo and C. Kleinstreuer, *J. Nanopart. Res.*, 2004, **6**, 577–588.
69. R. Prasher, P. Bhattacharya and P. E. Phelan, *J. Heat Transfer*, 2006, **128**, 588–595.
70. D. Song, D. Jing, W. Ma and X. Zhang, *Appl. Phys.*, 2019, **9**, 015103.
71. J. W. Gao, R. T. Zheng, H. Ohtani, D. S. Zhu and G. Chen, *Nano Lett.*, 2009, **9**, 4128–4132.
72. W. Yu and S. U. S. Choi, *J. Nanopart. Res.*, 2004, **6**, 355–361.
73. H. U. Kang, S. H. Kim and J. M. Oh, *Exp. Heat Transfer*, 2006, **19**, 181–191.
74. K. N. Shukla, T. M. Koller, M. H. Rausch and A. P. Fröba, *Int. J. Heat Mass Transf.*, 2016, **99**, 532–540.
75. J. Eapen, R. Rusconi, R. Piazza and S. Yip, *J. Heat Transf.*, 2010, **132**, 102402.
76. H. Zhu, Y. Lin and Y. Yin, *J. Colloid Interf. Sci.*, 2004, **277**, 100–103.
77. H. Bonnemann, S. S. Botha, B. Bladergroen and V. M. Linkov, *Appl. Organomet. Chem.*, 2005, **19**, 768–773.
78. M. S. Liu, M. C. C. Lin, I. T. Huang and C. C. Wang, *Int. Commun. Heat Mass Transf.*, 2005, **32**, 1202–1210.
79. D. Wen and Y. Ding, *IEEE Trans. Nanotechnol.*, 2006, **5**, 220–227.
80. M. Wan, R. R. Yadav, K. L. Yadav and S. B. Yadaw, *Exp. Therm. Fluid Sci.*, 2012, **41**, 158–164.
81. Y. Hwang, J.-K. Lee, J.-K. Lee, Y.-M. Jeong, S. Cheong, Y.-C. Ahn and S. H. Kim, *Powder Technol.*, 2008, **186**, 145–153.
82. L. Fedele, L. Colla, S. Bobbo, S. Barison and F. Agresti, *Nanoscale Res. Lett.*, 2011, **6**, 300.
83. AFA *vs.* Sonicators – Covaris, http://covaris.com/pre-analytical/afa-vs-sonicators/ (accessed 28 July 2021).
84. R. Arrigo, R. Teresi, C. Gambarotti, F. Parisi, G. Lazzara and N. Dintcheva, *Materials*, 2018, **11**, 383.
85. B. Pal, S. S. Mallick and B. Pal, *J. Nanosci. Nanotechnol.*, 2015, **15**, 3670–3676.
86. S. Koda, H. Mori, K. Matsumoto and H. Nomura, *Polymer*, 1994, **35**, 30–33.

87. M. T. Taghizadeh and T. Asadpour, *Ultrason. Sonochem.*, 2009, **16**, 280–286.
88. M. H. Dehghani, A. Zarei and M. Yousefi, *MethodsX*, 2019, **6**, 805–814.
89. *Colloid stability: the role of surface forces*, ed. T. F. Tadros, Wiley-VCH Verlag, Weinheim, Germany, 2007, vol. 1.
90. T. Tadros, in *Encyclopedia of Colloid and Interface Science*, ed. T. Tadros, Springer, Berlin, Heidelberg, 2013, pp. 1048–1049.
91. S. Iyahraja and J. S. Rajadurai, *Arabian J. Sci. Eng.*, 2016, **41**, 653–660.
92. F. Yu, Y. Chen, X. Liang, J. Xu, C. Lee, Q. Liang, P. Tao and T. Deng, *Prog. Nat. Sci.: Mater. Int.*, 2017, **27**, 531–542.
93. A. Ehle, S. Feja and M. H. Buschmann, *J. Thermophys. Heat Trans.*, 2011, **25**, 378–385.
94. R. Sadeghi, S. G. Etemad, E. Keshavarzi and M. Haghshenasfard, *Microfluid. Nanofluid.*, 2015, **18**, 1023–1030.
95. T. R. Ray, B. Lettiere, J. de Rutte and S. Pennathur, *Langmuir*, 2015, **31**, 3577–3586.
96. D.-E. A. Mansour and E. G. Atiya, in *Conference on Electrical Insulation and Dielectric Phenomena (CEIDP)*, Canada, 2016.
97. N. Ali, J. A. Teixeira and A. Addali, *J. Nanomater.*, 2018, **2018**, 1–33.
98. T. Luxbacher, *The Zeta Potential for Solid Surface Analysis - A practical guide to streaming potential measurement*, Anton-Paar GmbH, Austria, 1st edn, 2014.
99. S. Bhattacharjee, *J. Controlled Release*, 2016, **235**, 337–351.
100. D. Langevin, *et al.*, *NanoImpact*, 2018, **10**, 97–107.
101. B. J. Inkson, in *Materials Characterization Using Nondestructive Evaluation (NDE) Methods*, Woodhead Publishing, Cambridge, United Kingdom, 1st edn, 2016, pp. 17–43.
102. J. I. Goldstein, D. E. Newbury, P. Echlin, D. C. Joy, A. D. Romig, C. E. Lyman, C. Fiori and E. Lifshin, in *Scanning Electron Microscopy and X-Ray Microanalysis: A Text for Biologists, Materials Scientists, and Geologists*, Springer, Boston, 2nd edn, 1992, pp. 671–740.
103. B. J. Berne and R. Pecora, *Dynamic Light Scattering: With Applications to Chemistry, Biology, and Physics*, Dover Publications, 2000.
104. F. E. Berger Bioucas, C. Damm, W. Peukert, M. H. Rausch, T. M. Koller, C. Giraudet and A. P. Fröba, *J. Phys. Chem. B*, 2019, **123**, 9491–9502.
105. S. Bi, T. M. Koller, M. H. Rausch, P. Wasserscheid and A. P. Fröba, *Ind. Eng. Chem. Res.*, 2015, **54**, 3071–3081.
106. M. S. G. Knoll, C. Giraudet, C. J. Hahn, M. H. Rausch and A. P. Fröba, *J. Colloid Interf. Sci.*, 2019, **544**, 144–154.
107. S. W. Provencher, *Comput. Phys. Commun.*, 1982, **27**, 213–227.
108. A. G. Mailer, P. S. Clegg and P. N. Pusey, *J. Phys.: Condens. Matter*, 2015, **27**, 145102.
109. *Measurement of the transport properties of fluids*, ed. W. A. Wakeham, A. Nagashima and J. V. Sengers, International Union of Pure and Applied Chemistry, Blackwell Scientific Publications, 1991, vol. III, pp. 111–194.

110. P. Bhattacharya, S. Nara, P. Vijayan, T. Tang, W. Lai, P. E. Phelan, R. S. Prasher, D. W. Song and J. Wang, in *Heat Transfer*, ASMEDC, Anaheim, California, USA, 2004, vol. 2, pp. 51–56.
111. S. K. Das, N. Putra, P. Thiesen and W. Roetzel, *J. Heat Transfer*, 2003, **125**, 567–574.
112. H. E. Patel, T. Sundararajan and S. K. Das, *J. Nanopart. Res.*, 2010, **12**, 1015–1031.
113. D.-W. Oh, A. Jain, J. K. Eaton, K. E. Goodson and J. S. Lee, *Int. J. Heat Fluid Flow*, 2008, **29**, 1456–1461.
114. K. D. Antoniadis, G. J. Tertsinidou, M. J. Assael and W. A. Wakeham, *Int. J. Thermophys.*, 2016, **37**, 78.
115. P. G. Knibbe, *Int. J. Heat Mass Transfer*, 1986, **29**, 463–473.
116. M. J. Assael, E. Charitidou and C. A. Nieto de Castro, *Int. J. Thermophys.*, 1988, **9**, 813–824.
117. S. M. S. Murshed, K. C. Leong and C. Yang, *Int. J. Therm. Sci.*, 2008, **47**, 560–568.
118. Y. Hwang, H. S. Park, J. K. Lee and W. H. Jung, *Curr. Appl. Phys.*, 2006, **6**, e67–e71.
119. S. Lee, S. U.-S. Choi, S. Li and J. A. Eastman, *J. Heat Transfer*, 1999, **121**, 280–289.
120. C. H. Chon, K. D. Kihm, S. P. Lee and S. U. S. Choi, *Appl. Phys. Lett.*, 2005, **87**, 153107.
121. *ISO 22007-2, 2008: Plastics – Determination of thermal conductivity and thermal diffusivity – Part 2: Transient plane heat source (hot disc) method*, 1st edn, 2008.
122. R. J. Warzoha and A. S. Fleischer, *Int. J. Heat Mass Transf.*, 2014, **71**, 779–789.
123. R. J. Warzoha and A. S. Fleischer, *Int. J. Heat Mass Transf.*, 2014, **71**, 790–807.
124. Y. He, *Thermochim. Acta*, 2005, **436**, 122–129.
125. Q. Zheng, S. Kaur, C. Dames and R. S. Prasher, *Int. J. Heat Mass Transf.*, 2020, **151**, 119331.
126. B. Ma, N. Kumar, A. Kuchibhotla and D. Banerjee, in *2018 17th IEEE Intersociety Conference on Thermal and Thermomechanical Phenomena in Electronic Systems (ITherm)*, 2018.
127. C. W. Maranville, H. Ohtani, D. D. Sawall, J. T. Remillard, and J. M. Ginder, in *SAE 2006 World Congress & Exhibition*, United States, 2006.
128. J. Glory, M. Bonetti, M. Helezen, M. Mayne-L'Hermite and C. Reynaud, *J. Appl. Phys.*, 2008, **103**, 094309.
129. L. S. Sundar and K. V. Sharma, *Int. J. Nanopart.*, 2008, **1**, 66.
130. R. S. Vajjha and D. K. Das, *Int. J. Heat Mass Transf.*, 2009, **52**, 4675–4682.
131. M. T. Naik and G. R. Janardhana, *Int. J. Nanopart.*, 2010, **3**, 149.
132. B. Barbés, R. Páramo, E. Blanco, M. J. Pastoriza-Gallego, M. M. Piñeiro, J. L. Legido and C. Casanova, *J. Therm. Anal. Calorim.*, 2013, **111**, 1615–1625.

133. B. Barbés, R. Páramo, E. Blanco and C. Casanova, *J. Therm. Anal. Calorim.*, 2014, **115**, 1883–1891.
134. N. Shalkevich, W. Escher, T. Bürgi, B. Michel, L. Si-Ahmed and D. Poulikakos, *Langmuir*, 2010, **26**, 663–670.
135. X. Wang, X. Xu and S. U. S. Choi, *J. Thermophys. Heat Trans.*, 1999, **13**, 474–480.
136. C. H. Li and G. P. Peterson, *J. Appl. Phys.*, 2006, **99**, 084314.
137. C. H. Li and G. P. Peterson, *J. Appl. Phys.*, 2007, **101**, 044312.
138. C. H. Li, W. Williams, J. Buongiorno, L.-W. Hu and G. P. Peterson, *J. Heat Transfer*, 2008, **130**, 042407.
139. A. R. Challoner and R. W. Powell, *Proc. R. Soc. London, Ser. A*, 1956, **238**, 90–106.
140. M. H. Rausch, K. Krzeminski, A. Leipertz and A. P. Fröba, *Int. J. Heat Mass Transf.*, 2013, **58**, 610–618.
141. R. Mostert, *J. Chem. Phys.*, 1990, **92**, 5454.
142. M. Ya, M. M. Naziev, Bashirov and I. M. Abdulagatov, *Fluid Phase Equilib.*, 2004, **226**, 221–235.
143. F. Croccolo, L. García-Fernández, H. Bataller, A. Vailati and J. M. Ortiz de Zárate, *Phys. Rev. E*, 2019, **99**, 012602.
144. A. Michels, J. V. Sengers and P. S. Van der Gulik, *Physica*, 1962, **28**, 1216–1237.
145. H. A. Pakravan and M. Yaghoubi, *Int. J. Therm. Sci.*, 2011, **50**, 394–402.
146. R. Braun, S. Fischer and A. Schaber, *Wärme-und Stoffübertragung*, 1983, **17**, 121–124.
147. A. P. Fröba, M. H. Rausch, K. Krzeminski, D. Assenbaum, P. Wasserscheid and A. Leipertz, *Int. J. Thermophys.*, 2010, **31**, 2059–2077.
148. T. M. Koller, S. R. Schmid, S. J. Sachnov, M. H. Rausch, P. Wasserscheid and A. P. Fröba, *Int. J. Thermophys.*, 2014, **35**, 195–217.
149. V. Velagapudi, K. Konijeti and K. Aduru, *Therm. Sci.*, 2008, **12**, 27–37.
150. M. Corcione, *Energy Convers. Manage.*, 2011, **52**, 789–793.
151. F. Garoosi, *Powder Technol.*, 2020, **366**, 788–820.
152. J. C. Maxwell, in *A Treatise on Electricity and Magnetism*, Clarendon Press Series, Oxford, 2nd edn, 1873, vol. 1, pp. 360–365.
153. D. A. G. Bruggeman, *Ann. Phys.*, 1935, **416**, 636–664.
154. Q.-Z. Xue, *Phys. Lett. A*, 2003, **307**, 313–317.
155. W. Yu and S. U. S. Choi, *J. Nanopart. Res.*, 2003, **5**, 167–171.
156. C.-W. Nan, R. Birringer, D. R. Clarke and H. Gleiter, *J. Appl. Phys.*, 1997, **81**, 6692–6699.
157. Z. Liang and H.-L. Tsai, *Phys. Rev. E*, 2011, **83**, 041602.
158. X. Wang and D. Jing, *Int. J. Heat Mass Transf.*, 2019, **128**, 199–207.
159. M. F. Reedijk, J. Arsic, F. F. A. Hollander, S. A. de Vries and E. Vlieg, *Phys. Rev. Lett.*, 2003, **90**, 066103.
160. J. Eapen, J. Li and S. Yip, *Phys. Rev. E.*, 2007, **76**, 062501.
161. R. Prasher, P. Bhattacharya and P. E. Phelan, *Phys. Rev. Lett.*, 2005, **94**, 025901.

162. S. M. S. Murshed, in *International Conference on Mechanical Engineering*, London, 2011, p. 5.
163. Y. Ren, H. Xie and A. Cai, *J. Phys. D: Appl. Phys.*, 2005, **38**, 3958–3961.
164. P. Keblinski, S. R. Phillpot, S. U. S. Choi and J. A. Eastman, *Int. J. Heat Mass Transf.*, 2002, **45**, 855–863.
165. W. Evans, J. Fish and P. Keblinski, *Appl. Phys. Lett.*, 2006, **88**, 093116.
166. M. E. Meibodi, M. Vafaie-Sefti, A. M. Rashidi, A. Amrollahi, M. Tabasi and H. S. Kalal, *Int. Commun. Heat Mass Transf.*, 2010, **37**, 555–559.
167. W. Qu and J. Feng, in *2010 IEEE 5th International Conference on Nano/Micro Engineered and Molecular Systems*, IEEE, Xiamen, 2010.
168. J. P. Holman, T. W. Moore and V. M. Wong, *Ind. Eng. Chem. Fundam.*, 1965, **4**, 21–31.
169. D. R. Lide, *CRC Handbook of Chemistry and Physics, Internet Version 2005*, CRC Press, Boca Raton, FL, 2005.
170. J. E. Mark, *Polymer Data Handbook*, Oxford University Press, Inc., Oxford University Press, Inc., New York, 1999.
171. H. Xie, J. Wang, T. Xi, Y. Liu, F. Ai and Q. Wu, *J. Appl. Phys.*, 2002, **91**, 4568–4572.
172. S. Bhanushali, N. N. Jason, P. Ghosh, A. Ganesh, G. P. Simon and W. Cheng, *ACS Appl. Mater. Interfaces*, 2017, **9**, 18925–18935.
173. J. Jeong, C. Li, Y. Kwon, J. Lee, S. H. Kim and R. Yun, *Int. J. Refrig.*, 2013, **36**, 2233–2241.
174. H. Jiang, Q. Zhang and L. Shi, *J. Taiwan Inst. Chem. Eng.*, 2015, **55**, 76–81.
175. H. Xie, H. Lee, W. Youn and M. Choi, *J. Appl. Phys.*, 2003, **94**, 4967–4971.
176. H. Xie, J. Wang, T. Xi and Y. Liu, *Int. J. Thermophys.*, 2002, **23**, 571–580.
177. E. B. Haghighi, *et al.*, *Meas. Sci. Technol.*, 2013, **24**, 105301.
178. M. J. L. Ismay, E. Doroodchi and B. Moghtaderi, *Chem. Eng. Res. Des.*, 2013, **91**, 426–436.
179. R. Azizian, E. Doroodchi and B. Moghtaderi, *J. Heat Transfer*, 2016, **138**, 021301.
180. C. T. Wamkam, M. K. Opoku, H. Hong and P. Smith, *J. Appl. Phys.*, 2011, **109**, 024305.
181. P. Bhattacharya, S. Nara, P. Vijayan, T. Tang, W. Lai, P. E. Phelan, R. S. Prasher, D. W. Song and J. Wang, *Int. J. Heat Mass Transf.*, 2006, **49**, 2950–2956.

CHAPTER 7

Modelling and Mechanisms of Nanofluid Thermal Conductivity

R. FULMER AND S. VAFAEI*

Mechanical Engineering Department, Bradley University, Peoria, IL 61606, USA
*Email: svafaei@mail.bradley.edu

7.1 Introduction

Nanofluids are synthesis has increased due to the need for thermal management in modern technology. Traditional cooling fluids are unable to provide the necessary cooling of devices due to their low thermal conductivity. Therefore, particles with high thermal conductivity are added to a fluid to increase the thermal conductivity of the mixture.[1] The nanoparticles used to create nanofluids can be metals, metal oxides, or carbon-based materials such as diamond, graphene, carbon nanotubes, *etc.*[2] Nanoparticles are generally <100 nm in size, but sometimes can form larger clusters in nanofluids.[3] Sometimes a surfactant can be added to the nanofluid to prevent clustering and improve dispersion of the nanoparticles. The most common base fluids used to create nanofluids are water, ethylene glycol (EG), propylene glycol, and various oils.[4] The increased thermal conductivity of these mixtures leads to improved thermal management. More detail on this can be found in ref. 5.

There have been several proposed mechanisms on the increase in thermal conductivity of a nanofluid mixture. Increased thermal diffusion from highly conductivity particles, was initially thought to be the mechanism for thermal conductivity enhancement in a nanofluid, leading to greater overall thermal diffusion in the mixture.[6] Normally, nanoparticles do not remain stationary in a nanofluid, there is random motion of particles due to Brownian

Soft Matter Series No. 16
Fundamentals and Transport Properties of Nanofluids
Edited by S. M. Sohel Murshed

Published by the Royal Society of Chemistry, www.rsc.org

motion. This Brownian motion can lead to microconvection in the base fluid, which increases thermal conductivity.[7] In a nanofluid mixture, the base fluid forms a dense layer of liquid molecules around the nanoparticles. This layer is a few liquid molecules thick, and has higher thermal conductivity than the bulk liquid. This higher thermal conductivity layer increases thermal conductivity.[8] Additionally, clustering of the nanoparticles can affect the thermal conductivity of a nanofluid. The contact between particles in the clusters can promote heat transfer within the clusters.[9] There is controversy about which of these mechanisms are most dominant in the thermal conductivity of nanofluids and which mechanisms must be considered to predict thermal conductivity.

The first attempt to calculate the thermal conductivity of a solid–liquid mixture was the effective medium theory (EMT) proposed by Maxwell[10] and further expanded on by Hamilton and Crosser[11] and Timofeeva *et al.*[12] The EMT laid the groundwork for further investigation into nanofluid thermal conductivity, but fails to consider several important parameters, such as particle size, temperature, and particle motion. EMT also fails to consider the interaction between nanoparticles. To improve thermal conductivity modelling, Prasher *et al.*[13] used the work on the EMT to create a model that also considers the effect of the larger clusters that are formed in a nanofluid as a result of nanoparticle interactions. Xuan *et al.*[14] and Feng *et al.*[15] also considered nanoparticle aggregation along with Brownian motion and the nanolayer, respectively. Aggregation models consider additional mechanisms of heat transfer in a nanofluid. It is difficult to quantify the aggregation of nanoparticles and typically the information is not readily available, making the models difficult to use in some cases. Corcione,[16] Chon *et al.*,[17] Koo and Kleinstreuer,[18] Vajjha and Das,[19] and Patel *et al.*[20] all considered the effect of Brownian motion. This parameter is important since it is a primary mechanism of thermal conductivity enhancement. Many of these models rely on empirically determined constants that possibly limit the range of validity. Leong *et al.*,[21] Xie *et al.*,[22] and Tinga *et al.*[23] created models considering the effects of the nanolayer. The nanolayer effectively increases the concentration of the nanoparticles by creating an additional layer of thermally conductive material. The properties of the nanolayer are not well understood, so several approximations are needed in these models for the nanolayer properties. An additional way to study the mechanisms of heat transfer in a nanofluid is using molecular dynamics simulations. Evans *et al.*,[24] Vladkov and Barrat,[25] Lin *et al.*,[26] and Mirmohammadi *et al.*[27] determined nanofluid thermal conductivity models using molecular dynamics. These simulations are beneficial to understanding the solid and fluid interactions in a way that cannot be achieved experimentally. Since the mechanisms of nanofluid heat transfer are very complex, many current researchers are determining models based on correlation with experimental data. Moghaddam *et al.*,[28] Çolak[29] and, Kamel *et al.*[30] all determined correlations for nanofluid thermal conductivity from experimental data. Since these models are based on experimental data this guarantees the model is reflective of the observed behavior of the nanofluids. The models may not be accurate for nanofluid conditions that were not considered in determining the model.

There have been many proposed theoretical models for nanofluid thermal conductivity based on these mechanisms. These models are created based on the laws of heat transfer along with results from experiments and molecular dynamics simulations.[31] None of the proposed models have shown accuracy in all cases based on experimental results. The accuracy of the models can vary greatly based on the data set used in the comparison.[8] In order to explain the mechanisms of thermal conductivity enhancement in nanofluids, theoretical models considering each of the mechanisms and models from molecular dynamics simulations will be explained in this chapter. The models will also be compared to experimental data to evaluate the accuracy of each model based on the average absolute error. This will allow for the accuracy of the models to be evaluated.

7.2 Effective Medium Theory (EMT)

The EMT represents a basic attempt to model the thermal conductivity of a nanofluid and predict the effects of nanoparticle concentration on the thermal conductivity of the base fluid. The EMT was originally developed by Maxwell[10] in 1873. Maxwell[10] was considering electrical conductivity in a heterogeneous mixture when developing the model, but the model has since also been found to be applicable to thermal conductivity. It has been determined that thermal conductivity behavior is analogous electrical conductivity in many cases. Maxwell[10] calculated the flow of electrical current in a mixture and based on the premise that the current must be continuous where the two elements of the mixture meet. Maxwell[10] solved for the continuity equation at this interface between the two elements in the mixture. Based on this continuity equation, the conductivity of a mixture was solved for several types of mixtures based on the shape of the components. The Maxwell model[10] neglects the interaction of particles with molecules and each other. It is assumed that small spherical particles on the microscale range are dispersed in a base fluid uniformly, in a way that they are far enough apart not to interact with one another. These assumptions are only valid when the volume fraction of particles is low and the overall particle shape is spherical. Micro and milli sized particles are not commonly used in thermal management applications nowadays since they can cause clogging of micro channels. Practically speaking, Maxwell assumed a homogenous distribution of static, spherical, and non-interacting particles in a base fluid. Maxwell's model[10] can be used to calculate the thermal conductivity of a solid–liquid mixture, k_{nf}, based on the thermal conductivity of the solid particles, k_p, the thermal conductivity of the liquid, k_{bf}, and the volume fraction of the solid particles, ϕ. The Maxwell model[10] is given as:

$$k_{nf} = \frac{k_p + 2k_{bf} + 2\phi(k_p - k_{bf})}{k_p + 2k_{bf} - \phi(k_p - k_{bf})} k_{bf} \tag{7.1}$$

Even though Maxwell's model is a very simplistic approach by which to predict the thermal conductivity of nanofluids, it provides a solid basis for better theoretical models. Hamilton and Crosser[11] attempted to improve the prediction of nanofluid thermal conductivity of the Maxwell model by developing a new EMT model that considers the shape of solid particles. The effect of shape can be calculated by adding an empirical shape factor, n, to the model. The empirical shape factor takes into consideration the sphericity, ψ, of the solid particles. The sphericity is defined as the ratio of the surface area of a sphere and the surface area of the actual particle if the sphere and particle have an equal volume. The empirical shape factor can be calculated from:

$$n = \frac{3}{\psi} \tag{7.2}$$

Typically, n is 3 for spherical particles and 6 for cylindrical particles. As the sphericity of the particle decreases, which corresponds to a particle with a higher surface area to volume ratio, the shape factor increases. The effects of nanoparticle shape become important when the thermal conductivity of particles is much higher than that of the base fluid. It has been proposed that the effects of nanoparticle shape can be important if the thermal conductivity of the particles is round 100 times greater than that of the base fluid. The effects of nanoparticle shape decrease as the ratio of the thermal conductivity of nanoparticles to that of the base fluid decreases.[11] A model considering the shape factor developed by Hamilton and Crosser[11] is given as:

$$k_{\mathrm{nf}} = k_{\mathrm{bf}} \frac{k_{\mathrm{p}} + (n-1)k_{\mathrm{bf}} - (n-1)(k_{\mathrm{bf}} - k_{\mathrm{p}})\phi}{k_{\mathrm{p}} + (n-1)k_{\mathrm{bf}} + (k_{\mathrm{bf}} - k_{\mathrm{p}})\phi} \tag{7.3}$$

In the case of spherical particles, the Hamilton and Crosser model[11] becomes the same as the Maxwell model.[10] The predicted thermal conductivity of nanofluids by Hamilton and Crosser[11] increases as the shape factor increases. Therefore, it is predicted that reducing the sphericity of the particles will increase the thermal conductivity of the nanofluids. This implies that the thermal conductivity of nanofluids increases as the ratio of the surface area to volume of the nanoparticles increases. Timofeeva *et al.*[12] proposed that the EMT is a valid method for predicting the thermal conductivity of nanofluids because the effect of thermal diffusion through a solid–liquid mixture has a greater impact on the enhancement of the thermal conductivity of nanofluids than any other effects, including the effects of particle motion. It was estimated that the change in thermal conductivity in an Al_2O_3–water nanofluid predicted using the EMT was four orders of magnitude higher than the change in thermal conductivity due to particle motion. Based on this result, it was assumed that particle motion exhibits negligible effect on thermal conductivity in nanofluids. In their determination of a thermal conductivity model Timofeeva *et al.*[12] started from the work of Maxwell[10] and sought to simplify the proposed model. This simplification was done by assuming that the thermal conductivity of nanoparticles is significantly higher than that of the base fluid, which leads to eqn (7.4), which is a simplified version of the Maxwell model:[10]

$$k_{nf} = k_{bf}(1 + 3\phi) \tag{7.4}$$

In this model, it is assumed that the thermal conductivity of the nanoparticles is infinite compared to that of the base fluid. As a result, the thermal conductivity of the nanoparticles disappears. This model is valid when the thermal conductivity of nanoparticles is much greater than that of the base fluid, which is generally true in nanofluids. Under this assumption, the model proposed by Timofeeva *et al.*[12] cannot account for the effects of different thermal conductivities of nanoparticles. The next EMT model for the thermal conductivity of nanofluids was proposed by Jeffrey,[32] whom like many other researchers began from the work of Maxwell.[10] Jeffrey[32] thought that Maxwell's model could be improved by considering the interaction between spherical particles. Jeffrey[32] assumed a static homogenous distribution of spherical particles in a matrix of uniform thermal conductivity. This matrix represented the base fluid and the concentration of particles was assumed to be low in determining this model. An average of the interactions between particles was then calculated to determine the bulk properties of the particle–liquid suspension. Consequently, the average heat flux through the suspension could be determined, allowing for the calculation of thermal conductivity. The model determined by Jeffery[32] is given as:

$$\begin{aligned} k_{nf} = k_{bf}\Bigg(1 + 3\frac{k_p - k_{bf}}{k_p - 2k_{bf}}\phi + \phi^2\Bigg(3\left(\frac{k_p - k_{bf}}{k_p - 2k_{bf}}\right)^2 + \frac{3}{4}\left(\frac{k_p - k_{bf}}{k_p - 2k_{bf}}\right)^3 \\ + \frac{9}{16}\left(1\frac{k_p - k_{bf}}{k_p - 2k_{bf}}\right)^3\left(\frac{\frac{k_p}{k_{bf}} + 2}{2\frac{k_p}{k_{bf}} + 3}\right) + \frac{3}{64}\left(\frac{k_p - k_{bf}}{k_p - 2k_{bf}}\right)^4 + \cdots\Bigg)\Bigg) \end{aligned} \tag{7.5}$$

The EMT has also been used in combination with experimental data to develop models to predict the thermal conductivity of nanofluids. Sundar *et al.*[33] measured the thermal conductivity of Fe_3O_4–water nanofluids and found that the existing EMT models were unable to accurately match the experimental data they collected. This led to them determining a new correlation for the thermal conductivity of nanofluids. The model by Timofeeva *et al.*[12] was used as the basis for this new model since it allows for a simple correlation to be determined from the data. Based on the experimental data considered in this model, the range of validity was between 20 °C and 60 °C and the volume concentration was <2%. This model is given as:

$$k_{nf} = k_{bf}\,(1 + 10.5\phi)^{0.1051} \tag{7.6}$$

Davis[34] determined an additional model for the thermal conductivity of nanofluids based on the EMT. In their analysis, Davis[34] considered a composite material with a continuous phase and spherical inclusions, with the continuous phase representing the base fluid and the spherical inclusions representing the nanoparticles. It was then considered that there was a body placed in contact with the composite material. In the analysis, the volume of the composite material was assumed to be infinite, and the temperature of the body was kept constant. There was a temperature difference between the body and the composite material to drive a heat transfer between the materials. The heat flux caused by the body in the composite was calculated to determine the thermal conductivity of the composite material. The model proposed by Davis[34] is given as:

$$k_{\mathrm{nf}} = k_{\mathrm{bf}}\left(1 + \frac{3\left(\frac{k_{\mathrm{p}}}{k_{\mathrm{bf}}} - 1\right)}{\left(\frac{k_{\mathrm{p}}}{k_{\mathrm{bf}}} + 2\right) - \left(\frac{k_{\mathrm{p}}}{k_{\mathrm{bf}}} - 1\right)\phi}\left(\phi + f\left(\frac{k_{\mathrm{p}}}{k_{\mathrm{bf}}}\right)\phi^2 + 0(\phi^3)\right)\right) \tag{7.7}$$

In this model the function, $f\left(\frac{k_{\mathrm{p}}}{k_{\mathrm{bf}}}\right)$, is based on an infinite series and considers the ratio between the thermal conductivity of the nanoparticles and that of the base fluid. As the ratio approaches infinity, the series converges to 0.504. In most cases for common nanofluid conditions the function will generate a value of between 0.4 and 0.504.

Figure 7.1 presents a comparison between the thermal conductivity predicted by Maxwell,[10] Hamilton and Crosser[11] and Timofeeva *et al.*[12] EMT

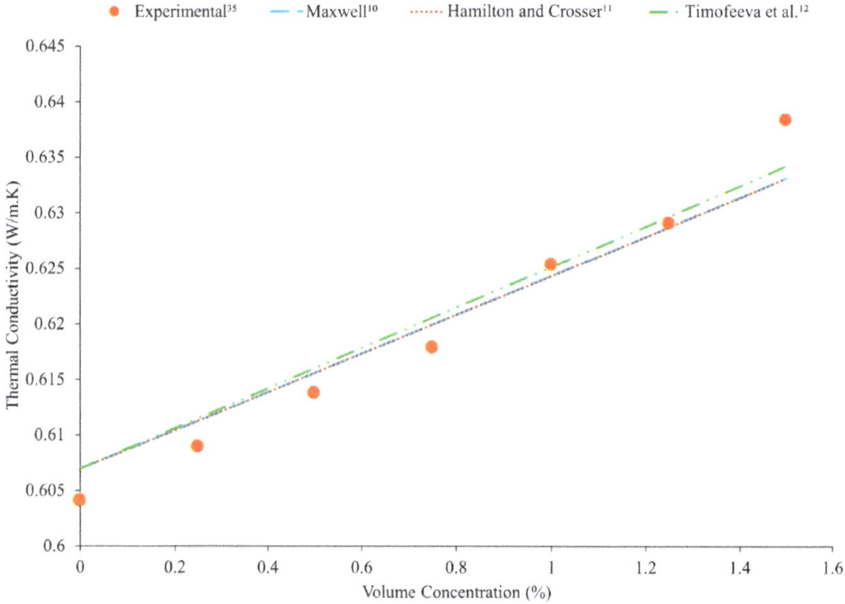

Figure 7.1 Comparison of EMT models to experimental data for spherical CuO–water nanofluid thermal conductivity as a function of volume concentration at 25 °C.

models and experimental data for a CuO–water nanofluid.[35] This figure shows the variation in the thermal conductivity of nanofluids as a function of volume concentration. Spherical conditions are used for the Hamilton and Crosser[11] model based on the appearance of the particles in a field-emission scanning electron microscopy (FESEM) image. All three models are accurate for the given experimental data, with the average absolute error for all three models being <0.358%. The most accurate model was that of Timofeeva *et al.*,[12] with an average absolute error of 0.347%. Since spherical conditions are used in both the Maxwell[10] and Hamilton and Crosser models,[11] the same results are observed.

Theoretical prediction of the thermal conductivity of a Fe_3O_4–water nanofluid using the Maxwell,[10] Hamilton Crosser,[11] and Timofeeva *et al.*[12] models was investigated and compared with the experimental data of Afrand *et al.*.[36] As the Fe_3O_4 nanoparticles were spherical, spherical conditions were used in the Hamilton and Crosser[11] model. It was observed that the absolute average error of predictions of the Maxwell,[10] Hamilton Crosser,[11] and Timofeeva *et al.*[12] models were all 21.61%. However, the EMT models all under predicted thermal conductivity compared to the experimental data and the level of accuracy decreased as the concentration of the nanoparticles increased. This indicates that EMT models have a varying level of accuracy when compared to different data sets. Experimental results have shown that there are significant differences in the thermal conductivity of nanofluids with different nanoparticle materials.

In order to better judge the applicability of the Hamilton and Crosser[11] model, nanofluids containing nanoparticles with non-spherical shapes also had to be considered. Experimental data for AlO(OH) and a mixture of water–ethylene glycol (50 : 50 mass : mass) containing brick- and blade-shaped nanoparticles were compared to the EMT models.[37] It can be seen in Figure 7.2 that all three models were able to predict the experimental data relatively accurately, but the most accurate model was the Hamilton Crosser[11] model, with an absolute average error of 1.19%. The Maxwell[10] and Timofeeva *et al.*[12] models exhibited absolute average errors of 1.60% and 1.71%, respectively. The shape factor was 3.7 for the Hamilton and Crosser[11] model, based on the particle shape and dimensions. For a given data set, the shape factor proposed by Hamilton and Crosser[11] can increase the accuracy of the prediction of EMT models. The brick-shaped nanoparticles had a higher shape factor than a spherical particle, and the experimental data for the nanofluid with brick-shaped nanoparticles was greater than the Maxwell model[10] prediction for spherical particles. For given experimental data, Hamilton and Crosser[11] predicted an increase in thermal conductivity as the shape factor of the particle increased. In the case of the blade-shaped nanoparticles, the Hamilton and Crosser[11] model significantly over-predicted the thermal conductivity of the nanofluid. Figure 7.3 compares the nanofluid made with blade-shaped nanoparticles with a shape factor of 8.6 for the EMT models. The absolute average error for the Hamilton and Crosser[11] model was 18.84%, while the Maxwell[10] and Timofeeva *et al.*[12] models only exhibited errors of 1.57% and 1.39%, respectively. For this

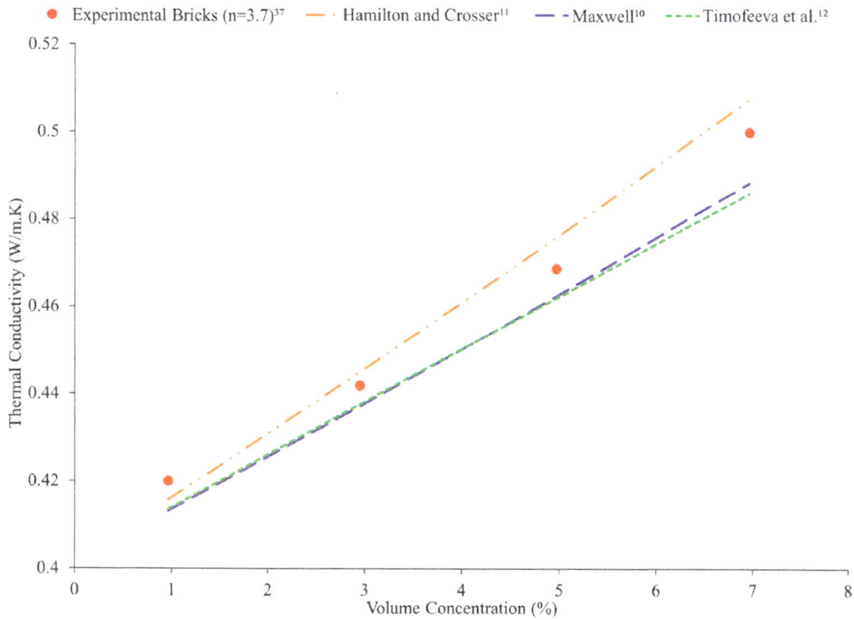

Figure 7.2 Comparison of EMT models to experimental data for brick-shaped, $n = 3.7$, AlO(OH)–50% water–50% ethylene glycol nanofluid thermal conductivity as a function of volume concentration at 21 °C.

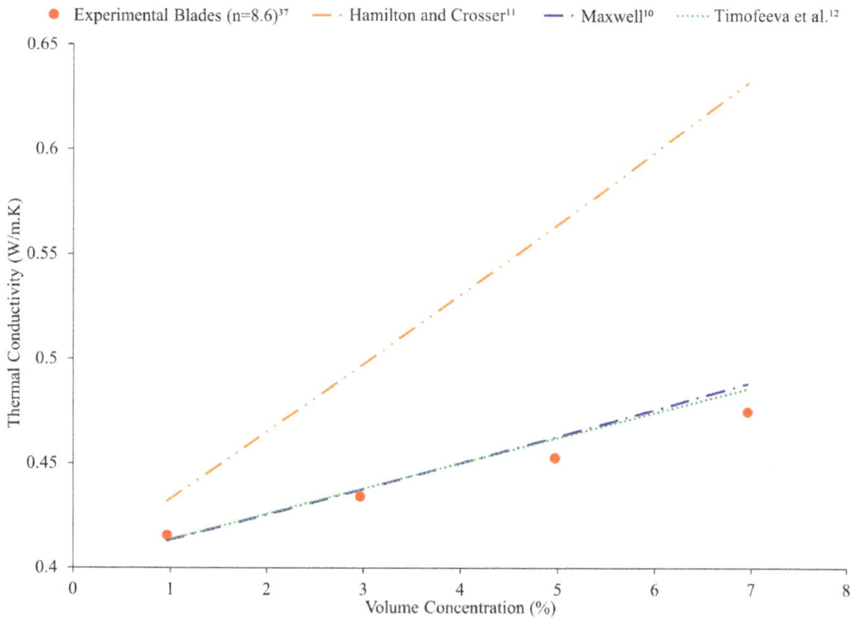

Figure 7.3 Comparison of the prediction of nanofluid thermal conductivity by EMT models to experimental data. The nanofluid was a mixture of water–ethylene glycol (50 : 50) containing AlO(OH) blade-shaped, $n = 8.6$, nanoparticles, at 21 °C.

Table 7.1 Average percentage errors of EMT models.

Average percentage error EMT models			
Ref.	Maxwell[10]	Hamilton and Crosser[11]	Timofeeva *et al.*[12]
Al_2O_3–water, 0.05 vol%, 29 nm[38]	4.73%	4.73%	4.73%
CuO–water, 0.1 vol%, 27 nm[39]	6.23%	6.23%	6.21%
Al_2O_3–water, 0.2 vol%, 50 nm[39]	4.87%	4.87%	4.85%
Al_2O_3–EG, 23 °C, 11 nm[12]	2.28%	2.28%	1.74%
CuO–water, 25 °C[35]	0.36%	0.36%	0.35%
TiO_2–water, 25 °C, 150 nm[40]	3.80%	3.80%	2.44%
Fe_3O_4–water, 1 vol%, 30 nm[41]	7.33%	7.33%	7.30%
Fe_3O_4–water, 20 °C, 20–30 nm[36]	21.62%	21.62%	21.62%
AlO(OH)–EG–water(50 : 50) 21 °C platelets $n = 5.7$[37]	1.62%	10.47%	1.50%
AlO(OH)–EG–water(50 : 50) 21 °C blades $n = 8.6$[37]	1.55%	18.84%	1.39%
AlO(OH)–EG–water(50 : 50) 21 °C cylinders $n = 4.9$[37]	4.29%	3.32%	4.57%
AlO(OH)–EG–water(50 : 50) 21 °C bricks $n = 3.7$[37]	1.60%	1.19%	1.71%
ZrO_2–water, 0.05 vol%, 30 nm[29]	1.01%	1.01%	0.94%
Cu–EG, 1 vol%, 7 nm[42]	17.93%	17.93%	17.94%
SiO_2–water, 1 vol%, 45 nm[43]	7.37%	7.37%	5.41%

dataset, considering the nanoparticle shape as described by Hamilton and Crosser[11] provided inferior results to the other EMT models that do not consider shape. It can thus be concluded that considering the effects of shape in the Hamilton and Crosser[11] model may increase the level of accuracy of prediction, but not always. Although particle shape would affect thermal conductivity, that effect is not yet entirely captured by the EMT models.

In Table 1, the average absolute errors for the EMT models compared to several sets of experimental data is given.

7.3 Aggregation Models

In a nanofluid, the nanoparticles tend to agglomerate and produce a cluster if they are not well designed. The aggregation of nanoparticles will place the nanoparticles in close contact with one another, and enable energy transfer from particle to particle, which potentially further increases the thermal conductivity of the nanofluids.[44] Prasher *et al.*[13] attempted to create a model for the thermal conductivity of nanofluids, considering the effects of aggregation. The model considers several features of the aggregates formed, such as the radius of gyration that represents the radius of a circle that can contain the nanoparticle aggregate. This model also breaks the clusters into two parts, the backbone and dead ends. The backbone of the aggregate runs the entire length of the chain of nanoparticles while the dead ends are simply branches that do not span the entire chain. Additionally, the model is based on the work conducted by Maxwell[10] in the EMT, except that

Prasher *et al.*[13] replaced the thermal conductivity of nanoparticles, k_p, with the thermal conductivity of the nanoparticle aggregate, k_a. The volume concentration of the nanoparticles, ϕ, was also replaced with the volume concentration of the nanoparticle aggregates, ϕ_a. The nanoparticle aggregates are thought to have a higher thermal conductivity than the nanoparticles themselves due to particle contact. The model is given as:

$$k_{nf} = \frac{k_a + 2k_{bf} + 2\phi_a(k_a - k_{bf})}{k_a + 2k_{bf} - \phi_a(k_a - k_{bf})} k_{bf} \tag{7.8}$$

The simulations used in the development of the model assumed that the thermal conductivity of the nanoparticle was 100 times greater than that of the base fluid. This assumption is reasonable for the base fluid and particle combinations that are frequently used in nanofluids. It was also shown in both the simulations and the model that the expected thermal conductivity will increase as the size of the nanoparticle aggregates increases. Xuan *et al.*[14] also considered the effects of nanoparticle aggregation on the thermal conductivity of nanofluids. Their model was generated based on the EMT and Brownian motion, with the additional consideration of the formation of particle clusters. Brownian motion is thought to play an important role in the formation of nanoparticle clusters since random motion leads to particle collisions, which contribute to the formation of clusters. It is also considered important to evaluate the effects of Brownian motion and aggregation together since the formation of nanoparticle clusters affects the Brownian velocity. As the radius of a cluster increases the Brownian velocity decreases. Therefore, it is important to use cluster dimensions when the combination of Brownian motion and aggregation is taken into consideration. The model proposed by Xuan *et al.*[14] is given as:

$$k_{nf} = \left(\frac{k_p + 2k_{bf} + 2\phi(k_p - k_{bf})}{k_p + 2k_{bf} - \phi(k_p - k_{bf})} + \frac{\rho_p \phi c_p}{2k_{bf}} \sqrt{\frac{k_b T}{3\pi r_c \mu_{nf}}} \right) k_{bf} \tag{7.9}$$

where r_c is the radius of the cluster form of particles, ρ_p is the nanoparticle density, c_p is the nanoparticle specific heat, k_b is the Boltzmann constant, T is the nanofluid temperature, and μ_{nf} is the nanofluid viscosity. Based on this model, thermal conductivity will decrease as the size of the nanoparticle clusters increases due to the reduced motion of the larger clusters. In this model, the combination of EMT, clustering, and Brownian motion were taken into consideration. As particle clustering increases the contribution of Brownian motion to the overall thermal conductivity enhancement decreases. Feng *et al.*[15] considered the effect of nanoparticle clustering and the nanolayer. It was proposed that the nanoparticle must be considered in combination with the ordered liquid layer surrounding it. The combined properties of the nanoparticle and the liquid layer will form an equivalent particle that must be considered rather than just the particle itself. Work conducted by Yu and Choi[45] was the basis for determining the effect of the nanolayer on the thermal conductivity of the mixture. Additionally, there are two parts of the nanofluid that must be considered, non-aggregating particles and aggregating particles. One dimensional heat conduction was used to evaluate the thermal conductivity between the nanoparticles in aggregates in contact. The final evaluation of the thermal conductivity was considered as the sum of two parts. One part of the total thermal conductivity was provided by the aggregating nanoparticles and the non-aggregating nanoparticles were responsible for the other part. The two parts of the thermal conductivity are weighted by approximating the ratio between the nanoparticles in aggregates and the total number of nanoparticles in the mixture. The weight factor, which represents the ratio between aggregating particles and total particles, is used to determine how much of the total thermal conductivity the aggregating nanoparticles and the nonaggregating nanoparticles are each responsible for. The weight factor is approximated based on the effective volume fraction, which is a volume fraction based on the combination of the particle and nanolayer volumes. This model is given as:

$$k_{nf} = \left((1-\phi_e)\frac{k_{pe} + 2k_{bf} + 2(k_{pe} - k_{bf})(\beta_{nl})^3\phi}{k_{pe} + 2k_{bf} - (k_{pe} - k_{bf})(\beta_{nl})^3\phi} + \phi_e\left(\left(1 - \frac{3}{2}\phi_e\right)\right.\right.$$

$$\left.\left. + \frac{3\phi_e}{\alpha}\left(\frac{1}{\alpha}\ln\left(\frac{r_p + \delta_{nl}}{(r_p + \delta_{nl})(1-\alpha)} - 1\right)\right)\right)\right) k_{bf} \tag{7.10}$$

$$\phi_e = \phi\beta_{nl}{}^3 \tag{7.11}$$

$$\alpha = 1 - \frac{k_{bf}}{k_{pe}} \tag{7.12}$$

where ϕ_e is the equivalent volume fraction combining the nanoparticle and nanolayer, k_{pe} is the particle equivalent thermal conductivity considering both the nanoparticle and nanolayer thermal conductivity, and α relates the ratio of the thermal conductivity of the base fluid to the equivalent thermal conductivity of the nanoparticles. Additionally, r_p is the nanoparticle radius, δ_{nl} is the nanolayer thickness, and β_{nl} is a dimensionless nanolayer parameter based on the nanoparticle radius and nanolayer thickness. In this model, the nanolayer thickness is 1 nm and the nanolayer thermal conductivity is $2k_{bf}$. The contribution of the effect of aggregating particles increases as the volume fraction increases and the particle size decreases. Wang *et al.*[46] also considered the effect of nanoparticle aggregation on the thermal conductivity of nanofluids. It was considered that the nanoparticle clusters that form will vary in size and that the distribution of these sizes can be approximated using fractals. The model was developed based on the EMT, except rather than considering the thermal conductivity of the nanoparticles the thermal conductivity of the cluster was considered. The thermal conductivity of the cluster was modeled as a function of the cluster radius. The combination of the cluster thermal conductivity function and a function to predict the distribution of particle radius can be used to develop the thermal conductivity model given as:

$$k_{\text{nf}} = k_{\text{bf}}\left(\frac{(1-\phi)+3\phi\int_0^\infty \frac{k_{\text{cl}}(r)n(r)}{k_{\text{cl}}(r)+2k_{\text{bf}}}\text{d}r}{(1-\phi)+3\phi\int_0^\infty \frac{k_{\text{bf}}n(r)}{k_{\text{cl}}(r)+2k_{\text{bf}}}\text{d}r}\right) \tag{7.13}$$

where $k_{\text{cl}}(r)$ is the function for the thermal conductivity of the cluster as a function of radius and $n(r)$ is the function for the distribution of the nanoparticle cluster radius.

In Figure 7.4 the aggregation models are compared to experimental data.[47] The nanofluid was CuO–gear oil and the temperature was 30 °C, with nanoparticles aggregated in the nanofluid. It was observed that as the concentration of nanoparticles increased the size of the aggregates increased. Both the models by Xuan *et al.*[14] and Feng *et al.*[15] displayed a high degree of accuracy, with average absolute errors of 0.42% and 1.07%, respectively. The model by Prasher *et al.*[13] was not accurate, with an average absolute error of 32.91%. Xuan *et al.*[14] considered aggregation in combination with Brownian motion and Feng *et al.*[15] considered aggregation in combination with the nanolayer, while Prasher *et al.*[13] only considered the effect of aggregation. Considering aggregation in combination with Brownian motion was more accurate than considering

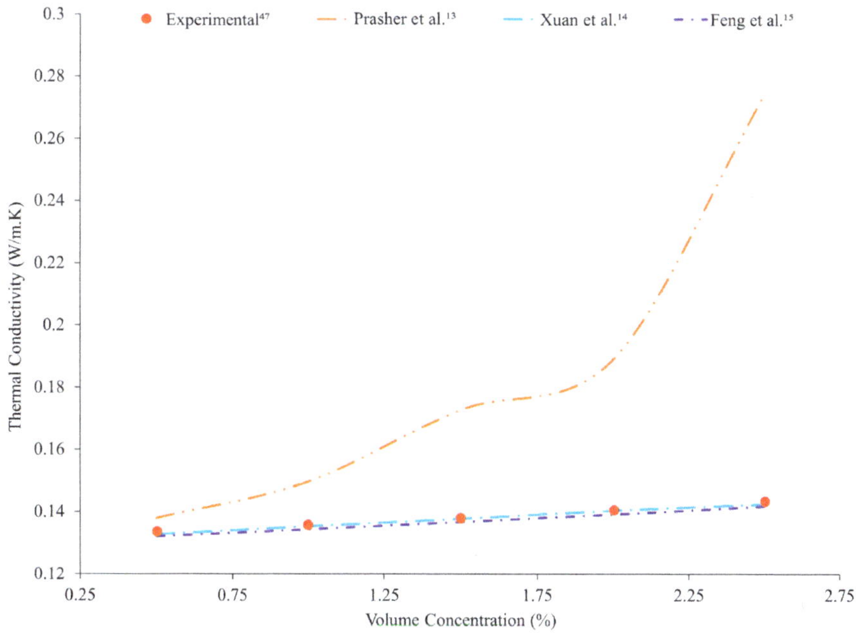

Figure 7.4 Comparison of aggregation models to experimental data for 40 nm CuO–gear oil nanofluid thermal conductivity as a function of volume fraction at a temperature of 30 °C.

aggregation with the nanolayer, which may suggest that the role of Brownian motion is more important. Both the models by Xuan *et al.*[14] and Feng *et al.*[15] were highly accurate, so it is difficult to definitively say whether Brownian motion or the nanolayer should be considered with aggregation.

In Figure 7.5, the aggregation models are compared to experimental data for MgO–water nanofluids.[48] It was observed that the size of the clusters of nanoparticles in the nanofluid decreased as the concentration of nanoparticles increased. For a given data, the model of Prasher *et al.*[13] was the most accurate model, with an average absolute error of 2.02%. Moreover, the models of Xuan *et al.*[14] and Feng *et al.*[15] showed absolute average errors of 2.79% and 2.85%, respectively. For this data set, the model of Prasher *et al.*[13] demonstrated the highest predicted thermal conductivity, which led to the most accurate results. In Figure 7.4, it can be seen that the model of Prasher *et al.*[13] also predicted a higher thermal conductivity than the other models, but in Figure 7.4 the Prasher *et al.*[13] model was the least accurate. Based on the given data sets analyzed, the model of Prasher *et al.*[13] tended to generate a higher predicted thermal conductivity than the models of Xuan *et al.*[14] and Feng *et al.*[15] This higher prediction can lead to more accurate or less accurate predictions depending on the experimental data set being analyzed.

In Figure 7.6, the aggregation models are compared to experimental data for Al_2O_3–Therminol 66 nanofluids.[49] The size of the aggregates that formed in the nanofluid were assumed to be constant as the temperature of the

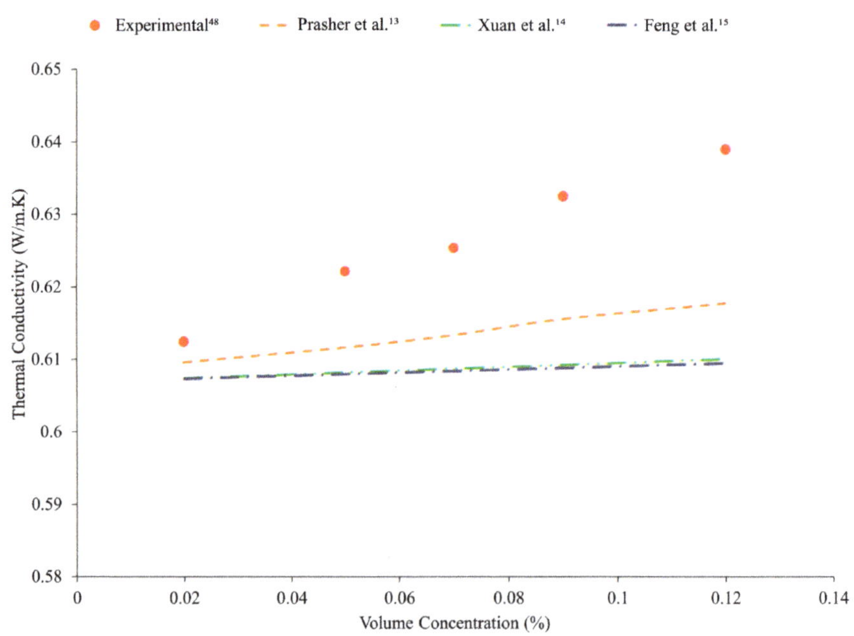

Figure 7.5 Comparison of clustering models to experimental data for 20 nm MgO–water nanofluid thermal conductivity as a function of volume fraction at a temperature of 25 °C.

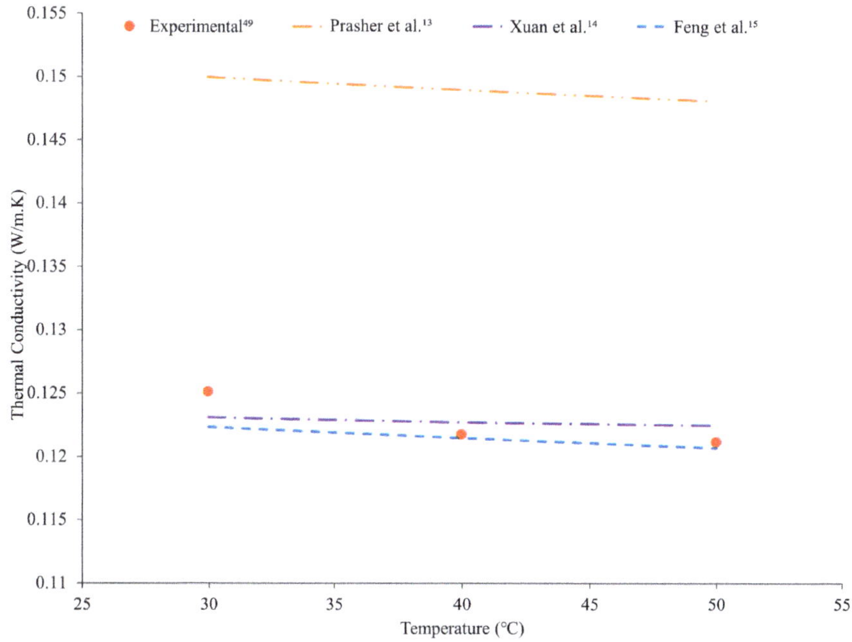

Figure 7.6 Comparison of clustering models to experimental data for 45 nm Al_2O_3–Therminol 66 nanofluid thermal conductivity as a function of temperature at a volume fraction of 0.01.

nanofluid was varied. The pattern seen in this figure is very similar to that in Figure 7.4, where the models of Xuan *et al.*[14] and Feng *et al.*[15] were very accurate, with absolute average errors of 1.15% and 0.98%, respectively. The model proposed by Prasher *et al.*[13] significantly overpredicted the thermal conductivity, with an absolute average error of 21.43%. For all three data sets, the models of Xuan *et al.*[14] and Feng *et al.*[15] were able to accurately predict the thermal conductivity with <3% absolute average error. Additionally, the data sets were all based on different nanoparticle and base fluid combinations. This suggests that these models are valid for a wide range of nanofluid compositions. The accuracy of these models could be because of considering the effects of aggregation in combination with another nanofluid heat transfer mechanism.

Table 2 shows the average absolute errors for the aggregation models compared to several sets of experimental data.

7.4 Brownian Models

In the EMT, the nanoparticles are assumed to be stationary, which is not accurate, since nanoparticles and molecules are constantly moving inside the nanofluids. In fact, the nanoparticles interact with each other, and molecules transfer energy as a result of these interactions, which can

Table 7.2 Average percentage error aggregation models.

Average percentage error aggregation models			
Ref.	Prasher *et al.*[13]	Xuan *et al.*[14]	Feng *et al.*[15]
ZnO–EG, 30 °C, 40 nm[50]	7.09%	13.83%	15.12%
CuO–gear oil, 30 °C, 40 nm[47]	32.91%	0.42%	1.07%
CuO–water, 25 °C, 40 nm[48]	4.20%	4.37%	4.46%
MgO–water, 25 °C, 40 nm[48]	2.02%	2.79%	2.85%
Al_2O_3–Therminol 66, 1 vol%, 45 nm[49]	21.43%	1.15%	0.98%

enhance the thermal conductivity. It was also proposed that the particle motion in the nanofluid might generate micro convection inside the nanofluid and consequently enhance the thermal conductivity of the nanofluid.[51] Considering, the particle motion inside the nanofluids, researchers have developed models to predict the thermal conductivity of nanofluids. For instance, Corcione[16] derived an empirical correlation for the thermal conductivity of a nanofluid that incorporated the effects of Brownian motion based on a wide range of previously published experimental data. The data sets considered nanofluids with nanoparticles ranging from 10 nm to 150 nm in diameter, mixed with water and ethylene glycol and nanoparticles, such as Cu, CuO, TiO_2, and Al_2O_3. Based on these data sets, regression analysis was performed to determine the correlation for the thermal conductivity of nanofluids, given as:

$$k_{\mathrm{nf}} = k_{\mathrm{bf}}\left(1 + 4.4\mathrm{Re}^{0.4}\mathrm{Pr}^{0.66}\left(\frac{T}{T_{\mathrm{fr}}}\right)^{10}\left(\frac{k_{\mathrm{p}}}{k_{\mathrm{bf}}}\right)^{0.03}\phi^{0.66}\right) \tag{7.14}$$

In this model, Re is the Reynolds number, Pr is the Prandtl number of the base fluid, T is the temperature of the nanofluid, and T_{fr} is the freezing point of the base fluid. The Prandtl number is defined as:

$$\mathrm{Pr} = \frac{c_{\mathrm{bf}}\mu_{\mathrm{bf}}}{k_{\mathrm{bf}}} \tag{7.15}$$

where c_{bf} is the specific heat of the base fluid and μ_{bf} is the dynamic viscosity of the base fluid. The Brownian motion of the nanoparticles is considered in the Reynolds number since the Reynolds number of the nanoparticles is a function of the Brownian velocity. For a nanoparticle, the Reynolds number is defined as:

$$\mathrm{Re} = \frac{\rho_{\mathrm{bf}} u_{\mathrm{B}} d_{\mathrm{p}}}{\mu_{\mathrm{bf}}} \tag{7.16}$$

where ρ_{bf} is the density of the base fluid, u_{B} is the mean Brownian velocity, and d_{p} is the nanoparticle diameter. In order to calculate the Reynolds number, the Brownian velocity is calculated as the ratio between the nanoparticle diameter and the time required for the nanoparticle to move a distance equal to the nanoparticle diameter.[44] Therefore, the Brownian velocity of the nanoparticles can be calculated as:

$$u_{\mathrm{B}} = \frac{2k_{\mathrm{b}}T}{\pi\mu_{\mathrm{bf}}d_{\mathrm{p}}^2} \tag{7.17}$$

where k_{b} is the Boltzmann constant. Eqn (7.16) and (7.17) can then be combined to calculate the Reynolds number using:

$$\mathrm{Re} = \frac{2\rho_{\mathrm{bf}}k_{\mathrm{b}}T}{\pi\mu_{\mathrm{bf}}^2 d_{\mathrm{p}}} \tag{7.18}$$

It was observed that this model is valid for nanofluids when the volume concentration is between 0.2% and 9%, and temperature is in the range of 21–51 °C, and nanoparticle size is in the range of 10–150 nm. Chon *et al.*[17] also determined a model for the thermal conductivity of nanofluids considering the effects of Brownian motion. This model was created from data for Al_2O_3–water nanofluids with particle diameters of 11 nm, 47 nm, and 150 nm. The thermal conductivities of the nanofluids were measured at volume concentrations of 1% and 4% over a temperature range of 21–71 °C. The Buckingham Pi theorem was then used on the experimental data to determine the model:

$$k_{\mathrm{nf}} = \left(1 + 64.7\phi^{0.746}\left(\frac{d_{\mathrm{bf}}}{d_{\mathrm{p}}}\right)^{0.369}\left(\frac{k_{\mathrm{p}}}{k_{\mathrm{bf}}}\right)^{0.746}\mathrm{Pr}^{0.9955}\mathrm{Re}^{1.2321}\right)k_{\mathrm{bf}} \tag{7.19}$$

where d_{bf} is the diameter of a molecule of the base fluid. In this model, the Prandtl number is defined as in eqn (7.15). The Reynolds number, however in this case is defined as:

$$\mathrm{Re} = \frac{\rho_{\mathrm{bf}}k_{\mathrm{b}}T}{3\pi\mu_{\mathrm{bf}}^2 l_{\mathrm{bf}}} \tag{7.20}$$

where l_{bf} is the mean free path of the base fluid. In the case of water, a mean free path of 0.17 nm was used. Koo and Kleinstreuer[18] proposed that the thermal conductivity of a nanofluid is based on two parts, static and Brownian motion. To calculate the static part, the Maxwell model[10] was used since Maxwell assumed the particles were static in the base fluid. The Brownian motion part would consider the effect of nanoparticle motion on the thermal conductivity of the nanofluid. A model by Koo and Kleinstreuer[18] considered the effects of particle motion, using the same methods as Corcione[16] and Chon *et al.*[17] The Koo and Kleinstreuer[18] model also considered the effect that particle motion will have on the motion of the base fluid. When the nanoparticles move due to Brownian motion, they will carry some of the base fluid with them. This can lead to micro mixing of the base fluid, with this mixing of the base fluid also increasing the heat transfer in the base fluid, contributing to enhancement in thermal conductivity. The model proposed by Koo and Kleinstreuer[18] is given as:

$$k_{nf} = \left(\frac{k_p + 2k_{bf} - 2\phi(k_{bf} - k_p)}{k_p + 2k_{bf} + \phi(k_{bf} - k_p)} \right) k_{bf} + \left(5 \times 10^4 \beta \phi \rho_{bf} c_{bf} \sqrt{\frac{k_b T}{\rho_p d_p}} f(T, \phi, etc.) \right) k_{bf} \tag{7.21}$$

where ρ_p is the density of the nanoparticle and β and $f(T,\phi,etc.)$ are determined based on the nanofluid characteristics. In this model, $f(T,\phi,etc.)$ considers the nanofluid temperature and concentration and is given as:

$$f(T,\phi) = (-6.04\phi + 0.4705)T + (1722.3\phi - 134.63) \tag{7.22}$$

β is based on the nanofluid concentration and nanoparticle materials properties. Therefore, there are different correlations for β, determined based on sets of experimental data collected for nanofluids containing different nanoparticle materials. β as defined for CuO and Al_2O_3 is given in Table 7.3.

Similar to the Koo and Kleinstreuer model,[18] Vajjha and Das[19] determined a thermal conductivity model considering Brownian motion. In this model, only the empirical correlations for β and $f(T, \phi, etc.)$ were changed to increase their range of validity. The new correlations were based on 133 data points from ZnO, CuO, and Al_2O_3 nanofluids with nanoparticle diameters of 29 nm, 53 nm, and 77 nm. The new correlation for $f(T,\phi,etc.)$ is given as:

$$f(T, \phi) = (0.028217\phi + 0.003917)\left(\frac{T}{T_0}\right) + (-0.030669\phi - 0.00391123) \tag{7.23}$$

T_0 is a reference temperature and is equal to 273 K. The new correlations for β are given in Table 7.4.

Patel *et al.*[20] considered the effect of Brownian motion by including the nanoparticle diameter in the thermal conductivity of the nanofluid. Since the velocity of Brownian motion is related to the particle size, the particle

Table 7.3 Correlation for Al_2O_3 and CuO nanoparticles.

Nanoparticle	β	Volume fraction	Temperature
Al_2O_3	$0.0011\ (100\phi)^{-0.7272}$	$\phi \geq 1\%$	$300\ K \leq T \leq 325\ K$
CuO	$0.0017\ (100\phi)^{-0.0841}$	$\phi \geq 1\%$	$300\ K \leq T \leq 325\ K$

Table 7.4 Correlation for Al_2O_3, ZnO, and CuO nanoparticles.

Nanoparticles	β	Volume fraction	Temperature
Al_2O_3	$8.4407\ (100\phi)^{-1.07304}$	$1\% \leq \phi \leq 10\%$	$298\ K \leq T \leq 363\ K$
ZnO	$8.4407\ (100\phi)^{-1.07304}$	$1\% \leq \phi \leq 7\%$	$298\ K \leq T \leq 363\ K$
CuO	$9.8810\ (100\phi)^{-0.9446}$	$1\% \leq \phi \leq 6\%$	$298\ K \leq T \leq 363\ K$

diameter can be an effective parameter on Brownian motion. It was observed that the thermal conductivity of the nanofluid increased as the size of the nanoparticle decreased. The Brownian velocity of 11 nm nanoparticles was approximated to be two orders of magnitude higher than for 150 nm nanoparticles, which could be the reason for the increased thermal conductivity in nanofluids with smaller particles. The increase in thermal conductivity with reduced particle size could also be attributed to a higher surface area to volume ratio for the smaller nanoparticles. Non-linear regression analysis was performed on experimental data for water and ethylene glycol based CuO and Al_2O_3 nanofluids with a diameter from 11 nm to 150 nm to determine the model. The analysis was performed using the least squares method to generate the model:

$$k_{nf} = k_{bf}\left(1 + 0.135\left(\frac{k_p}{k_{bf}}\right)^{0.273}\phi^{0.467}\left(\frac{T}{20}\right)^{0.547}\left(\frac{100}{d_p}\right)^{0.234}\right) \tag{7.24}$$

In this model, T is the temperature in Celsius as opposed to other models presented in this chapter where Kelvin is used. The particle diameter, d_p, in this model is calculated in nm. It was observed that this model is valid for nanofluids with spherical particles, ranging from 10 nm to 150 nm, the thermal conductivity of nanoparticles, ranging from 20 W m^{-1} K to 400 W m^{-1} K, the thermal conductivity of the base fluid, ranging from 0.1 W m^{-1} K to 0.7 W m^{-1} K, volume concentration, ranging from 0.1% to 3%, and temperature, ranging from 20 °C to 50 °C. Kumar *et al.*[52] created a model for the thermal conductivity of a nanofluid mixture considering the particle motion, based on Brownian velocity. It was thought that the Brownian velocity would account for the strong dependence of the thermal conductivity of the nanofluid on temperature since the Brownian motion would increase with an increase in temperature. The model began by considering the effect of stationary particles in the base fluid. It was considered that the total heat flux in the nanofluid was a combination of the heat flux of the particles and the base fluid. Based on this and the geometry of the base fluid molecules and nanoparticles, a model for the thermal conductivity of a nanofluid with stationary particles was determined. The model was then modified by replacing the thermal conductivity of the nanoparticles in the model with a constant multiplied by nanoparticle velocity. This was done because the effect of the thermal conductivity of the nanoparticles increases as the velocity of the nanoparticle increases. This model is given as:

$$k_p = k_{bf}\left(1 + C_u u_B \frac{\phi r_{bf}}{k_{bf}(1-\phi)r_p}\right) \tag{7.25}$$

where C_u is a constant, u_B is the Brownian velocity defined as in eqn (17), and r_{bf} is the radius of the base fluid molecules. There is not a determined C_u, but a value of around 2.9 to 3 was observed to be applicable for some data sets.

In Figure 7.7, the level of accuracy of Brownian models was examined according to the experimental data, given for CuO–water nanofluids.[20] The Koo and Kleinstreuer model exhibited an absolute average error of 14%, while the other models all exhibited an absolute average error of <1.56%. The Koo and Kleinstreuer model significantly overestimated the thermal conductivity of the nanofluids and the level of accuracy of the model decreased as the temperature increased.

In Figure 7.8 the Brownian models are compared to experimental data for CuO–water nanofluids.[53] The model of Koo and Kleinstreuer[18] predicts a higher thermal conductivity than the other models, but in this case, the Koo and Kleinstreuer[18] model is the most accurate, with an absolute average error of 1.20%. The other models have an absolute average error of at least 8.83%.

In Figure 7.9, the Brownian models are compared to the experimental data of Al_2O_3–water nanofluids.[54] For the given experimental data, both the models of Patel *et al.*[20] and Vajjha and Das[19] exhibited absolute average errors of 1.53% and 2.93%, respectively. The model of Corcione[16] exhibited an absolute average error of 17.94% and significantly overpredicted the thermal conductivity. In this case, the behavior of the model of Corcione[16] deviated significantly from the models of Chon *et al.*,[17] Vajjah and Das,[19] and Patel *et al.*,[20] which was not observed for the other data sets considered

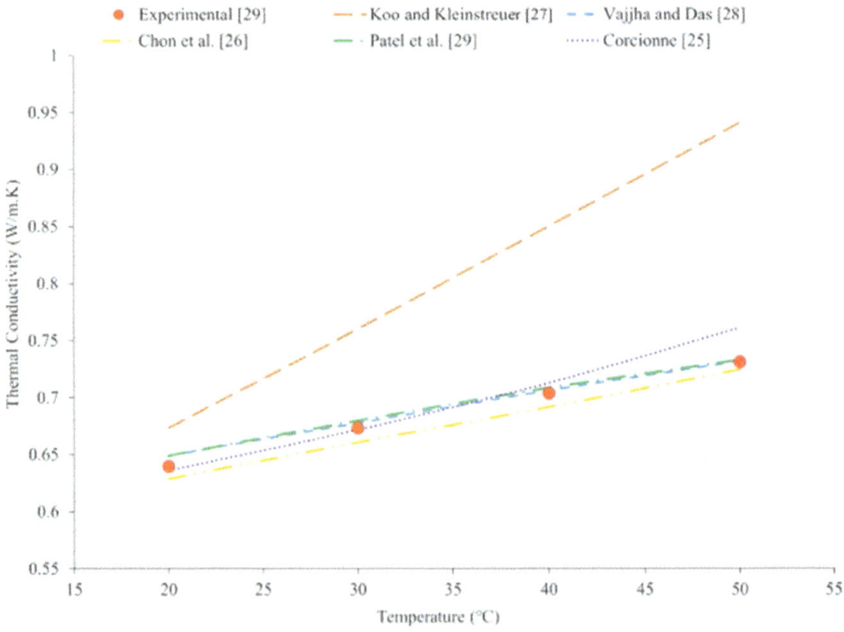

Figure 7.7 Comparison of Brownian models to experimental data for 31 nm CuO–water nanofluid thermal conductivity as a function of temperature at a volume fraction of 0.02.

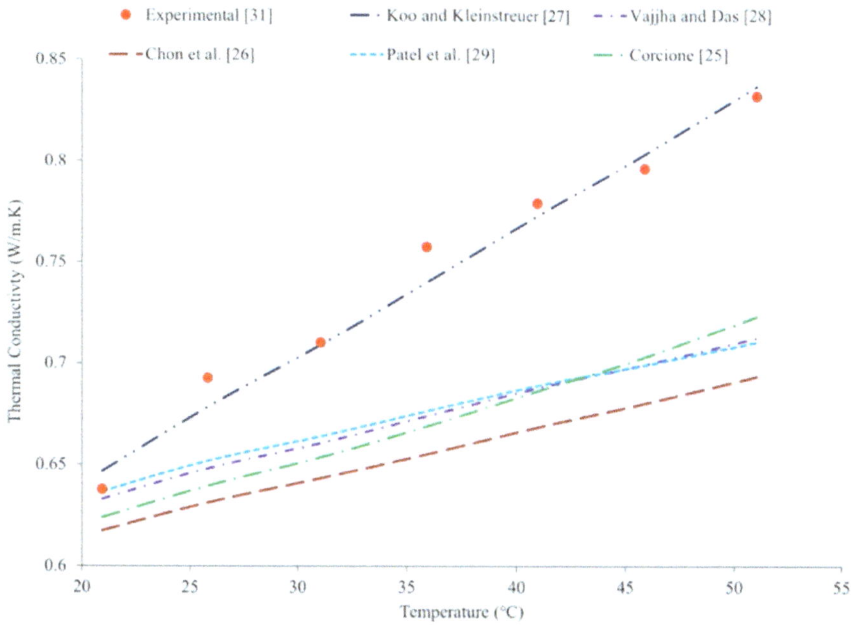

Figure 7.8 Comparison of Brownian models to experimental data for 28.6 nm CuO–water nanofluid thermal conductivity as a function of temperature at a volume fraction of 0.01.

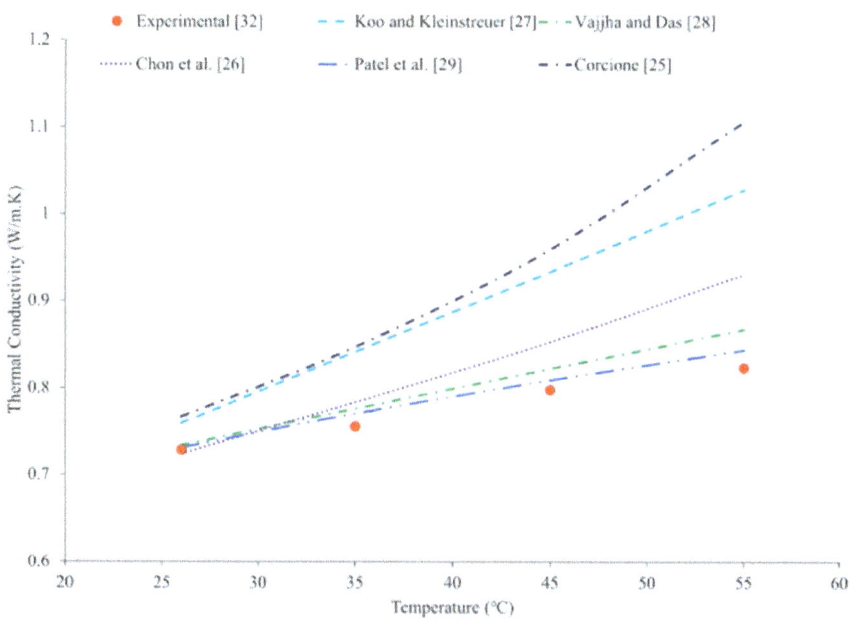

Figure 7.9 Comparison of Brownian models to experimental data for 5 nm Al_2O_3–water nanofluid thermal conductivity as a function of temperature at a volume fraction of 0.04.

Table 7.5 Average percentage error Brownian motion models.

Average percentage error Brownian models					
Ref.	Corcione[16]	Chon *et al.*[17]	Koo and Kleinstreuer[18]	Vajjha and Das[19]	Patel *et al.*[20]
Al_2O_3–water, 0.05 vol%, 29 nm[38]	3.43%	4.21%	3.85%	3.57%	2.71%
Al_2O_3–water, 0.2 vol%, 29 nm[38]	6.02%	7.67%	3.36%	1.80%	5.51%
Al_2O_3–water, 5 vol%, 5 nm[54]	18.25%	6.58%	6.13%	1.70%	1.18%
Al_2O_3–water, 4 vol%, 5 nm[54]	17.94%	6.04%	12.15%	2.93%	1.53%
Al_2O_3–water, 1.5 vol%, 50 nm[55]	4.09%	1.71%	8.25%	2.63%	2.74%
CuO–water, 1 vol%, 28.6 nm[53]	9.50%	11.50%	1.20%	9.07%	8.83%
CuO–water, 4 vol%, 28.6 nm[53]	4.24%	7.74%	11.95%	5.42%	6.67%
CuO–water, 2 vol%, 31 nm[20]	1.56%	1.54%	14.0%	0.65%	0.85%
CuO–water, 3 vol%, 31 nm[20]	1.82%	1.50%	16.33%	1.30%	0.88%
SiO_2–water, 1 vol%, 45 nm[43]	2.93%	3.17%	8.84%	2.30%	2.36%
Al_2O_3–water, 0.1 vol%, 20 nm[56]	0.88%	1.36%	1.09%	4.93%	0.49%

in Figures 7.7 and 7.8. This confirms that the level of accuracy of the models is significantly influenced by the nanofluid conditions.

Table 7.5 shows the average absolute errors for the Brownian motion models compared to several sets of experimental data.

7.5 Nanolayer Models

It was also suggested that in a nanofluid, the base fluid molecules form an ordered layer around the nanoparticles and the formation of this layer can enhance the thermal conductivity. The molecular structure of the surrounding layer will be more ordered than the bulk liquid. This could increase the thermal conductivity of the surrounded layer, and consequently the nanofluid as a whole. Ordered molecular structures, such as crystalline solids, exhibit much higher thermal transport than the molecular structures found in liquids.[44] This layer of more thermally conductive liquid could increase the effective volume of the nanoparticles by converting a portion of the volume of liquid into a material with higher conductivity. This layer could also act as a thermal bridge between the nanoparticles and the base fluid to facilitate energy transfer. Leong *et al.*[21] created a model for the thermal conductivity of a nanofluid considering the effect of the nanolayer. They assumed that in the nanofluid there are three separate parts, the nanoparticles, the nanolayer, and the base fluid itself. This model also considers spherical noninteracting particles. This model is given as:

$$k_{nf} = \frac{\begin{array}{c}(k_p - k_{nl})\phi k_{nl}\left[2\left(1+\dfrac{\delta_{nl}}{2r_p}\right)^3 - \left(1+\dfrac{\delta_{nl}}{r_p}\right)^3 + 1\right] \\ + (k_p + 2k_{nl})\left(1+\dfrac{\delta_{nl}}{2r_p}\right)^3\left[\phi\left(1+\dfrac{\delta_{nl}}{2r_p}\right)^3 (k_{nl} - k_{bf}) + k_{bf}\right]\end{array}}{\left(1+\dfrac{\delta_{nl}}{2r_p}\right)^3 (k_p + 2k_{nl}) - (k_p - k_{nl})\phi\left[\left(1+\dfrac{\delta_{nl}}{2r_p}\right)^3 + \left(1+\dfrac{\delta_{nl}}{r_p}\right)^3 - 1\right]} \tag{7.26}$$

where k_{nl} is the nanolayer thermal conductivity, δ_{nl} is the nanolayer thickness, and r_p is the nanoparticle radius. Currently, there is no method to experimentally determine the nanolayer thickness or the exact thermal conductivity of the nanolayer. The nanolayer thickness is believed to only be the thickness of a few liquid molecules. The typical thickness of a nanolayer is around 1 nm. Additionally, the thermal conductivity of the nanolayer is thought to be a multiple of the thermal conductivity of the base liquid. Leong *et al.*[21] found that, based on comparisons to experimental data, a nanolayer thermal conductivity of 2 or 3 times the base liquid was able to provide accurate results. Xie *et al.*[22] also considered the effect of the nanolayer on the thermal conductivity of a nanofluid. The effect of the nanolayer was considered to be much more important for mixtures with nanoscale particles than fluid mixtures with larger particles. The high surface area to volume ratio of the nanoparticles would create a higher volume of the ordered liquid layer than would be found in a mixture containing larger particles at the same concentration. The properties of the nanolayer are thought to be dependent on the nanoparticle material, the base fluid, and how the base fluid and nanoparticles interact with one another. The thermal conductivity of the nanolayer cannot be definitively determined, however it is thought to be between the thermal conductivity of the base fluid and the solid particle. Although the nanolayer has some properties that resemble the properties of a solid, it is still a liquid and does not have as high a thermal conductivity as the solid particles. The proposed model of the thermal conductivity of a nanofluid is given as:

$$\frac{k_{nf} - k_{bf}}{k_{bf}} = 3\Theta\phi + \frac{3\Theta^2\phi^2}{1 - \Theta\phi} \tag{7.27}$$

$$\Theta = \frac{\left(\dfrac{k_{nl} - k_{bf}}{k_{nl} + 2k_{bf}}\right)\left[\left(1 + \dfrac{\delta_{nl}}{r_p}\right)^3 - \left[\dfrac{(k_p - k_{nl})(k_{bf} + 2k_{nl})}{(k_p + 2k_{nl})(k_{bf} - k_{nl})}\right]\right]}{\left(1 + \dfrac{\delta_{nl}}{r_p}\right)^3 + 2\left[\left(\dfrac{k_{nl} - k_{bf}}{k_{nl} + 2k_{bf}}\right)\left(\dfrac{k_p - k_{nl}}{k_p + 2k_{nl}}\right)\right]} \tag{7.28}$$

Based on this model, it is expected that the thermal conductivity increases as the size of the nanoparticle decreases, since the nanolayer has a more significant impact on smaller particles. This model also assumes that there is a continuous distribution of thermal conductivity in the nanofluid mixture. This means that the outer surface of the nanolayer has the thermal conductivity of the base fluid, while the inner surface of the nanolayer has the thermal conductivity of the nanoparticles. Based on this assumption, there is a continuous transition from the high thermal conductivity particles to the lower thermal conductivity base fluid. A nanolayer thickness of 2 nm was used in evaluations of the model. Yu and Choi[45] modified the model by Maxwell[10] to include the effect of the nanolayer. This model considers three components of the nanofluid composition, the nanoparticles, the base fluid, and the nanolayer. The Maxwell model[10] can be modified by considering an equivalent thermal conductivity of the particles that includes the effect of the nanolayer rather than just the particles, as is used in the classical models. It is thought that the nanolayer will act as a thermal bridge to join the solid particles and the base fluid. In this model, the heat transfer in the nanofluid is assumed to be diffusive. This assumption is made since the distance between the suspended nanoparticles is greater than the mean free path of the base fluid molecules. Particles are assumed to be spherical in this model. The nanolayer is considered to be more ordered to have a higher thermal conductivity than the base fluid. Considering the nanolayer, a modified Maxwell model[10] by Yu and Choi[45] is given as:

$$k_{nf} = \frac{k_{pe} + 2k_{bf} + 2(k_{pe} - k_{bf})(\beta_{nl})^3 \phi}{k_{pe} + 2k_{bf} - (k_{pe} - k_{bf})(\beta_{nl})^3 \phi} k_{bf} \tag{7.29}$$

$$\beta_{nl} = 1 + \frac{\delta_{nl}}{r_p} \tag{7.30}$$

$$k_{pe} = \frac{[2(1-\gamma) + (\beta_{nl})^3(1+2\gamma)]\gamma}{-(1-\gamma) + (\beta_{nl})^3(1+2\gamma)} k_p \tag{7.31}$$

$$\gamma = \frac{k_{nl}}{k_p} \tag{7.32}$$

where k_{pe} is the equivalent thermal conductivity of the particles, considering both the effect of the nanoparticles and the nanolayer, γ is the ratio between the thermal conductivity of the nanolayer and that of the nanoparticles, and β_{nl} is a dimensionless nanolayer parameter based on the nanolayer thickness and nanoparticle radius. Because of the higher surface area to volume ratio of smaller particles, there is a significant increase in the effect of the nanolayer for smaller nanoparticles. For large particles the model will reduce to the classic Maxwell model.[10] In the analysis of the model of Yu and Choi, a nanolayer thickness of 2 nm was used. There was no clear estimation of the thermal conductivity of the nanolayer, but it was found that the model was most accurate when $k_{nl} > 10k_{bf}$. Additionally, Tinga *et al.*[23] created a model that was intended to determine the complex dielectric constant, a property quantifying the force between charges in a material, of a mixture containing inclusions. Similar to Maxwell,[10] the model that was created for electrical properties was modified to be used to model the thermal conductivity of a nanofluid. The model originally considered an air–water–cellulose mixture, where the cellulose was the inclusion in the mixture. In order to modify the model for use in nanoparticles, the cellulose was replaced with solid particles, the water shell was replaced with the nanolayer, and the host medium of air was replaced by the base fluid. The modified model is given as:

$$k_{nf} = \left(1 + \frac{3\phi\left[(\beta_{nl}^3 - 1)(2k_{nl} + k_p)(k_{nl} - k_{bf}) - (k_{nl} - k_p)(2k_{nl} + k_{bf})\right]}{(2k_{bf} + k_{nl})(2k_{nl} + k_p) - \frac{2}{\beta_{nl}^3 - 1}(k_{nl} - k_{bf}) - (k_{nl} - k_p) - 3\phi k_{nl}(k_p - k_{bf})}\right) k_{bf} \tag{7.33}$$

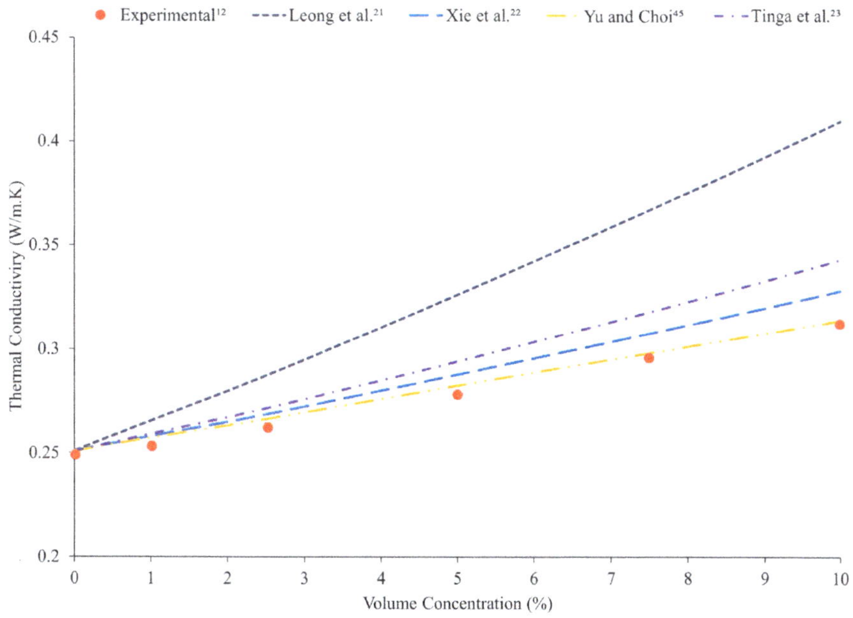

Figure 7.10 Comparison of nanolayer models to experimental data for 40 nm Al_2O_3–ethylene glycol nanofluid thermal conductivity as a function of volume fraction at a temperature of 23 °C.

In Figure 7.10 the nanolayer models are compared to experimental data for Al_2O_3–water nanofluids.[12] The models of Yu and Choi[45] and Xie *et al.*[22] both demonstrated a high degree of accuracy, with average absolute errors of 1.08% and 2.89%, respectively. The models of Tinga *et al.*[23] and Leong *et al.*[21] were less accurate, with average absolute errors of 4.90% and 14.56%, respectively. In the case of the Yu and Choi[45] model, the thermal conductivity of the nanolayer was taken as $10k_{bf}$ while for that of Leong *et al.*[21] it was taken as $2k_{bf}$. Obviously, there is not a clear method by which to determine the thermal conductivity of a nanolayer. Despite the Yu and Choi model[45] assuming a higher thermal conductivity of the nanolayer, the model of Leong *et al.*[21] predicted a higher overall thermal conductivity of the nanofluid. Therefore, the correlation between the thermal conductivity of the nanolayer and the overall thermal conductivity cannot yet by fully understood using nanolayer models.

In Figure 7.11, the layering models are compared with experimental data.[20] A CuO–water nanofluid was used at a volume concentration of 0.5%, with 31 nm in diameter nanoparticles. The Leong *et al.*[21] model predicts the highest thermal conductivity, but this time it is the most accurate model, with an average absolute error of 1.28%. The other three models are all still relatively accurate, with average absolute errors of <2.67%. Unlike in Figure 7.9, there was less variation between the models. This indicates that the nanofluid conditions determine if there is deviation between the models.

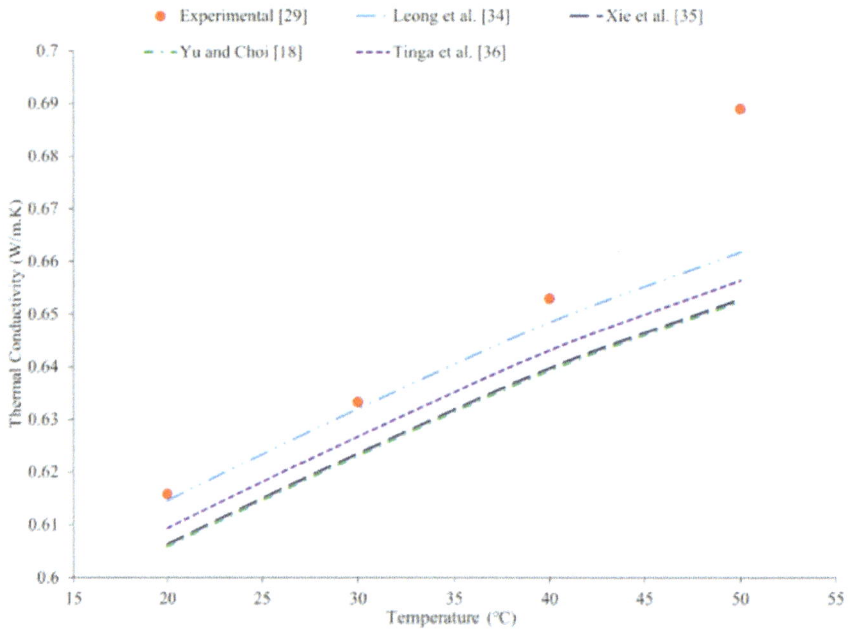

Figure 7.11 Comparison of Brownian models to experimental data for 31 nm CuO–water nanofluid thermal conductivity as a function of temperature at a volume fraction of 0.005.

For certain data sets, any of the nanolayer models may predict similar thermal conductivity. Additionally, the models are all very inaccurate for the final data point at the highest temperature. The layering models may also not fully consider the effect of temperature on the thermal conductivity of the nanofluid.

A shown in Figure 7.12, the nanolayer models were then compared to experimental data for TiO_2–water nanofluids.[40] For this data set, the most accurate models were the models of Leong *et al.*[21] and Tinga *et al.*,[23] with average absolute errors of 2.43% and 2.13%, respectively. It has been consistently shown that the models of Leong *et al.*[21] and Tinga *et al.*[23] predict a higher thermal conductivity than the models of Xie *et al.*[22] and Yu and Choi.[45] This higher thermal conductivity can either increase the accuracy or decease the accuracy of the prediction based on the data set. Since the models have different parameters for the nanolayer properties, and the accuracy of each model varies by data set, it is difficult to determine which set of parameters is correct. The accuracy of the models for a given nanolayer thickness and thermal conductivity of the nanolayer will vary based on the chosen data. Therefore, despite there being clear trends in the behavior of the models it is impossible to declare one model to be generally more accurate than the others.

In Table 6, the average absolute errors for the nanolayer models are compared to several sets of experimental data.

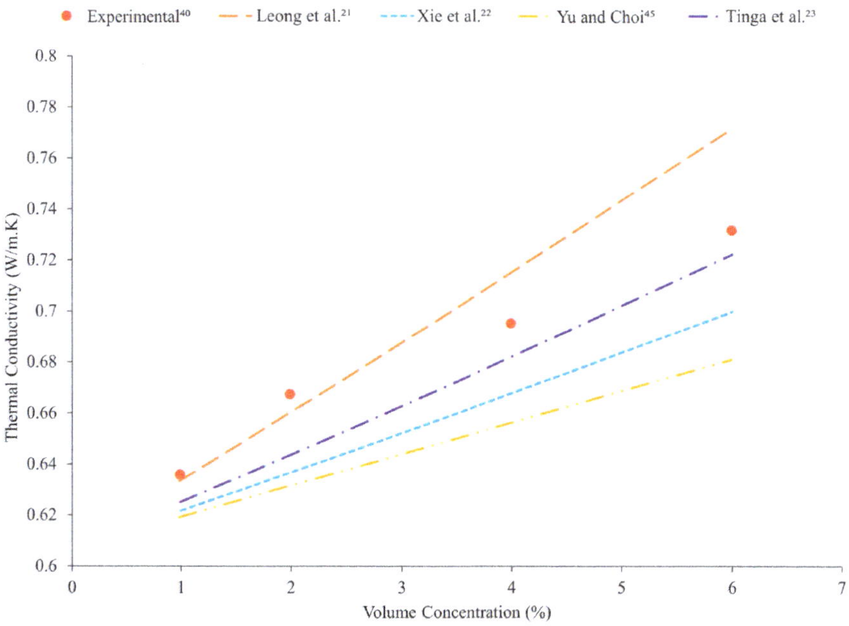

Figure 7.12 Comparison of nanolayer models to experimental data for 150 nm TiO_2–water nanofluid thermal conductivity as a function of volume concentration at a temperature of 25 °C.

Table 7.6 Average percentage errors of nanolayer models.

Average percentage error nanolayer models				
Ref.	Leong *et al.*[21]	Xie *et al.*[22]	Yu and Choi[45]	Tinga *et al.*[23]
Al_2O_3–water, 0.05 vol%, 29 nm[38]	4.61%	4.74%	4.74%	4.69%
Al_2O_3–water, 0.2 vol%, 29 nm[38]	8.44%	8.94%	14.74%	8.74%
Al_2O_3–EG, 23 °C, 40 nm[12]	14.56%	2.89%	1.08%	4.90%
Al_2O_3–EG, 23 °C, 11 nm[12]	15.06%	1.89%	9.39%	4.13%
Al_2O_3–water, 0.1 vol%, 50 nm[39]	3.08%	3.13%	3.14%	3.11%
TiO_2–water, 25 °C, 150 nm[40]	2.434%	3.80%	5.15%	2.13%
CuO–water, 0.5 vol%, 31 nm[20]	1.28%	2.61%	2.67%	2.10%
CuO–EG, 0.5 vol%, 31 nm[20]	3.79%	5.16%	5.21%	4.95%
ZrO_2–water, 0.05 vol%, 30 nm[29]	0.96%	0.78%	0.99%	0.32%
Al_2O_3–water, 1.30 vol%, 30 nm[56]	7.20%	2.83%	7.98%	21.40%

7.6 Molecular Dynamics Models

Molecular dynamics is a method of computer simulation used to calculate the interactions between atoms and molecules. These computer simulations provide a method to understand interactions that cannot be measured experimentally. Molecular dynamics simulations are based on solving classical equations of motion for an atomic system.[57] These simulations can also be used to study the behavior of nanofluid mixtures, with several researchers having used molecular dynamics simulations as a basis for predicting the thermal conductivity of a nanofluid. Evans *et al.*[24] used molecular dynamics simulations to research the mechanisms of the enhancement of thermal conductivity. They found that Brownian motion does not have much effect on the enhancement of thermal conductivity, and that the results were more consistent with the EMT. The parameters of the simulation included nanoparticles with a density 4.5 times higher than the base fluid. In the simulation, 8 particles were placed in a mixture of 50 000 fluid atoms, which is equivalent to a volume concentration of around 3.3%. In the simulations, the nanoparticles remained separate and did not form clusters. The simulations were also ran considering three different strengths of solid–fluid interactions for cases with no wetting, weak wetting, and wetting. As the strength of the solid–fluid interactions between the particle and the base fluid increased, based on the three cases considered in the simulation, the predicted thermal conductivity increased as well. This is because there is an interfacial thermal resistance where the solid and fluid meet, and this resistance is reduced by strengthening the solid–fluid interactions. There is a temperature drop across this interfacial thermal resistance, and this interfacial thermal resistance is further quantified by determining an equivalent matrix. The equivalent matrix represents the distance in the solid particle required to promote a temperature drop equal to what is caused by the interfacial thermal resistance. The nanofluid thermal conductivity model:

$$k_{\mathrm{nf}} = \left(1 + 3\phi\frac{\varepsilon - 1}{\varepsilon + 2}\right)k_{\mathrm{bf}} \tag{7.34}$$

where ε is the ratio between the particle radius and the equivalent matrix thickness. Based on this model, the thermal conductivity of the nanofluid decreases as the size of the particles decreases due to the increased thermal resistance at the interfacial boundary. Vladkov and Barrat[25] also used molecular dynamics simulations to study the thermal conductivity of a nanofluid. They took the effect of the interfacial resistance seriously in their model. Simulations were conducted on a particle containing 555 atoms surrounded by 30 000 liquid atoms. This corresponds to a particle with a radius of 1.5 nm and a total system with a side length of 10 nm. Simulations were then used to determine the heat flux through the system, with this heat flux used to determine thermal conductivity. The determined thermal conductivity model is given as:

$$k_{\mathrm{nf}} = \left(\frac{\left(\frac{k_{\mathrm{p}}}{k_{\mathrm{bf}}}(1 + 2\Omega) + 2\right) + 2\phi\left(\frac{k_{\mathrm{p}}}{k_{\mathrm{bf}}}(1 - \Omega) - 1\right)}{\left(\frac{k_{\mathrm{p}}}{k_{\mathrm{bf}}}(1 + 2\Omega) + 2\right) - \phi\left(\frac{k_{\mathrm{p}}}{k_{\mathrm{bf}}}(1 - \Omega) - 1\right)}\right)k_{\mathrm{bf}} \tag{7.35}$$

$$\Omega = \frac{R_{\mathrm{k}}k_{\mathrm{bf}}}{r_{\mathrm{p}}} \tag{7.36}$$

where R_k is the Kapitza resistance, which is the thermal resistance at the interface between the particle and base fluid. Based on the simulations the enhancement in thermal conductivity in nanofluids is primarily due to nanoparticles working collectively to form clusters rather than individual particles. The effect of the clusters is significant since the interfacial resistance limits the possible thermal conductivity enhancement from individual particles. Lin *et al.*[26] investigated the thermal conductivity of nanofluids by performing molecular dynamics simulations on Cu–ethylene glycol based nanofluids. They believed that previous molecular dynamics simulations were performed on unrealistic nanofluids and only focused on the effect of interfacial resistance. In order to correct this the effect of the nanolayer on thermal conductivity enhancement was considered in this model. The simulated nanofluid contained 1200 ethylene glycol molecules and the nanoparticle concentration was 3%. The model determined from the simulation combined the EMT and nanolayer models based on the work of Mawxell[10] and Yu and Choi.[45] The model is based on the heat conduction model in spherical coordinates and combines three parts in the nanofluid. The three parts are the nanoparticle, an ordered liquid layer around the particle and a base fluid layer around the particle. The determined model is given as:

$$k_{nf} = k_{bf} + \frac{(k_p - k_{bf})\left[\dfrac{(1-\Xi_1)(1-\Xi_2)}{1+2\rho_{nf}\Xi_1\Xi_2}\right]\phi_p + (k_{nl} - k_{bf})\left[\dfrac{(1-\Xi_1)}{1+2\rho_{nf}\Xi_1\Xi_2}\right](\phi_{pl} - \phi_p)}{\left[\dfrac{(1-\Xi_1)(1-\Xi_2)}{1+2\rho_{nf}\Xi_1\Xi_2}\right]\phi_p + \left[\dfrac{(1-\Xi_1)}{1+2\rho_{nf}\Xi_1\Xi_2}\right]\left(\phi_{pl} - \phi_p\right) + (1-\phi_{pl})} \tag{7.37}$$

$$\Xi_1 = \frac{k_p - k_{nl}}{k_p + 2k_{nl}} \tag{7.38}$$

$$\Xi_2 = \frac{k_{nl} - k_{bf}}{k_{nl} + 2k_{bf}} \tag{7.39}$$

where ϕ_{pl}, is the volume fraction of the nanoparticle plus the nanolayer and ϕ_p represents the volume fraction of only the nanoparticles. Using this model, the thermal conductivity enhancement is greater for smaller sized nanoparticles. Mirmohammadi *et al.*[27] used molecular dynamics simulations to investigate the effects of nanoparticle concentration, temperature, and nanoparticle shape on the thermal conductivity of nanofluids. In the molecular dynamics simulations, water was considered as the base fluid in a mixture with silver nanoparticles. Three nanoparticle shapes, spheres, cylinders, and rectangular prisms, were considered with several different dimensions. In the simulations, several parameters considering the effect of the particle shape were investigated. The elongation and sphericity of the particles were found to not have a strong correlation with thermal conductivity. The surface area to volume ratio showed a strong correlation with thermal conductivity, making it the most relevant parameter to be considered. The specific nanofluid conditions used in this simulation had a volume concentration from 0.14% to 1.4% and a temperature from

7 °C to 62 °C. In the simulations, the thermal conductivity of the nanofluids increased with an increase in temperature and nanoparticle concentration. These results were also supported by experimental data collected for silver–water nanofluids. Dimensional analysis was used to non-dimensionalize the data collected from the molecular dynamics simulation, which was used to determine the model of the thermal conductivity of the nanofluids:

$$k_{nf} = k_{bf}\left(1 + 9.89\left(\frac{T}{300}\right)^{4.75}(\phi)^{0.65}\left(0.30855\frac{SA}{V}\right)^{0.23}\right) \tag{7.40}$$

In this model, SA represents the surface area of the nanoparticle and V represents the volume of the nanoparticles. Increasing the surface area to volume ratio increases the thermal conductivity. The relationship between the surface area to volume ratio and thermal conductivity is asymptotic and for high surface area to volume ratios the thermal conductivity approaches a constant. This model is valid for volume concentrations from 0.14% to 1.4% and temperatures from 7 °C to 62 °C.

Figure 7.13 compares the experimental data for Al_2O_3–water nanofluids to the molecular dynamics models.[38] For this data set, the most accurate model is the model of Mirmohammadi *et al.*,[27] with an absolute average error of 1.15%, the three other models all have an absolute average error of between 4.80% and 4.90%. The model of Mirmohammadi *et al.*[27] remains accurate

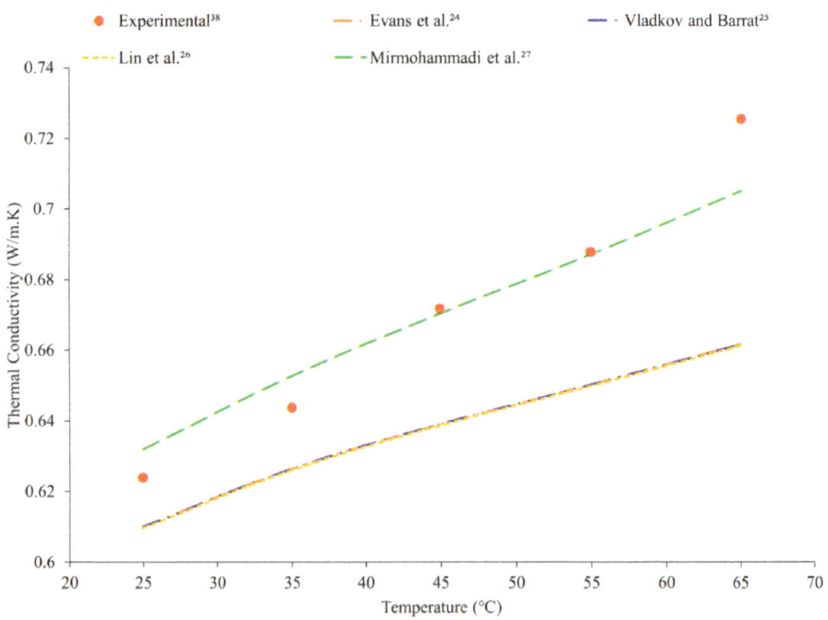

Figure 7.13 Comparison of molecular dynamics models to experimental data for 29.2 nm Al_2O_3–water nanofluid thermal conductivity as a function of temperature at a volume fraction of 0.0005.

for the entire data set, while the three other models become increasingly less accurate as the temperature increases. The model of Mirmohammadi *et al.*[27] is the only one that directly considers the effect of temperature, which might increase the level of accuracy of the model.

Figure 7.14 compares molecular dynamics models to experimental data for Al_2O_3–ethylene glycol nanofluids.[20] For a given data, the most accurate models are those of Evans *et al.*[24] and Vladkov and Barrat,[25] with average absolute errors of 1.49% and 1.52%, respectively. The least accurate model is that of Mirmohammadi *et al.*,[27] with an absolute average error of 15.9%. The model of Lin *et al.*[26] is moderately accurate, with an absolute average error of 3.90%. As shown in both Figures 7.13 and 7.14, the model by Mirmohammadi *et al.*[27] can be seen to predict higher thermal conductivity values than the other models. The level of accuracy for the model of Mirmohammadi *et al.*[27] is very different, as observed in Figure 7.13 and 7.14. This has been seen with other methods in this chapter, where the accuracy of the models varies significantly according to the dataset, suggesting that there is not yet a model that is valid for all datasets.

In Figure 7.15, the molecular dynamics models are compared to experimental data for Al_2O_3–water nanofluids.[39] In this case, all four of the models have a similar level of accuracy, with an absolute average error of between 2.34% and 2.92%. At low temperatures, the models of Evans *et al.*,[24] Vladkov and Barrat,[25] and Lin *et al.*[26] are more accurate, while at higher temperatures the model of Mirmohammadi *et al.*[27] is more accurate. The range of

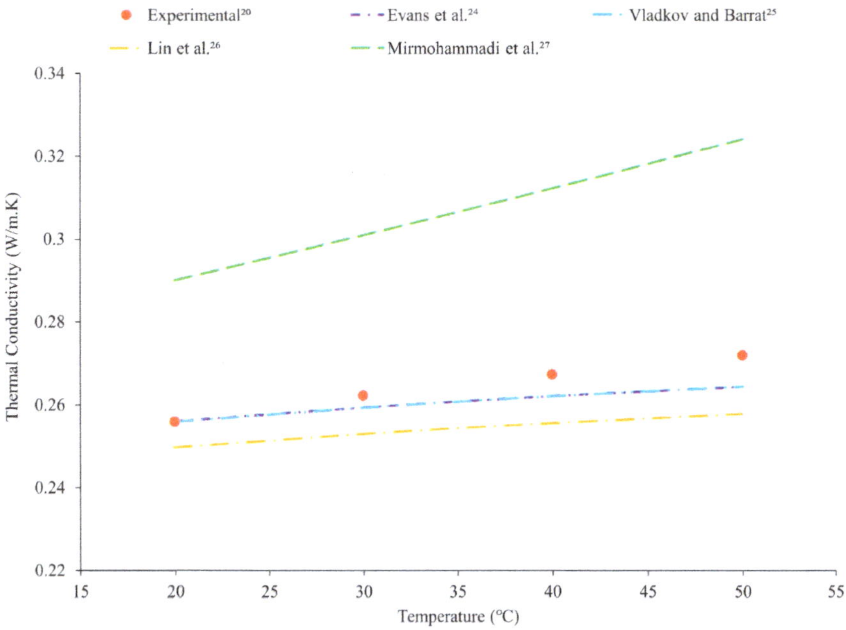

Figure 7.14 Comparison of molecular dynamics models to experimental data for 150 nm Al_2O_3–ethylene glycol nanofluid thermal conductivity as a function of temperature at a volume fraction of 0.01.

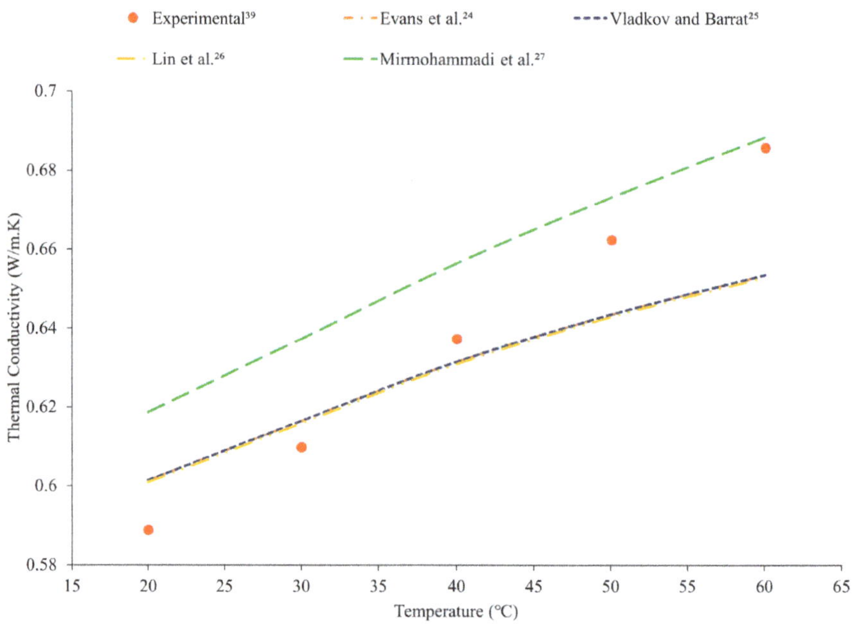

Figure 7.15 Comparison of molecular dynamics models to experimental data for 50 nm Al_2O_3–water nanofluid thermal conductivity as a function of temperature at a volume fraction of 0.0005.

Table 7.7 Average percentage error molecular dynamics models.

Average percentage error molecular dynamics models				
Ref.	Evans *et al.*[24]	Vladkov and Barrat[25]	Lin *et al.*[26]	Mirmohammadi *et al.*[27]
Al_2O_3–EG, 23 °C, 40 nm[12]	3.07%	2.97%	8.78%	37.35%
Al_2O_3–water, 0.05 vol%, 29 nm[38]	4.80%	4.81%	4.87%	1.15%
Al_2O_3–water, 0.05 vol%, 50 nm[39]	2.34%	2.34%	2.36%	2.92%
Al_2O_3–EG, 1 vol%, 150 nm[20]	1.49%	1.52%	3.90%	15.93%
Al_2O_3–water, 23 °C, 20 nm[12]	1.21%	1.31%	4.97%	59.98%
Al_2O_3–water, 1.5 vol%, 50 nm[55]	3.50%	3.62%	6.04%	29.21%

temperature may affect the accuracy of the model for a given data set. Based on the data examined in this case, none of the molecular dynamics models can account for the effect of temperature in all cases. Under certain conditions the models can be accurate though.

Table 7.7 shows the average absolute errors for the molecular dynamics models compared to several sets of experimental data.

7.7 Recent Developments

Since the mechanisms of heat transfer in a nanofluid are very complex and depend on many parameters it has been difficult for researchers to determine models that consider all parameters accurately. As a result, much of the research in the last couple years on modelling the thermal conductivity of nanoparticles is focused on determining correlations with experimental data. This can allow equations to be determined that can accurately provide results for a wide range of nanofluid conditions and mean that researchers will not have to choose which of the available models will be applicable to the nanofluid being studied.[28] Moghaddam *et al.*[28] determined a model for the thermal conductivity of nanofluids based on 472 experimental data points. The nanofluids considered were water and ethylene glycol containing nanoparticle materials such as Al, Al_2O_3, Au, CuO, Fe, Fe_3O_4, $Mg(OH)_2$, MgO, TiO_2, and ZnO. The nanoparticle sizes ranged from 5–150 nm and the nanofluid temperatures ranged from 5–70 °C. Regression analysis was performed on the data to generate a correlation considering volume concentration, nanofluid temperature, and nanoparticle diameter. The model is given as:

$$k_{nf} = k_{bf}\left(1.0139(1+\phi)^{2.964}\left(1+\frac{T}{70}\right)^{0.155}\left(1+\frac{d_p}{0.0275}\right)^{-0.01}\right) \tag{7.41}$$

In this model the nanofluid temperature, T, is in Celsius and the nanoparticle diameter, d_p, is in nm. Çolak[29] determined a model based on experimental data for ZrO_2–water nanofluids. ZrO_2 nanoparticles were chosen since at the time the research was conducted there were not any other correlations found for ZrO_2 nanofluids in the literature. The nanofluids were tested at volume concentrations from 0.0125% to 0.2% and temperatures from 10 °C to 65 °C. The data was then analyzed using an artificial neural network and the least squares method to determine the correlation given as:

$$k_{nf} = (0.0000859T + 0.03354)(17.01 + \phi) \tag{7.42}$$

where the nanofluid temperature, T, is given in Celsius. Additionally, Kamel *et al.*[30] proposed a correlation for Al_2O_3–water nanofluids based on experimental data for nanofluids with a volume concentration from 0.01% to 0.5% and a temperature from 35 °C to 50 °C. The correlation determined based on the experimental data is given as:

$$k_{nf} = k_{bf}(0.9845 - 0.000827T - 0.01939\phi + 0.00000293T^2 + 0.00123T\phi) \tag{7.43}$$

Figure 7.16 compares the recent nanofluid thermal conductivity models to experimental data for Al_2O_3–water nanofluids.[54] All models were relatively accurate with all percentage errors <3%. The most accurate model was the model of Moghaddam *et al.*,[28] with an average absolute error of 0.85%, followed by the models of Kamel *et al.*[30] and Çolak,[29] with average absolute errors of 1.72% and 2.82%, respectively. The model of Moghaddam *et al.*[28] is based on the largest data set, which could explain its high level of accuracy.

Figure 7.17 compares the models to the experimental data of the thermal conductivity of Al_2O_3–water nanofluids.[30] In this case, the model of Çolak[29]

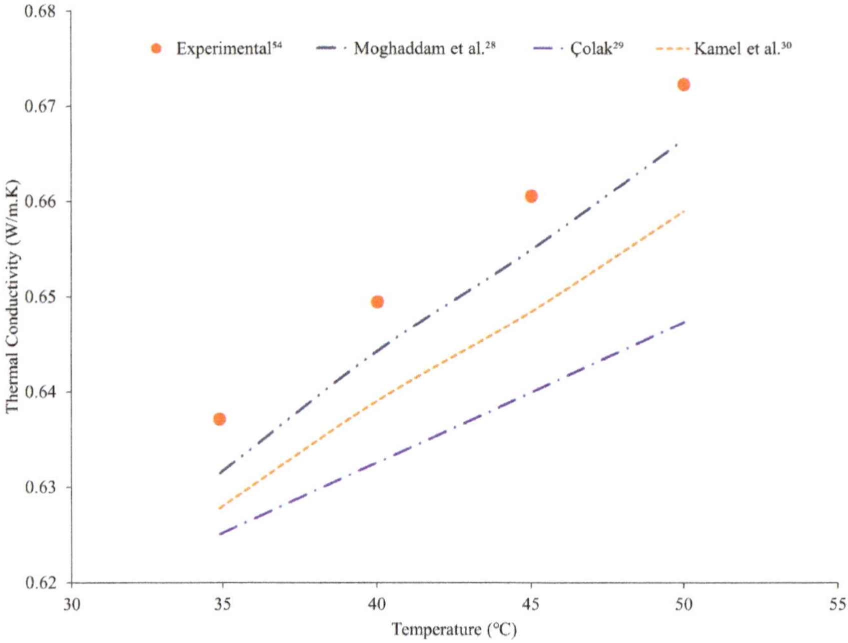

Figure 7.16 Comparison of recent models to experimental data for 20 nm Al_2O_3–water nanofluid thermal conductivity as a function of temperature at a volume fraction of 0.001.

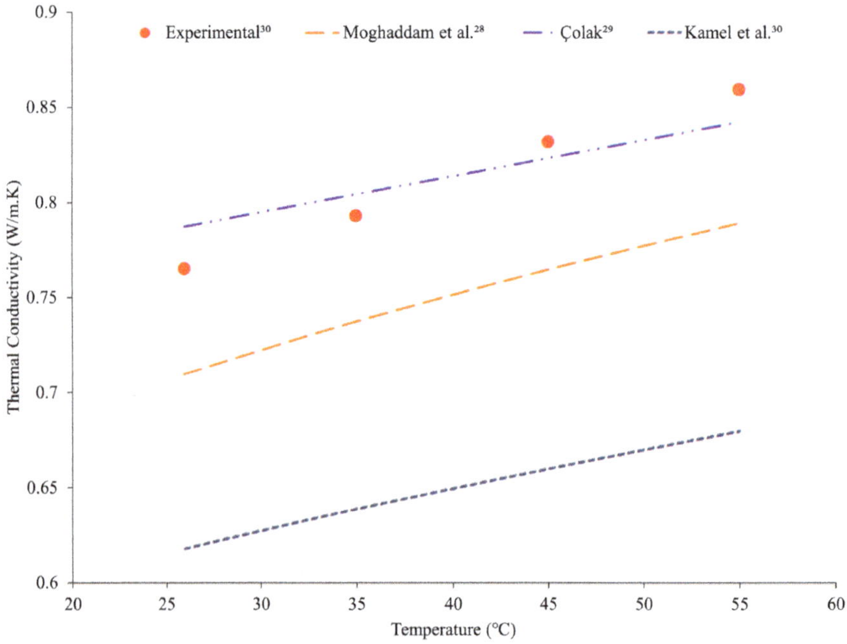

Figure 7.17 Comparison of recent models to experimental data for 5 nm Al_2O_3–water nanofluid thermal conductivity as a function of temperature at a volume fraction of 0.05.

Table 7.8 Average percentage errors of the most recent models.

Average percentage errors of new models			
Ref.	Moghaddam *et al.*[28]	Çolak[29]	Kamel *et al.*[30]
Al_2O_3-water, 0.05 vol%, 29 nm[38]	2.84%	5.01%	3.13%
Al_2O_3-water, 0.2 vol%, 29 nm[38]	7.10%	8.51%	7.74%
Al_2O_3-water, 5 vol%, 5 nm[54]	7.63%	1.85%	20.10%
Al_2O_3-water, 4 vol%, 5 nm[54]	5.99%	2.16%	16.63%
Al_2O_3-water, 1.5 vol%, 50 nm[55]	1.63%	2.32%	4.30%
CuO-water, 1 vol%, 28.6 nm[53]	12.76%	10.84%	14.90%
CuO-water, 4 vol%, 28.6 nm[53]	9.03%	4.33%	17.77%
CuO-water, 2 vol%, 31 nm[20]	3.37%	2.10%	8.00%
CuO-water, 3 vol%, 31 nm[20]	3.04%	3.91%	10.00%
SiO_2-water, 1 vol%, 45 nm[43]	5.24%	3.40%	6.68%
Al_2O_3-water, 0.1 vol%, 20 nm[30]	0.85%	2.82%	1.72%

had the highest level of accuracy, with an average absolute error of 1.85%. The models by Moghaddam *et al.*[28] and Kamel *et al.*[30] were both inaccurate, with average absolute errors of 7.63% and 20.10%, respectively. Even though the model by Çolak[29] was based on ZrO_2 nanofluids, the model was accurate for Al_2O_3 nanofluid data. This suggests that models based on experimental can be accurate for nanofluid conditions other than those used to determine the model.

In Table 7.8, the average absolute errors for the most recent models are compared to several sets of experimental data.

Table 7.9 presents the thermal conductivity of nanofluid models discussed in the chapter.

7.8 Conclusion

In this chapter, the primary methods used to understand the thermal conductivity of nanofluids, the EMT, aggregation, Brownian motion, nanolayer, and molecular dynamics, were explained. Using these methods, many models to predict the thermal conductivity of nanofluids have been created; some of these models were presented and compared to experimental data. From the comparison to experimental data, it was shown that the level of accuracy of the model was highly dependent on the given data. Based on this it is difficult to state which mechanism of thermal conductivity enhancement in a nanofluid is most significant, as none of the models demonstrate accuracy in all cases. Although there are existing models that can provide accurate results under certain conditions, the mechanisms of the thermal conductivity of nanofluids cannot be fully captured by theoretical modelling. Recently, much of the work to model the thermal conductivity of nanofluids has been based on experimental results. This work has produced accurate models in some cases, but there is still not a model for all cases.

Table 7.9 Nanofluid thermal conductivity models.

Model	Equation	Notes
Maxwell[10]	$k_{nf}=\frac{k_p+2k_{bf}+2\phi(k_p-k_{bf})}{k_p+2k_{bf}-\phi(k_p-k_{bf})}k_{bf}$	• Valid for spherical particles at low concentration, generally <2% • Assumes static and even particle distribution of particles with no particle interactions
Hamilton and Crosser[11]	$k_{nf}=k_{bf}\frac{k_p+(n-1)k_{bf}-(n-1)(k_{bf}-k_p)\phi}{k_p+(n-1)k_{bf}+(k_{bf}-k_p)\phi}$	• Based on static and homogenous particle distribution • Considers shape of particle agglomerates with $n=3$ used for spherical agglomerates and $n=6$ for cylindrical agglomerates
Timofeeva *et al.*[12]	$k_{nf}=k_{bf}(1+3\phi)$	• Suspended particles are assumed to be immobile • Valid in low concentrations
Prasher *et al.*[13]	$k_{nf}=\frac{k_a+2k_{bf}+2\phi_a(k_a-k_{bf})}{k_a+2k_{bf}-\phi_a(k_a-k_{bf})}k_{bf}$	• Model based on EMT • Considers the concentration and thermal conductivity of aggregates rather than particles
Xuan *et al.*[14]	$k_{nf}=\left(\frac{k_p+2k_{bf}+2\phi(k_p-k_{bf})}{k_p+2k_{bf}-\phi(k_p-k_{bf})}+\frac{\rho_p\phi c_p}{2k_{bf}}\sqrt{\frac{k_BT}{3\pi r_a\mu}}\right)k_{bf}$	• Combination of EMT and Brownian motion • Considers the radius of the cluster rather than the particle in the Brownian motion calculation
Feng *et al.*[15]	$k_{nf}=\left((1-\phi_e)\frac{k_{pe}+2k_{bf}+2(k_{pe}-k_{bf})(\beta_{nl})^3\phi}{k_{pe}+2k_{bf}-(k_{pe}-k_{bf})(\beta_{nl})^3\phi}+\phi_e\left(\left(1-\tfrac{3}{2}\phi_e\right)+\frac{3\phi_e}{\alpha}\left(\frac{1}{\alpha}\ln\left(\frac{r_p+\delta_{nl}}{(r_p+\delta_{nl})(1-\alpha)}-1\right)\right)\right)\right)k_{bf}$ $\phi_e=\phi\beta_{nl}^3\ \alpha=1-\frac{k_{bf}}{k_{pe}}$	• Combination of nanolayer and clustering. • Considers two volume fractions, one for the particles and one for the clusters
Corcione[16]	$k_{nf}=k_{bf}\left(1+4.4Re^{0.4}Pr^{0.66}\left(\frac{T}{T_{fr}}\right)^{10}\left(\frac{k_p}{k_{bf}}\right)^{0.03}\phi^{0.066}\right)$ $\mathrm{Re}=\frac{2\rho_{bf}k_{bf}T}{\pi\mu_{bf}^2d_p}\ \mathrm{Pr}=\frac{c_{bf}\mu_{bf}}{k_{bf}}$	• Based on line of best fit taken over large set of experimental data

Chon *et al.*[17]

$$k_{nf} = \left(1 + 64.7\phi^{.746}\left(\frac{d_{bf}}{d_{np}}\right)^{.369}\left(\frac{k_{np}}{k_{bf}}\right)^{.746} Pr^{.9955} Re^{1.2321}\right)k_{bf}$$

$$\mathrm{Re} = \frac{\rho_{bf} k_b T}{3\pi\mu_{bf}^2 l_{bf}} \quad \mathrm{Pr} = \frac{c_{pbf}\mu_{bf}}{k_{bf}}$$

- Model based on Buckingham–Pi theorem and large set of Al_2O_3–water nanofluid data
- Assumes $d_{bf} = 0.384$ nm and $l_{bf} = 0.17$ nm
- Micro mixing caused by Brownian motion has a significant impact on thermal conductivity enhancement
- Thermal conductivity enhancement has a static part and a part caused by Brownian motion

Koo and Kleinstreuer[18]

$$k_{nf} = \left[\frac{k_p + 2k_{bf} - 2\phi(k_{bf} - k_p)}{k_p + 2k_{bf} + \phi(k_{bf} - k_p)}\right]k_{bf} + \left[5 \times 10^4 \beta\phi\rho_{bf}c_{bf}\sqrt{\frac{k_b T}{\rho_p d_p}} f(T, \phi, etc.)\right]k_{bf}$$

$$f(T,\phi) = (-6.04\phi + 4705)\ T + (1722.3\phi - 134.63)$$

Nanoparticle	β	Volume fraction	Temperature
Al_2O_3	$0.0011\ (100\phi)^{-0.7272}$	$\phi 1\%$	$300K \leq T \leq 325K$
CuO	$0.0017\ (100\phi)^{-0.0841}$	$\phi 1\%$	$300K \leq T \leq 325K$

Vajjha and Das[19]

$$k_{nf} = \left[\frac{k_p + 2k_{bf} - 2\phi(k_{bf} - k_p)}{k_p + 2k_{bf} + \phi(k_{bf} - k_p)}\right]k_{bf} + [5 \times 10^4 \beta\phi\rho_{bf}c_{bf}\sqrt{\frac{k_b T}{\rho_p d_p}} f(T, \phi, etc.)])k_{bf}$$

$$f(T,\phi) = (.028217\phi + .003917)\left(\frac{T}{T_0}\right) + (-.030669\phi - .00391123)$$

Nanoparticle	β	Volume Fraction	Temperature
Al_2O_3	$8.4407(100\phi)^{-1.07304}$	$1\% \leq \phi \leq 10\%$	$298K \leq T \leq 363K$
ZnO	$8.4407(100\phi)^{-1.07304}$	$1\% \leq \phi \leq 7\%$	$298K \leq T \leq 363K$
CuO	$9.8810(100\phi)^{-.9446}$	$1\% \leq \phi \leq 6\%$	$298K \leq T \leq 363K$

- Based on same equation as Koo and Kleinstreuer with different $f(T,\phi,etc.)$ and β.
- Assumes thermal conductivity is temperature dependent

Table 7.9 (*Continued*)

Patel *et al.*[20]	$k_{nf} = k_{bf}\left(1 + 0.135\left(\frac{k_p}{k_{bf}}\right)^{.273} \phi^{.467}\left(\frac{T}{20}\right)^{.547}\left(\frac{100}{d_p}\right)^{.234}\right)$	• Based on regression analysis over a large set of experimental data • Assumes thermal conductivity is temperature dependent
Leong *et al.*[21]	$k_{nf} = \frac{(k_p - k_{nl})\phi k_{nl}\left[2\left(1+\frac{\delta_{nl}}{2r_p}\right)^3 - \left(1+\frac{\delta_{nl}}{r_p}\right)^3 + 1\right] + (k_p + 2k_{nl})\left(1+\frac{\delta_{nl}}{2r_p}\right)^3\left[\phi\left(1+\frac{\delta_{nl}}{2r_p}\right)^3 (k_{nl} - k_{bf}) + k_{bf}\right]}{\left(1+\frac{\delta_{nl}}{2r_p}\right)^3 (k_p + 2k_{nl}) - (k_p - k_{nl})\phi\left[\left(1+\frac{\delta_{nl}}{2r_p}\right)^3 + \left(1+\frac{\delta_{nl}}{r_p}\right)^3 - 1\right]}$	• Considers that the nanolayer has a significant impact on the thermal conductivity enhancement of the nanofluid • Nanolayer thermal conductivity is considered a multiple of base fluid thermal conductivity
Xie *et al.*[22]	$\frac{k_{nf} - k_{bf}}{k_{bf}} = 3\Theta\phi + \frac{3\Theta^2\phi^2}{1 - \Theta\phi}$ $\Theta = \frac{\left(\frac{k_{nl} - k_{bf}}{k_{nl} + 2k_{bf}}\right)\left[\left(1+\frac{\delta_{nl}}{r_p}\right)^3 - \left[\frac{(k_p - k_{nl})(k_{bf} + 2k_{nl})}{(k_p + 2k_{nl})(k_{bf} - k_{nl})}\right]\right]}{\left(1+\frac{\delta_{nl}}{r_p}\right)^3 + 2\left[\left(\frac{k_{nl} - k_{bf}}{k_{nl} + 2k_{bf}}\right)\left(\frac{k_p - k_{nl}}{k_p + 2k_{nl}}\right)\right]}$	• Considers thermal conductivity and size of nanolayer to be 2 nm • Assumes continuous thermal conductivity variation in nanolayer transition between base fluid and solid particles
Tinga *et al.*[23]	$k_{nf} = \left(1 + \frac{3\phi\left[(\beta_{nl}^3 - 1)(2k_{nl} + k_p)(k_{nl} - k_{bf}) - (k_{nl} - k_p)(2k_{nl} + k_{bf})\right]}{(2k_{bf} + k_{nl})(2k_{nl} + k_p) - \frac{2}{\beta_{nl}^3 - 1}(k_{nl} - k_{bf})(k_{nl} - k_p) - 3\phi k_{nl}(k_p - k_{bf})}\right)k_{bf}$	• Simplification of model to calculate dielectric constant to be applied to thermal conductivity • Assumes water as interfacial layer and air as the host medium
Evans *et al.*[24]	$k_{nf} = \left(1 + 3\phi\frac{\varepsilon - 1}{\varepsilon + 2}\right)k_{bf}$	• Based on EMT • Considers ratio between particle and equivalent matrix thickness
Vladkov and Barrat[25]	$k_{nf} = \left(\frac{\left(\frac{k_p}{k_{bf}}(1 + 2\Omega) + 2\right) + 2\phi\left(\frac{k_p}{k_{bf}}(1 - \Omega) - 1\right)}{\left(\frac{k_p}{k_{bf}}(1 + 2\Omega) + 2\right) - \phi\left(\frac{k_p}{k_{bf}}(1 - \Omega) - 1\right)}\right)k_{bf}$ $\Omega = \frac{R_k k_{bf}}{r_p}$	• Modified version of the Maxwell [2] model • Considers the ratio between Kapitza length to the particle radius

Lin *et al.*[26]	$k_{nf}=k_{bf}+\dfrac{(k_p-k_{bf})\left[\dfrac{(1-\Xi_1)(1-\Xi_2)}{1+2\rho_{nf}\Xi_1\Xi_2}\right]\phi_p+(k_{nl}-k_{bf})\left[\dfrac{(1-\Xi_1)}{1+2\rho_{nf}\Xi_1\Xi_2}\right](\phi_{pl}-\phi_p)}{\left[\dfrac{(1-\Xi_1)(1-\Xi_2)}{1+2\rho_{nf}\Xi_1\Xi_2}\right]\phi_p+\left[\dfrac{(1-\Xi_1)}{1+2\rho_{nf}\Xi_1\Xi_2}\right]\left(\phi_{pl}-\phi_p\right)+(1-\phi_{pl})}$ $\Xi_1=\dfrac{k_p-k_{nl}}{k_p+2k_{nl}}\ \Xi_2=\dfrac{k_{nl}-k_{bf}}{k_{nl}+2k_{bf}}$	• Based on simulation with Cu–ethylene glycol nanofluid • Considers several nanolayers with different properties
Mirmohammadi *et al.*[27]	$k_{nf}=k_{bf}\left(1+9.89\left(\dfrac{T}{300}\right)^{4.75}(\phi)^{0.65}\left(0.30855\dfrac{SA}{V}\right)^{0.23}\right)$	• Model based on dimensional analysis with constants determined from molecular dynamics simulation • Considers surface to volume ratio
Moghaddam *et al.*[28]	$k_{nf}=k_{bf}\left(1.0139(1+\phi)^{2.964}\left(1+\dfrac{T}{70}\right)^{0.155}\left(1+\dfrac{d_p}{0.0275}\right)^{-0.01}\right)$	• Based on regression analysis from 472 data points for water and ethylene glycol. • Nanoparticles in range of 5–150 nm and temperatures 5–70 °C considered
Çolak[29]	$k_{nf}=(0.0000859\ T+0.03354)(17.01+\phi)$	• Based on ZrO_2–water nanofluid • Volume concentration range of 0.0125–0.2% and temperatures of 10– 65 °C
Kamel *et al.*[30]	$k_{nf}=k_{bf}\,(0.9845-0.000827\ T-0.01939\ \phi+0.00000293\ T^2+0.0012\ T\phi)$	• Based on Al_2O_3– water nanofluid • Volume concentration range of 0.01–0.5% and temperatures of 35–50 °C
Jeffrey[32]	$k_{nf}=k_{bf}\left(1+3\dfrac{k_p-k_{bf}}{k_p-2k_{bf}}\phi+\phi^2\left(3\left(\dfrac{k_p-k_{bf}}{k_p-2k_{bf}}\right)^2+\dfrac{3}{4}\left(\dfrac{k_p-k_{bf}}{k_p-2k_{bf}}\right)^3+\tfrac{9}{16}\left(\dfrac{k_p-k_{bf}}{k_p-2k_{bf}}\right)^3\left(\dfrac{\dfrac{k_p}{k_{bf}}+2}{2\dfrac{k_p}{k_{bf}}+3}\right)+\tfrac{3}{64}\left(\dfrac{k_p-k_{bf}}{k_p-2k_{bf}}\right)^4+\cdots\right)\right)$	• Modification of the EMT to consider interactions between particles
Sundar *et al.*[33]	$k_{nf}=k_{bf}\,(1+10.5\phi)^{0.1051}$	• Model based on Fe_3O_4–water experimental data • Valid for volume fractions of 0–2% and temperatures of 20–60 °C

Table 7.9 (*Continued*)

Davis[34]	$k_{nf} = k_{bf}\left(1 + \frac{3\left(\frac{k_p}{k_{bf}} - 1\right)\phi}{\left(\frac{k_p}{k_{bf}} + 2\right) - \left(\frac{k_p}{k_{bf}} - 1\right)\phi}\left(\phi + f\left(\frac{k_p}{k_{bf}}\right)\phi^2 + 0(\phi^2)\right)\right)$	• EMT model determined by calculating the heat flux caused by a solid body in a composite material.
Yu and Choi[45]	$k_{nf} = \frac{k_{pe} + 2k_{bf} + 2(k_{pe} - k_{bf})(\beta_{nl})^3\phi}{k_{pe} + 2k_{bf} - (k_{pe} - k_{bf})(\beta_{nl})^3\phi}k_{bf}\ \beta_{nl} = 1 + \frac{\delta_{nl}}{r_p}$ $k_{pe} = \frac{[2(1-\gamma) + (\beta_{nl})^3(1+2\gamma)]\gamma}{-(1-\gamma) + (\beta_{nl})^3(1+2\gamma)}\ \gamma = \frac{k_{nl}}{k_p}$	• Modification of the EMT to include effect of nanolayer • Replaces particle thermal conductivity with an equivalent thermal conductivity including the nanolayer
Wang *et al.*[46]	$k_{nf} = k_{bf}\frac{(1-\phi) + 3\phi\int_0^\infty \frac{k_{cl}(r)n(r)}{k_{cl}(r) + 2k_{bf}}\mathrm{d}r}{(1-\phi) + 3\phi\int_0^\infty \frac{k_{bf}n(r)}{k_{cl}(r) + 2k_{bf}}\mathrm{d}r}$	• Uses fractals to predict distribution of nanoparticle cluster size • Cluster thermal conductivity is a function of cluster radius
Kumar *et al.*[52]	$k_p = k_{bf}\left(1 + C_u u_B\frac{\phi r_{bf}}{k_{bf}(1-\phi)r_p}\right)$	• Model considering the thermal conductivity enhancement of nanoparticles to be proportional to nanoparticle velocity

Nomenclature

k	Thermal conductivity [W m^{-1} K]
C_u	Constant relating Brownian velocity to thermal conductivity [J m^{-2} K]
d	Diameter [m]
n	Empirical shape factor [–]
k_b	Boltzmann constant [m^2 kg^{-1} s^{-2} K]
Re	Reynolds number
Pr	Prandtl number
T	Temperature [K]
T_{fr}	Freezing temperature [K]
T_0	Reference temperature [K]
c	Specific heat capacity [J kg^{-1} K]
u_B	Brownian velocity [m s^{-1}]
r	Radius [m]
SA	Surface area [m^2]
V	Volume [m^3]
$k_{cl}(r)$	Thermal conductivity of nanoparticle cluster [W m^{-1} K]
$n(r)$	Cluster radius distribution function [–]

Greek Symbols

Ψ	Sphericity [–]
μ	Dynamic viscosity [kg m^{-1} s]
ρ	Density [kg m^{-3}]
β	Empirical correlation [–]
β_{nl}	Dimensionless nanolayer parameter [–]
δ_{nl}	Nanolayer thickness [m]
γ	Ratio of nanolayer thickness to nanoparticle radius [–]
α	Dimensionless particle equivalent conductivity parameter [–]
ε	Ratio between equivalent the matrix thickness and particle radius [–]
Ω	Ratio between the Kapitza length and particle radius [–]
Ξ	Dimensionless nanolayer conductivity parameter [–]

Subscripts

bf	Base fluid
nf	Nanofluid
p	Nanoparticle
nl	Nanolayer
pe	Equivalent nanoparticle and nanolayer
a	Aggregate
e	Equivalent of aggregate

References

1. S. Simpson, A. Schelfhout, C. Golden and S. Vafaei, *Appl. Sci.*, 2019, **9**(1), 87.

2. K. S. Suganthi and K. S. Rajan, *Renewable Sustainable Energy Rev.*, 2017, **76**, 226.
3. V. Y. Rudyak and A. V. Minakov, *Eur. Phys. J. E: Soft Matter Biol. Phys.*, 2018, **41**(1), 1.
4. J. Bayat and A. H. Nikseresht, *Heat Mass Transfer*, 2011, **47**(9), 1089.
5. K. Apmann, R. Fulmer, A. Soto and S. Vafaei, *Materials*, 2021, **14**(5), 1291.
6. J. Buongiorno, *J. Heat Transfer*, 2006, **128**(3), 240.
7. S. Dong and X. Chen, *J. Therm. Anal. Calorim.*, 2017, **129**(2), 1255.
8. A. Soto, T. Gaster, C. Golden, and S. Vafaei, *In ASTFE Digital Library*, 2020. Begel House Inc.
9. D. Chauhan and N. Singhvi, *J. Thermoplast. Compos. Mater.*, 2017, **30**(3), 289.
10. J. C. Maxwell, *A Treatise on Electricity and Magnetism*, Clarendon press, Oxford, 1873.
11. R. L. Hamilton and O. K. Crosser, *Ind. Eng. Chem. Fundam.*, 1962, **1**(3), 187.
12. E. V. Timofeeva, A. N. Gavrilov, J. M. McCloskey, Y. V. Tolmachev, S. Sprunt, L. M. Lopatina and J. V. Selinger, *Phys. Rev. E*, 2007, **76**(6), 061203.
13. R. Prasher, W. Evans, P. Meakin, J. Fish, P. Phelan and P. Keblinski, *Appl. Phys. Lett.*, 2006, **89**(14), 143119.
14. Y. Xuan, Q. Li and W. Hu, *AIChE J.*, 2003, **49**(4), 1038.
15. Y. Feng, B. Yu, P. Xu and M. Zou, *J. Phys. D: Appl. Phys.*, 2007, **40**(10), 3164.
16. M. Corcione, *Energy Convers. Manage.*, 2011, **52**(1), 789.
17. C. H. Chon, K. D. Kihm, S. P. Lee and S. U. Choi, *Appl. Phys. Lett.*, 2005, **87**(15), 153107.
18. J. Koo and C. Kleinstreuer, *J. Nanopart. Res.*, 2004, **6**(6), 577.
19. R. S. Vajjha and D. K. Das, *Int. J. Heat Mass Transfer*, 2009, **52**(21–22), 4675.
20. H. E. Patel, T. Sundararajan and S. K. Das, *J. Nanopart. Res.*, 2010, **12**(3), 1015.
21. K. C. Leong, C. Yang and S. M. Murshed, *J. Nanopart. Res.*, 2006, **8**(2), 245.
22. H. Xie, M. Fujii and X. Zhang, *Int. J. Heat Mass Transfer*, 2005, **48**(14), 2926.
23. W. R. Tinga, W. A. Voss and D. F. Blossey, *J. Appl. Phys.*, 1973, **44**(9), 3897.
24. W. Evans, J. Fish and P. Keblinski, *Appl. Phys. Lett.*, 2006, **88**(9), 093116.
25. M. Vladkov and J. L. Barrat, *Nano Lett.*, 2006, **6**(6), 1224.
26. Y. S. Lin, P. Y. Hsiao and C. C. Chieng, *Appl. Phys. Lett.*, 2011, **98**(15), 153105.
27. S. A. Mirmohammadi, M. Behi, Y. Gan and L. Shen, *Phys. Rev. E*, 2019, **99**(4), 043109.
28. H. A. Moghaddam, A. Ghafouri and R. F. Khouzestani, *J. Braz. Soc. Mech. Sci. Eng.*, 2021, **43**(6), 1.
29. A. B. Çolak, *Int. J. Energy Res.*, 2021, **45**(2), 2912.
30. M. S. Kamel, O. Al-Oran and F. Lezsovits, *Period. Polytech., Chem. Eng.*, 2021, **65**(1), 50.
31. N. H. Abu-Hamdeh, E. Almatrafi, M. Hekmatifar, D. Toghraie and A. Golmohammadzadeh, *J. Mol. Liq.*, 2021, **327**, 114832.

32. D. J. Jeffrey, *Proc. R. Soc. London. A. Math. Phys. Sci.*, 1973, **335**(1602), 355.
33. L. S. Sundar, M. K. Singh and A. C. Sousa, *Int. Commun. Heat Mass Transfer*, 2013, **44**, 7.
34. R. H. Davis, *Int. J. Thermophys.*, 1986, 7(3), 609.
35. K. Singh and S. Kumar, *Int. Res. J. Eng. Technol.*, 2020, 7(8), 400.
36. M. Afrand, D. Toghraie and N. Sina, *Int. Commun. Heat Mass Transfer*, 2016, **75**, 262.
37. E. V. Timofeeva, J. L. Routbort and D. Singh, *J. Appl. Phys.*, 2009, **106**(1), 014304.
38. E. C. Okonkwo, I. Wole-Osho, D. Kavaz, M. Abid and T. Al-Ansari, *Sustainable Energy Technol. Assess.*, 2020, **37**, 100636.
39. R. Gangadevi and B. K. Vinayagam, *J. Therm. Anal. Calorim.*, 2019, **136**(1), 199.
40. M. I. Pryazhnikov, A. V. Minakov, V. Y. Rudyak and D. V. Guzei, *Int. J. Heat Mass Transfer*, 2017, **104**, 1275.
41. M. Goharkhah, A. Salarian, M. Ashjaee and M. Shahabadi, *Powder Technol.*, 2015, **274**, 258.
42. T. Khamliche, S. Khamlich, M. K. Moodley, B. M. Mothudi, M. Henini and M. Maaza, *J. Mol. Liq.*, 2021, **323**, 114975.
43. R. Ranjbarzadeh, A. Moradikazerouni, R. Bakhtiari, A. Asadi and M. Afrand, *J. Cleaner Prod.*, 2019, **206**, 1089.
44. P. Keblinski, S. R. Phillpot, S. U. Choi and J. A. Eastman, *Int. J. Heat Mass Transfer*, 2002, **45**(4), 855.
45. W. Yu and S. U. Choi, *J. Nanopart. Res.*, 2003, **5**(1), 167.
46. B. X. Wang, L. P. Zhou and X. F. Peng, *Int. J. Heat Mass Transfer*, 2003, **46**(14), 2665.
47. M. Kole and T. K. Dey, *Exp. Therm. Fluid Sci.*, 2011, **35**(7), 1490.
48. M. Motevasel, A. R. Nazar and M. Jamialahmadi, *Heat Mass Transfer*, 2018, **54**(1), 125.
49. G. Colangelo, E. Favale, P. Miglietta, M. Milanese and A. de Risi, *Energy*, 2016, **95**, 124.
50. M. Kole and T. K. Dey, *Thermochim. Acta*, 2012, **535**, 58.
51. S. P. Jang and S. U. Choi, *Appl. Phys. Lett.*, 2004, **84**(21), 4316.
52. D. H. Kumar, H. E. Patel, V. R. Kumar, T. Sundararajan, T. Pradeep and S. K. Das, *Phys. Rev. Lett.*, 2004, **93**(14), 144301.
53. S. K. Das, N. Putra, P. Thiesen and W. Roetzel, *J. Heat Transfer*, 2003, **125**(4), 567.
54. M. H. Esfe, M. Afrand, W. M. Yan and M. Akbari, *Int. Commun. Heat Mass Transfer*, 2015, **66**, 246.
55. P. K. Das, N. Islam, A. K. Santra and R. Ganguly, *J. Mol. Liq.*, 2017, **237**, 304.
56. M. Kong and S. Lee, *Adv. Mech. Eng.*, 2020, **12**(8), 1687814020952277.
57. M. P. Allen, *Comput. soft matter: from synth. polym. proteins*, 2004, **23**(1), 1.

CHAPTER 8

Specific Heat Capacity and Density of Nanofluids

T. J. CHOI,[a] S. H. KIM[b] AND S. P. JANG*[a,b]

[a] School of Aerospace and Mechanical Engineering, Korea Aerospace University, Goyang, Gyeonggi-do 10540, Republic of Korea; [b] Smart Drone Convergence Education & Research Center, Korea Aerospace University, Goyang, Gyeonggi-do 10540, Republic of Korea
*Email: spjang@kau.ac.kr

8.1 Specific Heat of Nanofluids

Nanofluids are engineered suspensions of well-dispersed nanoparticles in base fluids such as water, ethylene glycol, propylene glycol, synthetic oil, and other conventional fluids. Since their invention by Choi and Eastman[1] in 1995, nanofluids have attracted tremendous interest from researchers due to their superior thermal properties. Although researchers initially focused on the thermal conductivity, viscosity, and convective heat transfer coefficient of nanofluids,[2–10] recent studies have found that changes in the specific heat and density properties of nanofluids play a very important role in heat transfer applications, such as in heat exchangers and thermal storage devices. Pak and Cho[11] were the first to report the specific heat of nanofluids, which has since been investigated by many researchers. Therefore, in this chapter, the experimental characteristics and various prediction correlations of the specific heat of nanofluids are introduced.

Soft Matter Series No. 16
Fundamentals and Transport Properties of Nanofluids
Edited by S. M. Sohel Murshed

Published by the Royal Society of Chemistry, www.rsc.org

8.1.1 Experimental Characteristics of the Specific Heat of Nanofluids

According to the literature,[12–57] it has been clearly shown that key parameters that affect the specific heat of nanofluids are the volume fraction of nanoparticles (or the weight fraction of nanoparticles); the type of base fluid, such as water, synthetic oil, molten salt, and so on; the nanoparticle size; and suspension characteristics. We can classify the specific heat of nanofluids according to the base fluid and summarize it based on the volume fraction or weight fraction of the nanoparticles as there is insufficient data reported on the effect of the nanoparticle size and the suspension characteristics of nanofluids on the specific heat of nanofluids.

8.1.1.1 Water-based Nanofluids with Oxide Nanoparticles

Figure 8.1 shows the specific heats of water-based nanofluids reported by various researchers.[12–21] As shown in Figure 8.1, the specific heat of water-based nanofluids generally decreases as the volume fraction of the nanoparticles increases, regardless of the material of the nanoparticles. This is because the specific heat of the nanoparticle material, such as Al_2O_3, TiO_2, SiO_2, CuO, and SWCNT (single-walled carbon nanotubes), is lower than that of water. Zhou and Ni,[12] O'Hanley *et al.*,[13] and Moldoveanu and Minea[14] investigated the specific heat of water-based Al_2O_3 nanofluids. They reported that the specific heat of water-based nanofluids with Al_2O_3 decreased with an increase in the volume fraction of the nanoparticles because the specific heat of the Al_2O_3 nanoparticles ($c_{p,Al_2O_3} = 779.2\ J\ kg^{-1}\ K^{-1}$) is lower than that of water ($c_{p,\ water} = 4180.6\ J\ kg^{-1}\ K^{-1}$). Several researchers[15,16] have reported that the specific heat of water-based TiO_2 nanofluids also exhibits the same characteristics as the volume fraction of the TiO_2 nanoparticles ($c_{p,TiO_2} = 687.6\ J\ kg^{-1}\ K$). Moreover, it was reported

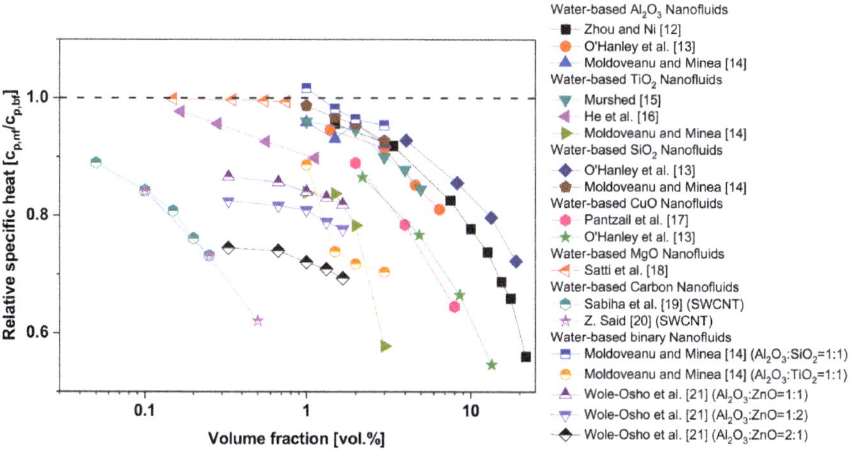

Figure 8.1 Relative specific heats of water-based nanofluids.

that the specific heat of water-based nanofluids with SiO_2, CuO, MgO, SWCNTs, and binary nanoparticles decreased with an increase in the volume fraction of the nanoparticles.[13,14,17–21] These results can be explained using the mixing law as the specific heats of the nanoparticle materials, such as Al_2O_3, TiO_2, SiO_2, CuO, MgO, SWCNTs, and binary oxide nanoparticles, are lower than that of water.

8.1.1.2 Molten Salt-based Nanofluids

Figure 8.2 shows the relative specific heats of molten salt-based Al_2O_3 nanofluids. In early research, specific heat studies of molten salt-based Al_2O_3 nanofluids were actively conducted because Al_2O_3 nanoparticles can be widely used to manufacture nanofluids with high thermal conductivity at high temperatures. Based on the literature,[22–24] there are two trends of the specific heat of molten salt-based Al_2O_3 nanofluids depending on the weight fraction of the nanoparticles. Malik[22] and Hu *et al.*[23] experimentally investigated the specific heat of molten salt-based Al_2O_3 nanofluids, noting that the specific heat increased with the concentration of the nanoparticles. The abnormal enhancement of the specific heat was explained using molecular dynamics simulations that consider the increase in the surface energy of the nanoparticles. In contrast, other studies[24–30] revealed that the specific heat of molten salt-based Al_2O_3 nanofluids has an optimum value at a certain weight fraction. Chieruzzi *et al.*[24] reported that the specific heat of molten salt-based Al_2O_3 nanofluids with a 0.5 wt% and 1.5 wt% is lower than that of the base fluid because the nanoparticles have a lower specific heat than the base fluid ($c_{p,Al_2O_3} = 779.2\ J\ kg^{-1}\ K$,[31] $c_{p,molten\ salt} = 1648.0\ J\ kg^{-1}\ K$).[24]

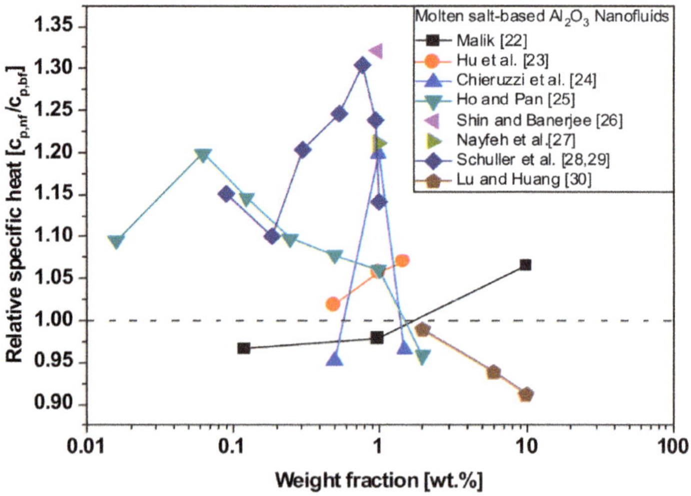

Figure 8.2 Relative specific heats of molten salt-based Al_2O_3 nanofluids.

These results are similar to those of the water-based nanofluids containing oxide nanoparticles presented in Section 8.1.1.1. However, they experimentally showed that the specific heat of molten salt-based nanofluids containing 1 wt% of Al_2O_3 significantly increased by approximately 20% compared to the base fluid. The different behavior of molten salt-based nanofluids containing 1 wt% of Al_2O_3 was explained by considering the formation of a solid-like nanolayer on the surface of the nanoparticles. Similarly, Ho and Pan,[25] Shin and Banerjee,[26] Nayfeh *et al.*,[27] and Schuller *et al.*[28,29] also reported that the specific heat of molten salt-based Al_2O_3 nanofluids significantly increased at specific weight fractions. It was also explained that the specific heat of the nanofluids could be improved by surface phenomena such as the surface energy, interfacial thermal resistance, and fractal-like fluid nanostructures, even though the specific heat of the nanoparticles was lower than that of the base fluid.

Figure 8.3 shows the relative specific heats of molten salt-based SiO_2 nanofluids reported by several researchers.[24,32–35] Most of them[24,32–35] reported that the specific heat of molten salt-based SiO_2 nanofluids greatly improved at 1 wt%.

In general, conventional theories such as the mixing law can predict that the specific heat of molten salt-based nanofluids decreases as the weight fraction of the nanoparticles increases. However, in ref. 32–35, it was reported that the specific heat of molten salt-based nanofluids could be enhanced by a high specific surface area,[36,37] interfacial thermal resistance between the nanoparticles and molecules of molten salt,[32,38] and the semi-solid layering of salt molecules.[39,40]

Figure 8.4 shows the specific heats of molten salt-based TiO_2 nanofluids.[24,41–44] Chieruzzi *et al.*[24] stated that the specific heat of nanofluids containing TiO_2 also

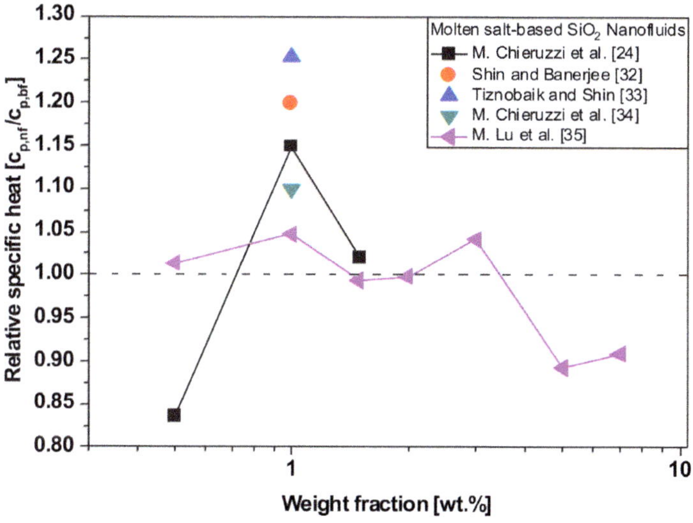

Figure 8.3 Relative specific heats of molten salt-based SiO_2 nanofluids.

has an optimal weight fraction; however, they suggested the specific heat cannot exceed that of the base fluid. In contrast, Lasfargues *et al.*,[41,42] Awad *et al.*,[43] and Jung[44] showed that the specific heat of nanofluids with TiO_2 can exceed that of the base fluid. Moreover, they also reported that the specific heat of the molten salt-based TiO_2 nanofluids increased slightly according to the weight fraction. It was explained that the improvement in the specific heat of the molten salt-based nanofluids was based on the enhancement in enthalpy due to the aggregation of the nanoparticles and the interfacial interaction at the nanolayer between the nanoparticles and molecules of the molten salt.[41,42]

Finally, other studies[45–49] investigated the specific heat of molten salt-based nanofluids with multi-walled carbon nanotubes (MWCNTs) and graphite. As shown in Figure 8.5, it was experimentally reported that the specific heat improved according to the weight fraction of the nanoparticles, although the mixing law predicted that the specific heat of molten salt-based MWCNT nanofluids is lower than that of the base fluid. However, it was suggested that there may be other causes that could increase the specific heat.

According to the literature, it is known that an improvement in the specific heat of molten salt-based nanofluids cannot be explained using only conventional theory such as the mixing law based on the specific heats of the nanoparticles and base fluid. Moreover, we can estimate there is another mechanism by which the specific heat of molten salt-based nanofluids can be improved, although understanding the exact mechanism is limited due to ongoing research.

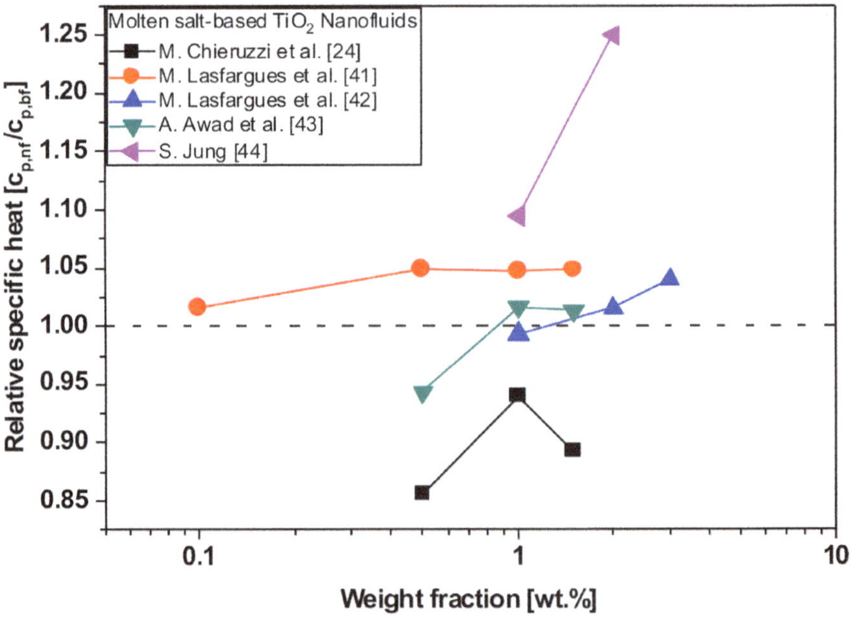

Figure 8.4 Relative specific heats of molten salt-based TiO_2 nanofluids.

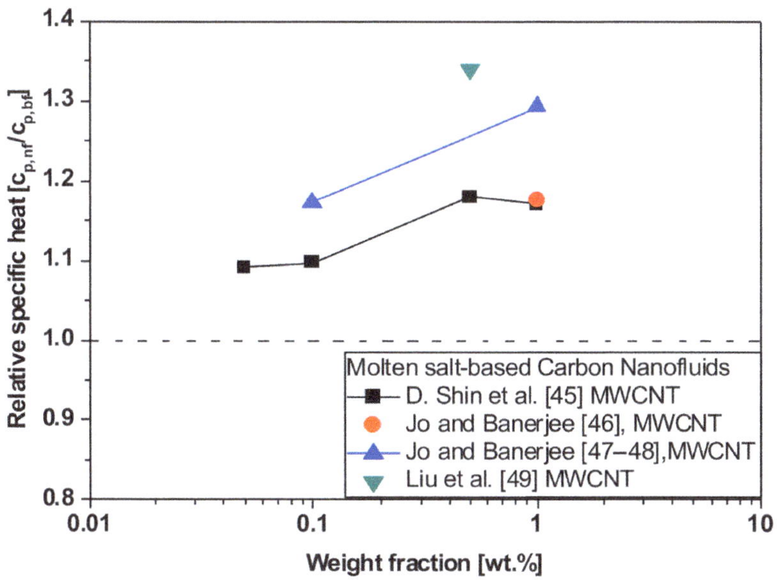

Figure 8.5 Relative specific heats of molten salt-based MWCNT nanofluids.

8.1.1.3 *Heat Transfer Fluid-based Nanofluids*

Figure 8.6 shows the relative specific heats of various heat transfer fluid-based nanofluids. As shown in Figure 8.6, the specific heat of the nanofluids decreases as the volume fraction of the nanoparticles increases, regardless of the type of nanoparticles when base fluids such as an ethylene glycol and water mixture (EGW), ethylene glycol (EG), propylene glycol and water mixture (PGW), and oil (engine oil or synthetic oil) are used. For example, Kulkarni *et al.*,[50] Vajjha and Das,[51] and Hassan *et al.*[52] investigated the specific heat of EGW-based Al_2O_3 nanofluids with respect to their volume fraction. They reported that the specific heat of EGW-based Al_2O_3 nanofluids decreased as the particle volume fraction increased. Namburu *et al.*[53] and Vajjha *et al.*[51,54] investigated the specific heats of EGW-based SiO_2 and ZnO nanofluids with respect to the volume fraction. They reported that the specific heats of the EGW-based SiO_2 and ZnO nanofluids also decreased as the particle volume concentration increased. Choudhary *et al.*[55] and Teng and Yu[56] investigated the specific heats of EGW-based nanofluids with MgO and MWCNT nanoparticles, respectively. They reported that the specific heats of these nanofluids gradually decreased with the volume fractions of MgO and MWCNTs. Moreover, Murshed,[57] Barbes *et al.*,[58] and Akilu *et al.*[59] investigated the specific heats of EG-based nanofluids with Al_2O_3, TiO_2, SiO_2, and Al nanoparticles. They reported that the specific heats of these nanofluids decreased with an increase in the volume fractions of the Al_2O_3, TiO_2, SiO_2, and Al nanoparticles because of the lower specific heat values of these nanoparticles compared with that of the basefluid. Similarly, Satti *et al.*[60] reported that the specific heats of PGW-based nanofluids containing Al_2O_3,

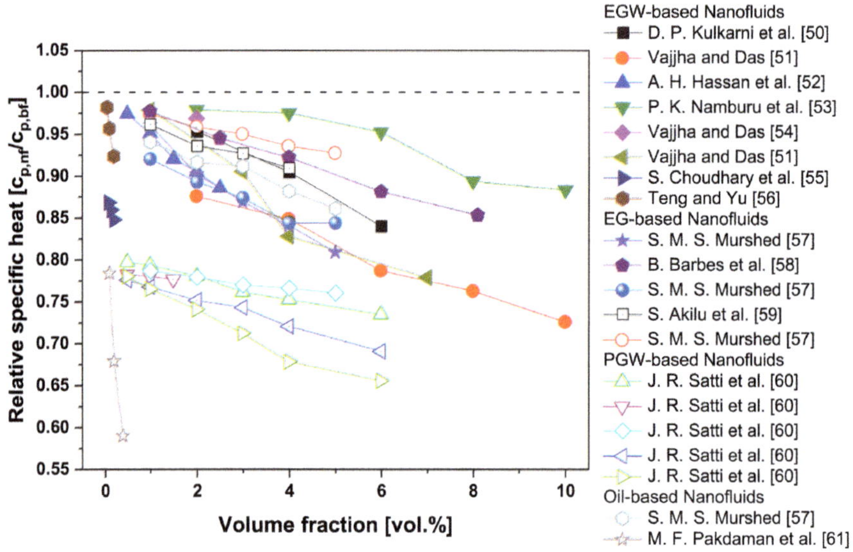

Figure 8.6 Relative specific heats of various heat transfer fluid-based nanofluids.

TiO_2, SiO_2, ZnO, and CuO nanoparticles decreased with an increase in the volume fractions of these nanoparticles. Furthermore, Murshed[57] and Pakdaman *et al.*[61] observed the same phenomenon in oil-based nanofluids containing Al and MWCNT nanoparticles. The specific heats of the various heat transfer fluid-based nanofluids, except for molten salt-based nanofluids, decreased with an increase in the volume fractions of the nanoparticles. These results are similar to those of the water-based nanofluids containing oxide nanoparticles presented in Section 8.1.1.1.

8.1.2 Mechanisms and Correlations of the Specific Heat of Nanofluids

To explain change in the specific heat of nanofluids according to the volume fraction or the weight fraction of nanoparticles, various mechanisms[11,51,62–64] have been proposed, including the mixing theory for ideal gas mixtures, thermal equilibrium correlation, surface energy of nanoparticles, interfacial thermal resistance, and the formation of nanolayers. Although it is currently difficult to determine which of the proposed mechanisms is appropriate, it has been shown that the various mechanisms proposed fall into two main categories. The first category of mechanisms is based on the mixing theory using the properties of the nanoparticles and base fluid. The second category of mechanisms that can explain the specific heat enhancement of molten salt-based nanofluids is based on the consideration of the high specific surface energy, high interfacial thermal resistance, and semi-solid layering of salt molecules (fractal-like fluid nanostructures).

8.1.2.1 Three Correlations Based on Mixing Theory using the Properties of the Nanoparticles and Base fluid

Three correlations for predicting the specific heat of nanofluids are widely used in the literature. Eqn (8.1), commonly referred to as correlation l, is simply based on the mixing theory of an ideal gas mixture.[65] Pak and Cho[11] were the first to propose correlation I to predict the specific heat of nanofluids, as given by:

$$c_{\mathrm{p,nf}} = \phi c_{\mathrm{p,np}} + (1-\phi)c_{\mathrm{p,bf}}, \tag{8.1}$$

where c_{p} and ϕ are the specific heat and volume fraction of nanoparticles, respectively. The subscripts nf, np, and bf represent nanofluids, nanoparticles and base fluid, respectively. In eqn (8.1), the specific heat of the nanofluids is expressed only in terms of the specific heats of the nanoparticles and base fluid without considering other factors such as the nanolayer on the surface of the nanoparticles,[39,66] the interfacial thermal resistance between the nanoparticles and base fluid,[38,67] and the high specific surface area.[36,68] Eqn (8.1) has been widely used to evaluate the heat transfer of nanofluids in numerous experiments and theoretical simulations.[11,65,69] However, it overestimates the specific heat of the nanofluids.[12,14,17] To overcome this challenge, Xuan and Roetzel[62] proposed a modified prediction correlation, as defined in eqn (8.2), based on the mixing theory by assuming thermal equilibrium between the nanoparticles and the base fluid:[38,48,49]

$$c_{\mathrm{p,nf}} = \frac{\phi(\rho c_{\mathrm{p}})_{\mathrm{np}} + (1-\phi)(\rho c_{\mathrm{p}})_{\mathrm{bf}}}{\phi \rho_{\mathrm{np}} + (1-\phi)\rho_{\mathrm{bf}}}, \tag{8.2}$$

where ρ represents density. Several studies[12–20,39,62–64] have reported that eqn (8.2) (correlation II) can be used to predict the specific heat of the nanofluids better than correlation I. Moreover, Vajjha and Das[51] proposed a new correlation (correlation III) as a function of temperature based on their experimental data, as expressed by:

$$\frac{c_{\mathrm{p,nf}}}{c_{\mathrm{p,bf}}} = \frac{(A\cdot T) + \left(B\frac{c_{\mathrm{p,np}}}{c_{\mathrm{p,bf}}}\right)}{C+\phi}, \tag{8.3}$$

where *A*, *B*, and *C* are the curve fit coefficients for each nanofluid. The curve fit coefficients for the EGW-based (60 : 40 wt%) Al_2O_3, SiO_2, and ZnO nanofluids are presented in Table 8.1.[51] Correlation III can be applied in the volume faction and temperature ranges of $0<\phi\leq 0.1$ and $315\ \mathrm{K}<T<363\ \mathrm{K}$, respectively, although correlations I and II are commonly employed to predict the specific heat of nanofluids without the range limitation.

However, the correlations defined in eqn (8.1)–(8.3) cannot be used to estimate the specific heat enhancement of molten salt-based nanofluids at 1 wt%, as reported in many studies.[24,26,28,29,32–34] Previous experimental data,

Table 8.1 Coefficients of correlation III for EGW-based (60 : 40 wt%) nanofluids.

Particle	*A*	*B*	*C*	Concentration
Al_2O_3	8.911×10^{-4}	0.5179	0.4250	$0<\phi\leq10$ (vol%)
SiO_2	1.769×10^{-3}	1.1937	0.8021	$0<\phi\leq10$ (vol%)
ZnO	4.604×10^{-4}	0.9855	0.2990	$0<\phi\leq7$ (vol%)

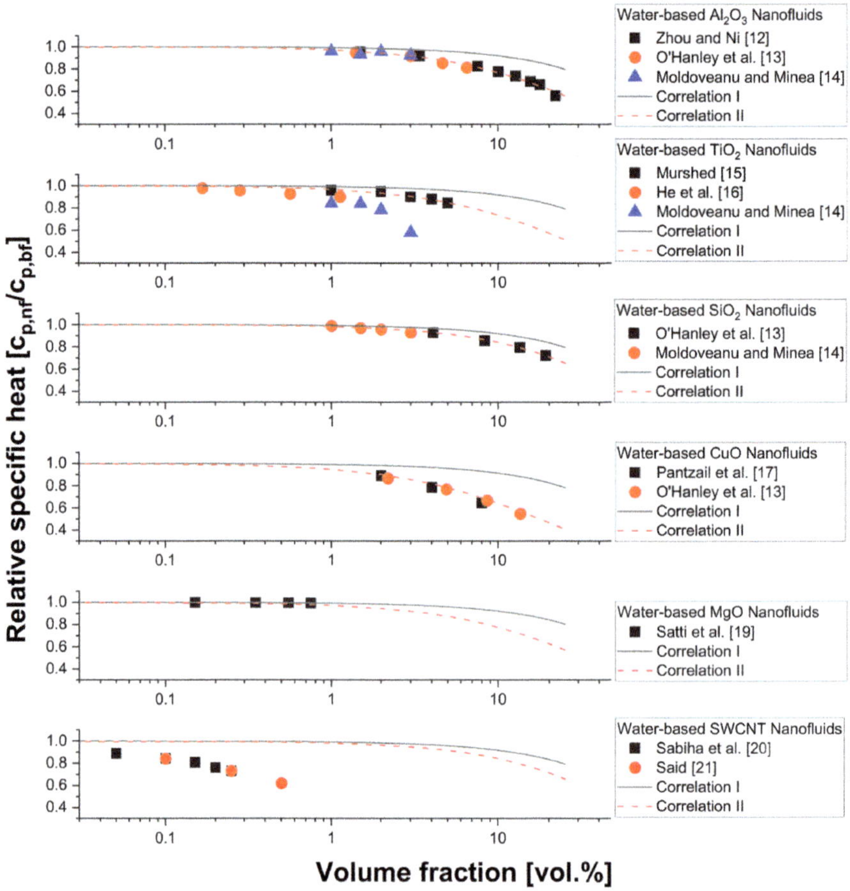

Figure 8.7 Comparison between the experimental data and correlations I and II.

except for those of molten salt-based nanofluids, were compared with the results predicted by correlations I and II, as shown in Figure 8.7. As shown in Figure 8.7, the specific heat calculated by correlation I is higher than all the experimental data, while most experimental data[12–17,19] on water-based nanofluids with various nanoparticles such as Al_2O_3, TiO_2, SiO_2, CuO, and MgO show good agreement with correlation II. However, it was reported that the experimental specific heat of water-based SWCNT[20,21] nanofluids does not follow any correlation.

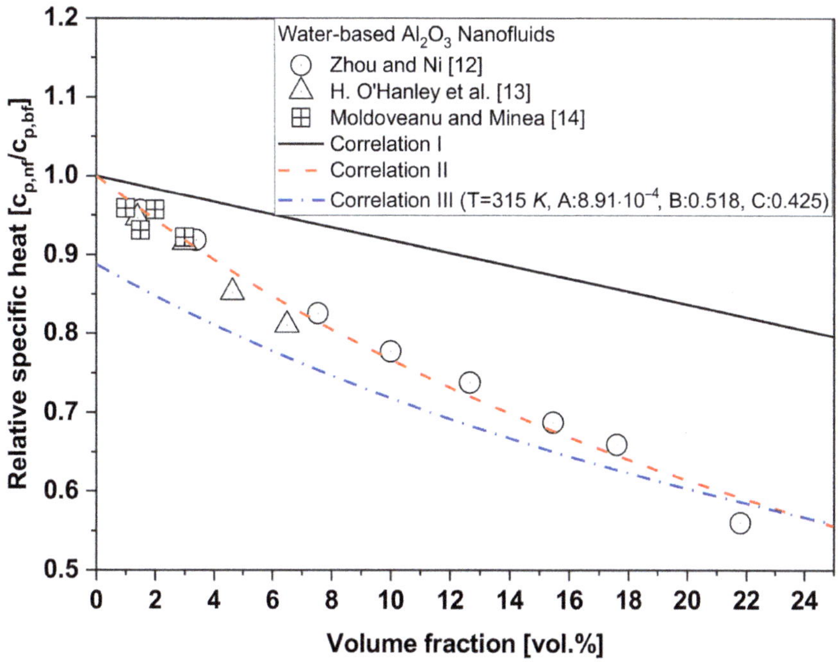

Figure 8.8 Comparison between the experimental data of water-based Al_2O_3 nanofluids and correlations I, II, and III.

Moreover, with the constant values of A, B, and C in correlation III presented by Vajjha and Das,[51] when the basefluid was EGW, the specific heats of water-based Al_2O_3 nanofluids and water-based SiO_2 nanofluids were estimated as shown in Figures 8.8 and 8.9. As shown in Figure 8.8, the experimental results[12–14] follow correlation II. The specific heat predicted by correlation I is still higher than the experimental data, while that predicted by correlation III is lower than the experimental data. Moreover, the slope of correlation III slightly changes with temperature.

Figure 8.9 shows that the experimental results[13,14] indicate good agreement with correlations II and III up to10 vol%, although the curve fit coefficients of different fluids are used. However, the specific heat calculated using correlation I is still higher than the experimental data. Based on the above comparisons, correlation II can be used to predict the specific heats of many nanofluids, except those of molten salt-based nanofluids. However, to estimate the specific heat of molten salt-based nanofluids with nanoparticles of 1 wt% concentration, another correlation that considers other factors is required.

8.1.2.2 *Action of the Nanoparticles*

The second category of mechanisms that can be used to explain the anomalous enhancement in the specific heat of molten salt-based nanofluids[24,26,32,36–38,40,70] involves three major factors.

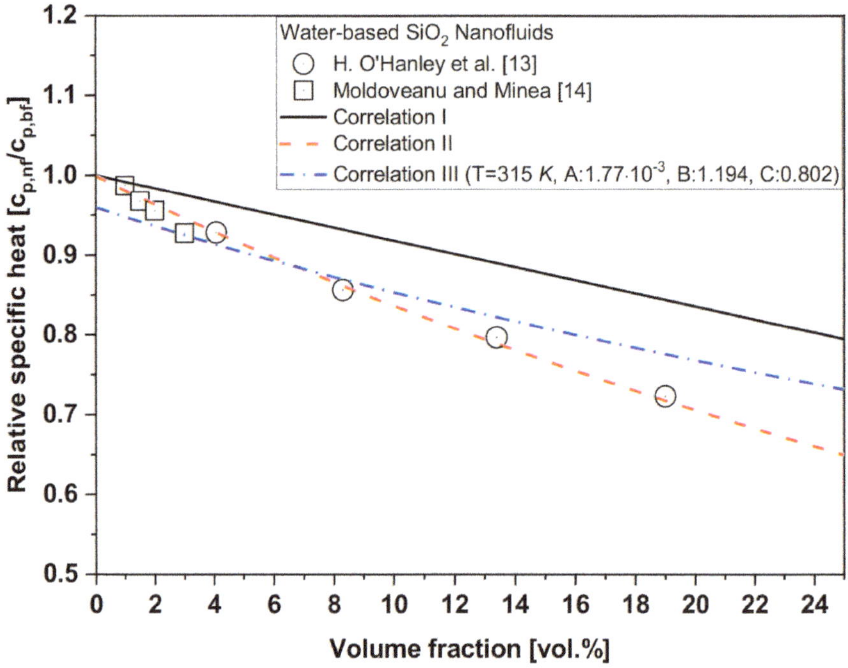

Figure 8.9 Comparison between the experimental data of water-based SiO_2 nanofluids and correlations I, II, and III.

First, the specific heat of molten salt-based nanofluids can be improved by increasing the specific surface energy.[24,36,37] The high surface energy of the nanoparticles induces a phase transformation in the solvent material near to the nanoparticles. To elaborate, it has been suggested that the layers of the adhered liquid molecules have a smaller intermolecular spacing on the surface than the liquid molecules in the bulk phase.[32,38,40,70] Therefore, the adhered layers of liquid molecules near to the surface of the nanoparticles can be expected to exhibit semi-solid properties.[32] As a result, the specific heat of the base fluid near to the surface of the nanoparticle with high energy is higher than that of the bulk base fluid. Second, the high interfacial thermal resistance between the nanoparticles and base fluid, known as the Kapitza resistance, can improve the specific heat of molten salt-based nanofluids[25,33,43] as the interfacial thermal resistance acts as additional thermal energy storage.[33] Third, the semi-solid layering of salt molecules near to the nanoparticles can improve the specific heat of molten salt-based nanofluids.[25,26,30,33,34,41] Many studies have demonstrated that a fractal-like fluid nanostructure is formed at the surface of the nanoparticles in molten salt-based nanofluids. Interestingly, it has been shown that nanostructures are not observed in other base fluids because they are formed near to the nanoparticles due to ionic interactions between the salts and nanoparticles.[26,71] It has been reported that nanostructures, so-called nanolayers or semi-solid layers, can significantly enhance the specific heat of molten salt-based nanofluids. In the literature, previous studies have suggested the possibility of the additional specific heat capacity of molten salt-based nanofluids using three mechanisms.[24–26,30,32–41,43,71] Shin and Banerjee[26,71] proposed a new theoretical correlation:

$$c_{\mathrm{p,nf}} = \frac{(\phi\rho c_{\mathrm{p}})_{\mathrm{np}} + (\phi\rho c_{\mathrm{p}})_{\mathrm{ns}} + \left(1 - \phi_{\mathrm{np}} - \phi_{\mathrm{ns}}\right)(\rho c_{\mathrm{p}})_{\mathrm{bf}}}{(\phi\rho)_{\mathrm{np}} + (\phi\rho)_{\mathrm{ns}} + \left(1 - \phi_{\mathrm{np}} - \phi_{\mathrm{ns}}\right)\rho_{\mathrm{bf}}}, \tag{8.4}$$

where the subscript ns denotes the nanostructure. They reported that the density and specific heat of the nanostructure near to the nanoparticles in salt molecules increased *via* three mechanisms. As shown in Figure 8.10, eqn (8.4) (correlation IV) predicts that the specific heat of molten salt-based nanofluids increases according to the weight fraction of the nanoparticles, assuming that the density and specific heat of the nanostructure near to the nanoparticles are 2 to 10 times those of the basefluid. However, eqn (8.4) cannot be used to predict why the specific heat of molten salt-based nanofluids exhibits an optimal value at a certain weight fraction. So, the effect of the nanostructure on the specific heat of the nanofluids should be investigated.

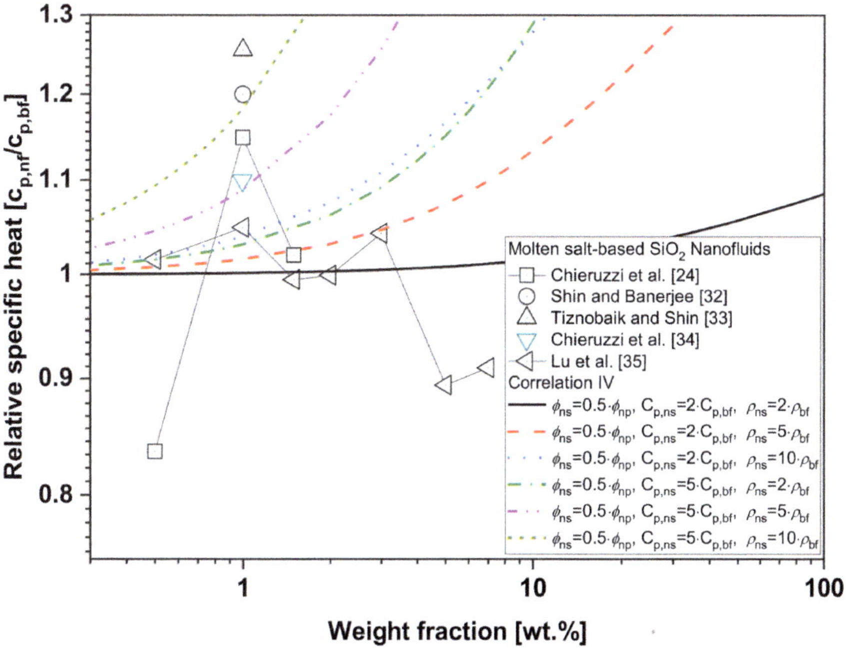

Figure 8.10 Comparison between the experimental data of molten salt-based SiO_2 nanofluids and correlation IV.

8.2 Density of Nanofluids

The thermal and physical properties of nanofluids are key parameters in heat transfer systems with high efficiency. Numerous studies have investigated the thermal conductivity, convective heat transfer coefficient, viscosity, and specific heat of nanofluids.[2–70] However, research on the density of nanofluids remains limited, although density is one of the critical properties of working fluids. Therefore, the experimental features, mechanisms, and correlations of the density of nanofluids reported in the literature are summarized.

8.2.1 Experimental Features of the Density of Nanofluids

Previous studies[72–77] have reported the effect of the volume fraction of the nanoparticles on the density of the nanofluids, as shown in Figure 8.11. The density of nanofluids increases linearly with the volume fraction regardless of the type of nanoparticle and base fluid, but their slopes are different. This phenomenon can be explained using the mixing law, which predicts a linear increase in density with the volume fraction of the nanoparticles. It has also been shown that the density gradients of nanofluids are dependent on the types of nanoparticles ($Al_2O_3 = 3600$ kg m^{-3},[76] $ZnO = 5600$ kg m^{-3},[76] $TiN = 5220$ kg m^{-3},[77] $Sb_2O_5 : SnO_2 = 6800$ kg m^{-3})[76] and base fluids.

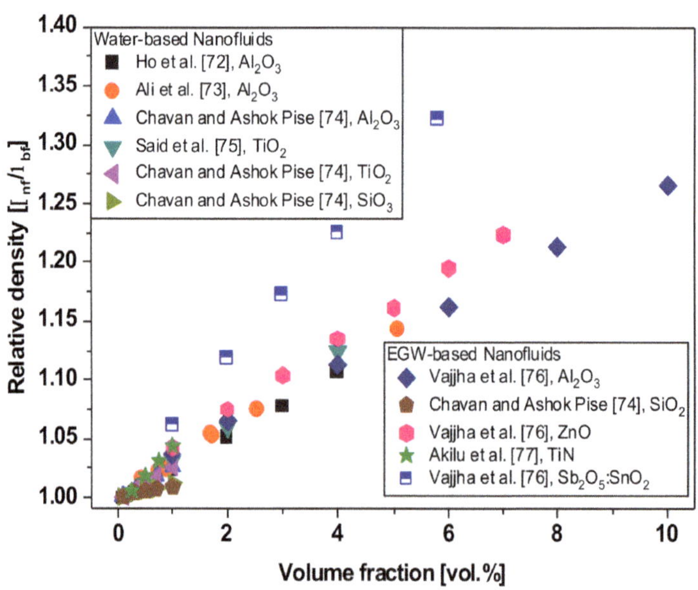

Figure 8.11 Relative density of nanofluids according to the volume fraction of the nanoparticles.

8.2.2 Mechanisms and Correlations for the Density of Nanofluids

To explain the change in the density of nanofluids according to the volume fraction of nanoparticles, Pak and Cho[11] calculated the density of nanofluids using the mixing law given as:

$$\rho_{nf} = \phi\rho_{np} + (1 - \phi)\rho_{bf}. \tag{8.5}$$

Since the density of nanofluids was first determined by Pak and Cho[11] using the mixing law, many studies have estimated the density of nanofluids using the mixing law defined in eqn (8.5).[73–78] Figure 8.12 shows a comparison between the experimental data and theoretical results for the density of the nanofluids. Excellent agreement can be observed between the measured values and the mixing law within ±5%. In the literature, it has been sufficiently verified that the density of nanofluids can be predicted using the mixing law.

8.3 State-of-the-art Technology

8.3.1 PCM-filled MWCNT Nanofluids

In general, nanoparticles improve the density of nanofluids but reduce their specific heat. Therefore, it is difficult to dramatically improve the heat-transfer

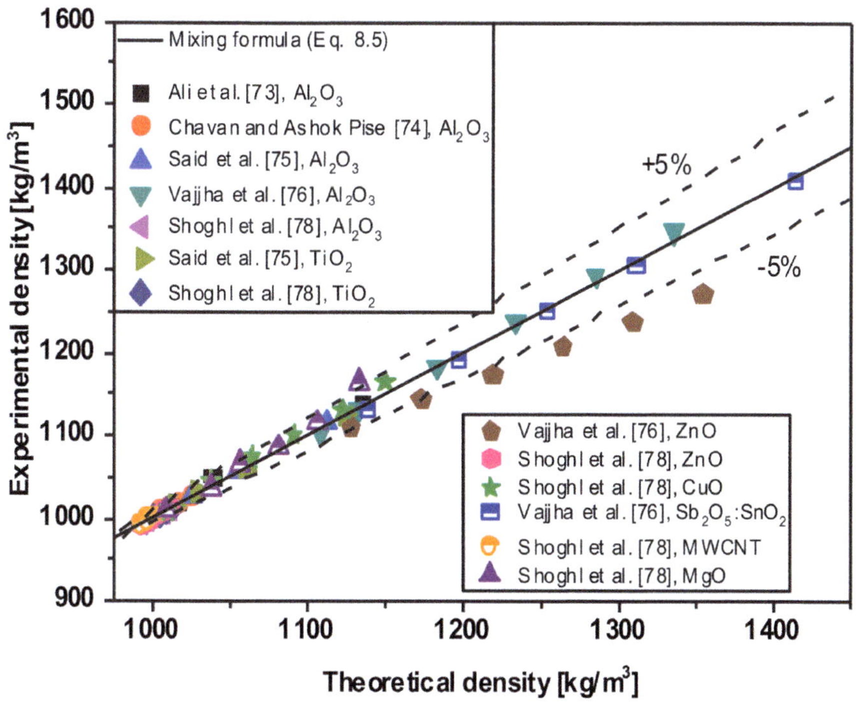

Figure 8.12 Comparison between the experimental data and theoretical results on the density of nanofluids.

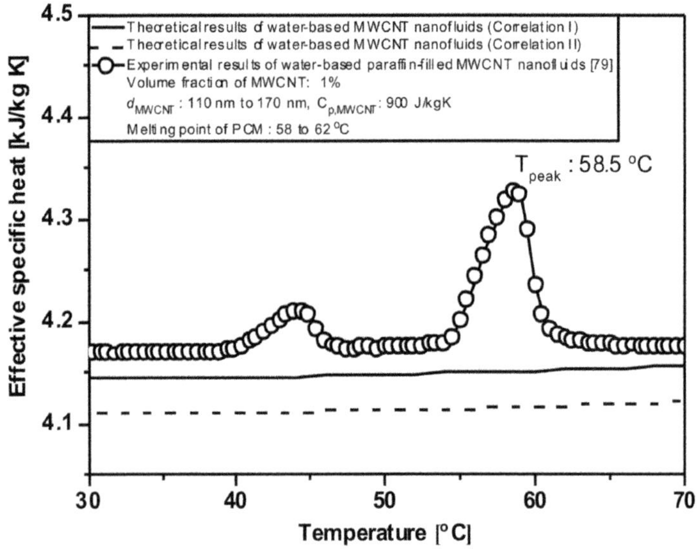

Figure 8.13 Effective specific heat of water-based paraffin-filled MWCNT nanofluids.

capacity of nanofluids. In this section, state-of-the-art technology that can overcome these limitations is introduced, namely phase change material (PCM)-filled MWCNT nanofluids.[79] This novel concept involves using MWCNTs filled with a PCM, such as paraffin wax that exhibits high latent heat, for manufacturing nanofluids. The PCM-filled MWCNT nanofluids are expected to improve not only the effective specific heat near to the melting point of paraffin wax but also the density, as the density of the MWCNTs is approximately 2100 kg m^{-3}. Choi *et al.*[79] were the first to report that the specific heat and density of water-based nanofluids can be simultaneously improved by dispersing paraffin-filled MWCNT nanoparticles. In particular, it was experimentally shown that the measured specific heat of paraffin-filled MWCNT nanofluids exhibited a maximum value near to the melting point of paraffin wax. Moreover, it was presented that for MWCNTs with larger diameters, the paraffin-filled MWCNTs exhibited a considerably greater improvement in specific heat because they contained more paraffin wax in their cavity than MWCNTs with smaller diameters. It was suggested that the temperature range for enhancing the specific heat of working fluids can be designed by selecting one of a variety of PCM materials with the desired melting point (Figure 8.13).

8.3.2 Supercritical Fluid-based MWCNT Nanofluids

In the field of solar energy, many investigators have researched volumetric solar receivers to improve the thermal efficiency of solar thermal

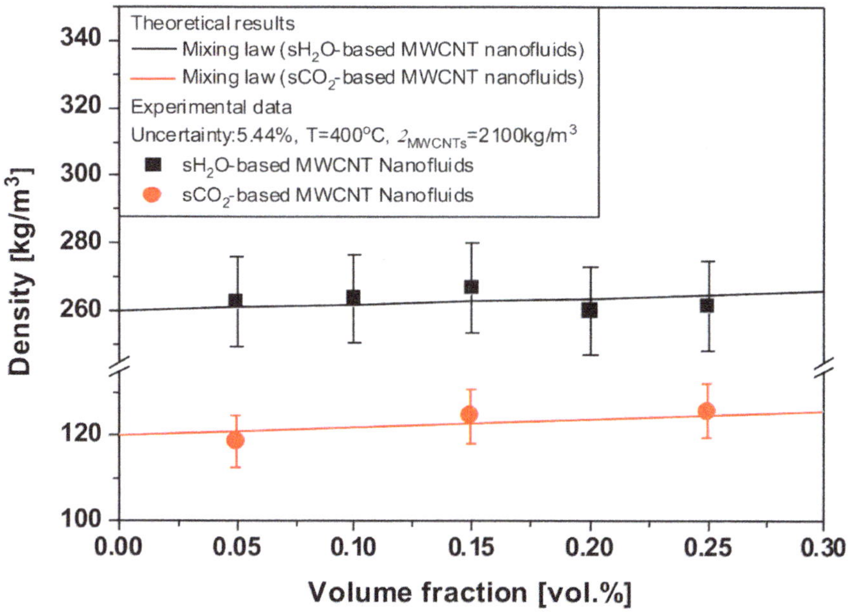

Figure 8.14 Density of supercritical fluid-based MWCNT nanofluids.

collection.[80–83] The concept behind volumetric solar receivers is to absorb solar thermal energy directly into the working fluid. Conventional working fluids in volumetric solar receivers, such as water, EG, and synthetic oil, not only have a low extinction coefficient but are also not suitable for concentrated solar power (CSP) systems owing to the operating temperature limitations. To overcome this challenge, several studies have investigated the use of nanofluids as working fluids in volumetric solar receivers, as it has been theoretically reported that the operating temperature of nanofluids in the supercritical state can exceed 1000 °C.[84,85] Recently, Choi and Jang[86] experimentally investigated the thermophysical and optical properties of supercritical water-based MWCNT nanofluids and supercritical carbon dioxide-based MWCNT nanofluids. They reported that the density of supercritical fluid-based nanofluids with MWCNTs linearly increased with the volume fraction, as shown in Figure 8.14. Moreover, it was demonstrated that the density of supercritical nanofluids was similar to that predicted using eqn (8.5).[11]

Acknowledgements

This work was supported by a National Research Foundation of Korea (NRF) grant funded by the Korean government (MSIP) (NRF2017R1A2B2011617) and partially supported by the 2021 Korea Aerospace University Faculty Research Grant (Prof. Seok Pil Jang).

References

1. S. U. S. Choi and J. A. Eastman, presented in part at *ASME International Mechanical Engineering Congress & Exposition*, San Francisco, USA, November, 1995.
2. Q. Li and Y. Xuan, *Sci. China, Ser. E: Technol. Sci.*, 2002, **45**(4), 408.
3. S. P. Jang and S. U. S. Choi, *Appl. Phys. Lett.*, 2004, **84**(21), 4316.
4. K. S. Hwang, S. P. Jang and S. U. S. Choi, *Int. J. Heat Mass Transfer*, 2009, **52**, 193.
5. H. A. Mintsa, G. Roy, C. T. Nguyen and D. Doucet, *Int. J. Therm. Sci.*, 2009, **48**(2), 363.
6. W. Yu, H. Xie, L. Chen and Y. Li, *Powder Technol.*, 2010, **197**(3), 218.
7. J. Sarkar, *Renewable and Sustainable Eng. Rev.*, 2011, **15**(3), 3271.
8. A. M. Hussein, K. V. Sharma, R. A. Bakar and K. Kadirgama, *Renewable Sustainable Energy Rev.*, 2014, **29**, 734.
9. B. Ilhan and H. Erturk, *Int. J. Heat Mass Transfer*, 2017, **111**, 500.
10. D. G. Subhedar, B. M. Ramani and A. Gupta, *Case Stud. Therm. Eng.*, 2018, **11**, 26.
11. B. C. Pak and Y. I. Cho, *Exp. Heat Transfer*, 1998, **11**(2), 151.
12. S. Q. Zhou and R. Ni, *Appl. Phys. Lett.*, 2008, **92**(9), 093123.
13. H. O'Hanley, J. Buongiorno, T. McKrell and L. W. Hu, presented in part at *ASME International Mechanical Engineering Congress & Exposition*, Denver, USA, 2011.
14. G. M. Moldoveanu and A. A. Minea, *J. Mol. Liq.*, 2019, **279**, 299.
15. S. M. S. Murshed, *Heat Transfer Eng.*, 2012, **33**(8), 722.
16. Q. He, S. Wang, M. Tong and Y. Liu, *Energy Convers. Manage.*, 2012, **64**, 199.
17. M. N. Pantzali, A. G. Kanaris, K. D. Antoniadis, A. A. Mouza and S. V. Paras, *Int. J. Heat Fluid Flow*, 2009, **30**(4), 691.
18. I. Wole-Osho, E. C. Okonkwo, D. Kavaz and S. Abbasoglu, *Powder Technol.*, 2020, **363**, 699.
19. Z. K. Kadhim, M. S. Kassim and A. Y. A. Hassan, *Int. J. Mech. Eng. Technol.*, 2016, 7, 11.
20. M. A. Sabiha, R. M. Mostafizur, R. Saidur and S. Mekhilef, *Int. J. Heat Mass Transfer*, 2016, **93**, 862.
21. Z. Said, *Int. Commun. Heat Mass Transfer*, 2016, **78**, 207.
22. D. R. Malik, PhD thesis, Texas A & M University, 2010.
23. Y. Hu, Y. He, Z. Zhang and D. Wen, *Energy Convers. Manage.*, 2017, **142**, 366–373.
24. M. Chieruzzi, G. F. Cerritelli, A. Miliozzi and J. M. Kenny, *Nanoscale Res. Lett.*, 2013, **8**, 448.
25. M. X. Ho and C. Pan, *Int. J. Heat Mass Transfer*, 2014, **70**, 174.
26. D. Shin and D. Banerjee, *Int. J. Heat Mass Transfer*, 2014, **74**, 210.
27. Y. Nayfeh, S. M. M. Rizvi, B. El Far and D. Shin, *Nanomaterials*, 2020, **10**(11), 2131.
28. M. Schuller, F. Little, D. Malik, M. Betts, Q. Shao, J. Luo, W. Zhong, S. Shankar and A. Padmanaban, Texas A & M University, College Station,

TX (United States), Texas A & M Engineering Experiment Station, 2012, No. DOE/GO18154.
29. M. Schuller, Q. Shao and T. Lalk, *Int. J. Therm. Sci.*, 2015, **91**, 142–145.
30. M. C. Lu and C. H. Huang, *Nanoscale Res. Lett.*, 2013, **8**, 292.
31. D. Lide and H. P. R. Frederikse, *CRC Handbook of Chemistry and Physics*, CRC Press, Cleveland, Ohio, 1978.
32. D. Shin and D. Banerjee, *J. Heat Transfer*, 2011, **133**, 024501.
33. H. Tiznobaik and D. Shin, *Int. J. Heat Mass Transfer*, 2013, **57**, 542.
34. M. Chieruzzi, A. Miliozzi, T. Crescenzi, L. Torre and J. M. Kenny, *Nanoscale Res. Lett.*, 2015, **10**, 273.
35. M. Liu, J. Severino, F. Bruno and P. Majewski, *J. Energy Storage*, 2019, **22**, 137.
36. B. X. Wang, L. P. Zhou and X. F. Peng, *Int. J. Thermophys.*, 2006, **27**, 139.
37. L. Wang, Z. Tan, S. Meng, D. Liang and G. Li, *J. Nanopart. Res.*, 2004, **3**, 483.
38. L. Xue, P. Keblinski, S. R. Phillpot, S. U. S. Choi and J. A. Eastman, *Int. J. Heat Mass Transfer*, 2004, **47**, 4277.
39. L. Li, Y. Zhang, H. Ma and M. Yang, *J. Nanopart. Res.*, 2010, **12**(3), 811.
40. S. H. Oh, Y. Kauffmann, C. Scheu, W. D. Kaplan and M. Ruhle, *Science*, 2005, **310**, 661.
41. M. Lasfargues, Q. Geng, H. Cao and Y. Ding, *Nanomaterials*, 2015, **5**, 1136.
42. M. Lasfargues, A. Bell and Y. Ding, *J. Nanopart. Res.*, 2016, **18**(6), 1.
43. A. Awad, H. Navarro, Y. Ding and D. Wen, *Renewable Energy*, 2018, **120**, 275.
44. S. Jung, PhD thesis, Texas A & M University, 2012.
45. D. Shin, B. Jo, H. E. Kwak and D. Banerjee, in *International Heat Transfer Conference*, Washington DC, USA, Jan., 2010.
46. B. Jo and D. Banerjee, presented in part at *ASME International Mechanical Engineering Congress & Exposition*, Denver, USA, 2011.
47. B. Jo and D. Banerjee, *Acta Mater.*, 2014, **75**, 80.
48. B. Jo and D. Banerjee, *Int. J. Therm. Sci.*, 2015, **98**, 219.
49. S. Liu, D. Wu, J. Liu, Y. Nian and P. Qiu, *Renewable Power Generation Conference*, Beijing, China, Sep., 2013.
50. D. P. Kulkarni, R. S. Vajjha, D. K. Das and D. Oliva, *Appl. Therm. Eng.*, 2008, **28**(14–15), 1774.
51. R. S. Vajjha and D. K. Das, *J. Heat Transfer*, 2009, **131**(7), 071601.
52. A. H. Hassan, M. A. M. Hassan and M. H. Shedid, *J. Therm. Sci. Eng. Appl.*, 2021, **14**(2), 021001.
53. P. K. Namburu, D. P. Kulkarni, A. Dandekar and D. K. Das, *Micro Nano Lett.*, 2007, **2**(3), 67.
54. R. S. Vajjha and D. K. Das, *Int. J. Heat Mass Transfer*, 2012, **55**(15–16), 4063.
55. S. Choudhary, A. Sachdeva and P. Kumar, *Renewable Energy*, 2020, **147**, 1801.
56. T. P. Teng and C. C. Yu, *Exp. Therm. Fluid Sci.*, 2013, **49**, 22.
57. S. M. S. Murshed, *J. Exp. Nanosci.*, 2011, **6**(5), 539.

58. B. Barbes, R. Paramo, E. Blanco, M. J. Pastoriza-Gallego, M. M. Pineiro, J. L. Legido and C. Casanova, *J. Therm. Anal. Calorim.*, 2013, **111**(2), 1615.
59. S. Akilu, A. T. Baheta, K. V. Sharma and M. A. Said, *AIP Conf. Proc.*, 2017, **1877**, 090001.
60. J. R. Satti, D. K. Das and D. Ray, *Int. J. Heat Mass Transfer*, 2016, **94**, 343.
61. M. F. Pakdaman, M. A. Akhavan-Behabadi and P. Razi, *Exp. Therm. Fluid Sci.*, 2012, **40**, 103.
62. Y. Xuan and W. Roetzel, *Int. J. Heat Mass Transfer*, 2000, **43**, 3701.
63. J. Buongiorno, *J. Heat Transfer*, 2006, **128**, 240.
64. J. Avsec and M. Oblak, *Int. J. Heat Mass Transfer*, 2007, **50**, 4331.
65. J. M. Smith and M. C. Van Ness, *Introduction to Chemical Engineering Thermodynamics*, McGraw-Hill, New York, 1987.
66. D. Shin and D. Banerjee, *J. Heat Transfer*, 2011, **133**(2), 024501.
67. D. Shin and D. Banerjee, *Int. J. Heat Mass Transfer*, 2011, **54**(5–6), 1064.
68. L. Wang, Z. Tan, S. Meng, D. Liang and G. Li, *J. Nanopart. Res.*, 2001, **3**(5), 483.
69. S. E. B. Maiga, C. T. Nguyen, N. Galanis and G. Roy, *Superlattices Microstruct.*, 2003, **125**, 567.
70. P. Keblinski, S. R. Phillpot, S. U. S. Choi and J. A. Eastman, *Int. J. Heat Mass Transfer*, 2002, **45**(4), 855.
71. D. Shin, H. Tiznobaik and D. Banerjee, *Appl. Phys. Lett.*, 2014, **104**(12), 121914.
72. C. J. Ho, W. K. Liu, Y. S. Chang and C. C. Lin, *Int. J. Therm. Sci.*, 2010, **49**, 1345.
73. M. Ali, O. Zeitoun and S. Almotairi, *Int. J. Therm. Sci.*, 2013, **63**, 115.
74. D. Chavan and A. Pise, *Mater. Today*, 2019, **16**, 504.
75. Z. Said, M. Sajid, A. Kamyar and R. Saidur, *Earth Environ. Sci.*, 2013, **16**, 012002.
76. R. S. Vajjha, D. K. Sas and B. M. Mahagaonkar, *Pet. Sci. Technol.*, 2009, **27**(6), 612.
77. S. Akilu, A. T. Baheta and K. V. Sharma, *J. Therm. Anal. Calorim.*, 2020, **140**, 1999.
78. S. N. Shoghl, J. Jamali and M. K. Moraveji, *Exp. Therm. Fluid Sci.*, 2016, **74**, 339.
79. T. J. Choi, S. H. Kim, S. P. Jang, L. Lin and M. A. Kedzierski, *Energy*, 2020, **210**, 118523.
80. S. H. Lee and S. P. Jang, *Int. J. Heat Mass Transfer*, 2015, **80**, 58.
81. S. H. Lee, T. J. Choi and S. P. Jang, *Energy*, 2016, **115**, 404.
82. R. Capuano, T. Fend, H. Stadler, B. Hoffschmidt and R. Pitz-Paal, *Renewable Energy*, 2017, **114**, 556.
83. R. Zhang, J. Qu, M. Tian, X. Han and Q. Wang, *Int. J. Energy Res.*, 2018, **42**(7), 2456.
84. D. C. H. Aita, MS thesis, Delft University, 2014.
85. S. H. H. J. Smit, MS thesis, Delft University, 2016.
86. T. J. Choi, PhD thesis, Korea Aerospace University, 2021.

CHAPTER 9

Dielectric and Magnetic Properties of Nanofluids

M. TIMKO,*[a] P. KOPCANSKY,[a] M. RAJNAK,[a] M. KARPETS,[a] K. PAULOVICOVA,[a] O. V. KOVALCHUK[b,c,d] AND L. A. BULAVIN[e]

[a] Institute of Experimental Physics, Slovak Academy of Sciences, Watsonova 47, 040 01 Kosice, Slovakia; [b] Kyiv National University of Technologies and Design, Nemirovich-Danchenko str. 2, Kyiv 01011, Ukraine; [c] National Technical University of Ukraine "Igor Sikorsky Kyiv Polytechnic Institute", 3, prospect Peremohy, 03056 Kyiv, Ukraine; [d] Institute of Physics, NAS of Ukraine, 46, prospect Nauky, 03680 Kyiv, Ukraine; [e] Faculty of Physics, Taras Shevchenko National University of Kyiv, Volodymyrska Str. 64, Kyiv 01601, Ukraine
*Email: timko@saske.sk

9.1 Dielectric and Magnetic Properties of Nanofluids

9.1.1 Preparation of Magnetic Nanofluids based on Non-polar Carrier Liquid

Magnetic nanofluids (MNFs) are well-known colloidal stable systems of nanosized sub-domain magnetic particles dispersed in a non-magnetic carrier liquid. The preparation of stable MNFs in our laboratory consists of two consecutive steps:

(a) the synthesis of magnetic nanoparticles;
(b) and the subsequent stabilization/dispersion of the nanoparticles in a non-polar solvent medium.[1,2]

Soft Matter Series No. 16
Fundamentals and Transport Properties of Nanofluids
Edited by S. M. Sohel Murshed

Published by the Royal Society of Chemistry, www.rsc.org

For the synthesis of magnetic nanoparticles (spinel ferrite type), the bottom-up chemical method is often applied. Ferrite particles are obtained from the co-precipitation technique of aqueous solutions of trivalent and bivalent iron salts in a 2 : 1 molar ratio. The process of nucleation and growth of the spherical nanoparticles takes place at 80–82 °C in the presence of a concentrated NH_4OH solution (25%) and $C_{17}H_{33}COOH$ (90%) used as precipitating and stabilizing agents, respectively. The temperature and rate of the addition of an excess of the precipitating agent are chosen to ensure the formation of magnetite nanoparticles predominantly. The surface of the magnetic nanomaterial is sterically functionalized *via* the chemisorption of chemically pure oleic acid to prevent both agglomeration and/or oxidation of the nanoparticles and ensure the hydrophobicity of their surface. The resultant black precipitate (containing a monolayer of surfactant covered nanoparticles) was separated using a permanent magnet and then the nanoparticles were washed with deionized water and acetone (flocculation removed the free oleic acid). The stabilized magnetic nanoparticles with hydrophobic character were homogenously dispersed (~120 °C) in a carrier liquid of non-polar character (GTL-based transformer oil) to obtain a stable MNF without any phase separation and sedimentation.[2–4]

This new generation gas-to-liquid (GTL) transformer oil is based on iso-paraffin hydrocarbons extracted from natural gas (CH_4). Electrical insulation liquids based on these hydrocarbons are cleaner, chemically stable, and the thermal conductivity values of these fluids are higher, indicating their improved heat transfer properties. In the field of electrotechnics and power engineering, these composite materials represent a way of reducing the dielectric and thermal losses of various devices and are promising cooling and insulating media.[4]

The typical magnetization loops of a prepared ferrofluid sample measured at room temperature and at 2 K are shown in Figure 9.1 (for a sample of a ferrofluid consisting of the transformer oil Mogul Trafo CZ-A and iron oxide nanoparticles stabilized by oleic acid).

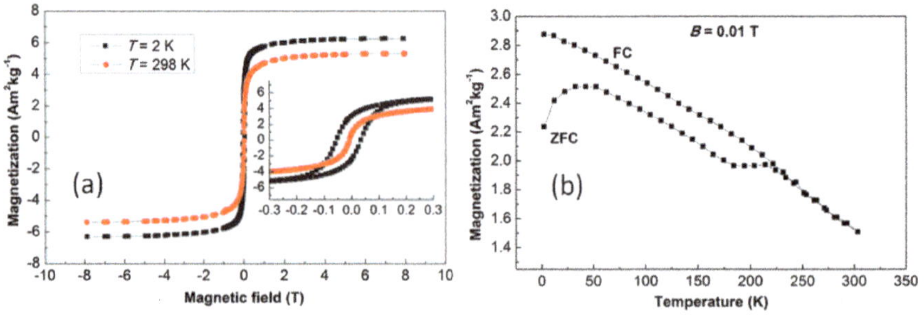

Figure 9.1 Magnetic properties of a ferrofluid with a particle volume fraction of $\Phi_V = 1\%$. (a) Magnetization curves. (b) ZFC and FC curves measured at 0.01 T. Reproduced from ref. 12 with permission from Elsevier, Copyright 2017.

The high temperature paramagnetic nature of the depicted curve and the hysteresis (inset graph) revealed at low temperature confirm the superparamagnetic and blocked state of the nanoparticles, respectively. The temperature-dependent magnetization measured in the zero field cooled (ZFC) and field cooled (FC) regimes is presented in Figure 9.1(b). One can clearly see the area of the blocking temperatures at around 50 K, while the high temperature kink in the ZFC curve reflects the melting of the oil. The pronounced ZFC kink separating the FC and ZFC curves is associated with the presence of greater particles that respond to the magnetic field *via* Brownian rotation when the oil undergoes transition from the frozen to the liquid state. These particles prefer to respond to the field *via* physical rotation, so once the surrounding material (frozen oil) allows that rotation, the magnetization increases markedly. The FC curve exhibits a typical continuous increase in magnetization with decreasing temperature (as the moments acquire a higher degree of orientation), so the two curves remain separated below the transition from the liquid to the frozen state.

9.1.2 Studies of Magnetic Fluids Under an Electric Field

Physical properties of magnetic fluids (ferrofluids) have been studied by numerous research groups, leading to many technical applications. In particular, transformer oil-based ferrofluids and their enhanced thermal properties and higher breakdown voltage (compared to pure transformer oil) have promoted very intensive study of these materials. Among various physical properties, the dependence of dielectric permittivity of ferrofluids on the applied magnetic field, the so-called magneto-dielectric effect, has been studied by different authors. The results of the studied effects of the homogeneous magnetic field on the permittivity have revealed that the field-induced magnetodielectric effect is distinctly frequency dependent and should be attributed to structural anisotropy, which affects not only the interfacial polarization of the nanoparticles, but also the strength of the hopping process observed in the samples. Both these effects influence the low-frequency permittivity values.[5]

However, experimental results concerning the dispersion of the magneto-dielectric anisotropy as a function of the electric field frequency, its dependence on the value of the magnetic field strength and on the physical parameters of ferrofluids have been presented.[6]

Previous study in our laboratory showed that the dielectric response of a magnetic fluid (MF) composed of purified magnetite nanoparticles dispersed in an inhibited mineral transformer oil (MOGUL CZ-A, PARAMO) is dependent on the strength and frequency of the electric field.[7] For explanation of the low frequency relaxation process, the Schwarz model was applied.[8] This model considers the fact that electrically-charged colloidal nanoparticles by fixed or adsorbed ions are surrounded by counterions, so forming an electric double layer (EDL) on the surface of each particle. These counterions are free to move laterally but not transversally on the particle surface. During the action of the external electric field, the system is polarized and this polarization process is correlated with the corresponding relaxation process. The position of the relaxation maximum shifts towards higher frequencies with increasing magnetic volume fraction of the nanoparticles, and so with the corresponding number of double layers. It was also proven[9,11] that the low frequency relaxation process is slowed down when the inter-electrode gap is increased in units of micrometers and when a direct current bias field is applied to the sample. This effect is related to the formation of aggregates. The isothermal dielectric spectroscopy of a transformer oil based ferrofluid was employed to detect the predicted electric field induced particle cluster formation. In earlier studies[7,10] on transformer oil-based ferrofluids, the low-frequency relaxation process was assigned to EDL polarization on the particle surfaces. In the case of nonpolar carrier liquids, such as transformer oil, it is possible to consider the fixed adsorbed OH^- and bound layer NH_4^+ as the potential EDL components. So, in this type of arrangement, one can describe the low-frequency relaxation process using the Schwarz theory and the relaxation time can be given by:

$$\tau = \frac{R^2}{2ukT} = \frac{R^2}{2u_0kT}\exp\left(\frac{E_a}{kT}\right)$$

where R is the radius of the nanoparticles, k is the Boltzmann constant, T is the absolute temperature and u is the mechanical mobility of the ions on the particle surface dependent on the activation energy E_a of the motion along the surface and the charge mobility u_0 in free solution. The low-frequency relaxation process was assigned to EDL polarization. Furthermore, the effect of electrode separation distance at a constant electric field intensity on the relaxation maxima was investigated. The broadening of the relaxation maximum towards lower frequencies was observed as the response to the increased electrode separation distance.[9] On the basis of the particle size dependent relaxation time, the broadening is associated with the cluster formation induced by the applied alternating current (AC) electric field. The broader relaxation maximum therefore indicates the increase in the magnetic solid fraction size distribution. A similar indication of the cluster formation is observed when the DC bias voltage is applied (see Figure 9.2).

The observed temporal hysteresis in the relaxation time reflects the particle cluster formation and decay behavior. From Figure 9.2, it can be seen that the relaxation maximum amplitude increases for an applied bias voltage of 1 V only and is shifted to lower frequencies. This is associated with the initial formation of particle clusters as a consequence of the DC electric field application. However, this relaxation maximum shifts to lower frequencies when the amplitude of the applied DC electric field increases. This is a consequence of the dominant polarization force and the DC electric field induces the dipoles in the particles and subsequent formation of larger clusters. However, this DC field polarizes and keeps the counterions on the opposite cluster sides and prevents their response to the AC field.

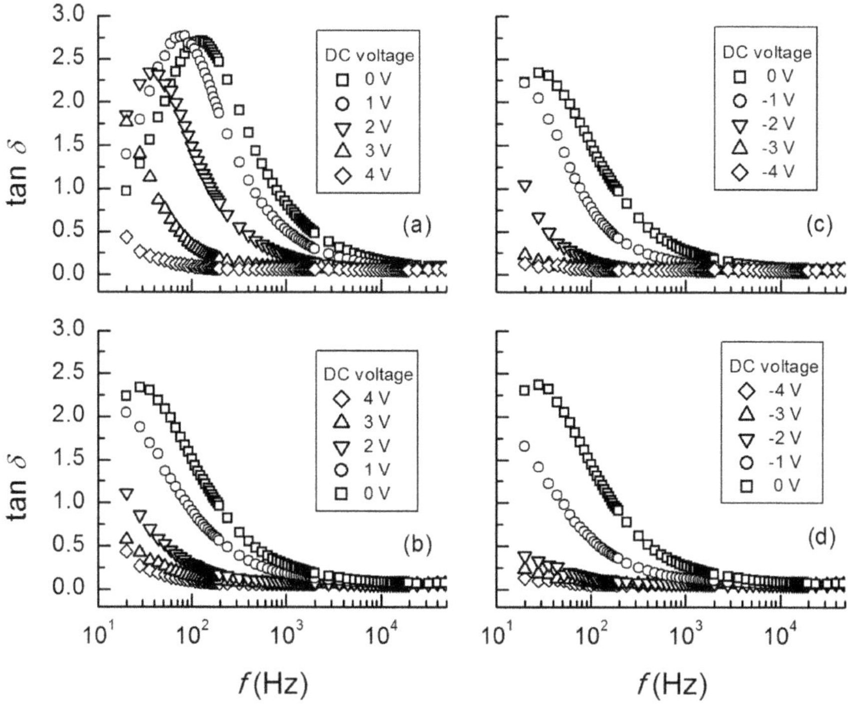

Figure 9.2 The dependence of the tan δ spectrum on the applied DC bias voltage. The measurements were conducted one after the other with the following sequence: (a) from 0 V up to 4 V and (b) from 4 V down to 0 V. The investigated transformer oil-based MF with a magnetic volume fraction of $\Phi V = 6.6\%$ was measured in a capacitor cell with an electrode separation distance of $d = 1.6$ μm and an applied test signal level of 32 mV (20 kV m^{-1}). The intensity of DC electric field was 0.625 MV m^{-1}, 1.25 MV m^{-1}, 1.875 MV m^{-1}, and 2.5 MV m^{-1} corresponding to applied DC bias voltages of 1 V, 2 V, 3 V, and 4 V, respectively. Reproduced from ref. 9 with permission from American Physical Society, Copyright 2014.

The DC electric field raises the energy barrier and reduces the mobility and moving range of the counterions, which is reflected in the lower-frequency position of the loss peak and decrease in amplitude.

The particle clusters induced by the external electric field can contribute to the process of the electric breakdown field strength enhancement in transformer oils. The clustering should be taken into account in connection with a charged nanoparticle model, in which the nanoparticles act as electron scavengers.[9]

The formation of visually observable pattern formation in a transformer oil based ferrofluid (TOFF) exposed to DC electric field was presented.[10] The presence of space charge and its motion towards electric field gradients was assumed to be a trigger of colloidal cloud formation, as it is possible to see in Figure 9.3.

Electrical forces on space charge and dielectric permittivity contrast in the ferrofluid components, each playing a role in inducing pattern formation and its dynamics. Subsequently, a spiky pattern forms from the cloud with the spikes oriented in the field gradient. The initial particle accumulation can be explained by the induced electric dipole–dipole interaction, subsequent dielectrophoresis and the related electrohydrodynamic flow, which push the particles towards the electrical equator. Moreover, the distortion of the cloud is the result of permittivity and conductivity gradients in the sample with the separated cloud, and the action of an anisotropic electrostatic force on the transition layer between the cloud and its surroundings. The observed macroscopic structural changes of the ferrofluid are a reversible phenomenon, independent on the voltage polarity and the orientation of the sample holder with regard to gravity force. In a similar way, magnetic particle assembly was observed in a case where the electrodes were not in direct contact with the ferrofluid sample. Instead, a submillimeter droplet was placed on a microscope slide and two needle electrodes were attached to the slide from outside the droplet (Figure 9.3(c and d)). Even though the large aggregates slowly disappeared after interruption of the electric field, one can suppose that the applied electric force acting on the surfactant has a deteriorative consequence on the ferrofluid stability.

The effect is accompanied by specific particle aggregation on a nanoscale, as concluded from small-angle neutron scattering (SANS) experiments. Electric field induced changes in a ferrofluid structure at the nanoscale were recently confirmed *via in situ* SANS experiments,[12] where the structure of magnetite nanoparticles and their assemblies as a result of magnetic and van der Waals interactions were studied. In Figure 9.4, an increase in the scattered intensity in the small q (scattering vector) region can be seen to emerge and intensify with increasing electric field strength.

This is direct evidence for the formation of particle aggregates in a transformer oil based ferrofluid (TOFF) induced by the above analyzed electrohydrodynamics and polarization forces between the particles.

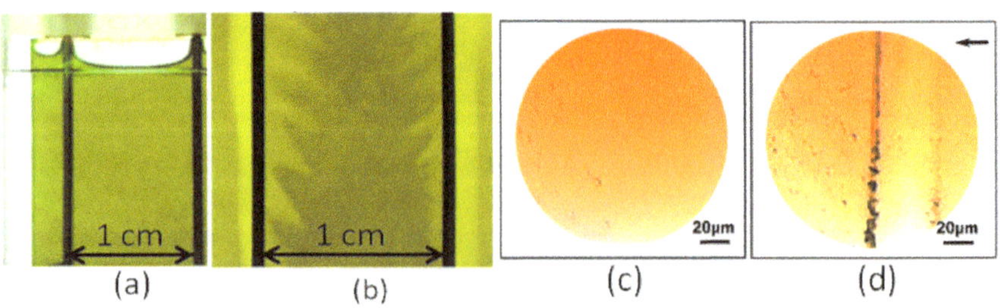

Figure 9.3 Ferrofluid with a particle volume fraction of $\Phi_v = 0.05\%$ in a glass cuvette with two electrodes under (a) zero electric field and (b) under the action of 5 kV cm^{-1}. Optical microscopy view focused on a submillimeter droplet of the ferrofluid ($\Phi_v = 1\%$) in a (c) zero electric field and (d) under the action of 2 kV mm^{-1}. The arrow indicates the field direction. Reproduced from ref. 12 with permission from Elsevier, Copyright 2017.

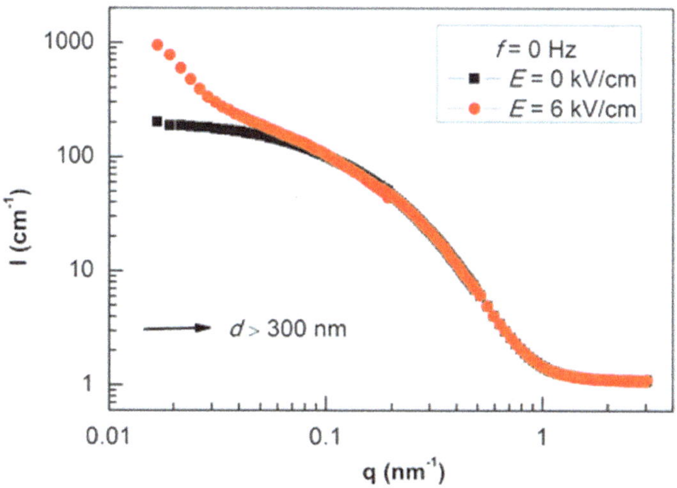

Figure 9.4 SANS profiles of a ferrofluid ($\Phi_v = 1\%$) exposed to zero and 6 kV cm^{-1} electric field intensity. Reproduced from ref. 12 with permission from Elsevier, Copyright 2017.

The dielectric contrast in the TOFF results in induced electric dipole–dipole interactions and subsequent aggregation. Thus, the scattering curve obtained at 6 kV cm^{-1} represents two subsystems of scattering objects in the sample. The dielectric contrast between the magnetic particles and the oil molecules plays a crucial role, leading to induced electric dipole–dipole interactions and subsequent aggregation. It was estimated that the mean aggregate size was >300 nm, as determined from the minimal q covered in the experiment according to $D \sim 2\pi/q$. However, in an applied DC electric field, 2D scattering patterns showed anisotropy corresponding to the elongated shape of the formed aggregates oriented along the electric field.

In comparison to SANS, small angle X-ray scattering (SAXS)[13] cannot be used to prove the proportional increase in the low q scattering intensity with increasing electric field, but rather stochastic variations. The observed intensity variations during SAXS measurements reflect the local structural nanofluid changes caused by induced electrohydrodynamics. The scattering profiles were compared for a magnetic nanofluid in direct electrical contact with electrodes and for a magnetic fluid externally exposed to an electric field. No scattering changes were observed for the non-contact mode, reflecting the unchanged initial structure of the nanofluid in the electric field. Moreover, stochastic variation in the scattering intensity under various voltages was detected in direct contact mode as a result of the small sized X-ray beam (small irradiated area) scattered from the flowing concentration inhomogeneity. This inhomogeneity is brought about by the conductivity and polarization of the particles and the resulting electrohydrodynamics.

The insulating and dielectric properties of TO were found to exhibit peculiar enhancement upon the addition of magnetic nanoparticles.[14] A comprehensive mechanism of the increased breakdown field strength is still unknown; however, a few models have been proposed. In the nanoparticle charging model,[13] the difference in the dielectric permittivity of the dispersed nanoparticles and the surrounding oil is considered to be an essential condition leading to polarization, charge trapping, and subsequent reduction of streamer velocity. Thus, the dielectric permittivity of MNFs is a crucial parameter that has been extensively studied.[8] In a previous study,[15] it was shown that a transformer oil-based magnetic nanofluid can exhibit DC bias tunable apparent negative permittivity behavior at low frequencies (Figure 9.5). This effect is a result of particle clustering and interface charge overlap in an applied electric field. Due to created percolative conductive paths, the nanofluid undergoes transition from capacitive to inductive reactance. This mechanism is feasible under a low-frequency electric field without dynamic perturbations.

The obtained theoretical results of Selyshchev *et al.*[16] showed that the homogeneous distribution of dielectric particles in a dielectric carrier became inhomogeneous under an applied electric field. The reason for the inhomogeneous development was the polarization of the particles and their interaction as dipoles. Transition of originally homogenous suspension to an inhomogeneous distribution has a threshold character. The critical value of the applied field increased with increasing temperature, with a decrease in the number of particles and their radius. The critical value of the applied field also increased with decreasing difference between the dielectric permittivity of the particles and medium. The observed distribution of

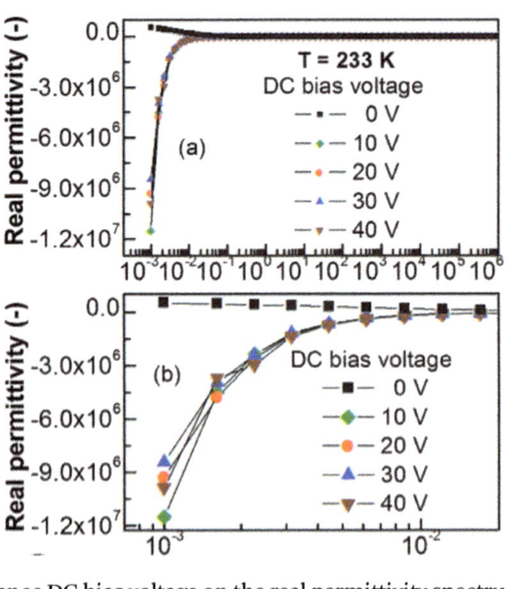

Figure 9.5 Influence DC bias voltage on the real permittivity spectrum of a ferrofluid at 233 K. (a) Whole spectrum with linear vertical scale and (b) low-frequency detail of the negative permittivity. Reproduced from ref. 15 with permission from American Physical Society, Copyright 2019.

nanoparticle concentration according to the proposed theoretical approach was in good agreement with the experimentally observed non-homogeneous distribution of ferrofluid particles under an external electric field. The current challenges in electric energy production and its distribution are the reduction in power loss, minimalization of the environmental impact of transformers and prolonging of their service life. These appeals are connected with insulating oils used in power transformer as cooling and insulating media. Besides this, novel mineral oils and natural ester-based oils are gaining increased interest due to their recyclability and biodegradability. Moreover, GTL technology has been introduced to produce isoparaffinic transformer oil that offers a high degree of compositional and performance consistency, high purity and excellent resistance to degradation, as well as being sulfur free. The current research on dielectric liquids aims to enhance their thermal and insulating properties by doping the liquids with various particles. It has been shown that nano-sized particles can significantly enhance both the thermal and insulating properties of the base liquid.

9.1.3 Dielectric Breakdown in Transformer Oil based Ferrofluids

Mineral oils have been used as insulating and cooling materials for transformers due to their thermal and isolative potential. Therefore, the development of a transformer oil with favorable dielectric and thermal properties is extensively required. Moreover, transformer oils doped with magnetic nanoparticles (nanofluids) have been produced to accomplish the above mentioned and required attributes. In recent years, nanofluids have attracted attention and some review articles concerning their thermal, electrical and isolating properties have been published.[17,18] Moreover, a high AC breakdown voltage is a very important requirement for insulation liquids. Experimental studies have shown that AC breakdown strength is affected by factors such as moisture, nanoparticle type, volume concentration and the surface modification of nanoparticles. An increasing number of recent papers on the insulating properties of nanofluids have reported on enhanced dielectric breakdown strength compared to a base liquid.

It has been demonstrated that the heat transfer properties of oils can be significantly improved using nanoparticle additives.[19,20] Ongoing research will increase our knowledge of the fundamental mechanisms through which nanoparticles interact with liquid matrices. A potential alternative in many cases is the replacement of mineral oil with appropriately modified nanoparticle liquids, which lead to considerable cost savings. Another important potential benefit of nanofluids is an increase in breakdown voltage, which should allow more compact electrical apparatus design.

Electrical breakdown in insulating liquids is a multifaceted phenomenon, with various theories having been proposed for its explanation. Enhanced electrical breakdown due to the addition of nanoparticles is still not fully understood. It is known that it depends not only on the type of the nanoparticles but also on their concentration. The breakdown voltage of magnetic fluids is usually increased upon the application of an external magnetic field.[22] In this work, the effect of volume nanoparticle concentration in the transformer oil based magnetic fluid and external magnetic field on dielectric breakdown voltage (DBV) was investigated. It was shown that the DBV for a nanofluid was twice that of pure transformer oil. It might be considered that the added conductive nanoparticles affect as electron scavengers in electrically stressed transformer oil based nanofluids and transforms fast electrons into slowly negatively-charged nanoparticles for electrical breakdown. After application of an external magnetic field, the DBV is 30% higher than that of the non-applied magnetic field. These phenomena indicate that the DVB of a transformer oil based nanofluid is increased because the magnetic nanoparticles are well-dispersed by the application of a magnetic field.

Furthermore, from the breakdown testing of magnetic fluids by Segal,[21] a greater enhancement of breakdown voltage (in regard to the base transformer oil) was proven for positive breakdown than for negative breakdown as a consequence of the polarity effect. The polarity effect is generally well-known, even for base transformer oils. It is associated with the fact that the mean velocity of positive streamer propagation is significantly greater than that of negative streamers.[23] Even AC breakdowns of dielectric oils are usually triggered by positive streamers, as they propagate much faster than negative streamers.[24] Recently, several researchers have accepted the charged nanoparticle model when explaining enhanced breakdown voltage due to magnetic nanoparticles.[11] The model is based on the charging dynamics of nanoparticles, which act as electron scavengers. According to the model, nanoparticle charge trapping reduces the speed of the streamer, leading to enhanced insulating performance of the oil. Thus, the speed of the streamer development plays a key role in understanding the enhanced breakdown. Then, considering that the speed of the streamer propagation in transformer oil is dependent on the electrical polarity, the charged nanoparticles can reduce that speed. Therefore, it is important to study the electrical breakdown of magnetic fluids under both electrical polarities. DC electric breakdown of MOGUL TRAFO CZ-A[9] transformer oil containing iron oxide nanoparticles stabilized by oleic acid showed dependence on the high potential polarity and magnetic nanoparticle concentration. Enhancement in BDV was observed for all of the used concentrations, for each voltage rise rate and both needle polarities, compared to pure transformer oil. Maximal BDV enhancement was observed for magnetic fluid containing 0.35 vol% of magnetic nanoparticles, measured under a voltage rise rate of 470 V s^{-1} with a negative potential applied to the needle electrode.[25] Thus, the average BDV increased by approximately 47% compared to the BDV of the clear oil. As reported by Devins,[26] electrical breakdown in transformer oils is a consequence of streamers, which are formed due to the ionization of oil molecules and their propagation between the electrodes. When a positive streamer is formed, the positive ions enhance the electric field strength in front of the streamer tip and induce further oil ionization resulting in streamer propagation and breakdown. For negative streamers, the positive ions reduce the electric field strength between the streamer and the anode.[27] The propagation of negative streamers is thus restrained and breakdown occurs at higher electric field strength.[20] According to the charged nanoparticle model,[9] nanoparticles can trap the electrons from field ionization if the nanoparticles are characterized by a short relaxation time constant, $\tau = (2\varepsilon 1 + \varepsilon 2)/(2\sigma 1 + \sigma 2)$, determined by the dielectric permittivity and conductivity of oil ($\varepsilon 1$ and $\sigma 1$) and the nanoparticles ($\varepsilon 2$ and $\sigma 2$), respectively.

It has been shown that in partial discharge activity, the inception voltage of the transformer oil increases linearly with increasing magnetite nanoparticle concentration.[14]

Similarly, the DC breakdown voltage was found to increase with magnetite nanoparticle concentration as well as with voltage rise rate, with these effects being dependent on voltage polarity. The nanoparticles immersed in the base liquid attract a space charge (counter ions) from the liquid and form an EDL.[8,28] These nanoparticles act as free charge scavengers even before their polarization, and before the electric field reaches the ionization threshold. Under an externally applied electric field, the nanoparticles undergo induced polarization, which leads to more efficient charge scavenging from the streamer, and the charge density in the EDL region increases. Moreover, the charges attracted to the nanoparticles decrease the net charge density at streamer tips. As a result, the electric field distribution at the streamer tip is homogenized. This behavior leads to suppressed propagation of positive streamers and subsequently to the reduction in the polarity effect on the electric breakdown. This consideration is adequate, as the presence of an EDL on the magnetic nanoparticles in studied magnetic fluids has been concluded in earlier papers by means of dielectric spectroscopy.[7,9]

Nanoparticle charge trapping and a reduction in net charge density at the streamer tip is considered to be the major reason for decreasing polarity effect with increased nanoparticle concentration. According to the charged nanoparticle model,[11] nanoparticles can trap the electrons from field ionization if the nanoparticles are characterized by a short relaxation time constant according to the dielectric permittivity and conductivity of the oil and nanoparticles. However, some studies have shown magnetite nanoparticles to have a negative impact on the breakdown performance of transformer oil. This can be associated with nanoparticle interface modification[29] or with aging processes in ferrofluids.[30] Moreover, the breakdown voltage of ferrofluids may be significantly increased by applying an external magnetic field.[22] The effect of magnetic nanoparticles on the breakdown performance of transformer oil is often interpreted in view of an electron-scavenging theory, which is based on the modeling of the electrodynamic processes in ferrofluids.[11]

A comparative experimental study of the insulating properties of conventional mineral oil based ferrofluids and ferrofluids based on a novel insulating oil produced using GTL technology was published recently.[31] These two transformer oils exhibited comparable breakdown voltage values but the magnetic fluids based on these oils behaved differently. It was shown that magnetic nanoparticles enhanced the breakdown voltage in the mineral oil in comparison to the GTL oil. This enhancement was attributed to charge trapping by the polarized nanoparticles. The better performance in the mineral oil was associated with its greater viscosity, which impeded the migration of the charged nanoparticles. Similarly, partial discharge testing revealed more intense activity in the GTL oil based ferrofluids.

9.1.4 Conclusions

Magnetic nanoparticles and transformer oil based magnetic fluids represent interesting targets of investigation due to their special basic physical properties and promising application potential as cooling and isolative media in high-power transformers. In this chapter we reviewed the obtained results concerning the selected dielectric, magnetic and structural properties of transformer oil based magnetic fluids reported in recent years, including our latest research results. The main idea behind the incorporation of magnetic nanoparticles in transformer oil is the improvement in material properties and to achieve better conditions for wide application in power engineering technology. The obtained results of implemented experiments are encouraging for the future outlook in this field of science.

Acknowledgements

This work was supported by Slovak Scientific Grant Agency VEGA (projects No. 0011 and 0043), by the Slovak Research and Development Agency under the contract No. APVV 15-0453 and 18-0160, by the European Structural Funds in the frame of the projects MODEX ITMS 313011T548, Project PROMATECH ITMS 26220220186, Project MVTS SAV FMF ERANET 2; COST NANOUPTAKE CA15119 and COST NANOConVEX CIG-15119.

References

1. R. E. Rosensweig, *Ferrohydrodynamics*, Dover Publications, 2013, ISBN 0486783006.
2. L. Vekas, *Adv. Sci. Tech.*, 2008, **54**, 127.
3. D. Bica, L. Vekas, M. V. Avdeev, O. Marinica, V. Socoliuc, M. Balasoiu and V. M. Garamus, *J. Magn. Magn. Mater.*, 2007, **311**(1), 17.
4. K. Paulovičová, J. Tóthová, M. Rajňák, M. Timko, P. Kopčanský and V. Lisý, *Acta Phys. Pol., A*, 2020, **137**, 908.
5. A. Spanoudaki and R. Pelster, *J. Magn. Magn. Mater.*, 2002, **252**, 71.
6. C. Cotae, *J. Magn. Magn. Mater.*, 1983, **39**, 85.
7. M. Rajnak, J. Kurimsky, B. Dolnik, K. Marton, L. Tomco, A. Taculescu, L. Vekas, J. Kovac, I. Vavra, J. Tothova, P. Kopcansky and M. Timko, *J. Appl. Phys.*, 2013, **114**, 034313.
8. L. Malaescu and C. N. Marin, *J. Colloid Interface Sci.*, 2002, **251**, 73.

9. M. Rajnak, J. Kurimsky, B. Dolnik, P. Kopcansky, N. Tomasovicova, E. A. Taculescu-Moaca and M. Timko, *Phys. Rev. E: Stat., Nonlinear, Soft Matter Phys.*, 2014, **90**, 032310.
10. M. Rajnak, V. I. Petrenko, M. V. Avdeev, O. I. Ivankov, A. Feoktystov, B. Dolnik, J. Kurimsky, P. Kopcansky and M. Timko, *Appl. Phys. Lett.*, 2015, **107**, 073108.
11. J. G. Hwang, M. Zahn, F. M. O'Sullivan, L. A. A. Pettersson, O. Hjortstam and R. S. Liu, *J. Appl. Phys.*, 2010, **107**, 014310.
12. M. Rajnak, M. Timko, P. Kopcansky, K. Paulovicova, J. Tothova, J. Kurimsky, B. Dolnik, R. Cimbala, M. V. Avdeev, V. I. Petrenko and A. Feoktystov, *J. Magn. Magn. Mater.*, 2017, **431**, 99.
13. M. Rajnak, V. M. Garamus, M. Timko, P. Kopcansky, K. Paulovicova, J. Kurimsky, B. Dolnik and R. Cimbala, *Acta Phys. Pol., A*, 2020, **137**, 942.
14. J. Kurimsky, M. Rajnak, R. Cimbala, J. Rajnic, M. Timko and P. Kopcansky, *J. Magn. Magn. Mater.*, 2020, **496**, 165923.
15. M. Rajnak, Z. Spitalsky, B. Dolnik, J. Kurimsky, L. Tomco, R. Cimbala, P. Kopcansky and M. Timko, *Rev. Phys. Appl.*, 2019, **11**, 024032.
16. P. A. Selyshchev, V. I. Petrenko, M. Rajnak, B. Dolnik, J. Kurimsky, P. Kopcansky, M. Timko and L. A. Bulavin, *J. Mol. Liq.*, 2019, **278**, 491.
17. M. Rafiq, Y. Lv and C. Li, *J. Nanomater.*, 2016, **2016**, 8371560.
18. Y. Lv, Y. Zhou, C. Li, Q. Wang and B. Qi, *IEEE Electr. Insul. Mag.*, 2014, **30**, 23.
19. T. Tanaka and T. Imai, *IEEE Electr. Insul. Mag.*, 2013, **29**(21), 10.
20. J. Gerhold, *IEEE Trans. Dielectr. Electr. Insul.*, 2002, **9**(6), 878.
21. V. Segal, A. Hjortsberg, A. Rabinovich, D. Nattrass and K. Raj, *Conference Record of the IEEE International Symposium on Electrical Insulation*, (Cat. No. 98CH36239), IEEE, n.d., 1998, pp. 619–622.
22. J.-C. Lee, H.-S. Seo and Y.-J. Kim, *Int. J. Therm. Sci.*, 2012, **62**, 29.
23. H. Yamada, S. Sakamoto and Y. Nakao, *J. Electrost.*, 1979, **7**, 155.
24. E. F. Kelley and R. E. Hebner, *Appl. Phys. Lett.*, 1981, **38**, 231.
25. P. Bartko, M. Rajňák, R. Cimbala, K. Paulovičová, M. Timko, P. Kopčanský and J. Kurimský, *J. Magn. Magn. Mater.*, 2020, **497**, 166007.
26. J. C. Devins, S. J. Rzad and R. J. Schwabe, *J. Appl. Phys.*, 1981, **52**, 4531.
27. Z. Wang, Y. Zhou, W. Lu, N. Peng and W. Chen, *Nanomater*, 2019, **9**, 627.
28. M. Dong, J. Dai, Y. Li, J. Xie, M. Ren and Z. Dang, *AIP Adv.*, 2017, 7, 025307.
29. M. Rajnak, J. Kurimsky, R. Cimbala, Z. Conka, P. Bartko, M. Suga, K. Paulovicova, J. Tothova, M. Karpets, P. Kopcansky and M. Timko, *J. Mol. Liq.*, 2020, **309**, 113243.
30. J. Kurimsky, M. Rajnak, P. Bartko, K. Paulovicova, R. Cimbala, D. Medved, M. Dzamova, M. Timko and P. Kopcansky, *J. Magn. Magn. Mater.*, 2018, **465**, 136.
31. J. Kurimsky, M. Rajnak, R. Cimbala, K. Paulovicova, Z. Rozynek, P. Kopcansky and M. Timko, *J. Mol. Liq.*, 2021, **325**, 115244.

Section III: Thermal Features of Nanofluids

CHAPTER 10

Convective Heat Transfer Using Nanofluids

M. H. BUSCHMANN

Institut für Luft- und Kältetechnik Dresden, 01309 Dresden, Germany
Email: Matthias.Buschmann@ilkdresden.de

10.1 Introduction

Convective heat transfer has proven to offer vast potential to increase the efficient use of energy in various processes in both industry and everyday life. However, the global transition to sustainable, green industry and the corresponding lifestyle demands more effective strategies. Standard methods for improving convective heat transfer,

(1) increasing the heat transfer area,
(2) increasing the Reynolds number, and
(3) structuring heat transfer surfaces,

appear to have reached their physical and technical limits. Moreover, most of them are associated with various drawbacks, such as high costs due to high material requirements, high pumping capacity, or increased production efforts, thus, alternatives are needed.

Let us first consider the structure of a classical Nusselt number correlation. According to the Buckingham Π theorem, the most general form is:[1]

$$\mathrm{Nu} = f(\mathrm{Geo}, \mathrm{Re}^n, \mathrm{Pr}^m) \tag{10.1}$$

Soft Matter Series No. 16
Fundamentals and Transport Properties of Nanofluids
Edited by S. M. Sohel Murshed

Published by the Royal Society of Chemistry, www.rsc.org

where $Nu = hl/k$ (h is the heat transfer coefficient, l is the characteristic length of the device, k is the thermal conductivity) indicates the Nusselt number. Geo is any geometry parameter that characterises the flow situation, such as a pipe, plate, bend, *etc.* The Reynolds number is denoted as $Re = \rho ul/\eta$ (ρ is the density, u is the flow velocity, η is the dynamic viscosity) and the Prandtl number as $Pr = \eta c_p/k$ (c_p is the heat capacity, k is the thermal conductivity). The powers n and m can be any real number. While the first similarity number can be understood as the ratio of inertia to viscous force, the second is derived from splitting up the Péclet number Pe = Re Pr into the Reynolds and Prandtl numbers. The Péclet number is the equivalent of the Reynolds number for a related heat transfer problem. While the three classical strategies mentioned above can more or less be described according to changes in Geo and Re, the third parameter Pr opens up a new way of manipulating heat transfer. The Prandtl number is a thermophysical property that characterises the working fluid. Changing this parameter would be limited entirely to the alteration of the thermophysical properties of the fluid.

For conventional heat carriers such as water and thermal oils, alteration of the thermophysical properties is possible through the dispersion of nanoparticles with sizes ranging from 10 to 100 nm. The main aim of creating such suspensions is to increase thermal conductivity. Therefore, the nanoparticles should have a thermal conductivity that is significantly higher than that of the base fluid. Assuming water as a base fluid with $k\,(20\ ^{\circ}C) = 0.6\ W\,m^{-1}\,K^{-1}$, this is true for most solid nanoparticles. However, suspending nanoparticles in a base fluid not only increases thermal conductivity, due to the higher density of solid nanoparticles it also increases the effective density of the nanofluid. Solids have lower specific heat capacities than the most common base fluid, water. Hence, the effective heat capacity of the nanofluid is also lower. The last remaining thermophysical property that is relevant for convective heat transfer is viscosity, which has been analysed intensely, as the viscosity of nanofluids is a complex, multifaceted parameter. Unfortunately, it seems to be that in most cases it increases,[10,11] which is especially true the higher the nanoparticle concentration. Therefore, it makes sense to add nanoparticles with significantly high thermal conductivity to the base fluid. This allows the nanoparticle concentration and thus the increase of the viscosity to remain as low as possible.

Adding solid particles to a liquid creates a two-phase suspension, where the question is if the flow of such a compound can still be referred to as a single-phase flow. If so, real-world applications would be hampered. Moreover, well-established pressure loss and heat transfer correlations, such as the one shown in eqn (10.1), could no longer be applied. In addition, numerical schemes employed for the design of heat transfer apparatus would have to change from single-phase to two-phase flow approaches. The next chapter will therefore discuss if additional physical mechanisms such as thermophoresis, Brownian motion of the nanoparticles, and container-like transport of thermal energy significantly contribute to the convective heat transfer of nanofluids.

10.2 A Closer Look at the Governing Equations

In a joint effort, teams from Spain, Sweden, Finland, Germany, and the US compiled their experimental data.[2] The general question of this study was if nanoparticles have their *own lives* that generate, besides conduction and convection, additional heat transfer effects. This question was tackled by two approaches. First the governing equations were non-dimensionalised to gain a more general form suitable for nanofluid flows with heat transfer and, second, by analysing the experimental findings with respect to the applicability of conventional Nusselt number correlations. Conventional Nusselt number correlations have been developed solely based on the assumption of a single-phase flow. In the case where nanofluid flow offers additional benefits, *e.g.*, the previously mentioned thermophoresis, these correlations would no longer be applicable.

To investigate nanofluid heat transfer problems, a general set of governing equations describing a two-phase flow comprising a base fluid and nanoparticles must be employed. Only such an approach would allow a ruling out of the two-phase flow character of nanofluid flows in a very general way. For that purpose, the following equations are considered:[2]

(1) continuity equation of the base fluid
(2) momentum balance of the base fluid
(3) thermal energy transport of the base fluid
(4) continuity equation of the solid phase (nanoparticles)
(5) momentum balance of the nanoparticles
(6) temperature transport equation of the particles

Additionally, thermal and momentum coupling of the two phases are taken into account.

We leave the details of the analysis aside and present just the obtained extension of eqn (10.1)

$$\mathrm{Nu}_{\mathrm{nf}} = f(\mathrm{Geo}, \mathrm{Re}_{\mathrm{nf}}^{n}, \mathrm{Pr}_{\mathrm{nf}}^{m}, \mathrm{St}, \mathrm{Fo}, \mathrm{Sc}_{\mathrm{B}}, \mathrm{Sc}_{\mathrm{T}}, \Pi_{\tau}, \mathrm{K}_{\mathrm{Pe}}, \mathrm{K}_{\mathrm{Nu}}) \tag{10.2}$$

where the subscript nf denotes a nanofluid. Eqn (10.2) indicates essentially two new groups of parameters. The first group – St, Fo, $\mathrm{Sc_B}$, $\mathrm{Sc_T}$ – are the Stokes, Fourier, and two Schmidt numbers related to Brownian motion and thermophoresis, respectively. A fourth similarity number Π_τ relates the time it takes for a nanoparticle to sediment inside the heat transfer device on the characteristic time scale of the device. The second group of parameters – $\mathrm{K_{Pe}}$, $\mathrm{K_{Nu}}$ – are weighting functions for the Péclet and Nusselt numbers, respectively. All of these parameters are related directly to the thermophysical properties of the nanoparticles and the forces acting on them while being convected through heat transfer apparatus. In the case of the flow of a pure liquid, eqn (10.2) reduces to eqn (10.1).

By carefully analysing the compiled experimental data and the thermophysical properties of the employed nanoparticles, the authors of the study[2] concluded the following points that are relevant to the convective heat transfer of nanofluids.

(1) The Stokes number St characterises the importance of the mass inertia of a nanoparticle for its movement in the flowing base fluid. This similarity number is in all experiments significantly smaller than unity. Therefore, the nanoparticles follow the local base fluid flow with almost no delay. Therefore, effects, *e.g.*, vortex streets, induced by nanoparticle movement are not expected.
(2) The Fourier number Fo characterises the instantaneous heat transfer between nanoparticles and the base fluid. For the analysed experiments, the Fourier numbers of orders of magnitude ranging between 10^9 and 10^{11} are observed. Consequently, a nanoparticle has an almost constant inner temperature and adapts immediately to the surrounding base fluid temperature. Therefore, a container-like transport of thermal energy by the nanoparticles does not seem possible.
(3) The Schmidt numbers Sc_B and Sc_T describe the ratio of diffusive momentum transport to diffusive mass transport. The analysed experiments indicate that the Schmidt numbers are much greater than unity. Therefore, mass diffusion due to either Brownian motion or thermophoresis is negligible, even though it takes place in the flow.
(4) The last remaining parameter of eqn (10.2) Π_τ denotes the ratio of the time a nanoparticle needs to sediment inside the heat transfer device to the characteristic time scale of the same. Again, analysing the experiments reveals that Π_τ ranges between 10^{-7} and 10^{-4}. The consequence is that a nanoparticle has, literally speaking, no time to sediment in heat transfer, it is simply washed away by the main current.

In 1961, Sergé and Silberberg[37] described for the first time an effect – shearphoresis driven radial transport of particles in a pipe flow – that was later named after them. Based on the idea of the Magnus effect, the authors derived an equation for the radial velocity of particles in a parabolic velocity profile. This correlation indicates that the radial velocity of a particle is proportional to the fourth power of the ratio of the particle diameter to the inner pipe radius. For nanofluids this ratio ranges between 10^{-2} ($d_{np} = 100$ nm, $R = 1$ mm) and 10^{-7} ($d_{np} = 10$ nm, $R = 20$ mm). The fourth power of this ratio is between 10^{-8} and 10^{-28} roughly. Therefore, the shear induced radial velocity of nanoparticles will be very small. It follows that the time required for a nanoparticle to be significantly transported by shearphoresis is much longer than the time required for it to be convected through a heat transfer device. Later in 2006, Buongiorno[38] published his seminal paper on "Convective Transport in Nanofluids", clearly showing that for turbulent flows (Section 2.5 Magnus effect), shear-induced transport should be negligible for nanoparticles in turbulent flows. The reason for this is that such tiny particles cannot be injected into the viscous sublayer – where sufficiently high shear rates would be present – due to their low inertia.

The main conclusion to be drawn here from this analysis is that flows of dilute, water-based nanofluids can be treated as single-phase flows. Their effective thermophysical properties must be either calculated (density, specific heat capacity, thermal expansion coefficient) based on the mixing rule or measured (thermal conductivity, dynamic viscosity, surface tension, *etc.*). Base fluid thermophysical properties do not describe nanofluid flows properly. Moreover, only the thermophysical properties of the nanofluid must be predicted according to the local temperature.

The well-established Nusselt number correlations for pure conductive, mixed convective, and pure convective heat transfer should be applicable. No anomalous or unexplainable increase or decrease of heat transfer can be expected. Additional physical effects that can affect heat transfer, such as thermophoresis, the Brownian motion of nanoparticles, the sedimentation of nanoparticles, *etc.*, are not identical to zero. However, for any heat transfer problem utilising nanofluids, the intensity of these effects should be analysed individually.

A detailed derivation of the above statements can be found in the study by Buschmann *et al.*[2] In this study, all time scales relevant to the above statements are also presented.

10.3 Experimental Validation of Base Fluid and Nanofluid Heat Transfer Performance

Despite a decade or so of intense research on nanofluids, several questions remain open. Therefore, newly developed nanofluids must first be tested experimentally to show their thermal performance, but also their stability over time under shear stress, aging, *etc.*, in heat transfer devices such as plate and tube bundle heat exchangers.

A standard test rig for carrying out such experiments is the seemingly simple pipe flow. Figure 10.1 presents a general sketch of such a pipe test rig.

Both a vertical and horizontal positioning of the pipe (see *e.g.*, Azizian *et al.*)[4] are possible. However, the horizontal version has become the *golden standard* of nanofluid heat transfer testing. The reasons for this are manifold. First, pipe bundle heat exchanges are among the most widely used heat transfer devices employed in industry. However, more important from an academic point of view is that analytical solutions based on first principles exist for laminar pipe flow. For turbulent flows, well-established Nussel number correlations are available that may be suitable for validation and comparison with conventional heat carriers.

The test rig shown in Figure 10.1 consists of two flow circuits. The inner one transports the working fluid, which is either the investigated nanofluid or its base fluid. The outer circuit consists of a heat exchanger and a thermostat and cools the working fluid back to a constant temperature provided at the inlet of the test section. The inner circuit consists of a pump, a flow metre, and a control valve. Practical experience tells us that nanofluids may agglomerate, sediment, or deteriorate in one way or another during the

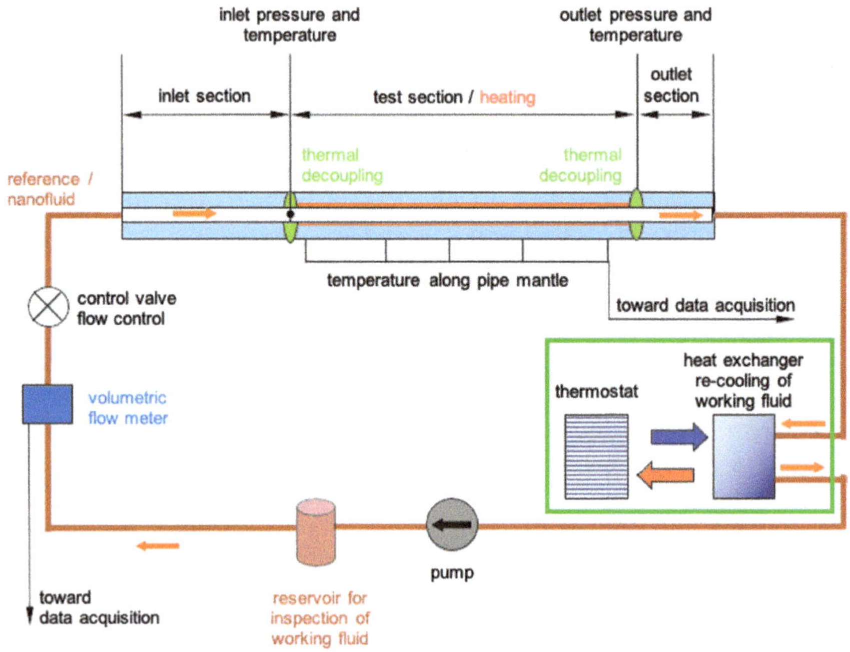

Figure 10.1 General setting of a horizontal pipe flow test rig. The red arrows indicate the direction of flow, for either reference fluid or nanofluid. The green rectangle indicates the outer flow circuit for back cooling of the working fluid. Sketch courtesy of ILK Dresden.

experiments. To monitor such processes, permanent *in situ* control of the nanofluid quality is needed. Hence, it is highly recommended to add a transparent reservoir wherein in particular the investigated nanofluid is collected. In the following, the three main parts of the test rig – the inlet, test, and outlet sections – are briefly discussed.

10.3.1 Inlet Section

The inlet section allows the establishment of a fully developed radial velocity profile, regardless of if the anticipated flow is laminar or turbulent. Therewith, this ensures an always reproducible flow situation at the entrance of the test section.

Although the inlet section is not heated, it still has to be insulated to prevent heat loss to the surroundings. Moreover, the inlet section must be thermally decoupled from the test section to avoid upstream oriented heat conduction in the pipe mantle. The same is true for the connection between the test and outlet sections.

Ideally, both the radial as well as the axial temperature profile are constant and equal the inlet temperature. The inlet temperature of the test section should not be measured at the inlet of this part of the test rig but rather at the beginning of the inlet section.

The length of the inlet section follows from the constraint that the velocity at the centre line of the pipe should differ by <1% from the ideal profile.[5] For laminar flow, the length of the test section is dependent on the Reynolds number and can be predicted according to:

$$x/d_i = 0.065\mathrm{Re}_x \tag{10.3}$$

where x denotes the streamwise coordinate, d_i the inner pipe diameter, and Re_x the Reynolds number built with the streamwise coordinate. The inlet section for turbulent flow is significantly shorter and is almost independent of the Reynolds number. A profile similar to the final one is already obtained after $10 \times d_i$. At 30 to $40 \times d_i$ the complete profile is established.[5]

10.3.2 Test Section

The test section serves for the combined development of temperature and velocity profile and, for this reason, it is heated. It may have two different sets of boundary conditions. Most nanofluid studies operate with *uniformly heated* pipes. The alternative is a *constant temperature* boundary condition. Figures 10.1, 10.2b and 10.3a show test sections that are heated from the outside employing heating coils. However, in most cases the pipe is utilised directly as ohmic resistance. Heating is carried out by applying a heating voltage over the entire test section length.

The general idea of a pipe flow test rig is to measure the temperature development along the test pipe and predict an experimental Nusselt number. For that purpose, the inlet and outlet temperature and the temperature development along the test pipe are measured. Figure 10.2a presents an example of Pt100 placed along the circumference of the mantle of the test pipe.[3]

For laminar pipe flow, analytical solutions in the form of series expansions exist going as far back as 1883 (Graetz,[6] constant wall temperature) and 1910 (Nusselt,[7] constant wall heat flux). More recent representations of these solutions have been reported by Jiji.[8] These correlations have the following general structure:

$$\mathrm{Nu}_{\mathrm{lam}} = f(X^*),\ X^* = (x/d_i)/\mathrm{Pe} \tag{10.4}$$

were $\mathrm{Nu}_{\mathrm{lam}}$ denotes the laminar Nusselt number, X^* the Graetz number and Pe the Péclet number (product of the Reynolds and Prandtl numbers).

The correlations in eqn (10.4) are based on first principles. However, they are relevant to pure liquids with constant thermophysical properties and without any particles. They are thus excellently suited to validate the quality of laminar pipe flow experiments as long as the changing thermophysical properties do not generate secondary motion. The latter is caused by the radial temperature distribution that causes a density gradient. Due to this density gradient, free convection appears in the flow, significantly enhancing heat transfer. Two counter rotating vortices stretch along the axial

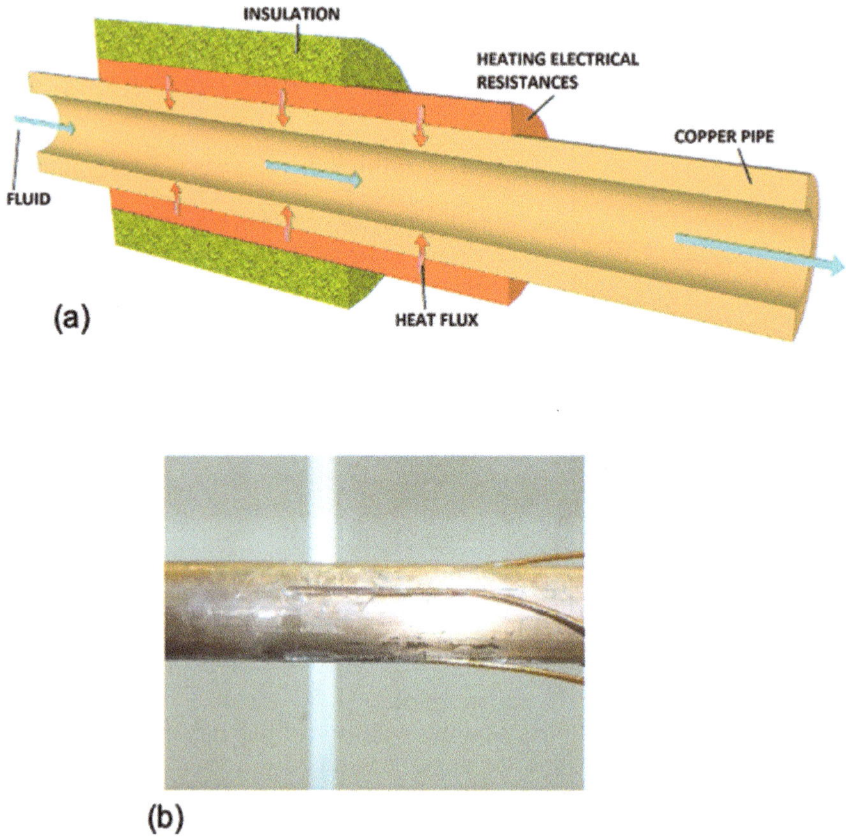

Figure 10.2 Details of pipe flow test rig. (a) Schematic of the pipe, heating, and insulation. (b) Thermocouples along the pipe mantle. Reproduced from ref. 3 with permission from the Authors, Copyright 2015 Elsevier Masson SAS.

coordinate of the pipe exchange hot liquid from the wall toward the core of the pipe.[3] Note that the flow is still laminar but the heat transfer is of a mixed type (conduction plus free convection).

In case the nanoparticles affect laminar pipe flow and heat transfer significantly by physical phenomena not prescribed by the governing equations that back eqn (10.4) – *e.g.* Brownian or thermophoretic motion of nanoparticles – the comparison of experimental results with eqn (10.4) should uncover this. However, it seems that such a strategy is only applicable if the flow is not affected by secondary motion.[3]

To the best of the author's knowledge, an analytical solution capturing secondary motion exists only for isothermally heated horizontal pipe flow.[9]

For fully developed turbulent flow, several Nusselt number correlations can be utilised. Among them is the well-established Dittus–Boelter equation:

$$\mathrm{Nu}_{\mathrm{tur}} = 0.023\mathrm{Re}_{\mathrm{di}}^{0.8}\mathrm{Pr}^{0.4},\ 0.6 \leq \mathrm{Pr} \leq 160,\ \mathrm{Re}_{\mathrm{di}} > 10^4 \tag{10.5}$$

Here, Nu_{tur} denotes the Nusselt number for convective turbulent heat transfer, Pr the Prandtl number, and Re_{di} the Reynolds number constructed with the inner diameter d_i of the pipe. Note that, no matter if the flow is laminar or turbulent, the thermophysical properties of the employed working fluid – base fluid/reference fluid or nanofluid – must be employed to calculate the similarity numbers Pr, Re, *etc.* correctly.

10.3.3 Outlet Section

The outlet section, although much shorter than its inlet counterpart, ensures that the flow leaves the measuring section straight and without interference from the subsequent set-up.

To calculate the local heat transfer coefficient along the test section, the inlet and the outlet temperature must be measured. While the inlet temperature is constant over the entire pipe diameter, a radial temperature profile characterises the flow at the outlet of the test section. In fact, the so-called *mixing cup temperature* should be measured at the outlet. Therefore, the temperature profile must be equalised by a flow mixer. Figure 10.3b shows such a flow mixer constructed as a multi-element twisted tape.

10.4 Comparison of Base Fluid and Nanofluid Heat Transfer Performance

The comparison of the convective heat transfers of a base fluid and a nanofluid is not an easy task. In order to compare two flow and heat transfer situations, fluid geometric, kinematic, dynamic, and thermodynamic similarity is necessary. Geometrical similarity is achieved by simply keeping the test facility unchanged. It can be assumed that, in most cases, kinematical similarity, as shown by *e.g.* the Mach number (ratio of fluid velocity and local speed of sound), is only negligibly changed. Here we postulate that, in order to accomplish dynamic and thermodynamic similarity, the products of the powers of the Reynolds and Prandtl numbers of both flow configurations – the base fluid and nanofluid flow – should be kept identical. For laminar flows, this is guaranteed by a constant Péclet number, which is a product of the Reynolds and Prandtl numbers.

Let us first consider the system of governing equations given in Section 10.2 that describe a coupled flow and heat transfer problem. If one intends to enhance heat transfer following classical routes (Section 10.1) by increasing, *e.g.*, the Reynolds number, the flow velocity u could then increase. Higher velocity enhances the local heat transfer h coefficient. The increased input of thermal energy into the flow alters the radial and the axial temperature profile of the fluid. Yet, an increase in the temperature also means changing all of the thermophysical properties of the working fluid and therewith its Prandtl number. The Reynolds and Prandtl numbers,

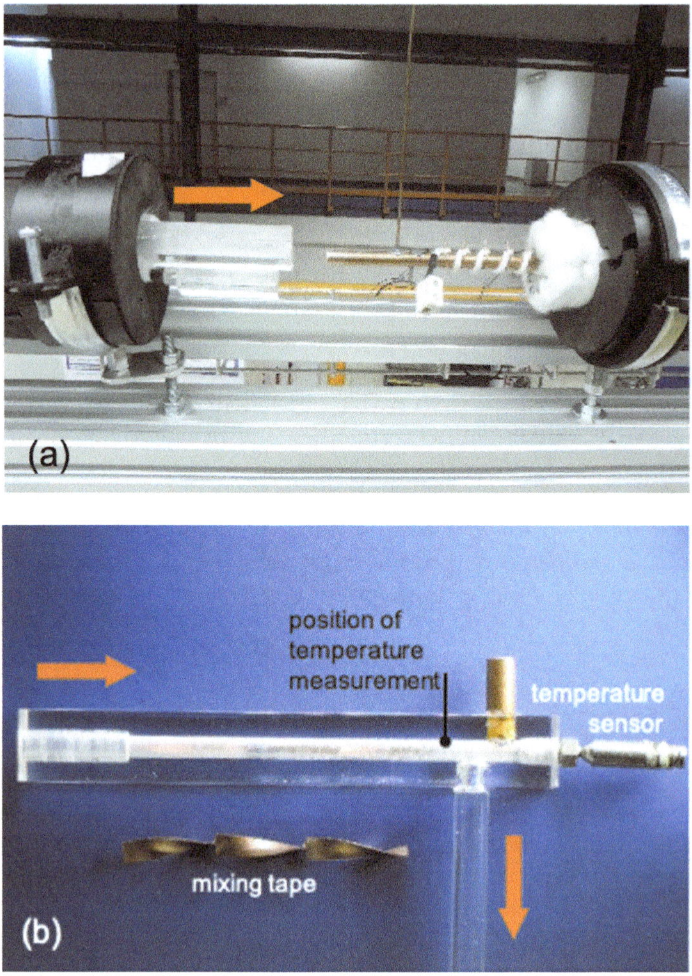

Figure 10.3 (a) Inlet and (b) outlet sections of the test section, where the arrows indicate the direction of flow. Photos courtesy of ILK Dresden.

or better, the Péclet number, are/is not independent. This basically expresses the coupling of the governing equations.

Employing nanofluids means changing all of the thermophysical properties of the working fluid. Consequently, the Prandtl and Reynolds numbers are different even if there is geometric and kinematic similarity. A prevalent feature of several nanofluid studies is the plotting of the experimentally obtained heat transfer coefficient or the thereof predicted Nusselt number solely *versus* the Reynolds number. As already shown in eqn (10.4), even for a fixed Reynolds number, the Nusselt number should be amplified by an increased Prandtl number. The *heat transfer improvements* seen in several studies may therefore just be an expression of Prandtl number effects caused by changed thermophysical properties.

The strategy of flow and heat transfer similarity has been, for example, applied for transitional flow in helical heat exchangers,[12] mixed convection in a laminar pipe flow with twisted tape,[13] convective heat transfer of nanofluids in a chevron plate heat exchanger,[14] and for laminar flow in horizontal pipes[15] (see also Section 10.5). In all these cases, the base fluid data collapse with the nanofluid results within the experimental uncertainty.

An alternative to the strategy of similarity is a very straightforward engineering approach. For industrial applications, often it is just the gains and losses that matter. Therefore, comparing, *e.g.*, the average heat transfer coefficient enhancements *versus* the Graetz number at fixed pumping power or entrance velocity are very valuable (see *e.g.* work by Utomo *et al.*[15]).

Numerical[16] and statistical[17] studies are underway to tackle more complex heat transfer problems, especially the dependency of fluid flow and heat transfer, as discussed above.

10.5 Nanofluid Heat Transfer in Laminar, Transient, and Turbulent Pipe Flow

The well-known paper by Choi and Eastman[18] on the enhancement of thermal conductivity of fluids by adding nanoparticles was published in 1995. Just three years later, Pak and Cho[19] published what was probably the first study on the hydrodynamics and heat transfer of fluids with submicron metal oxide particles. Titania and γ-alumina nanoparticles with primary diameters ranging from 13 to 27 nm were used, where the concentration of the water-based suspensions was between 1.0 and 4.3 vol%.

The main results of this study focused on turbulent flow ($10^4 < Re_{di} < 10^5$) are compiled in Fig. 12 in a study by Pak and Cho.[19] The data are first represented as Nu_{tur} *versus* Re_{di} and, second, following the similarity strategy (Section 10.4). For the utilised nanoparticle concentrations up to 3.16 vol%, the Prandtl number changes from the value of water of around 6.5 to nanofluid values of around 12.33. Therefore, the Nusselt number according to eqn (10.4) should increase simply because of the higher Prandtl numbers of the nanofluid. Plotting the heat transfer coefficient h or the Nusselt number *versus* the Reynolds number, as often proposed, leads to the erroneous conclusion that nanoparticles increase Nu and therewith heat transfer. Following the arguments in Section 10.4 delivers a different picture. As Fig. 12 in the study by Pak and Cho[19] clearly indicates, all data, regardless of the nanoparticle concentration (including 0.0 vol%) or material, collapse on one single curve. Through curve-fitting, the authors found that the correlation describing this curve is:

$$Nu_{tur} = 0.021\ Re_{di}^{0.8}\ Pr^{0.5}\quad 6.5 \leq Pr \leq 12.3,\ 10^4 < Re_{di} < 10^5 \tag{10.6}$$

Eqn (10.6) closely resembles the Dittus–Boelter equation (eqn (10.4)). However, for fixed $Re_{di}{}^{0.8}Pr^{0.5}$ values, the Nusselt numbers of the base fluid flow (water) and nanofluid flow are identical. Additionally, Pak and Cho[19] compared the heat transfer coefficients at constant averaged velocities. At a volume concentration of 3.0 vol%, the nanofluid values are 12% smaller than the respective values of pure water.

The findings by Pak and Cho[19] were confirmed by Sommers and Yerkes,[20] employing a similarity strategy for propanol-based alumina suspensions (10 nm; 0.5, 1.0, 3.0 wt%; $500 < Re_{di} < 5.200$; Fig. 9c[20]) and by Liao and Liu[21] for water-based carbon nanotube suspensions (0.5, 1.0 wt%; $500 < Re_{di} < 10.000$; Fig. 7[21]). An exception was a 1.0 wt% alumina suspension,[20] where a significant enhancement of the heat transfer coefficient for $1.800 < Re_{di} < 2.800$ was observed. The reason for this seems to be that the laminar-turbulent transition occurred at an earlier time. For high concentrations and/or high temperatures, up to 10–12% increased Nusselt numbers were observed for the carbon nanotubes.[21] However, due to their large length-to-diameter ratio, nanotubes are nanoobjects rather than nanoparticles. Effects related to the connection of distant fluid bales may have caused these increases.

The picture is even clearer for laminar flow in horizontal and vertical pipes. Rea *et al.*,[22] experimenting with water-based alumina and zirconia (50 nm; 0.6 to 6.0 vol%; $140 < Re_{di} < 1.900$), observed a perfect collapse of base fluid and nanofluid data with the theoretical curve for laminar pipe flow under constant heat flux boundary conditions. That in turn means that, due to the enhanced thermal conductivity of the nanofluids, the heat transfer coefficient must have also increased. For example, for a 6.0 vol% water–alumina nanofluid, Rea *et al.*[22] observed an increase in the heat transfer coefficient of 17% (entrance region) and 27% (developed region).

Employing a similarity strategy, comparable results were reported by Pohl *et al.*[23] for alumina (42 nm; 3.0 vol%; $220 < Re_{di} < 530$) and titania (30–80 nm; 5.25 vol%; $Re_{di} = 126$), by Utomo *et al.*[24] for alumina (10–20 nm, 100–200 nm; 9.0 wt%; $300 < Re_{di} < 1.070$) and titania (20–50 nm; 9.0 vol%; $530 < Re_{di} < 940$), and by Haghighi *et al.*[25] for alumina (10–15 nm), titania (20–30 nm), and ceria (50–100 nm). In the latter case, the concentration was 9.0 wt% and $140 < Re_{di} < 1.900$. Kim *et al.*[26] investigated alumina nanoparticles of different shapes (cylinders, bricks, and blades) with concentrations ranging from 0 to 5 vol%. When plotting the experimental Nusselt numbers of nanofluids together with the respective water values *versus* the Péclet ($1000 < Pe < 9000$) number, no differences were observed.

Experiments by Utomo *et al.*[24] revealed that the average heat transfer coefficient enhancement at various bases of comparison (equal mass flow rate, averaged velocity, Reynolds number, pumping power) were well described by correlation (eqn (10.4)). This is to some extent supported by work by Haghighi *et al.*,[25] who found that nanofluids do not show an advantage over pure water when compared at equal pumping power.

Employing micro-particle image velocimetry (PIV), Walsh *et al.*[27] analysed the laminar isothermal flow of alumina–water nanofluids (0.3, 3.1, and 6.3 vol%) in a straight pipe. Their most important finding was that the measured velocity profiles of the nanofluid flow could be properly described using the Hagen–Poiseuille profile. Even if these experiments are carried out without heat transfer, they hint at the Newtonian character of nanofluids. Moreover, thermophysical properties can be sufficiently characterised using the effective medium theory (see also ref. 2). Walsh *et al.*[27] repeated their experiments after two weeks and found that the suspension had changed its character to non-Newtonian. It is reasonable that sonication was sufficient to stabilise the suspension for the first experiments. However, after two weeks of downtime, agglomeration may have taken place and changed the rheology of the suspension. The authors argued that this phenomenon could be one of the reasons for the great variability of nanofluid experiments.

To find a rational explanation for the findings described above, we assume that, for many experiments, the Nusselt number of the base fluid flow coincides with the Nusselt number of the nanofluid flow when represented according to a similarity strategy. Keeping in mind the definition of the Nusselt number for pipe flow:

$$\mathrm{Nu} = hd_{\mathrm{i}}/k \tag{10.7}$$

where h denotes the heat transfer coefficient and k the thermal conductivity of the working fluid. The characteristic length scale of the experiment – the inner diameter of the pipe – is not changed. Under such circumstances, the base fluid and nanofluid flow can then be described using the same correlation $f(\mathrm{Re}^n\mathrm{Pr}^m)$. The following conclusion is then possible.

The ratio of the thermal conductivities $k_{\mathrm{nf}}/k_{\mathrm{bf}}$ is equal to the ratio of the heat transfer coefficients $h_{\mathrm{nf}}/h_{\mathrm{bf}}$ for a certain $\mathrm{Re}^n\mathrm{Pr}^m$ value. In other words, the heat transfer coefficient is increased by the same amount as the thermal conductivity.

Clearly, the temperature distribution of the nanofluid flow differs from that of the base fluid flow. However, that temperature is the sole parameter to explain the observed Nusselt number identities seems unlikely. It seems to be more probable that a combination of the changed thermophysical properties (due to the addition of nanoparticles) with the changed temperature alters the heat transfer. The same is obviously also true for different geometries of heat transfer devices, such as helical heat exchangers,[12] pipe flow with twisted tape,[13] and chevron plate heat exchangers.[14] In all these cases, the experimental Nusselt numbers of the reference fluid flow as well as of the nanofluid flow collapse on a single curve.

Nikulin *et al.*[28] performed a comprehensive experimental study on pipe flow covering a Reynolds number range from 200 to around 8000. The experiments covered laminar, transient, and turbulent pipe flow. The working fluid was either isopropanol or isopropanol with alumina nanoparticles (75–82 nm; 0.387, 0.992, 3.130, and 4.710 wt%). Similar to the previous work of Sommer and Yerkes,[20] an earlier transition from laminar to turbulent was observed. The authors analysed the pressure pulsation for the turbulent flow and observed a significant reduction when nanoparticles were added. Based on that observation, the authors argued that turbulence and therewith turbulent convection is hampered by the nanoparticles, with the most recent numerical simulations by Tschisgale and Kempe[29] confirming this finding. The simulations reveal that adding nanoparticles to a flow reduces turbulence significantly, which overcompensates for the positive effects resulting from increased thermal conductivity.

10.6 Magnetically Enhanced Heat Transfer in a Ferronanofluid Flow

As described in the previous chapter, the expected heat transfer enhancement related to classical nanofluids seems to be limited. Additional external forces could help to overcome this drawback. One of the few options is a magnetically enhanced or controlled heat transfer employing ferronanofluids. In fact, ferronanofluids were known long before the nanofluids that are currently known, with early patents dating back to that of Papell[30] in 1965. Among the earliest studies on the influence of magnetic fields on convective heat transfer in pipe flows was work by Fujita *et al.*[31] (1990) and Berkovski[32] (1996).

To create magnetically sensitive nanofluids – ferronanofluids – a change from classical nanoparticle materials to magnetic ones is necessary. Among others, iron, cobalt, nickel, magnetite, and hematite belong to this group. A comprehensive overview on ferronanofluid pipe flow experiments[33] indicated that the most widely employed nanoparticle material is magnetite. Magnetite is a naturally occurring mineral described by the molecular formulae Fe_3O_4 that represents the most stable compound of iron and oxygen.

To investigate heat transfer under magnetic fields, general assembly test rigs similar to that shown in Figure 10.1 are appropriate, but with the addition of a magnetic field being imposed on the flow. This can be done either by employing permanent magnets (see *e.g.*, work by Rudl *et al.*[34]) or electromagnetic systems. The latter allows the application of alternating magnetic fields with different frequencies and/or wave types (see *e.g.*, work by Tekir *et al.*[35]).

Employing ferronanofluids under magnetic fields allows not only for an enhancement of the heat transfer, but also the ability to control it. Simply put, it should be possible to increase or decrease heat transfer, both locally or temporarily. These options go well beyond the capabilities of classical nanofluids. However, the challenges related to such systems are also significantly higher than for heat exchangers operated using nanofluids. This is due to the fact that the relevant physical mechanisms acting in a flow are not completely understood at this point.

In addition to inertial, viscous, and gravitational forces, magnetic force must also be considered in magnetically influenced heat transfer problems. This magnetic force is a body force and can act with or against gravity.[34] It can be characterised by a magnetic number Mn, which is defined as the ratio between magnetic and viscous force.

Figure 10.4 describes the complex interplay between the forces acting on magnetic nanoparticles. According to Buschmann,[33] the ferromagnetic

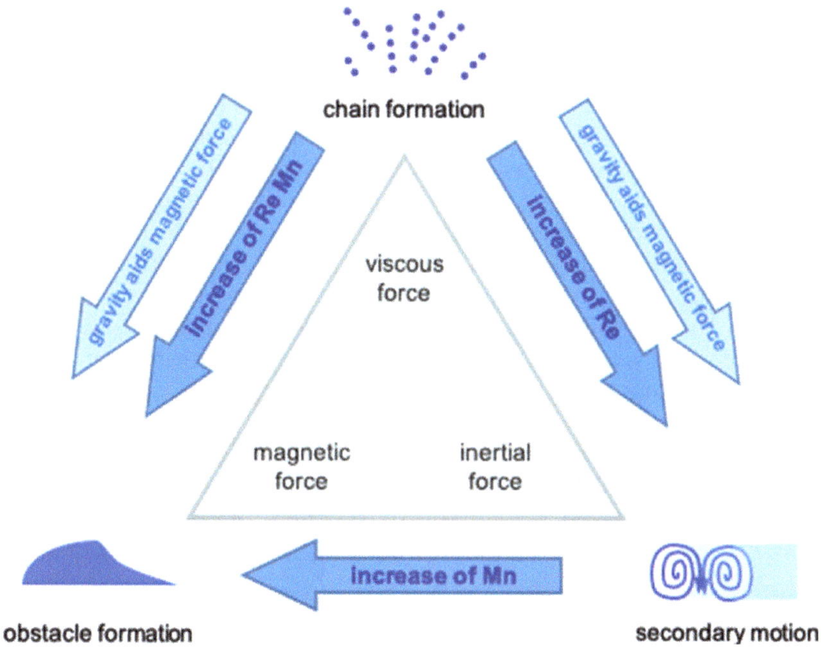

Figure 10.4 Interplay of viscous, inertial, magnetic, and gravitational forces for magnetically affected heat transfer problems. Reproduced from ref. 3 with permission from The Authors, Copyright 2020 Elsevier Masson SAS.

nanoparticles form three types of structures. For viscosity dominated flows – small Reynolds and magnetic number – ferronanoparticles may form chains. Asfer *et al.*[36] visualised such chains employing bright field microscopy. Increasing the magnetic force by keeping the Reynolds number constant led to secondary motion. Secondary motion refers to vortices or vortex systems generated solely by magnetic force. Numerical simulations[37] clearly reveal the existence of such secondary motion. Raising the magnetic force further attracts ferronanoparticles to the inner wall of the heat transfer device. Plugs or dune-like humps form.[36] These obstacles narrow the flow channel and increasing therewith the local velocity. Moreover, depending on the local Reynolds number, vortex shedding may occur on their downstream flanks. All three effects alter the flow situation in a manner that goes far beyond the changing of thermophysical properties of classical nanofluids. Heat transfer is either enhanced by heat bridges (chains, percolation effect), the exchange of wall bound and core fluid (secondary motion), or the creation of vortex streets (hump formation). However, it seems that these local effects are not the only ones that need to be considered. All three types of ferronanoparticle structures need a certain time to decay. Therefore, the heat transfer downstream of a magnetic field creating such structures locally should also be enhanced.[34]

10.7 Summary

The application of classical nanofluids and ferronanofluids is a new strategy by which to improve and/or control convective heat transfer. However, as with any novel method, the expected gains should not be overestimated. There are definitely no anomalous effects. Effects not understood so far will become explainable through theoretical, numerical and experimental studies. Therefore, further detailed investigations need to be carried out to develop nanofluids and ferronanofluids as robust heat transfer fluids for industrial applications. The focus should be placed on the development of nanofluids that exhibit a strong increase in thermal conductivity but only slightly increased viscosity. Another goal is the detailed theoretical analysis, as well as the experimental and numerical investigation, of magnetically-assisted heat transfer.

In summary, convective heat transfer using nanofluids has reached a certain level of applicability. Some possibilities are known. To go further, detailed fundamental research is needed to understand heat transfer on the nanometre length scale. It is therefore hoped that this will happen soon.

Acknowledgements

This study was carried out under the grants MF090026, MF 120019, 49VF170005, and 49VF180041 (Bundesministerium für Wirtschaft und Energie), the NanoUptake COST Actions CA15119 Overcoming Barriers to Nanofluids Market Uptake and the NANOConVEX COST Action IG15119 Nanofluids for convective heat transfer devices (European Cooperation in Science and Technology). My special thanks go to my students and the technicians, engineers, and scientists of ILK Dresden who supported our work in the field of nanofluids.

References

1. M. C. Ruzicka, *Chem. Eng. Res. Des.*, 2008, **86**, 835.
2. M. H. Buschmann, R. Azizian, T. Kempe, E. J. Juliá, R. Martínez-Cuenca, B. Sundén, Z. Wu, A. Seppälä and T. Ala-Nissila, *Int. J. Therm. Sci.*, 2018, **129**, 504.
3. L. Colla, L. Fedele and M. H. Buschmann, *Int. J. Therm. Sci.*, 2015, **97**, 26.
4. R. Azizian, E. Doroodchi, T. McKrell, J. Buongiorno, L. W. Hu and B. Moghtaderi, *Int. J. Heat Mass Transfer*, 2014, **68**, 94.
5. B. Eck, *Technische Strömungslehre*, Springer-Verlag, Berlin Heidelberg New York London Paris Tokyo, 9th edn, 1988.
6. L. Graetz, *Ann. Phys.*, 1883, **18**, 79.
7. W. Nusselt, *Z. Des. Ver. Dtsch. Ing.*, 1910, **54**, 1154.
8. L. M. Jiji, *Heat Convection*, Springer-Verlag, Berlin/Heidelberg, 2nd edn, 2009.

9. C. A. Hieber and S. K. Sreenivasan, *Int. J. Heat Mass Transfer*, 1974, **17**, 1337.
10. J. P. Meyer, S. A. Adio, M. Sharifpur and P. N. Nwosu, *Heat Transfer Eng.*, 2016, **37**, 387.
11. S. M. S. Murshed and P. Estellé, *Renewable Sustainable Energy Rev.*, 2017, **76**, 1134.
12. Z. Wu, L. Wang, B. Sundén and L. Wadsö, *Appl. Therm. Eng.*, 2016, **96**, 364.
13. M. H. Buschmann, *Heat Transfer Eng.*, 2017, **38**, 162.
14. D. Huang, Z. Wu and B. Sundén, *Int. J. Heat Mass Transfer*, 2015, **89**, 620.
15. A. T. Utomo, E. B. Haghighi, A. I. T. Zavareh, M. Ghanbarpourgeravi, H. Poth, R. Khodabandeh, B. Palm and A. W. Pacek, *Int. J. Heat Mass Transfer*, 2014, **69**, 77.
16. B. Kamenik, E. B. Elcioglu, A. Turgut, R. Mondragón, L. Hernández Lopez, J. P. Vallejo, L. Lugo, M. H. Buschmann and J. Ravnik, *3rd European Symposium on Nanofluids*, September 9–10, 2021, Iasi, Romania.
17. E. B. Elcioglu, B. Kamenik, A. Turgut, R. Mondragón, L. Hernández Lopez, J. P. Vallejo, L. Lugo, M. H. Buschmann and J. Ravnik, *3rd European Symposium on Nanofluids*, September 9–10, 2021, Iasi, Romania.
18. U. S. Choi and J. A. Eastman, *ASME International Mechanical Engineering Congress & Exposition*, November 12–17, 1995, San Francisco, CA.
19. B. C. Pak and Y. I. Cho, *Exp. Heat Transfer*, 1998, **11**, 151.
20. A. D. Sommer and K. L. Yerkes, *J. Nanopart. Res.*, 2009, **12**, 1003.
21. L. Liao and Z. H. Liu, *Int. J. Heat Mass Transfer*, 2009, **45**, 1129.
22. U. Rea, T. McKrell, L. W. Hu and J. Buongiorno, *Heat Mass Transfer*, 2009, **52**, 2042.
23. S. Pohl, S. Feja and M. H. Buschmann, *International Symposium Thermal and Materials Nanoscience and Nanotechnology*, May 29–June 3, 2011, Antalya, Turkey.
24. A. T. Utomo, H. Poth, P. T. Robbins and A. W. Pacek, *Int. J. Heat Mass Transfer*, 2012, **55**, 7772.
25. E. B. Haghighi, M. Saleemi, N. Nikkam, Z. Anwar, I. Lumbreras, M. Behi, S. A. Mirmohammadi, H. Poth, R. Khodabandeh, M. S. Toprak, M. Muhammed and B. Palm, *Exp. Therm. Fluid Sci.*, 2013, **49**, 114.
26. S. Kim, H. Yoo and C. Kim, *Korean J. Chem. Eng.*, 2012, **29**, 1321.
27. P. A. Walsh, V. M. Egan and E. J. Walsh, *Microfluid. Nanofluid.*, 2010, **8**, 837.
28. A. Nikulin, A. S. Moita, A. L. N. Moreira, S. M. S. Murshed, A. Huminic, Y. Grosu, A. Faik, J. Nieto-Maestre and O. Khliyeva, *Int. J. Heat Mass Transfer*, 2019, **130**, 1032.
29. S. Tschisgale and T. Kempe, *Int. J. Heat Mass Transfer*, 2021, **175**, 121392.
30. S. S. Papell, *US Pat.*, US3215572A, 1965.
31. T. Fujita, M. Mamiya and B. Jeyadevan, *J. Magn. Magn Mater.*, 1990, **85**, 203.

32. B. Berkovski, *Magnetic Fluids and Applications Handbook*, *Begell House inc.*, 1996.
33. M. H. Buschmann, *Int. J. Therm. Sci.*, 2020, **157**, 106426.
34. J. Rudl, C. Hanzelmann, S. Feja, A. Meyer, A. Potthoff and M. H. Buschmann, *Nanomaterials*, 2021, **11**, 824.
35. M. Tekir, E. Taskesen, B. Aksu, E. Gedik and K. Arslan, *Appl. Therm. Eng.*, 2020, **179**, 115624.
36. M. Asfer, B. Mehta, A. Kumar, S. Khandekar and P. K. Panigrahi, *Int. J. Heat Fluid Flow*, 2016, **59**, 74.
37. D. Sergé and A. Silberberg, *Nature*, 1961, **189**, 209.
38. J. Buongiorno, *J. Heat Transfer*, 2005, **128**, 240.

CHAPTER 11

Numerical Development of the Thermal Convection Characteristics of Nanofluids

W. AJEEB,[a] J. Z. ZHANG,[b,c] Z. WU,*[c,d,e] S. M. S. MURSHED[a] AND B. SUNDÉN[c]

[a] IDMEC, Instituto Superior Técnico, Universidade de Lisboa, Lisboa, Portugal; [b] School of Energy and Power Engineering, Shandong University, 250061 Jinan, China; [c] Department of Energy Sciences, Lund University, SE-22100 Lund, Sweden; [d] College of Electrical Engineering, Zhejiang University, 310027 Hangzhou, China; [e] Zhejiang University-Hangzhou Global Scientific and Technological Innovation Center, 311200 Hangzhou, China
*Email: zan.wu@energy.lth.se

11.1 Introduction

11.1.1 Numerical Simulation

Thermal convection is recognized as one of the most valuable heat transfer modes for a wide range of applications, such as energy production, chemical processing, thermal management of electronics, and other systems, which include cooling or heating processes. Nanofluids are an emerging type of thermal fluid that is under extensive investigation aiming to enhance their thermal convection performance toward higher energy efficiency and miniaturization of devices and systems.[1,2] To date, many types of nanofluids have been produced and investigated under certain operating conditions using

Soft Matter Series No. 16
Fundamentals and Transport Properties of Nanofluids
Edited by S. M. Sohel Murshed

Published by the Royal Society of Chemistry, www.rsc.org

experimental and numerical methods. However, it can be noted from the previous studies in the literature that a large number of nanofluid types exist due to the opportunity to use different kinds of nanoparticles (metallic and non-metallic nanoparticles) in different base fluids, leading to the formation of superior thermophysical properties for a particular application.[3–7] Moreover, the nanofluid field is still under development and new types of nanofluids are being discovered, such as ionic liquid-based nanofluids,[8] besides evaluating their performances under various conditions for possible applications. This leads to the importance of using flexible and low-cost investigation methods such as computational fluid dynamics (CFD), where numerical modelling is considered to be a very effective method to investigate the thermal convection performance of nanofluids.[9]

Furthermore, the numerical approach helps to explain the experimental data of convection heat transfer studies, as it is hard to measure spatial and temporal temperature and flow fields and heat fluxes by conducting experiments. Thus, CFD tools become an efficient prediction technique by avoiding the need for many prototype tests and thus saving money and time, when tackling the thermal convection characteristics of nanofluids for heat transfer applications.

The numerical approach can be explained starting with inputs (geometry, boundary conditions and materials properties). The inputs are defined according to the physics of the thermal convection problem. Then, a suitable mathematical model should be chosen among the available ones in the used CFD software (or building and inserting a new model into the CFD software) depending on relevant important physical principles and assumptions. Subsequently the needed variables (pressure, velocity, temperature, *etc.*) should be obtained at the selected points on the mesh of the geometry.[10] The mesh consists of a number of nodes that can be of different shapes and sizes according to the physical determination. The boundary of the operating conditions should be well defined and then a validation process should be carried out, where the obtained numerical results can be verified for consistency with the physical principles of the problem and the used numerical model (such as the performance at the boundaries). Furthermore, a comparison of the numerical results with available experimental data or theoretical correlations should be done in order to validate the employed numerical model. This may require a change in some assumptions of the physical problem or improvement in the mesh structure.[11] A scheme of the steps for performing and establishing a numerical approach using CFD tools is shown in Figure 11.1.

11.1.2 Nanofluid Properties in Modelling

An established numerical model of thermal convection systems involving nanofluids should be well validated, using proper modelling of the properties of nanofluids. Thermophysical properties of nanofluids, such as thermal conductivity, viscosity, specific heat, and density, are needed in any numerical approach to investigate the thermal performance of nanofluids. These properties

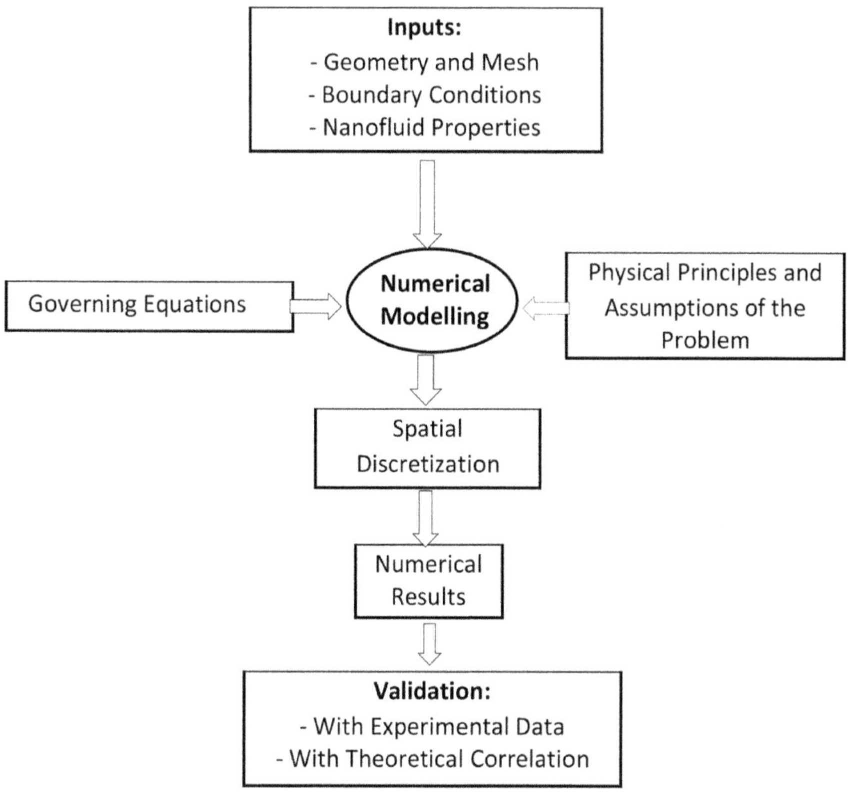

Figure 11.1 A scheme showing the use of the numerical method.

should be well defined by experimental investigations or based on validated theoretical correlations before they can be designed and tailored through the numerical modelling approach of the desired thermal system.

In the open literature, many numerical approaches have been established to investigate the behaviour of nanofluids in thermal convection. Most of the studies have considered constant properties of the nanofluids (independent of temperature) without defining the rheology of the nanofluid samples (such as Newtonian or non-Newtonian behaviour).[12–14]

These parameters, mainly rheological behaviour and temperature-dependent properties, have a considerable impact on heat transfer enhancement due to their influence on the hydraulic and thermal development of the flow, showing a higher impact for specific operating conditions, such as for flows in mini/microchannels.[15] Moreover, the pressure drop increase caused by an increase in viscosity (and change in rheology) due to nanoparticle addition should be defined. The latter implies the need for good tailoring of nanofluids with correlations considering temperature-dependent thermal and transport properties for the numerical modelling of thermal convection systems. In this regard, some researchers such as Kamali *et al.* have numerically studied multi-walled carbon nanotube

(MWCNT) nanofluids as single-phase and non-Newtonian fluids in a mini tube under constant wall heat flux.[16] The numerical model was constructed using a power-law non-Newtonian model at constant temperature.

Well-defined modelling of nanofluid properties is essential to determine the flow regime by finding the proper Reynolds number (Re) value. As is well known, the Reynolds number is important to determine flow regimes, where one of the most important inputs of the numerical model is the flow velocity at each desired Re. The generalized Re for Newtonian fluid flow through a pipe can be calculated using eqn (11.1),

$$\mathrm{Re} = \frac{\rho \cdot u \cdot D}{\mu} \tag{11.1}$$

where ρ is the density of the fluid, μ is the viscosity of the fluid, D is the tube diameter and u is the fluid velocity. In the case of non-Newtonian fluids, particularly power law fluids (shear thinning fluids), the Reynolds number was firstly introduced by Metzner and Reed.[17] Several studies[18–20] on the fluid flow of non-Newtonian nanofluids have applied a specific Re correlation, as shown in eqn (11.2):

$$\mathrm{Re} = \frac{\rho \cdot u^{2-n} \cdot D^{n}}{K} \tag{11.2}$$

where K is the consistency index and n is the shear thinning index of the power law non-Newtonian fluid.

11.2 Numerical Methods

Detailed information about CFD can be found in classic references.[11,21] There are basically two ways of modelling a flow field; the fluid is either treated as a collection of molecules or a continuous and indefinitely divisible continuum. In the latter approach, velocity, temperature, pressure, *etc.* are all defined at every point in space and time, and the conservation of mass, momentum and energy leads to a set of nonlinear partial differential equations (Navier–Stokes equations).[22]

The numerical methods for studying the convective heat transfer of nanofluids can be classified as single-phase and two-phase models, as shown in Figure 11.2.

CFD numerical simulation of a particular heat exchanger geometry is achieved by solving the main conservation laws for fluid flow, using the following three main steps:[23]

(a) Integrating the conservation equations,
(b) Converting the integral equations into algebraic equations using discretization methods (*e.g.*, finite volume method (FVM), finite element method (FEM)),
(c) Solve the obtained algebraic equations using numerical iterative methods.[10]

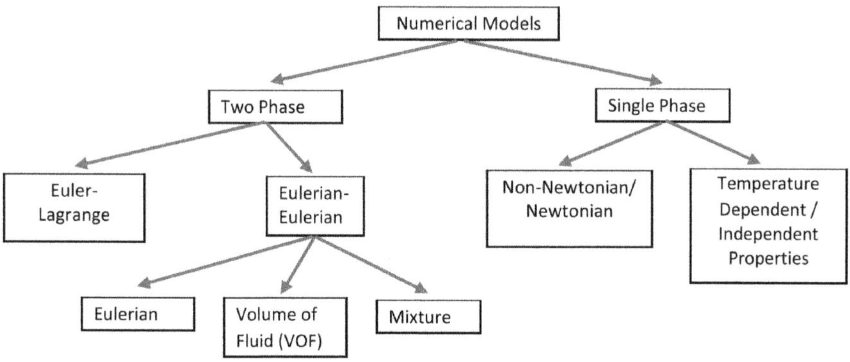

Figure 11.2 Numerical models used for studying the thermal convection of nanofluids.

The variables of the problem (pressure, velocity, *etc.*) are determined at the centre of each cell of the mesh in the domain by a process called discretization. Discretization has an error that can be reduced using a higher number of cells in the mesh (mesh refinement) or by increasing the order of accuracy of the interpolation. Moreover, a linearization process may be used to decrease the number of iterations, to account for imbalances in mass, momentum and energy within a certain selected tolerance.[24]

The most common governing equations for numerical studies on the thermal behaviour of nanofluids are the Navier–Stokes and energy equations. Successful implementation of the FVM in fluid flow and heat transfer problems has been noted, and this can be explained using several reasons. For instance, several CFD codes have been developed (*e.g.*, FLUENT, CFX) that allow smooth and easy implementation.[25] Also, the lower required computational specifications, especially for 2D approaches, as well as the available conservation equations, allow excellent control of the studied phenomena with the possibility of applying several options, such as temperature-dependent thermophysical properties.[26] Moreover, some assumptions are usually considered, for instance, that the fluid is Newtonian and incompressible, and single-phase, steady-state and constant thermophysical properties.

11.2.1 Singe-phase Model

11.2.1.1 Pressure–Velocity Coupling Methods

CFD tools such as ANSYS-Fluent provide several methods of pressure–velocity coupling: SIMPLE, SIMPLEC, PISO, and coupled. The semi-implicit method for the pressure-linked equation (SIMPLE) coupling algorithm is a popular algorithm used for single-phase problems to couple pressure and velocity. The SIMPLE algorithm has been modified into the SIMPLER algorithm using an additional pressure equation.[27] The SIMPLER method requires fewer iterations than the SIMPLE method and does not guess a pressure field, but extracts the pressure field from a given velocity field, which should be a corrected one.[10]

11.2.1.2 Discretization Methods

There are different types of discretization methods for governing equations. A common method known as the "second-order upwind method" is commonly used for viscous terms.[10] Single-phase problems can be solved using second-order upwind discretization for the flow equations and all scalar equations except those for turbulence quantities, which can be solved using first-order upwind discretization. For triangular and tetrahedral meshes, since the flow is never aligned with the mesh, one can generally obtain more accurate results using second-order discretization. Quad/hex meshes also enable better results using the second-order discretization, especially for complex flows.[27]

11.2.1.3 Solvers

Two solver technologies can be implemented:

1 pressure based
   and
2 density based. The density-based solver has been designed to approach high-speed compressible flows. While, in the case of incompressible flows, the pressure-based solver is the common one.[10]

11.2.1.4 Governing Equations

The physical principles of the used governing equations, the contained assumptions and the boundary conditions are the most essential parts of the numerical approach. The governing equations are based on the fundamental laws of fluid flow.

To solve a fluid flow and heat transfer problem, firstly the thermophysical properties of the working fluids need to be determined. For nanofluids, mixture models or correlations can be used to describe the thermophysical properties within an appropriate range. The governing equations such as the Navier–Stokes equations can be solved using the FVM at the cells (nodes) of the mesh in the fluid flow domain using the discretization process for a specific case. The computational domain is divided into many small sub-volumes by means of meshing.

11.2.1.4.1 For laminar Flow. So far, several numerical approaches have been used regarding the fluid flow and heat transfer performance of nanofluids. The governing equations used to simulate a nanofluid as a single-phase fluid under steady-state and laminar flow conditions can be briefly described as follows:

- Conservation of mass

  The conservation of mass is represented by the following continuity equation:

$$\nabla \cdot (\rho \boldsymbol{V}) = 0 \tag{11.3}$$

  For a fluid flow, the sum of the mass flowing in and the mass flowing out with time is equal to the change in mass as a result of the fluid density with time. The density (ρ) is assumed as a constant for incompressible fluid flows.[11]
- Conservation of momentum

  The conservation of momentum is represented by the following equation:

$$\nabla \cdot (\rho \boldsymbol{V}\boldsymbol{V}) = -\nabla \mathrm{P} + \nabla \cdot (\mu \nabla \boldsymbol{V}) \tag{11.4}$$

  This equation represents Newton's second law, where the change in momentum is equal to the sum of the forces on a fluid particle (such as pressure and viscous forces).
- Conservation of energy

The conservation of energy is represented by the following equation:

$$\nabla \cdot (\rho C_\mathrm{p} \boldsymbol{V}T) = \nabla \cdot (k \nabla T) \tag{11.5}$$

This energy equation represents the first law of thermodynamics, where the change in energy is equal to the added heat and the work done on a fluid particle.[11]

Moreover, the governing equations can be in differential or integral form. In the differential form (eqn (11.3)–(11.5)), the fundamental laws are applied to an infinitesimal fluid element, while in integral form the fundamental laws are applied to a finite volume in the flow domain.[11,24]

11.2.1.4.2 For Turbulent Flow. In turbulent flows there are additional terms arising from the rapid and random fluctuations of the mean value. These additional terms are the so-called Reynolds stresses, turbulent heat flux, turbulent diffusion flux, *etc.* Numerical turbulence models have been developed to express these terms (fluxes) based on the mean properties of the flow.[11] A common numerical code in the ANSYS software[27] uses two additional equations of conservation to describe the turbulence properties, particularly the turbulent kinetic energy equation (K') and the equation of dissipation (ε),[11] eqn (11.6) and (11.7), respectively. There are different types of turbulent models, such as K'–ε models (standard, RNG and realizable K'–ε turbulence models) and K'–ω models. Since the realizable model is a developed form of other K'–ε turbulence models, it satisfies certain mathematical constraints on the Reynolds stresses, consistent with the physics of turbulent flows.[27]

The realizable K'–ε model was proposed by Shih *et al.* in order to develop the traditional K'-ε model by considering a new eddy–viscosity formulation (C_μ as a variable).[28] The turbulence kinetic energy (K') and its rate of dissipation (ε) are obtained from the following transport equations:

$$\nabla \cdot (\rho \boldsymbol{V} K') = \nabla \cdot \left[\left(\mu + \frac{\mu_t}{\sigma_K}\right)\nabla K'\right] + G_K - \rho\varepsilon \tag{11.6}$$

$$\nabla \cdot (\rho \boldsymbol{V} \varepsilon) = \nabla \cdot \left[\left(\mu + \frac{\mu_t}{\sigma_\varepsilon}\right)\nabla \varepsilon\right] + \rho C_{1\varepsilon} S\varepsilon - \rho C_{2\varepsilon}\frac{\varepsilon^2}{K' + \sqrt{\upsilon\varepsilon}} \tag{11.7}$$

where,

$$C_{1\varepsilon} = \max\left[0.43, \frac{\eta}{\eta + 5}\right], \quad \eta = S\frac{K'}{\varepsilon}, \quad S = \sqrt{2S_{ij}S_{ij}} \tag{11.8}$$

and ε is the kinematic rate of dissipation, G_K represents the generation of turbulence kinetic energy due to mean velocity gradients, the quantity $G_K - \rho\varepsilon$ is the net source term in the equation, $C_{2\varepsilon}$ and $C_{1\varepsilon}$ are constants, σ_K and σ_ε are the turbulent Prandtl numbers for K' and ε, respectively, and S is the mean strain rate.[27]

11.2.2 Two-phase Models

11.2.2.1 Mixture Model

The mixture model is an efficient and simplified multiphase model, which can be adopted to model different phases moving with independent velocities. Slip velocities between the different phases are important for this model. The nanoparticles and base fluid are treated as a mixed fluid containing liquid and particle phases in a mixture model. The coupled influence of different phases in any control volume is identified by the volume fraction of the primary and secondary phases.[29] Compared with the single-phase simulation, an additional equation for the volume fraction of each phase must be solved. For a simple two-phase model, the sum of the volume fractions of the primary and secondary phases is unity in a control volume. Thus, only the volume fraction of the primary phase needs to be solved using iteration. The governing equations for the mixture model are listed below.

Continuity equation:

$$\frac{\partial}{\partial t}(\rho_m) + \nabla \cdot (\rho_m \boldsymbol{V}_m) = 0 \tag{11.9}$$

Momentum equation:

$$\begin{aligned}\frac{\partial}{\partial t}(\rho_m \boldsymbol{V}_m) + \nabla \cdot (\rho_m \boldsymbol{V}_m \boldsymbol{V}_m) &= -\nabla P + \rho_m \boldsymbol{g} + \boldsymbol{F} \\ &+ \nabla \cdot \left[\mu_m\left(\nabla \boldsymbol{V}_m + \nabla \boldsymbol{V}_m^T\right)\right] - \nabla \cdot \left(\sum_{i=1}^{n} \alpha_i \rho_i \boldsymbol{V}_{dr,i} \boldsymbol{V}_{dr,i}\right)\end{aligned} \tag{11.10}$$

Energy equation:

$$\frac{\partial}{\partial t}\sum_{i=1}^{n}(\alpha_i\rho_i E_i) + \nabla\cdot\sum_{i=1}^{n}[\alpha_i \boldsymbol{V}_i(\rho_i E_i + P)] = \nabla\cdot(k_{\text{eff}}\nabla T) + S_{\text{e}} \tag{11.11}$$

Volume fraction equation:

$$\frac{\partial}{\partial t}(\alpha_i\rho_i) + \nabla\cdot(\alpha_i\rho_i \boldsymbol{V}_{\text{m}}) = -\nabla\cdot\left(\alpha_{\text{p}}\rho_{\text{p}}\boldsymbol{V}_{\text{dr},i}\right) \tag{11.12}$$

The subscript m in these equations denotes the mixture, i denotes phase i, n is the number of different phases (usually 2 for nanofluid simulations), p denotes the secondary phase (nanoparticles). The density of the mixture, ρ_{m}, is calculated using the volume fraction weighted average mixture rule:

$$\rho_{\text{m}} = \sum_{i=1}^{n}\alpha_i\rho_i \tag{11.13}$$

For simulations of nanofluids using the mixture model, particles are considered as the second phase. Thus, the viscosity μ_{m} in the momentum equation is mainly obtained from the viscosity of the base fluid and the volume fraction of the nanoparticles. Many correlations have been developed to calculate this effective viscosity.[29] For example, the Brinkman model[30] has been used in some numerical studies to predict the nanofluid viscosity:

$$\mu_{\text{m}} = \frac{\mu_{\text{f}}}{\left(1-\alpha_{\text{p}}\right)^{2.5}} \tag{11.14}$$

The continuity, momentum, and energy equations are similar to the equations used in the single-phase method. However, the velocity field is shared by the nanoparticles and the base fluid. $\boldsymbol{V}_{\text{dr,p}}$ is the drift velocity between the base fluid and nanoparticles, while $\boldsymbol{V}_{\text{m}}$ is the mass-averaged velocity of the nanofluid:

$$\boldsymbol{V}_{\text{m}} = \frac{\sum_{i=1}^{n}\alpha_i\rho_i\boldsymbol{V}_i}{\rho_{\text{m}}} \tag{11.15}$$

$$\boldsymbol{V}_{\text{dr,p}} = \boldsymbol{V}_{\text{p}} - \boldsymbol{V}_{\text{m}} \tag{11.16}$$

Manninen *et al.*[31] proposed a method by which to calculate the velocity of the secondary phase (nanoparticles) $\boldsymbol{V}_{\text{p}}$, as given in eqn (11.17):

$$\boldsymbol{V}_{\text{p}} = \frac{\frac{\rho_{\text{d}}d_{\text{p}}^2}{18\mu_{\text{d}}}}{f_{\text{drag}}}\frac{\left(\rho_{\text{p}}-\rho_{\text{m}}\right)}{\rho_{\text{p}}}[\boldsymbol{g} - (\boldsymbol{V}_{\text{m}}\cdot\nabla)\boldsymbol{V}_{\text{m}}] \tag{11.17}$$

where the drag force is calculated using the correlation proposed by Schiller and Naumann:[32]

$$f_{\mathrm{drag}} = \begin{cases} 1 + 0.15\mathrm{Re}_{\mathrm{p}}^{0.687} & \mathrm{Re} \le 1000 \\ 0.0183\mathrm{Re}_{\mathrm{p}} & \mathrm{Re} > 1000 \end{cases} \tag{11.18}$$

Here, Re_{p} is the particle Re number, which is obtained from eqn (11.19):

$$\mathrm{Re}_{\mathrm{p}} = \frac{\rho_{\mathrm{m}} V_{\mathrm{m}} d_{\mathrm{p}}}{\mu_{\mathrm{m}}} \tag{11.19}$$

A widely adopted two-phase mixture model for nanofluid simulations is the Buongiorno model,[33] in which the Brownian diffusion and thermophoresis are included. The governing equations for this model are similar to the single-phase model mentioned before, while the source terms are different in the energy and volume fraction equations. For the volume fraction equation (shown in eqn (11.20)), the diffusion mass fluxes caused by Brownian diffusion and thermophoresis are included as source terms.

Volume fraction equation:

$$\frac{\partial \phi}{\partial t} + \boldsymbol{V}_{\mathrm{m}} \cdot \nabla \phi = \nabla \cdot \left(D_{\mathrm{B}} \nabla \phi + D_{\mathrm{T}} \frac{\nabla T}{T} \right) \tag{11.20}$$

where, D_{B} is the Brownian diffusion coefficient derived from the Einstein–Stokes equation:

$$D_{\mathrm{B}} = \frac{k_{\mathrm{B}} T}{3\pi \mu d_{\mathrm{p}}} \tag{11.21}$$

D_{T} is the thermal diffusion coefficient for thermophoresis defined as:

$$D_{\mathrm{T}} = 0.26 \frac{k}{2k + k_{\mathrm{p}}} \frac{\mu}{\rho} \phi \tag{11.22}$$

These two components also contribute to the energy equation, as shown below:

$$\rho c \left[\frac{\partial T}{\partial \tau} + \boldsymbol{V}_{\mathrm{m}} \cdot \nabla T \right] = \nabla \cdot (k \nabla T) + \rho_{\mathrm{p}} c_{\mathrm{p}} \left[D_{\mathrm{B}} \nabla \phi \cdot \nabla T + D_{\mathrm{T}} \frac{\nabla T \cdot \nabla T}{T} \right] \tag{11.23}$$

11.2.2.2 VOF Model

The volume of fluid (VOF) model is an efficient and widely adopted Eulerian–Eulerian multiphase flow model. Similar to the mixture model, the flow field of all phases is shared in the computational domain, which means that a single set of momentum equations are solved for these phases. Different phases are identified by the volume fraction of each phase in the computational domain. Unlike the mixture model, the VOF model assumes that all phases are immiscible fluids, and the interfaces between different phases are reconstructed based on the volume fraction in each computational domain. This model is more suitable for studies of bubble/droplet formation, boiling, and condensing flows, *etc.*, where the interfaces between different phases are of interest. For nanofluid simulations, the VOF method has also been employed by some researchers. The solid–liquid interfaces are not reconstructed, while only the volume fraction distributions are obtained to calculate the fluid properties in the computational cell. In this way, the VOF method adopted for nanofluid simulation is quite similar to the mixture model mentioned before. The differences in the heat transfer coefficient and frictional pressure drop between these two models are a result of the drift velocity, which is not included in the VOF method. The governing equations for the VOF methods are similar to the mixture model described above. As the slip velocity is not considered in the VOF method, the terms $V_{\mathrm{dr,I}}$ are zero for the VOF method.

11.2.2.3 Eulerian Two-phase Model

The Eulerian two-phase model has also been adopted in nanofluid simulations. For this model, the phases can be fluid, solid, or gas. Although the number of phases is only limited by the computational memory, two or three phases are usually considered for nanofluid simulations in the open literature. Unlike the aforementioned VOF and mixture models, the fluid flow field of each phase is not shared for the Eulerian two-phase model, which means that the governing equations for each phase are to be solved. The effect of each phase on the others is of importance for the accuracy of the Eulerian two-phase model, like the mass and momentum exchanges for different phases, the fluid–fluid or fluid–solid interface exchanges, drag and liquid forces, wall lubrication force, *etc.* Several models have been proposed and validated for these effects, which have also been included in widely adopted software, such as Ansys Fluent, CFX, OpenFOAM, COMSOL Multiphysics, *etc.*

The volume fraction of each phase is the key parameter for numerical simulations using the Eulerian two-phase model. The space occupied by each phase and the mass and momentum exchanges between phases can be represented by this parameter. The volume of phase i in the Eulerian two-phase model is calculated using the following equation.

$$V_i = \int_V \alpha_i \mathrm{d}V \tag{11.24}$$

$$\sum_{i=1}^{n} \alpha_i = 1 \tag{11.25}$$

The momentum equation for each phase needs to be solved for the Eulerian two-phase model. For phase i, the continuity equation is:

$$\frac{\partial}{\partial t}(\alpha_i \rho_i) + \nabla \cdot (\alpha_i \rho_i \boldsymbol{V}_i) = \sum_{i=1}^{n} (\dot{m}_{ij} - \dot{m}_{ji}) + S_i \tag{11.26}$$

where, $\boldsymbol{V}_i$ is the velocity vector of phase i and $\dot{m}_{ij}$ and $\dot{m}_{ji}$ are the mass transfer from phase i to phase j or from phase j to phase i, respectively. For nanofluid simulations, these two terms are usually zero as there is no mass transfer from the nanoparticles to the base fluid. S_i is an additional mass transfer source for phase i, which could be modified based on the particular case.

For the Eulerian two-phase model, the pressure field is shared by all phases. The momentum equation is as follows:

$$\frac{\partial}{\partial t}(\alpha_i\rho_i\boldsymbol{V}_i) + \nabla\cdot(\alpha_i\rho_i\boldsymbol{V}_i\boldsymbol{V}_i) = -\alpha_i\nabla P + \alpha_i\rho_i\boldsymbol{g} + \nabla\tau_i + \sum_{i=1}^{n} R_{ij} + \left(\boldsymbol{F}_i + \boldsymbol{F}_{\mathrm{vm},i} + \boldsymbol{F}_{\mathrm{td},i}\right) \tag{11.27}$$

where P is the pressure field, R_{ij} is the interface force, $\boldsymbol{F}_i$ is the external body force, $\boldsymbol{F}_{\mathrm{vm},i}$ is the virtual mass force, and $\boldsymbol{F}_{\mathrm{td},i}$ is the turbulent dispersion force, which is only valid for turbulent flows. The interface force R_{ij} is important for the Eulerian two-phase model. This force is determined by friction, pressure, cohesion, and other effects. In this model, the interphase force follows Newton's third law of motion, that is, $R_{ij} = -R_{ji}$. A simple interaction is usually adopted for this force:

$$\sum_{i=1}^{n} R_{ij} = \sum_{i=1}^{n} K_{ij}\left(\boldsymbol{V}_i - \boldsymbol{V}_j\right) \tag{11.28}$$

where K_{ij} is the interphase momentum exchange coefficient, which is described in detail later.

The stress–strain tensor for the ith phase τ_i is defined as:

$$\tau_i = \alpha_i\mu_i\left(\nabla\boldsymbol{V}_i + \nabla\boldsymbol{V}_i^T\right) + \alpha_i\left(\lambda_i - \frac{2}{3}\mu_i\right)\nabla\cdot\boldsymbol{V}_i\boldsymbol{I} \tag{11.29}$$

Similar to the momentum equation, the energy equation is solved separately based on the following equation:

$$\frac{\partial}{\partial t}(\alpha_i\rho_i h_i) + \nabla\cdot(\alpha_i\rho_i h_i\boldsymbol{V}_i) = -\alpha_i\frac{\mathrm{d}P_i}{\mathrm{d}t} + \tau_i : \nabla\boldsymbol{V}_i - \nabla q_i + S_i + \sum_{i=1}^{n} Q_{ij} \tag{11.30}$$

where, h_i is the specific enthalpy of the ith phase, q_i is the heat flux, S_i is a source term, and Q_{ij} is the intensity of heat exchange between phase i and phase j. In the numerical simulation, the heat exchange between two phases must comply with the local balance condition, that is $Q_{ij} = -Q_{ji}$.

For nanofluid simulations, the base fluid is always set as the fluid phase, while the nanoparticles are the solid phase. Hence, the fluid–solid Eulerian two-phase model is adopted for these simulations. The fluid–solid momentum equations are slightly different from the aforementioned equations. The equations for the fluid phase are described above, while the momentum equation for the *s*th solid phase is:

$$\frac{\partial}{\partial t}(\alpha_s \rho_s \boldsymbol{V}_s) + \nabla \cdot (\alpha_s \rho_s \boldsymbol{V}_s \boldsymbol{V}_s) = -\alpha_s \nabla P - \nabla P_s + \alpha_s \rho_s \boldsymbol{g} + \nabla \tau_s + \sum_{i=1}^{v} R_{ij} + \left(F_i + F_{\mathrm{vm},i} + F_{\mathrm{td},i}\right) \tag{11.31}$$

P_s is the pressure of the *s*th solid phase, which is defined as follows:

$$P_s = -\alpha_s \rho_s T_s + 2(1 + e_{ss}) {\alpha_s}^2 T_s \tag{11.32}$$

where e_{ss} is the coefficient of the restitution for particle collisions and T_s is the temperature of the nanoparticles. The value of e_{ss} is recommended as 0.9 for most cases, but this value may change for different types of particles.

As mentioned before, phase exchange coefficients are important for the numerical approach. In nanofluid simulations, the liquid–solid exchange coefficient K_{ls} is determined by the particle size and velocities of the liquid and solid phases, as shown in eqn (11.33).

$$K_{ls} = \frac{3}{4} C_D \frac{\alpha_s \alpha_l \rho_l |\boldsymbol{V}_s - \boldsymbol{V}_l|}{d_s} \alpha_l^{-2.65} \tag{11.33}$$

$$C_D = \frac{24}{\alpha_l \mathrm{Re}_s}\left[1 + 0.15(\alpha_l \mathrm{Re}_s)^{0.687}\right] \tag{11.34}$$

$$\mathrm{Re}_s = \frac{\rho_l d_s |\boldsymbol{V}_s - \boldsymbol{V}_l|}{\mu_l} \tag{11.35}$$

11.2.2.4 Euler–Lagrange Model

The combined Euler–Lagrange method can also be adopted for nanofluid simulations. In this model, the base fluid is treated as the continuum phase, which is solved with the Euler method, while the nanoparticles are considered as the discrete phase, which is calculated and tracked using the Lagrange method. This is also known as the discrete phase model (DPM). The governing equations of single-phase flow is adopted for the base fluid. The effect of nanoparticles on the nanofluid flow is considered using the source terms in these governing equations. Rashidi *et al.*[34] adopted this method in the simulation of an Al_2O_3–water nanofluid in an obstructed duct with two side-by-side obstacles. The fluid flow of the base fluid without the influence of discrete nanoparticles was solved initially. Then, the flow characteristics of nanoparticles were tracked based on the obtained initial flow field. Finally, a coupling iteration was conducted to include the effect of the discrete phase on the fluid flow with source terms in the governing equations.

The governing equations for the base fluid follow the Eulerian method, which are listed as follows, including continuity, momentum, and energy equations.

Continuity equation:

$$\frac{\partial}{\partial t}(\rho_l) + \nabla \cdot (\rho_l \boldsymbol{V}_l) = 0 \tag{11.36}$$

Momentum equation:

$$\frac{\partial}{\partial t}(\rho_l \boldsymbol{V}_l) + \nabla \cdot (\rho_l \boldsymbol{V}_l \boldsymbol{V}_l) = -\nabla P + \rho_l \boldsymbol{g} + \nabla \tau_l - S_{p,m} \tag{11.37}$$

Energy equation:

$$\frac{\partial}{\partial t}(\rho_l \boldsymbol{E}_l) + \nabla \cdot ((\rho_l E_l + P)\boldsymbol{V}_l) = \nabla \cdot (k_f \nabla T_f) + S_{p,e} \tag{11.38}$$

where $S_{p,m}$ and $S_{p,e}$ are the source terms due to the effect of discrete nanoparticles. $S_{p,m}$ is defined as:

$$S_{p,m} = \frac{1}{\delta V} \sum_{i=1}^{n} m_p \boldsymbol{F}_p \tag{11.39}$$

where n is the number of nanoparticles, m_p is the particle mass, and $\boldsymbol{F}_p$ is the total force of particles acting on the base fluid, which is determined from the Brownian motion, drag, Saffman and Magnus lift, pressure gradient force, thermophoretic force, and virtual mass force. The motion of the particles is determined using the Lagrangian method. The velocities of each nanoparticle follow Newton's second law, which is described in eqn (11.40):

$$m_p \frac{d\boldsymbol{V}_p}{dt} = \boldsymbol{F}_p \tag{11.40}$$

$S_{p,e}$ is the energy source term considering the effect of nanoparticles on energy transport. This term is defined as:

$$S_{p,e} = \frac{1}{\delta V} \sum_{i=1}^{n} m_p C_{p,p} \frac{dT_p}{dt} \tag{11.41}$$

The temperature of the nanoparticles is determined by the energy equation for the particles, which is given by eqn (11.42):

$$m_p C_{p,p} \frac{dT_p}{dt} = Nu_p \pi d_p k_f (T_f - T_p) \tag{11.42}$$

where Nu_p is obtained using the correlation by Ranz and Marshall:[35]

$$Nu_p = 2.0 + 0.6 \mathrm{Re_p}^{0.5} \mathrm{Pr}^{0.3} \tag{11.43}$$

11.3 Results and Discussion

It is to be noted that an extensive review of literature studies on the numerical simulation of the thermal convection of nanofluids is not provided herein as it is not the aim of this chapter. Instead, this section attempts to present numerical results that are typical for nanofluid convection, based on the most relevant and representative numerical studies.

11.3.1 Forced Convection

So far, many numerical investigations have been performed on the thermal convection of nanofluids. This sub-section briefly presents the numerical results of convective heat transfer of nanofluids in single tubes, heat sinks and plate heat exchangers based on the single-phase model, in addition to numerical simulations of forced convection based on the two-phase model.

11.3.1.1 Single Tube

For nanofluid convection, the single tube is the mostly studied geometry. Some numerical studies on the thermal performance of nanofluids in a single tube are summarized in Table 11.1.

Arjun and Rakesh[36] studied numerically a Al_2O_3 nanofluid flow through a circular microchannel and found that the enhancement of heat transfer in laminar flow was greater than that in turbulent flow compared to the base fluid. The increase in the Re number caused an increase in Nu and pressure drop. Also, a numerical investigation conducted by Esmaeilnejad *et al.*[37] for Al_2O_3 and CuO nanofluids flowing through a rectangular microchannel observed a good enhancement in heat transfer for the two nanofluids compared to the base fluid.

Ajeeb *et al.*[38] presented an experimental study complemented by a CFD simulation of single-phase laminar flow of a well-characterized MWCNTs nanofluid in a uniformly heated tube. The nanofluid was produced by dispersing 0.25% and 0.5% volume fractions of MWCNTs in a mixture of distilled water (DW) and 30% ethylene glycol. The numerical approach was validated by comparing the data with that obtained experimentally. The results indicated an average enhancement in the heat transfer coefficient of 10% and 14% with corresponding rises in pressure drop of 9.8% and 16.2% for the 0.25% and 0.5% MWCNTs nanofluids, respectively. Moreover, the heat transfer coefficient, pressure drop and wall shear stress all increased with an increase in both the Reynolds number and particle volume fraction, while the friction factor was found to decrease with the Reynolds number. Later, Ajeeb *et al.*[39] numerically studied the laminar flow of Al_2O_3 nanofluids in a horizontal minitube of 3.5 mm diameter under constant heat flux boundary conditions (15 kW m^{-2}) and at an inlet temperature of 25 °C. The numerical approach was validated in comparison with experimental data under the same operation conditions. The thermophysical properties of the nanofluids were measured and used in the numerical approach, with the results showing that the heat transfer was enhanced by the addition of Al_2O_3 nanoparticles to the base fluid by 6.77%, 9.08%, 11.38%, 13.03% and 16.32%, respectively, for volume fractions of 0.05%, 0.1%, 0.15%, 0.2% and 0.3%. Moreover, an

Table 11.1 Numerical studies of heat transfer of nanofluids in a single tube.

Authors	Nanofluid	Heat exchanger geometry	Numerical model	Remarks
Ajeeb *et al.*[38]	MWCNTs/EG + DW 0.25, 0.5 vol%	Circular tube $L = 1.420$ m $D = 6$ mm	Navier–Stokes, 2D, single-phase Laminar	Enhancement in HTC of 10% and 14% with raising pressure drop of 9.8% and 16.2% for 0.25 vol% and 0.5 vol%, respectively.
Rashmi *et al.*[13]	CNTs/water 0.01 vol%	Concentric tube $D = 11.3$ mm	Navier–Stokes, 2D, single-phase Laminar	Enhancement of HTC of 68.6% at 30 g s^{-1} flow rate.
Ajeeb *et al.*[39]	Al_2O_3/EG + DW	$L = 2.4$ m $D = 3.5$ mm	Navier–Stokes, 2D, single-phase Laminar	Maximum enhancement in Nu of 16.3% with an increase in pressure drop of 18.96%.
Arjun and Rakesh[36]	Al_2O_3/water 0–5 vol%	Circular channel $D = 0.0005$ mm	Laminar and standard k–ε models Turbulent	Heat transfer enhancement up to 83% in laminar and turbulent regimes.
Ying *et al.*[40]	Al_2O_3/water 0.25 and 0.5 wt%	Tube $L = 1$ m $D = 2.25$ mm	Single-phase models two-phase models 2D, laminar	DPM, X-R DM, and T-A DM are the most accurate models for this study conditions.
Kamali and Binesh[16]	CNTs/water 1 vol%	Tube $D = 1.55$ mm	Navier–Stokes, 2D, single-phase Laminar	Enhancement in HTC about 30% at Re = 600.
Nikkhah *et al.*[42]	FMWCNTs/water 0.12 and 0.25 vol%	Microchannel —	Navier–Stokes, 2D, single-phase Laminar	The maximum value of Nu number occurs at $\varphi = 0.25\%$ and Re = 100.

increase in the friction factor and pressure drop was noted with the addition of nanoparticles, thus requiring extra pumping power.

Ying *et al.*[40] applied various single-phase and two-phase models for the convective heat transfer of aqueous Al_2O_3 nanofluids in laminar flow through a tube under constant wall heat flux, and compared the results with the available experimental data. The results showed the suitability of the numerical models to predict the behaviour of Al_2O_3 nanofluids, particularly for low particle concentrations. The Eulerian–Lagrangian model was recommended by Ying *et al.*[40] in the case where no prior experimental data is available for validation.

Rashmi *et al.*[13] studied experimentally and numerically CNTs/water–gum arabic nanofluids through a concentric tube heat exchanger. The results indicated an heat transfer enhancement of 68.6% at a flow rate of 30 $g\,s^{-1}$ in comparison with the base fluid. The authors conjectured the Newtonian behaviour of the CNT nanofluids at 0.01 vol%. The numerical results showed good agreement with the experimental data when a 2D Fluent 6.3 model was used under single-phase, steady-state and incompressible flow conditions to simulate the laminar flow of CNT nanofluids.

Furthermore, Kamali and Binesh[16] studied numerically an aqueous 1.0 wt% MWCNT nanofluid as a single-phase and non-Newtonian fluid in a horizontal tube with a diameter of 1.55 mm under constant wall heat flux. The numerical model was constructed using a new correlation for viscosity based on fitting the experimental viscosity data using the power-law non-Newtonian model at constant temperature. The numerical results were compared with the correlated experimental data reported in the literature[41] for the local HTC and fairly good agreement was observed. The latter renders the numerical approach an important tool for analysing the fluid flow and heat transfer phenomena even when addressing non-Newtonian fluid behaviour. Also, these results indicated a significant effect of the near-wall region on the HTC in the case of non-Newtonian behaviour of MWCNT nanofluids.

11.3.1.2 Heat Sink

A heat sink is a passive heat exchanger that transfers heat to a fluid medium. The addition of nanoparticles into the fluid medium has been studied in order to improve the cooling performance of heat sinks. Table 11.2 presents a summary of some relevant numerical studies on the thermal performance of nanofluids in heat sinks.

Elbadawy and Fayed[43] numerically studied the laminar flow of Al_2O_3 in a rectangular cross-sectional microchannel heat sink (MCHS), and a heat transfer enhancement of 13.12% was observed in comparison with water. A numerical study by Kamali *et al.*[44] investigated the performance of Al_2O_3/water and MWCNT/water nanofluids in a typical heat sink. The results indicated a reduction in thermal resistance by 18% and 1% for MWCNT and Al_2O_3 nanofluids, respectively, in comparison with the base fluid. Also, Ebrahimi *et al.*[45] studied the thermal performance of a MWCNT/water nanofluid in a

Table 11.2 Numerical studies of heat transfer of nanofluids in heat sink.

Authors	Nanofluids	Heat exchanger geometry	Numerical model	Remarks
Elbadawy and Fayed[43]	Al_2O_3/water 1% to 5 vol%	MCHS $W_c = 200$ μm $H_c = 700$ μm	Navier–Stokes, 3D, single-phase Laminar	An increase in HTC by 13.12% compared to water.
Kamali *et al.*[44]	MWCNT, Al_2O_3/water 1.0 vol%	MCHS $W_c = 280$ μm $H_c = 430$ μm	Navier–Stokes, 3D, single-phase Laminar	Thermal resistance was reduced by 18% and 1%, for MWCNTs and Al_2O_3 nanofluids, respectively.
Ebrahimi *et al.*[45]	CNTs/water 1.0 vol%	MCHS $W_c/W = 0.6$ $H/W_c = 8.2$	Navier–Stokes —	CNTs nanofluids caused a decrease in thermal resistance by 9%.
Kadja and Rezaiguia[46]	Al_2O_3/water, 1 and 4 vol%	MCHS $W_c = 200$ μm $H_c = 160$ μm	Navier–Stokes, 3D, single-phase Laminar	Heat transfer enhancement by 30% for the nanofluids considering dependent-temperature properties.
Arabpour *et al.*[14]	MWCNTs/kerosene 4 and 8 vol%	MCHS $H_c = 50$ μm	Navier–Stokes, 2D, single-phase Laminar	Heat transfer enhancement significantly affected by using oscillating heat fluxes.

microchannel heat sink and a reduction in thermal resistance was observed. In another numerical study by Kadja and Rezaiguia,[46] the performance of several heat sink geometries with Al_2O_3 nanofluids was considered and it was found that the rectangular cross-section is preferred for heat transfer enhancement and that less pumping power is required in comparison with other shapes, such as elliptical, rectangular, and trapezoidal. They also studied the sensitivity of the heat transfer to thermophysical properties dependent on temperature, realising that heat transfer increased up to 30% when temperature-dependent properties were considered. Another numerical study by Arabpour *et al.*[14] investigated kerosene/MWCNTs in a two-dimensional double-layer microchannel. The heat transfer enhancement increased with increasing Reynolds number and nanoparticle volume fraction.

11.3.1.3 Plate Heat Exchangers

A numerical study by Bahiraei and Monavari[47] reported the thermal performance of an Al_2O_3 nanofluid (1.0 vol% particles) in a microchannel plate heat exchanger (PHX) for five particle shapes (*i.e.*, platelet, brick, cylinder, blade, and oblate spheroid) at high inlet temperature and Re = 500 for the hot fluid. The results indicated a maximum overall heat transfer coefficient for the nanofluid containing platelet-shaped particles.

Ahmed *et al.*[48] numerically studied the thermal performance of a Cu/water nanofluid in a two-plate channel for Re between 100 and 1000 and particle concentration between 0 and 5 vol%. The numerical modelling assumed 2D, steady, incompressible, Newtonian and homogenous flow (single-phase). The results showed an increase in HTC with an increase in the concentration of the nanoparticles and increase in Re, with a slight increase in pressure drop. Another numerical study by Shirzad *et al.*[49] focused on different Al_2O_3, CuO and TiO_2/water nanofluids for a pillow plate heat exchanger when Re ranged between 1000 and 8000. The numerical simulations applied the realizable turbulent k–ε model for turbulent flow and assumed 3D single-phase, incompressible, steady-state and no-slip conditions. A considerable increase in heat transfer behaviour was observed for all the nanofluid samples in comparison to the base fluid. Moreover, the best heat transfer performance was attained for an Al_2O_3 nanofluid at low Re, and a TiO_2 nanofluid showed a higher thermal performance at high Re numbers.

Pantzali *et al.*[50] studied numerically and experimentally CuO/water nanofluids in a miniature plate heat exchanger with a particle concentration of 4 vol%, observing a 29.4% increase in overall heat transfer.

Table 11.3 lists the relevant numerical studies on the thermal performance of nanofluids in plate heat exchangers.

Despite numerical research being conducted, there is still a lack of knowledge on modelling the temperature-dependent properties and non-Newtonian rheology of nanofluids for convection thermal investigations. Thermophysical properties should be well addressed and validated numerical approach modelled. New correlations that describe the behaviour of

Table 11.3 Numerical studies of the heat transfer of nanofluids in a plate heat exchanger.

Authors	Nanofluids	Heat exchanger Geometry	Numerical model	Remarks
Shirzad *et al.*[49]	Al_2O_3, CuO and TiO_2/ water 2–5 vol%	Pillow PHX —	Navier–Stokes, 3D, single-phase, Turbulent k-ε model T-dependent properties	Better thermal behaviour of an Al_2O_3 nanofluid in terms of low Re.
Bhattad *et al.*[51]	Al_2O_3 and MWCNT hybrid/water	PHX $W = 100$ mm $L = 355$ mm $2b_c = 2.8$ mm	Navier–Stokes, 3D Standard turbulent k-ε model Constant properties	Enhancement in HTC of 39.16%, and increase in pumping power of 1.23%.
Pantzali *et al.*[50]	CuO/water 4 vol%	Miniature PHX —	Navier–Stokes, 3D, single phase Turbulent k–ω model Constant properties	Maximum enhancement in HTC is 29.41%.
Bahiraei and Monavari[47]	Al_2O_3/EG– water. Five particle shapes 1 vol%	Micro PHX $W_c = 0.25$ mm $H_c = 0.26$ mm	Navier–Stokes, 3D, single phase Laminar Newtonian, constant properties	The maximum overall HTC enhancement for nanofluids was when platelet-shaped particles were used.
Ahmed *et al.*[48]	Cu/water 0–5 vol%	Corrugated PHX $H_{max} = 16$ mm $H_{min} = 10$ mm	Navier–Stokes, 2D, single phase Laminar Newtonian, constant properties	Maximum enhancement in HTC of 43.9%.
Tiwari *et al.*[52]	CeO_2 and Al_2O_3/water	Chevron corrugated PHX	Navier–Stokes, 3D, single-phase Standard turbulent k-ε model Constant properties	Better heat transfer performance for s CeO_2/water nanofluid.

nanofluid properties considering temperature change and rheology are needed for numerical modelling of developing thermal convection systems. Then, a tool to clarify the importance of nanofluids in thermal convection should be established that can support the implementation of nanofluids in industrial applications.

11.3.1.4 Two-phase Models for Forced Convection

Hazeri-Mahmel *et al.*[53] studied the forced convection of Al_2O_3/water nanofluid in a horizontal pipe using the mixture model in the OpenFOAM software. For a 1.0% volume fraction of nanoparticles, the numerical heat transfer coefficient attained using a non-Newtonian mixture model was a better fit than the single-phase model using the experimental results. Zhang *et al.*[54] experimentally and numerically investigated the heat transfer and flow characteristics of a hybrid Al_2O_3–CuO/water nanofluid in a circular tube. The mean absolute relative deviation of the mixture model between numerical and experimental heat transfer coefficients was 10.2%.

Zhang *et al.*[55] numerically studied a fully-developed gas–liquid Taylor flow using CuO/water nanofluids in a vertical capillary. The temperature contours were similar for the VOF model and the mixture model in work by Fischer *et al.*[56] Naphon and Nakharintr[57] numerically studied the turbulent heat transfer and flow characteristics of nanofluids in a minichannel heat sink using single-phase, mixture, and VOF methods. Compared with the homogenous single-phase model, the two-phase models showed better prediction of the experimental data.

Chen *et al.*[58] adopted the Eulerian two-phase model to study the laminar and turbulent forced convection of nanofluids in minitubes with inner diameters of 1.02, 4.5, and 4.75 mm. Two different nanofluids with CuO and γ-Al_2O_3 nanoparticles were included in this work. The results showed that the numerical results of both the Eulerian two-phase model and the mixture model were a good fit to the experimental data in terms of average and local distributions of Nu.

Rashidi *et al.*[59] used the Euler–Lagrange model in the simulation of nanofluids passing a square duct. As shown in Figure 11.3, nanoparticles with a particle diameter of 30 nm were dispersed over the entire domain. With increasing particle diameters, the nanoparticles tended to accumulate within the vorticity regions. He *et al.*[60] investigated the laminar convective heat transfer characteristics of nanofluids using the Euler–Lagrange method. The results suggest that the convective heat transfer coefficient was more affected by thermal conductivity than by viscosity, and Brownian, lift and thermophoretic forces played a very minor role. Bahremand *et al.*[61] experimentally and numerically studied turbulent flow in helically coiled tubes under constant wall heat flux using silver–water nanofluids. It was concluded that the Euler–Lagrange method gave more accurate results than the homogeneous single-phase model.

(a)

(b)

(c)

(d)

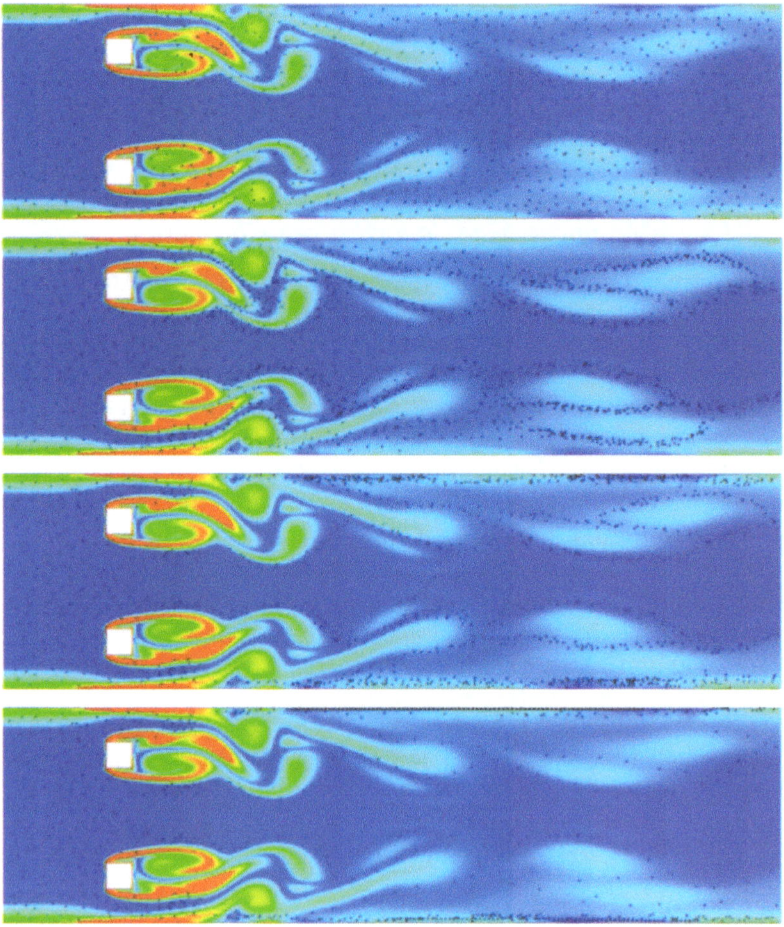

Figure 11.3 Dispersion of silver nanoparticles inside the duct for (a) $d_p = 30$ nm; (b) $d_p = 0.1$ μm; (c) $d_p = 0.25$ μm; (d) $d_p = 0.5$ μm, at Re = 100 and 1.0% volume fraction. Reproduced from ref. 59, https://doi.org/10.3390/app7040431, under the CC BY 4.0 license, https://creativecommons.org/licenses/by/4.0/.

11.3.2 Natural and Mixed Convection

In natural convection (also referred to as free convection), fluid motion is induced by the density differences in the fluid due to temperature gradients. Natural convection differs from forced convection as the former does not require any external sources such as a pump or fan, and the driving force for natural convection is gravity. Natural convection within enclosures finds applications in different fields such as electronics cooling, solar collector, thermal storage and food drying.

Mixed convection occurs when neither natural convection nor forced convection are negligible. Both pressure forces and buoyancy are present in mixed convection. The ratio between the buoyancy force due to density variations and the inertia force is governed by the Richardson number (Ri), *i.e.*, the Grashof number (Gr) divided by the square of the Reynolds number (Re). The convective heat transfer can then be classified as: (1) $Ri = Gr/Re^2 \ll 1$, forced convection; (2) $Ri \gg 1$, natural convection; and (3) $Ri \approx 1$, mixed convection or combined forced and natural convection.

Either single-phase model or two-phase models have been employed to study natural convection heat transfer of nanofluids in enclosures, such as square, rectangular and triangular partially-heated cavities and mixed convection in channels and tubes. Abu-Nada and Oztop[62] numerically investigated the effect of inclination angle on the natural convection of Cu–water nanofluids at 5.0 vol% and 10.0 vol% in a square cavity. The single-phase model was adopted in this numerical work, where the Maxwell–Garnett model[63] and the Brinkman model[30] were used to understand nanofluid thermal conductivity and viscosity, respectively. The governing equations for laminar and steady-state natural convection were solved using the finite-volume method for Rayleigh numbers ($Ra = Gr \times Pr$) in the range of 10^3 to 10^5. A second-order central difference scheme was adopted for the diffusion terms, while a second-order upwind difference scheme was used for the convective terms in the governing equations. The results indicated that the nanofluid can remarkably enhance heat transfer compared to a pure fluid, and the effect of nanoparticle fraction on the Nusselt number was more pronounced at low Ra than at high Ra. The effect of inclination angle on heat transfer enhancement was insignificant at low Ra but decreased the enhancement of the nanofluid. It needs to be noted that in this work, and similar numerical works, the nanofluids were assumed to be Newtonian. However, for volume fractions of as high as 10.0%, the validity of such a Newtonian assumption is questionable. Besides this, it should be noted that the "remarkable" heat transfer enhancement in this work and similar works is dependent on the viscosity and thermal conductivity models used in the simulations. There is ambiguity concerning the viscosity and thermal conductivity of nanofluids, not to mention the rheological behaviour of non-Newtonian nanofluids and its change with elapsed time.

Two-phase models have been less investigated in the literature. A few relevant studies using two-phase models are listed in Table 11.4.

Most two-phase models are based on the governing equations proposed by Buongiorno,[33] by considering two slip mechanisms, *i.e.*, Brownian motion and thermophoresis, while neglecting other slip mechanisms such as inertia, diffusiophoresis, the Magnus effect, fluid drainage and gravitational settling in the absence of turbulence. The Buongiorno model is a two-fluid (liquid and nanoparticles), four-equation (1 continuity equation for nanofluid, 1 continuity equation for nanoparticles, 1 momentum equation and 1 energy equation), non-homogeneous (slip between liquid and particles) and equilibrium (same temperature of the liquid and particles) model.

Corcione *et al.*[67] numerically studied the laminar natural convection of aqueous alumina nanofluids in a differentially heated square cavity based on the Buongiorno two-phase model. Empirical correlations for dynamic

Table 11.4 Numerical studies of thermal convection using two-phase models.

Authors	Nanofluid	Thermal conductivity	Viscosity	Numerical model	Validation
Akbari *et al.*[64]	Al_2O_3–water, 0–6 vol%; mixed convection	Jang and Choi[65]	Maiga *et al.*[66]	Volume of fluid; mixture model; Euler–Euler model	Experimental data from one source
Corcione *et al.*[67]	Al_2O_3–water, 0–6 vol%; natural convection	Corcione correlation[68]	Corcione correlation[68]	Two-phase mixture model	Numerical data from three sources and the Berkovsky–Polevikov correlation
Goodarzi *et al.*[69]	Cu–water, 0–4 vol%; mixed convection (laminar and turbulent)	Chon *et al.*[70]	Masoumi *et al.*[71]	Two-phase mixture model	Numerical and experimental data
Garoosi *et al.*[72]	Al_2O_3–water, 0–5 vol%; natural and mixed convection	Corcione correlation[68]	Corcione correlation[68]	Two-phase mixture model	Numerical data from two sources
Ziad Saghir *et al.*[73]	Al_2O_3–water, 0.01–5 vol%; natural convection	Ho *et al.*[74]	Ho *et al.*[74]	Eulerian–Eulerian model	Experimental data from one source
Izadi *et al.*[75]	Cu–water, 0–10 vol%; natural convection in porous media	Empirical correlation	Empirical correlation	Thermal nonequilibrium mixture model	Numerical data from one source
Qi *et al.*[76]	Cu–liquid metal, 1–5 vol%; natural convection	Hamilton and Crosser model[77]	Brinkman model[30]	Two-phase lattice Boltzmann method	Numerical and experimental data
Xu *et al.*[78]	Cu–water, 1–5 vol%; natural convection	Hamilton and Crosser model[77]	Brinkman model[30]	Two-phase lattice Boltzmann method	Numerical data from three sources

viscosity and thermal conductivity were proposed based on 14 sets of experimental data from literature, as follows:

$$\frac{\mu_n}{\mu_\mathrm{f}} = \frac{1}{1 - 34.87\left(d_\mathrm{p}/d_\mathrm{f}\right)^{-0.3}\varphi^{1.03}} \tag{11.44}$$

$$\frac{k_n}{k_\mathrm{f}} = 1 + 4.4\mathrm{Re}_\mathrm{p}^{0.4}\mathrm{Pr}_\mathrm{f}^{0.66}\left(\frac{T}{T_\mathrm{fr}}\right)^{10}\left(\frac{k_\mathrm{s}}{k_\mathrm{f}}\right)^{0.03}\varphi^{0.66} \tag{11.45}$$

where Re_p is the nanoparticle Reynolds number, defined as:

$$\mathrm{Re}_\mathrm{p} = \frac{2\rho_\mathrm{f}k_\mathrm{b}T}{\pi\mu_\mathrm{f}^2 d_\mathrm{p}} \tag{11.46}$$

Corcione[68] detailed how to use these two correlations properly. The two correlations for dynamic viscosity and the thermal conductivity of nanofluids were used in their numerical simulations and were later used in other numerical studies. The finite-difference method was used to solve the governing equations. The numerical approach was well validated by four different tests and also the usually-recommended Berkovsky–Polevikov correlation.[79] Based on the benchmarking, it was suggested that the two-phase mixture approach is more accurate than the single-phase homogeneous method.

Effects of the temperature difference between the hot and cold walls (ΔT) and the nanoparticle size (d_p) are presented in Figure 11.4, where E indicates

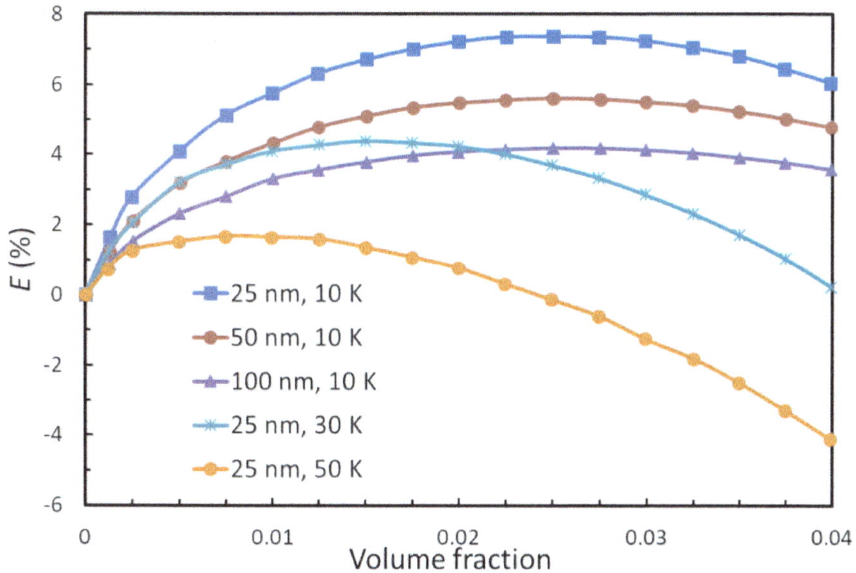

Figure 11.4 Effect of Al_2O_3 nanoparticle diameter and temperature difference on E (heat transfer enhancement of the nanofluid relative to the performance of the base fluid) as a function of the volume fraction of nanoparticles. The numerical values were extracted from ref. 67.

the ratio of the heat transfer rates flowing through the nanofluid and the base fluid, respectively. The heat transfer enhancement ratio E increases as ΔT increases and d_p increases. Besides this, there is an optimal particle loading at which the heat transfer enhancement peaks, and the optimal value increases as the temperature difference, ΔT, increases. It is to be noted that this dependence on particle size is because both the viscosity and thermal conductivity are predicted by the two correlations.

Garoosi *et al.*[72] performed a numerical study on both the natural and mixed convection of water-based alumina nanofluids in a laterally heated square cavity. Heat transfer and nanoparticle distribution were simulated within the following parameter ranges: $10^2 \leq \text{Ra} \leq 10^7$, $0 \leq \text{volume fraction} \leq 0.05$, $25\ \text{nm} \leq d_p \leq 145\ \text{nm}$, $0.01 \leq \text{Ri} \leq 1000$, $10^2 \leq \text{Gr} \leq 10^4$. The viscosity and thermal conductivity correlations proposed by Corcione[68] were also used in this work. An optimal nanoparticle volume fraction was observed at each Ra and Ri, where a maximum of the heat transfer rate was achieved. At low Ra and high Ri, the nanoparticle distribution was non-uniform, while at high Ra and low Ri values, the nanoparticle distribution was almost uniform, for both natural and mixed convection. As shown in Figure 11.5, at $\text{Gr} = 100$, the nanoparticle distribution was almost uniform at $\text{Ri} = 1$. As Ri increased to 100 at the same Gr value, the nanoparticle loading was obviously higher close to the cold wall (the right side) than that close to the hot wall (the left side) due to thermophoresis.

Laminar mixed convection of alumina–water nanofluids was investigated by Akbari *et al.*[64] using the single-phase and three two-phase models (volume of fluid, mixture, and Eulerian two-phase models). The numerical results indicate that the flow field is similar for these four different models, while the thermal characteristics are different between the single-phase and the two-phase models.

11.4 Conclusions and Outlook

The convective heat transfer of nanofluids may find applications in a variety of heating and cooling processes, while numerical simulation can be seen as an effective method to investigate the thermal convection performance of nanofluids, for example, optimization of geometry, concentration of nanoparticles and operating ranges. Numerical modelling can also be used to improve the understanding of the underlying physics of the addition of nanoparticles on the heat transfer performance of nanofluids. Single-phase and two-phase models have been considered to numerically investigate various convective heat transfer mechanisms, including forced, natural, and mixed convection. A few remarks and the outlook of this chapter are summarized as follows:

- Single-phase models may give a relatively good prediction for forced convection, especially for turbulent flow.
- The two-phase mixture model based on the Buongiorno model[33] is the most studied among all two-phase models, allowing deep insight regarding the effects of Brownian motion and thermophoresis.

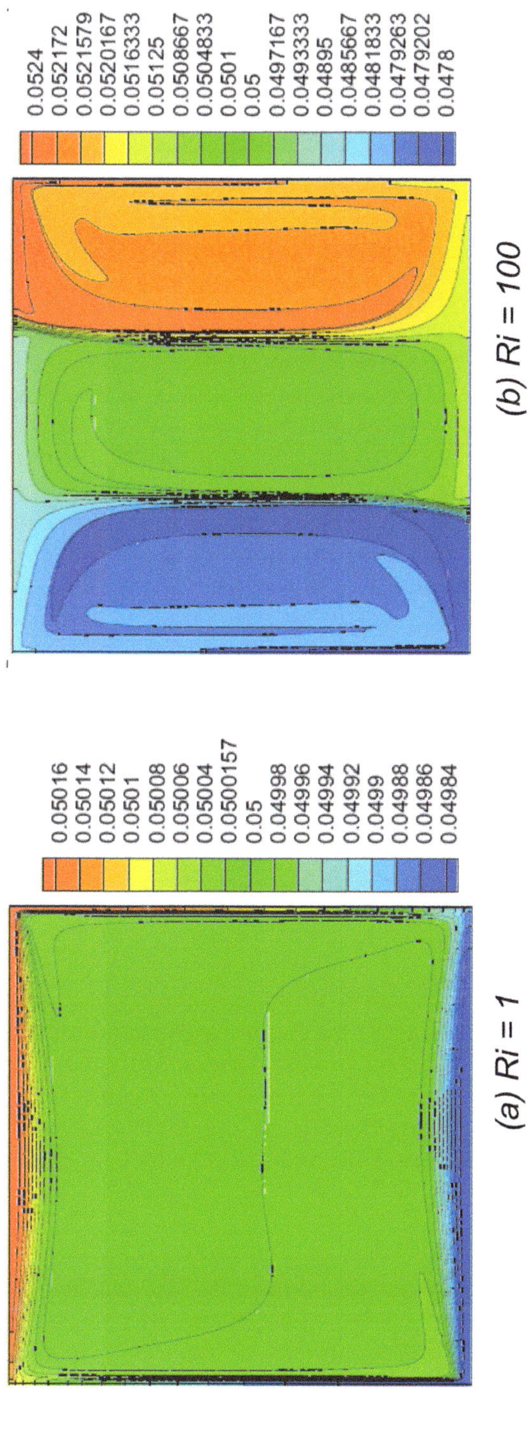

Figure 11.5 Contours of Al_2O_3 nanoparticle distribution at $d_p = 25$ nm and Gr = 100, at (a) Ri = 1, and (b) Ri = 100. Reproduced from ref. 72 with permission from Elsevier, Copyright 2014.

- Two-phase models are able to provide a better prediction than the single-phase model at low Grashof numbers and high Richardson numbers.
- Regarding the comparison of nanofluids with their corresponding base fluids, the heat transfer enhancement and the increase in pressure drop in numerical simulations is dependent on the selection of nanofluid properties, especially viscosity and thermal conductivity. Therefore, some careful consideration should be given when interpreting the benefit of nanofluids in numerical modelling.
- Solid benchmarking is a must for either the single-phase or two-phase modelling of nanofluids, by comparing simulation results with well-interpreted experimental and/or numerical data from different sources in the literature.
- One needs to take a closer look at the rheology of nanofluids. Nanofluids tend to present non-Newtonian behaviour either at relatively high-volume fractions (for example, >5.0%) or for non-spherical nanoparticles or for non-Newtonian base fluids.[80] However, this possible non-Newtonian rheological behaviour is very much ignored in the literature.
- Besides this, the nanoparticle morphology in nanofluids, such as aggregate-laden nanofluids,[81] is yet to be covered in numerical works, although nanoparticle clustering has attracted much interest in the nanofluid community, particularly in thermal conduction. A possible way to address nanoparticle morphology in a numerical simulation (*e.g.*, single-phase, two-phase mixture, and VOF models) is to use thermal conductivity and viscosity relationships that include the effect of nanoparticle clustering/agglomeration. However, inevitably additional parameters need to be introduced to properly characterize nanoparticle aggregates. For the Eulerian and Euler–Lagrange models, the size distribution of clusters and the relevant liquid/cluster forces are non-trivial challenges on the way.

Acknowledgements

This work was supported by Fundação para a Ciência e a Tecnologia (FCT), through IDMEC, under LAETA, project UIDB/50022/2020 and through project PTDC/NAN-MAT/29989/2017, by the Swedish Research Council through project 2017-04724, and additionally by the Joint Funds of the National Natural Science Foundation of China (U20A20300).

References

1. V. Trisaksri and S. Wongwises, *Renewable Sustainable Energy Rev.*, 2007, **11**, 512.
2. M. Liu, M. Lin and C. Wang, *Nanoscale Res. Lett.*, 2011, **6**, 297.
3. Y. Xuan and Q. Li, *Int. J. Heat Fluid Flow*, 2000, **21**, 58.
4. A. Akbarinia, *Int. J. Heat Fluid Flow*, 2008, **29**, 229–241.

5. S. Mirmasoumi and A. Behzadmehr, *Int. J. Heat Fluid Flow*, 2008, **29**, 557.
6. Y. He, Y. Men, Y. Zhao, H. Lu and Y. Ding, *Int. J. Heat Fluid FlowInt. J. Heat Fluid Flow*, 2009, **29**, 1965.
7. S. Kumar, S. K. Prasad and J. Banerjee, *Appl. Math. Model.*, 2010, **34**, 573.
8. A. Adriana and S. M. S. Murshed, *Renewable Sustainable Energy Rev.*, 2017, **91**, 584.
9. S. M. Vanaki, P. Ganesan and H. A. Mohammed, *Renewable Sustainable Energy Rev.*, 2016, **54**, 1212.
10. ANSYS FLUENT 12.*0 User's Guide*, ANSYS, 2009.
11. S. Patankar, *Numerical Heat Transfer and Fluid Flow*, Hemisphere Publishing, New York, 1980.
12. A. A. A. Arani, O. A. Akbari, M. R. Safaei, A. Marzban, A. A. Alrashed, G. R. Ahmadi and T. K. Nguyen, *Int. J. Heat Mass Transfer*, 2017, **113**, 780.
13. W. Rashmi, M. Khalid, A. F. Ismail, R. Saidur and A. K. Rashid, *J. Exp. Nanosci.*, 2013, **10**, 545.
14. A. Arabpour, A. Karimipour and D. Toghraie, *J. Therm. Anal. Calorim.*, 2018, **131**, 1553.
15. G. L. Morini, *Heat Transfer Eng.*, 2006, **27**, 64.
16. R. Kamali and A. R. Binesh, *Int. Commun. Heat Mass Transfer*, 2010, **37**, 1153.
17. A. B. Metzner and J. C. Reed, *AIChE J.*, 1955, **4**, 434.
18. S. A. Sajadifar, A. Karimipour and D. Toghraie, *Eur. J. Mech. B/Fluids*, 2017, **61**, 25.
19. R. Shyam and R. P. Chhabra, *Int. J. Heat Mass Transfer*, 2013, **57**, 742.
20. M. Shojaeian and A. Kos, *Int. J. Heat Mass Transfer*, 2014, **70**, 664.
21. A. S. Mujumdar, *Drying Technol.*, 1989, **7**, 843.
22. A. N. Smith and P. M. Norris, in *Heat Transfer Handbook*, ed. A. Bejan and A. D. Kraus, John Wiley & Sons, Inc., New Jersey, 2003, ch. 18, pp. 1309–1358.
23. Cornell University, A Hands-on Introduction to Engineering Simulations: Computational Fluid Dynamics (CFD), Available at https://courses.ed11.org/courses/course-v1:CornellX+ENGR2000X+1T2018/course/.
24. J. Anderson, *Computational Fluid Dynamics: The Basics with Applications*, Cambridge University Press, Cambridge, 1995.
25. H. K. Versteeg and W. Malalasekera, *An Introduction to Computational Fluid Dynamics*, Person Education, Essex, 1995.
26. S. Özerinç, A. G. Yazcoglu and S. Kakaç, *Int. J. Therm. Sci.*, 2012, **62**, 138.
27. ANSYS FLUENT 18.2 User's Guide, ANSYS, 2017.
28. T. H. Shih, W. W. Liou, A. Shabbir, Z. Yang and J. Zhu, *Comput. Fluids*, 1995, **24**, 227.
29. O. Mahian, L. Kolsi, M. Amani, P. Estellé, G. Ahmadi, C. Kleinstreuer, J. S. Marshall, M. Siavashi, R. A. Taylor, H. Niazmand, S. Wongwises, T. Hayat, A. Kolanjiyil, A. Kasaeian and I. Pop, *Phys. Report.*, 2019, **790**, 1.
30. H. C. Brinkman, *J. Chem. Phys.*, 1952, **20**, 571.
31. M. Manninen, V. Taivassalo and S. Kallio, *On the Mixture Model for Multiphase Flow*, VTT Publications, 1996.

32. L. Schiller and Z. Naumann, *Vdi Zeitung*, 1935, **7**, 318.
33. J. Buongiorno, *ASME J. Heat Transfer*, 2006, **128**, 240.
34. M. M. Rashidi, A. Hosseini, I. Pop, S. Kumar and N. Freidoonimehr, *Appl. Math. Mech.*, 2014, **35**, 831.
35. W. E. Ranz and W. R. Marshall, *Chem. Eng. Prog.*, 1952, **48**, 141.
36. K. S. Arjun and K. Rakesh, *J. Eng. Sci. Technol.*, 2017, **12**, 265.
37. A. Esmaeilnejad, H. Aminfar and M. S. Neistanak, *Int. J. Therm. Sci.*, 2014, **75**, 76.
38. W. A. Ajeeb, M. S. A. Oliveira, N. Martins and B. Abreu, *J. Nanofluids*, 2019, **8**, 1.
39. W. Ajeeb, D. Roque and S. M. S. Murshed, *5th–6th Thermal and Fluids Engineering Conference (TFEC)*, 2021, 1.
40. Z. Ying, B. He, D. He, Y. Kuang, J. Ren and B. Song, *Adv. Powder Technol.*, 2020, **31**, 3050.
41. P. Garg, J. L. Alvarado, C. Marsh, T. A. Carlson, D. A. Kessler and K. Annamalai, *Int. J. Heat Mass Transfer*, 2009, **52**, 5090.
42. Z. Nikkhah, A. Karimipour, M. R. Safaei, P. Forghani-Tehrani, M. Goodarzi, M. Dahari and S. Wongwises, *Int. Commun. Heat Mass Transfer*, 2015, **68**, 69.
43. I. Elbadawy and M. Fayed, *Alexandria Eng. J.*, 2020, **59**, 1771.
44. R. Kamali, Y. Jalali and A. R. Binesh, *Micro Nano Lett.*, 2013, **8**, 319.
45. S. Ebrahimi, J. Sabbaghzadeh, M. Lajvard and I. Hadi, *Heat Mass Transfer*, 2010, **46**, 549.
46. M. Kadja and I. Rezaiguia, ASME International Mechanical Engineering Congress and Exposition, Proceedings (IMECE), 2013, 15.
47. M. Bahiraei and A. Monavari, *Appl. Therm. Eng.*, 2020, **179**, 115621.
48. M. A. Ahmed, N. H. Shuaib, M. Z. Yusoff and A. H. Al-Falahi, *Int. Commun. Heat Mass Transfer*, 2011, **38**, 1368.
49. M. Shirzad, S. Soheil, M. Ajarostaghi, M. Aghajani and K. Sedighi, *Adv. Powder Technol.*, 2019, **30**, 1356.
50. M. N. Pantzali, A. G. Kanaris, K. D. Antoniadis, A. A. Mouza and S. V. Paras, *Int. J. Heat Fluid Flow*, 2009, **30**, 691.
51. A. Bhattad, J. Sarkar and P. Ghosh, *Int. Commun. Heat Mass Transfer*, 2018, **91**, 262.
52. A. K. Tiwari, P. Ghosh, J. Sarkar, H. Dahiya and J. Parekh, *Int. J. Therm. Sci.*, 2014, **85**, 93.
53. N. Hazeri-Mahmel, Y. Shekari and A. Tayebi, *Int. Commun. Heat Mass Transfer*, 2021, **121**, 105119.
54. S. Zhang, L. Lu, T. Wen and C. Dong, *Appl. Therm. Eng.*, 2021, **188**, 116589.
55. J. Zhang, S. Li, X. Wang, B. Sundén and Z. Wu, *Int. Commun. Heat Mass Transfer*, 2020, **116**, 104665.
56. M. Fischer, D. Juric and D. Poulikakos, *ASME J. Heat Transfer*, 2010, **132**, 10.
57. P. Naphon and L. Nakharintr, *Int. J. Heat Mass Transfer*, 2015, **82**, 388.
58. Y. Chen, Y. Li and Z. Liu, *Int. J. Heat Mass Transfer*, 2014, **78**, 993.

59. S. Rashidi, J. A. Esfahani and R. Ellahi, *Appl. Sci.*, 2017, **7**, 431.
60. Y. He, Y. Men, Y. Zhao, H. Lu and Y. Ding, *Appl. Therm. Eng.*, 2009, **29**, 1965.
61. H. Bahremand, A. Abbassi and M. Saffar-Avval, *Powder Technol.*, 2015, **269**, 93.
62. E. Abu-Nada and H. F. Oztop, *Int. J. Heat Fluid Flow*, 2009, **30**, 669.
63. C. W. Nan, R. Birringer, D. R. Clarke and H. Gleiter, *J. Appl. Phys.*, 1997, **81**, 6692.
64. M. Akbari, N. Galanis and A. Behzadmehr, *Int. J. Therm. Sci.*, 2011, **50**, 1343.
65. S. P. Jang and S. U. S. Choi, *ASME J. Heat Transfer*, 2007, **129**, 617.
66. S. E. B. Maïga, C. T. Nguyen and N. Galanis, *Superlattices Microstruct.*, 2004, **35**, 543.
67. M. Corcione, M. Cianfrini and A. Quintino, *Int. J. Therm. Sci.*, 2013, **71**, 182.
68. M. Corcione, *Energy Convers. Manage.*, 2011, **52**, 789.
69. M. Goodarzi, M. R. Safaei, K. Vafai, G. Ahmadi, M. Dahari, S. N. Kazi and N. Jomhari, *Int. J. Therm. Sci.*, 2014, **75**, 204.
70. C. H. Chon, K. D. Kihm, S. P. Lee and S. U. S. Choi, *J. Appl. Phys.*, 2005, **87**, 153107.
71. N. Masoumi, N. Sohrabi and A. Behzadmehr, *J. Phys. D: Appl. Phys.*, 2009, **42**, 055501.
72. F. Garoosi, S. Garoosi and K. Hooman, *Powder Technol.*, 2014, **268**, 279.
73. M. Ziad Saghir, A. Ahadi, A. Mohamad and S. Srinivasan, *Int. J. Therm. Sci.*, 2016, **109**, 148.
74. C. J. Ho, W. K. Liu, Y. S. Chang and C. C. Lin, *Int. J. Therm. Sci.*, 2010, **49**, 1345.
75. M. Izadi, G. Hoghoughi, R. Mohebbi and M. Sheremet, *J. Mol. Liq.*, 2018, **261**, 357.
76. C. Qi, L. Liang and Z. Rao, *Int. J. Heat Mass Transfer*, 2016, **94**, 316.
77. R. L. Hamilton and O. K. Crosser, *Ind. Eng. Chem. Fundam.*, 1962, **1**, 187.
78. D. Xu, Y. Hu and D. Li, *Case Stud. Therm. Eng.*, 2019, **13**, 100358.
79. F. P. Incropera, D. P. Dewitt, T. L. Bergman and A. S. Lavine, *Fundamentals of Heat and Mass Transfer*, John Wiley & Sons, Inc., New Jersey, 6th edn, 2007.
80. M. Hothar, B. Sundén and Z. Wu, *J. Mol. Liq.*, 2021, 118064.
81. Z. Wu and B. Sundén, *Int. J. Heat Mass Transfer*, 2016, **93**, 1107.

CHAPTER 12

Boiling Features and Performances of Nanofluids

E. FABRE,[a] S. MANCIN*[b] AND S. M. S. MURSHED[a]

[a] IDMEC, Instituto Superior Técnico, University of Lisbon, 1049-001 Lisbon, Portugal; [b] Department of Management and Engineering, University of Padova, Stradella S. Nicola, 3, Vicenza 36100, Italy
*Email: simone.mancin@unipd.it

12.1 Introduction

Boiling is an important phase-change heat transfer process that has attracted much attention due to the increasing demand for more efficient and compact devices.[1] Developments in technology have led to highly demanding equipment with high heat flux that needs adequate cooling to function properly. In addition, the conventional method to increase cooling in devices is to increase the heat transfer surface, which is not desirable and viable anymore in high-tech industries. Moreover, a two-phase liquid system (such as boiling) offers cooling performance that is much better than that of single-phase heat transfer modes.[2] Moreover, boiling processes are capable to dissipate high heat flux at low wall superheats.

Considering the current energy crises and the non-stop search for the improvement of energy consumption in thermal systems, boiling plays a vital role in many processes, such as in refrigeration, power generation, nuclear fuels, *etc.*[3]

In principle, heterogeneous boiling is a phase change (liquid to vapor) process that promotes large heat transfer from a hot solid surface in contact with a saturated liquid. Boiling occurs when the temperature of the solid

Soft Matter Series No. 16
Fundamentals and Transport Properties of Nanofluids
Edited by S. M. Sohel Murshed

Published by the Royal Society of Chemistry, www.rsc.org

surface is sufficiently higher than the saturation temperature of the liquid. According to the mode of boiling, it can be classified into two different processes: pool and flow boiling. Pool boiling refers to when the heated surface is immersed in a pool of initially quiescent liquid, while in flow boiling the fluid flows and boils over a heated surface or inside a channel, which can then be categorized as external or internal flow boiling, respectively.[4,5]

Regarding the enhancement of boiling performance, there are three major target features: increasing the heat transfer coefficient (HTC), increasing the critical heat flux (CHF), and decreasing of the wall superheat at the onset of nucleate boiling (ONB). HTC is related to the efficiency of the process and the CHF is the thermal limit of the boiling process at which the maximum heat flux is exchanged and beyond this point the HTC suddenly drops, and boiling crisis occurs. After the occurrence of the CHF, the temperature in the heater rises exponentially and may result in damage or burnout. Therefore, the improvement of both HTC and CHF is essential to improve the heat transfer process, resulting in more compact and efficient cooling systems. In this sense, the search for improved heat transfer fluids for applications in boiling processes has been the subject of many studies over the past few decades. In this scenario, nanofluids are recognized as a great potential alternative. They are defined as colloidal dispersions of single or multiple types of nanoparticles in base fluids (water, ethylene glycol, light oils, *etc*). These nanoparticles have at least one dimension on the nanometer scale (normally <100 nm), a high relative surface area and excellent thermal conductivity, normally orders of magnitude superior to those of common base fluids. Hence, blending nanoparticles in conventional liquids will result in greater thermal conductivities, even using very low concentrations. Because of their nano size, they can be stabilized better than particles with a micro size that have been studied in the past. Besides this, microparticles are associated with the clogging of channels, erosion of pumps and pressure drop, issues that can be avoided using nanoparticles.[3]

Since nanofluids were introduced in 1995 at the Argonne National Laboratory in the USA,[6] these fluids have been exhaustively investigated. For their use in boiling processes, searching for the keywords "nanofluid(s)" and refined by "boiling" in the Web of Science Core Collection database led to 1216 results, distributed according to years shown in Figure 12.1.

However, even before the concept of nanofluids upsurged, in 1984, a study led by Yang and Maa[7] evaluated the effect of dispersions of particles of alumina (size of 50 nm to 1000 nm) in working fluids in the enhancement of pool boiling heat transfer.[7] The authors noted that the addition of low concentrations of nanoparticles (0.1–0.5 wt%) to water significantly increased boiling heat transfer.[7]

Many techniques can be used for the enhancement of boiling heat transfer, which can be classified into two groups: active and passive methods. The active method involves the use of external energy to promote additional motion of the fluid, such as mechanical agitators, rotators, *etc.* This method increases the cost of the system and may not meet the features

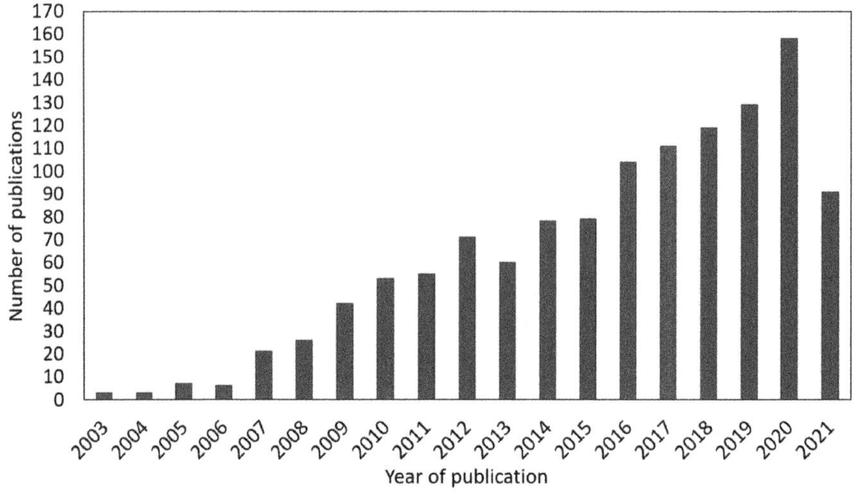

Figure 12.1 Number of publications mentioning "nanofluid" and "boiling" in the Web of Science Core Collection database on 29th September 2021.

of compact design frequently requested for devices. However, the passive method does not require any extra energy and only focuses on the change in the properties of the working fluid and/or the morphology of the heating surfaces. Changes in the properties of working fluids can be achieved using nanoparticles, surfactants, or polymers, while the surface of the heater can be modified with coatings or adding extended and tailored structures to improve roughness or porosity.[5] The use of nanofluids can be combined with other methods to achieve better performance and investigate the effects of different variables on the main boiling heat transfer parameters.

In this context, this chapter covers a description of the fundamentals of boiling heat transfer, approaching both pool boiling and flow boiling regimes, and then presents the application of nanofluids for boiling heat transfer comprising the main variables that can influence this process, namely roughness, the effect of the addition of dispersant agents (*i.e,* surfactants) into nanofluids, and the wettability of the heater surface.

12.2 Fundamentals on Boiling Heat Transfer

The phase change heat transfer process that occurs at the solid surface (heated) in contact with a liquid when the temperature of the heated surface is above the saturation temperature of the liquid is referred to as boiling. The surface heat flux is given by Newton's law of cooling:[8]

$$q'' = h(T_S - T_{Sat}) = h\Delta T_e \tag{12.1}$$

where ΔT_e is the wall superheat (difference between the surface temperature T_S and the liquid saturation temperature T_{Sat}) and h is the boiling heat transfer coefficient (HTC). The boiling process is characterized by the formation of vapor bubbles at the heated wall, which grow and depart from the surface promoting motion of the liquid. The formation of bubbles and consequently the boiling heat transfer itself are a function of the wall superheat, the surface nature, and the properties of the liquid.[8]

There are two modes *via* which boiling may occur: pool and flow boiling. In pool boiling there is a pool of stagnant liquid in which the heated surface is immersed. Heat transfer occurs due to natural convection and the motion caused by bubble growth and detachment. In flow boiling, there is a forced movement of liquid promoted by external power in addition to natural convection and induced circulation by bubble dynamics.[8]

12.2.1 Pool Boiling

Pool boiling is well represented by the typical boiling curve originally obtained by Nukiyama[9] in a power-controlled heating experiment using a platinum (Pt) wire as a heater, submerged in a pool of saturated water. Figure 12.2 shows a schematic representation of this boiling curve, where the heat flux is presented on the vertical axis and the wall superheat ΔT_e on the horizontal axis. During the heat (power) controlled pool boiling process, three regimes can be identified, namely: natural convection, nucleate boiling, and film boiling.

In the first region, up to point A, the heat transfer can be mainly ascribed to single-phase natural convection and small temperature differences exist between the heating wall and the liquid, a phase change process then occurs at the interface, referred to as evaporation. As the heat flux increases, some isolated bubbles start to form along the surface (at point A), which is referred to as the ONB from which the nucleate pool boiling regime initiates.

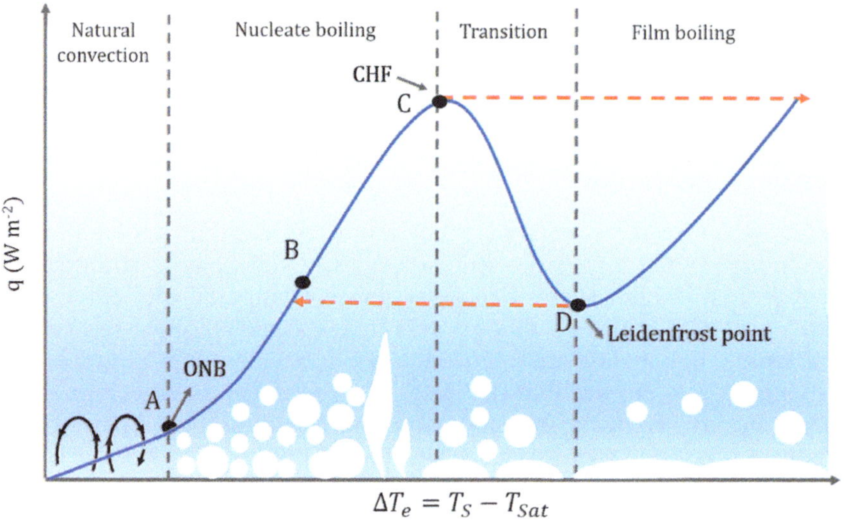

Figure 12.2 Schematic representation of a typical pool boiling curve.

From point A to C, heat is mainly transferred from the heating wall through the formation and ascending of bubbles. Initially, isolated bubbles appear at specific spots at the heating surface (termed as active nucleation sites), and grow and depart from the surface as a consequence of buoyancy forces. At this stage, significant heat is transferred by the mixing of the fluid, and h and q'' are greatly enhanced. As the boiling proceeds, at point B the frequency of bubbles departure is higher, the convective bubble motion is very intense and there are more active nucleation sites. As the heat flux continues to increase, coalescence of the bubbles is observed, and the bubbles detachment is replaced by columns or jets of vapor. Such coalescence of bubble impairs the contact of the liquid with the heating wall, hence reducing convective heat transfer. The curve approaches point C, where the heat flux is at a maximum. Point C is referred to as the CHF, as after this point the high density of vapor at the surface creates a film layer in which the temperature rises abruptly, leading to damage or even burnout of the heating wire. For that reason, it is desirable that the boiling system operates under or even at CHF but should never exceed it. Besides this, an additional increase in heat flux leads to a sudden transition following the red dashed line to the film boiling region. In fact, the surface is completely covered by a vapor blanket and film boiling is the new phase change mechanism. In this region, besides conduction, the contribution of heat transfer by radiation becomes more relevant, and heat flux rapidly increases. For a temperature-controlled process, if the temperature keeps increasing, the heat flux decreases and the boiling curve reaches the lowest point D, the so-called Leidenfrost point, the last point at which a stable vapor film can exist. Then, if the temperature keeps increasing, the heater surface is completely blanketed by the vapor film and the boiling curve follows an upward trend again (beyond point D). It must be underlined that the visualization of the curve from C to D is only possible in an experiment with wall temperature control. A transition region is observed when an unstable film layer or a vapor blanket is formed at the heating surface. The Leidenfrost point (point D) is reached as a result of the much lower thermal conductivity of the vapor in the stagnant film layer decreasing the HTC up to a minimum q'' is observed.[4,8]

12.2.2 Flow Boiling

Boiling with forced convection is much more complex than pool boiling and may involve much higher heat transfer rates, which is applied for thermal management in high energy density systems.[5] Compared with single-phase convective heat transfer, this process may provide higher temperature uniformity, higher HTC and reduced flows.[5]

In this process, the heat transfer is a result of the combination of two effects, the motion created by the formation and detachment of the bubbles from the heating surface and the directed heat transfer promoted by the forced convection.[4,8] According to the type of flow, flow boiling can be divided in external flow boiling, such as the flow over a heated plate or cylinder, or internal flow boiling, which may occur inside heated channels or microchannels.[8]

Regarding external flow boiling, the total heat flux can be estimated considering the equations of pure forced convection combined with pool boiling.[8,10] In comparison with pool boiling, the maximum heat flux (CHF) associated with external flow boiling is superior, being for example around 35 $MW\,m^{-2}$ for the boiling of water at 1 atm compared with the reported value for pool boiling under the same conditions being 1.3 $MW\,m^{-2}$.[8,11]

Internal flow boiling depends on the orientation of the tube through which the liquid is flowing, the operating conditions (*i.e.*, vapor quality, mass velocity), the properties of the fluid, the heating surface, the channel geometry, and the heat flux. As in pool boiling, different regimes can be identified as the heat flux increases. Figure 12.3 presents an illustration of flow boiling in an upward vertical flow inside a heated tube. Considering that a subcooled liquid enters the tube, the first stage is characterized by conventional single-phase forced convection. When the temperature at the surface of the tube increases above the saturation temperature of the liquid, a few bubbles nucleate at the heating surface. Under this subcooled regime, the bubbles are located only near the tube wall as bubbles tend to condensate in the subcooled bulk. As the temperature of the wall increases, all the fluid reaches saturated conditions, and the bubbly flow initiates, where the bubbles are formed at the surface, detach, and move through the radius of

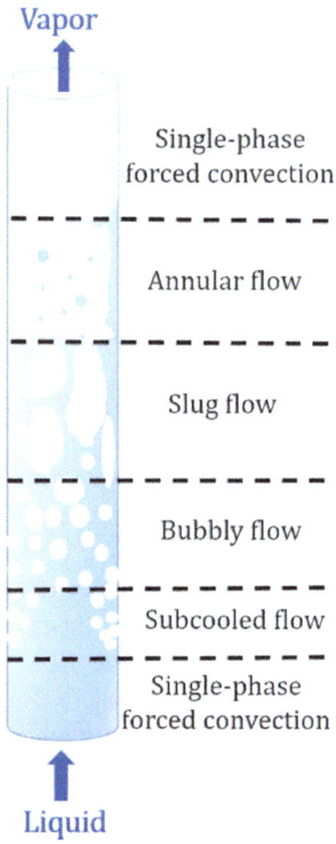

Figure 12.3 Illustration of internal flow boiling regimes.

the tube. As boiling proceeds, coalescence of the bubbles is observed, and the slug flow region is observed. The system evolves to the annular flow regime in which a liquid film is formed on the wall of the tube while vapor is moving very fast in the center. As the vapor quality increases, this state becomes unstable, dry spots start to appear at the surface, and liquid is entrained in the vapor core. As most of the tube wall dries, incipient dryout occurs at which point the heat transfer coefficient suddenly falls, and the tube temperature dramatically increases. As the boiling progresses, all the film liquid is exhausted and only vapor flows, characterizing the last stage, a second single-phase forced convection process. Because of the large difference in the densities of the liquid and vapor, the velocity of the flow is much higher under this regime.[8] Under the vapor regime, the wall temperature significantly rises and that is the CHF of the flow boiling.

12.2.3 Correlations to the Boiling Heat Transfer Coefficient and CHF

There are many correlations available in the open literature to predict boiling heat transfer for the different regions shown in the typical pool boiling curve. For the first regime, conventional single-phase convection equations can be applied. Considering thermal engineering applications, nucleation pool boiling reaches higher values of HTC and hence many attempts have been made to develop models that accurate describe this boiling regime, but a mathematical equation is yet to be proposed that is completely reliable. Nucleate boiling is influenced, among the other parameters, by the number of active nucleation sites and the frequency of bubble departure and diameter, which are all dependent on the features of the heating surface and the properties of the liquid.

One of the earliest and most applied correlations for nucleate pool boiling heat flux was proposed by Rohsenow,[12] expressed as:

$$\frac{c_{\mathrm{p}}(T_{\mathrm{S}}-T_{\mathrm{Sat}})}{h_{\mathrm{fg}}}=C_{\mathrm{sf}}\left[\frac{q''}{\mu h_{\mathrm{fg}}}\sqrt{\frac{\sigma}{g(\rho-\rho_{\mathrm{v}})}}\right]^{1/3}\left(\frac{c_{\mathrm{p}}\mu}{k}\right)^{n} \tag{12.2}$$

where c_{p}, μ, ρ, and k are the specific heat capacity, viscosity, density and thermal conductivity of the saturated liquid, respectively. The latent heat of the fluid is represented by h_{fg}, σ is the surface tension, ρ_{v} is the density of the saturated vapor and g is the acceleration of gravity. The constants C_{sf} and n are dependent of the surface–liquid combination and a list of some values determined experimentally can be found in the literature.[13]

The equation proposed by Rohsenow has its limitations and can only be applied for systems with smooth surfaces. The equation is thus better used for estimation of the wall superheat, which may have errors of ±25%,[4] much lower than the estimated values for heat flux that can be as high as ±100%.[8] Such high deviations can be attributed to many parameters, including different surface roughness treatments, surface functionalization, liquid purity, *etc.*[4]

As mentioned before, CHF is the maximum heat flux that is possible to dissipate in nucleate boiling, which cannot be exceeded so as to avoid catastrophic failure of the heater and systems and also to maintain safe operation. It is also the point of departure from the nucleate boiling regime (DNB) to film boiling. In 1958, a correlation for the prediction of critical heat flux (q''_{CHF}) was proposed based on dimensional analysis by Kutateladze,[13] and a hydrodynamic stability analysis of vapor columns by Zuber,[14] represented as follows:

$$q''_{\mathrm{CHF}} = C\rho_{\mathrm{v}}^{\frac{1}{2}} h_{\mathrm{fg}} [g\sigma(\rho - \rho_{\mathrm{v}})]^{\frac{1}{4}} \tag{12.3}$$

where C is a constant weakly dependent on the geometry of the heater surface,[8] in the range of 0.138–0.157.[15] For horizontal cylinders and spheres $C = 0.131$ and for horizontal plates $C = 0.149$.[8] Eqn (12.3) is used when the ratio of the heated surface length and bubble diameter is significantly large, otherwise a correction factor must be applied.[8] Many correction factors can be found in the work of Lienhard.[16]

In the transition region, the Leidenfrost point or minimum heat flux (q''_{min}) is identified, which corresponds to the turning point at which there is the formation of a stable vapor film layer on the heater surface. The value of q''_{min} can be estimated by:

$$q''_{\mathrm{min}} = 0.09\rho_{\mathrm{v}} h_{\mathrm{fg}} \left[\frac{g\sigma(\rho - \rho_{\mathrm{v}})}{(\rho + \rho_{\mathrm{v}})^2}\right]^{\frac{1}{4}} \tag{12.4}$$

This expression was proposed by Zuber,[14] and the numerical coefficient of 0.09 was experimentally determined by Berenson.[17] It is important to note that eqn (12.4) is not a function of the wall superheat, and that the surface roughness has a negligible effect on the minimum heat flux. However, it is significantly affected by pressure, and the equation below presents errors of around 50% at low or moderate pressures, but at higher pressures the accuracy drops.[4,8]

During the film boiling regime, the HTC is an association of convection and radiation heat transfer since conductive effects can be neglected due to the low thermal conductivity of the vapor. However, since radiation contributes to increasing film thickness, the relationship between them is given by Bromley, considering that the effective radiation coefficient ($\bar{h}_{\mathrm{rad}}$) is lower than the effective convective coefficient ($\bar{h}_{\mathrm{conv}}$):[8]

$$\bar{h} = \bar{h}_{\mathrm{conv}} + \frac{3}{4}\bar{h}_{\mathrm{rad}} \tag{12.5}$$

$\bar{h}_{\mathrm{conv}}$ can be obtained from the following equation:

$$\frac{\bar{h}_{\mathrm{conv}} D}{k_{\mathrm{v}}} = C\left[\frac{g(\rho - \rho_{\mathrm{v}})h'_{\mathrm{fg}} D^3}{\nu_{\mathrm{v}} k_{\mathrm{v}} (T_{\mathrm{S}} - T_{\mathrm{Sat}})}\right]^{\frac{1}{4}} \tag{12.6}$$

here the constant C has a value of 0.62 for cylinders and 0.67 for spheres. Some observations are worth pointing out. For instance, the convective heat transfer coefficient (h) is independent of the material of the heated surface and the effect of pressure is accounted for only on its impact on the physical properties of liquid and vapor.[8,18] The latent heat used in eqn (12.6) is a corrected value that takes into account the sensible energy necessary to guarantee temperature of the vapor is over the saturation temperature, estimated by:[4,18]

$$h'_{\mathrm{fg}} = h_{\mathrm{fg}} + 0.40c_{\mathrm{p,v}}(T_{\mathrm{S}} - T_{\mathrm{Sat}}) \tag{12.7}$$

All the vapor properties in eqn (12.6) and (12.7) must be calculated at the film temperature, which is the arithmetic average temperature of the hot surface and the boiling liquid.[4,18]

The effective radiation coefficient can be calculated as:[8]

$$\bar{h}_{\mathrm{rad}} = \frac{\varepsilon\sigma\left(T_{\mathrm{S}}^{4} - T_{\mathrm{Sat}}^{4}\right)}{T_{\mathrm{S}} - T_{\mathrm{Sat}}} \tag{12.8}$$

where ε is the emissivity and σ is the Stefan–Boltzmann constant.

Regarding flow boiling, the general heat transfer can be calculated *via* the combination of the heat flux from nucleate pool boiling, q''_{w} with the single-phase convection heat flux q''_{c}, $q''_{\mathrm{c}} = h_{\mathrm{c}}(T_{\mathrm{S}} - T_{\mathrm{Sat}})$, assuming that the bulk of the liquid is stationary:[4]

$$q'' = q''_{\mathrm{w}} + q''_{\mathrm{c}} \tag{12.9}$$

The equation above is more accurate for boiling systems with subcooled liquid.

12.3 Nanofluids Applied in Boiling Heat Transfer

Boiling heat transfer performance depends on the heating surface features as well as the properties of the fluid. As mentioned before, two methodologies can be applied to improve boiling efficiency: active and passive methods. Of particular interest are passive methods, such as surface modification or the use of additives such as nanoparticles and surfactants. Changing the surface roughness may improve the number of nucleation active sites that favor bubble formation and departure, affecting boiling heat transfer. Wettability of the heating surface can be enhanced, directly affecting the CHF. Besides this, the hydrophilicity of the surface can be chemically changed to improve the CHF.[5,15]

In the case of nanofluids, as the boiling process occurs, some fractions of suspended nanoparticles deposit on the solid surface, changing the roughness and wettability and consequently influencing the course of boiling. Many studies in the literature have attempted to assess the impact of using nanofluids in pool boiling and flow boiling performance. Moreover, the addition of surfactants alters the superficial tension, with the bubble dynamics also being worthy of investigation. The main parameters evaluated are the HTC and CHF, which are indicative of the boiling efficiency and the extent of the boiling process. The number of studies on this matter is enormous, with Table 12.1 illustrating some of the most relevant studies on the boiling of nanofluids over the last decade, together with their major findings.[5,15] Figures 12.4 and 12.5 show examples from different research groups of boiling curves and their corresponding HTC along the heat flux, respectively. A more detailed approach of the effect of different parameters on boiling is presented in the next sections, starting with a description of the methods used in the characterization of nanofluids.

12.4 Characterization of Nanofluids

Nanofluids are suspensions of nanoparticles (particles that have at least one dimension <100 nm) in common base liquids such as water, ethylene glycol, thermal and engine oil, *etc.* These fluids are known to exhibit improved thermophysical properties in comparison to their base fluids, such as thermal conductivity, thermal diffusivity, and heat transfer coefficient.

Regarding the usage of nanofluids in pool boiling, the most relevant characterization includes properties such as thermal conductivity, viscosity, and stability.

Thermal conductivity (TC) is a key property regarding the thermal energy process. It is closely related to the efficiency of the system, associated with the cooling/heating performance. It is well known that nanofluids have improved thermal conductivities attributed to the much higher thermal conductivity of the nanoparticles. For instance, the addition of only 5 vol% of TiO_2 nanoparticles has been shown to increase the thermal conductivity of deionized water by 33% at room temperature.[43] Furthermore, dispersion of 5 vol% of Al_2O_3 in water resulted in an improvement in thermal conductivity up to 29%,[44] and the same volume concentration of CuO up to 60%.[45]

In general, the parameters that mainly influence the thermal conductivity of nanofluids are the nature of the nanoparticles and the base fluid, concentration, and temperature. The type of nanoparticles is very important to the extent of the improvement in thermal conductivity, but the choice of the base fluid is also worthy of attention. The most used base fluids for the preparation of nanofluids are water (TC $\approx$ 0.60 W m^{-1} K^{-1})[46] and ethylene glycol (TC $\approx$ 0.25 W m^{-1} K^{-1}),[47] but engine and thermal oil and some refrigerants have also been investigated. Aside from that, the concentration of nanoparticles plays a significant role in thermal conductivity, with higher nanoparticle concentrations usually leading to higher thermal conductivity. Figure 12.6 presents the experimental data of the effect of nanoparticle concentration on the thermal conductivity of different nanofluids reported in the literature. Regarding temperature, it has been reported to increase the thermal conductivity of nanofluids.[37,48–50]

The viscosity of nanofluids is a rheological property that indicates the resistance of the fluid and is related to the energy demanded in the process. Higher viscosity indicates more pumping and mixing power required, and

Table 12.1 Summary of boiling heat transfer studies.

Researchers	Type of boiling	Heater	Nanofluids	Key findings
Sarafraz *et al.*[19]	External flow boiling	Vertical stainless steel cylinder	CuO–water	(1) Deterioration of the HTC of the nanofluid was reported when the concentration of NPs increased. (2) HTC decreased with time due to deposition of NPs on the heater surface.
Raveshi *et al.*[20]	Pool boiling	Copper block	Al_2O_3–water/EG	(1) HTC improved with increasing NP concentration of up to 0.75 vol% and then deteriorated. (2) HTC enhanced by 64%.
Amiri *et al.*[21]	Pool boiling	Stainless steel cylinder	MWCNT functionalized (covalent and non-covalent)-water	(1) HTC increased for covalent MWCNT nanofluids and decreased for non-covalent MWCNT nanofluids. (2) CHF increased for all nanofluids. (3) As the size of the NPs increased, HTC and CHF decreased due to lower surface area.
Peng *et al.*[22]	Pool boiling	Copper block	Cu-refrigerant R113.	(1) Addition of surfactant (SDS, CTAB and Span-80) increased the HTC to a maximum concentration. (2) HTC enhanced by 67% at an optimal SDS concentration of 2000 ppm (3) HTC enhanced with a decrease in NP concentration.
Lee *et al.*[23]	Internal flow boiling	Stainless steel tube	Graphene oxide–water	(1) CHF enhanced up to 100%. (2) Deposition of NPs during boiling tests reduced the contact angle in the inner surface and enhanced the wettability.
Sarafraz *et al.*[24]	Pool boiling	Stainless steel cylinder	CuO–water	(1) HTC reduced for nanofluids without surfactant, but increased for nanofluids with surfactant. (2) Roughness of the surface depends on the NP concentration (deposition on the surface). (3) Boiling contact angle decreased by increasing nanofluid concentration.
Shahmoradi *et al.*[25]	Pool boiling	Flat plate	Al_2O_3–water	(1) Augmentation of CHF with increasing NP concentrations was reported. (2) HTC decreased with NP concentration except for a very low concentration (0.001 vol%).
Kole *et al.*[26]	Pool boiling	Cylindrical polished copper/ uncoated constantan wire	ZnO–EG	(1) HTC enhanced up to 22% when NP concentration increased to 1.6 vol%. Higher concentrations resulted in deterioration of the HTC. (2) CHF increased to 117% with a NP loading of 2.6 vol%.

Song *et al.*[27]	Pool boiling	Stainless steel plate	SiC–water	(1) CHF ameliorated 105% for 0.01 vol% of NPs, attributed to the better wettability promoted by NP deposition on the heater surface. (2) CHF varied non-linearly with SiC concentration.
Heris[28]	Pool boiling	Cylindrical cartridge	CuO–water/EG	(1) HTC enhanced with CuO concentration. (2) HTC ameliorated 55% with 0.5 wt% of CuO.
Kim *et al.*[29]	Internal flow boiling	Bare round tube	Al_2O_3–water	(1) CHF improved by 80% due to NP deposition on the tube. (2) CHF increased with increasing mass flow. (3) NPs concentration had no effect on CHF.
Wen *et al.*[30]	Pool boiling	Brass plate	Al_2O_3–water	(1) Boiling heat transfer is related to NP size and the geometry of the heater surface. (2) On a smooth surface the roughness increased due to NP deposition and the HTC increased. On a rough surface the surface geometry did not change and HTC remained the same.
Tang *et al.*[31]	Pool boiling	Horizontal flat square copper surface	Al_2O_3-R141b	(1) Change in the surface angle had no significant effect. (2) Addition of NPs enhanced HTC in most cases.
Xu *et al.*[32]	Internal flow boiling	Single-microchannel	Al_2O_3–water	(1) NP deposition was not observed. (2) HTC enhanced with nanofluid.
Shoghl *et al.*[33]	Pool boiling	Rod heater	ZnO–water and CuO–water	(1) Presence of surfactant increased HTC for all nanofluid concentrations, but HTC was lower than water without surfactant addition. (2) Boiling of nanofluids (without surfactant) made the surface smoother.
Li *et al.*[34]	Pool boiling	Copper rod	CuO–water	(1) Different modified surfaces were tested, and the presence of fins were found to improve HTC. (2) Highest HTC was reported for the surface with the largest number of fins and the smallest space between the fins. (3) Increasing in NP concentration enhanced HTC.
Okawa *et al.*[35]	Pool boiling	Copper block	TiO_2–water	(1) CHF improved by 90%. (2) The lower the contact angle, the higher the CHF.
Ahmed *et al.*[36]	Pool boiling	Horizontal flat copper surface	Al_2O_3–water	(1) Only the lowest NP concentration enhanced HTC, attributed to the effect of thermal conductivity of nanofluids being more dominant than the effect of NP deposition on the surface.

Table 12.1 (*Continued*)

Researchers	Type of boiling	Heater	Nanofluids	Key findings
				(2) Effect of NP deposition was evaluated by running only water on the NP-deposited surfaces and an ameliorated HTC was registered in relation to nanofluid running.
Ali *et al.*[37]	Pool boiling	Copper block	TiO_2–water	(1) HTC improved up to 150% with a NP concentration of 15 wt%.
Lee *et al.*[38]	Internal flow boiling	SS316 tube	Fe_3O_4–water	(1) CHF enhanced ($\approx$57%) with the addition of NPs, ascribed to the deposition of NPs on the surface, which improved wettability and re-wetting.
Norouzipour *et al.*[39]	Pool boiling	Copper block	SiO_2–water	(1) Under almost all the conditions tested the HTCs of nanofluids were lower than those of pure water, attributed to the filling of boiling surface cavities. (2) The effect of NP size on HCT was evaluated, and the higher the NP diameter, the fewer active nucleation sites become occupied by NPs, resulting in higher HCT.
Yagnem *et al.*[40]	Pool boiling	Copper block	Al_2O_3–water, CuO–water and Al_2O_3 + CuO–water	(1) The CHF of hybrid nanofluids was higher than for a single-particle nanofluid. (2) Maximum enhancement in the CHF was 49.8% for the highest hybrid nanofluid concentration of 0.1 vol%. (3) HTC was dependent of the concentration of NPs in hybrid nanofluids and increased for 0.01 and 0.03 vol% and then reduced for higher concentrations.
Kamel *et al.*[41]	Pool boiling	Horizontal copper tube	CeO_2–water	(1) HTC increased with the addition of NPs in relation to pure water up to 70%. (2) HTC increased from concentrations of 0.001–0.007 (maximum) vol% and then decreased until 0.04 vol%. (3) HTC was associated with the roughness of the surface and nanofluids properties.

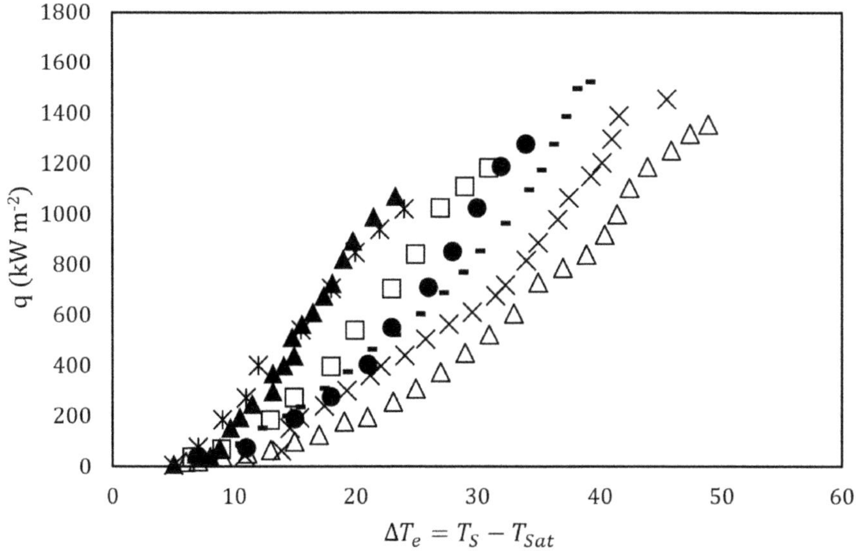

Figure 12.4 Boiling curves from different research studies: * water,[42] ▲ Al_2O_3–water (0.001 vol%),[25] □ Al_2O_3–water (0.005 vol%),[42] ● Al_2O_3–water (0.01 vol%),[42] – CNT–water (0.01 vol%),[21] × Al_2O_3–water (0.01 vol%),[25] and △ CNT–water (0.1 vol%).[21]

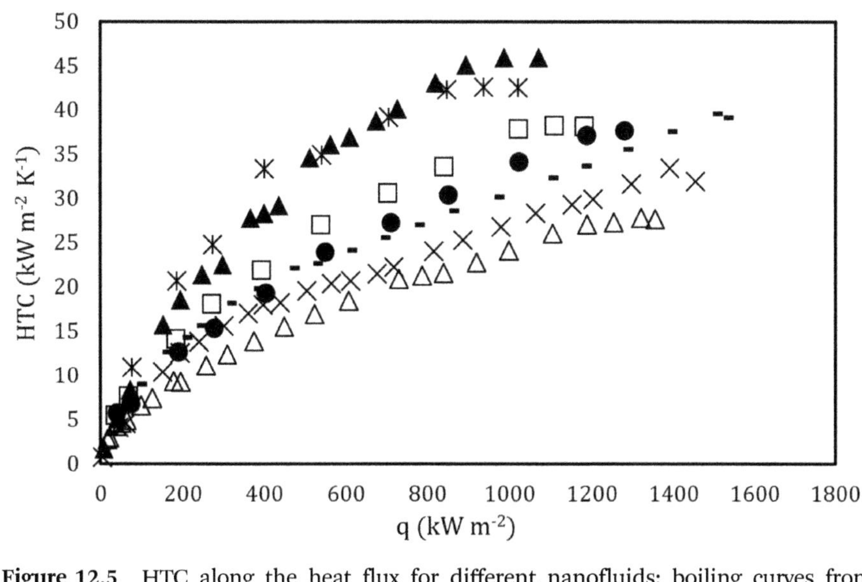

Figure 12.5 HTC along the heat flux for different nanofluids: boiling curves from different research studies: * water,[42] ▲ Al_2O_3–water (0.001 vol%),[25] □ Al_2O_3–water (0.005 vol%),[42] ● Al_2O_3–water (0.01 vol%),[42] – CNT–water (0.01 vol%),[21] × Al_2O_3–water (0.01 vol%),[25] and △ CNT–water (0.1 vol%).[21]

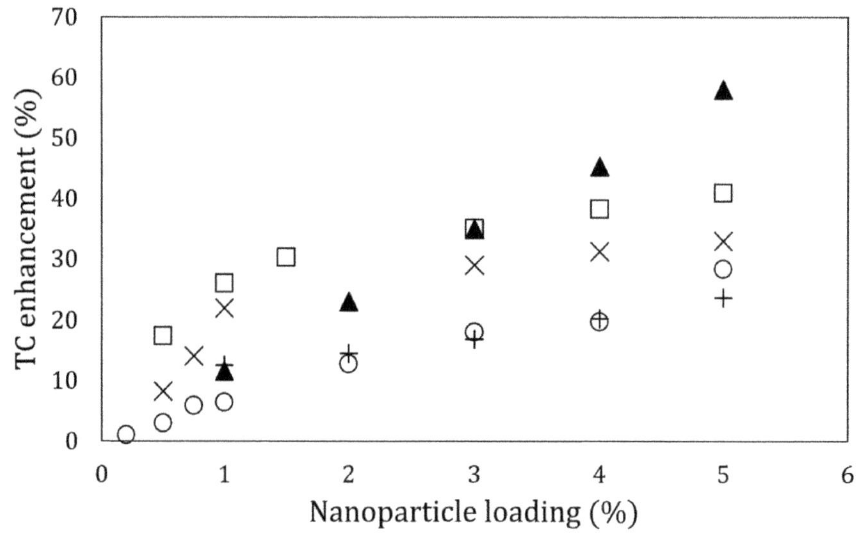

Figure 12.6 Effect of nanoparticle volume concentration on the thermal conductivity of nanofluids: ▲ CuO–water,[45] □ Fe_3O_4–water,[51] × TiO_2–water,[43] + Al_2O_3–water,[44] and ○ Al_2O_3–water.[52]

hence the heat transfer process becomes more expensive and less effective. Therefore, it is of great importance to examine the effect of the dispersion of nanoparticles on the viscosity of liquids. In this regard, the effective viscosity of suspensions has been reported to increase with the increasing nanoparticle volume concentration of the base fluids.[46,53–55] In relation to temperature, its effect on viscosity is acknowledged as being very relevant and can be even more remarkable than the effect of nanoparticle concentration.[48,56] As is shown in Figure 12.7, viscosity reduces with an increase in temperature. This reduction may be attributed to the increase in intermolecular distance at higher temperatures.[50] In addition, the effect of temperature is usually more pronounced for higher nanoparticle loadings.[48,57]

Maintaining a stable dispersion for a long time period is one of the biggest challenges when working with nanofluids. Nanoparticles interact between each other and have a high tendency to agglomerate, with subsequent sedimentation. Nanoparticle settling reduces the thermal conductivity of nanofluids and may clog channels or pumps.[50] Some approaches have been reported to increase nanofluid stability, such as by changing the pH of the medium, adding surfactants or increasing dispersion times.[50,60]

12.5 Experimental Investigations of the Pool Boiling of Nanofluids

12.5.1 Pool Boiling Apparatus

Generally, the pool boiling experimental setup comprises a test section, an auxiliary heater to maintain the fluid in its saturation temperature, a

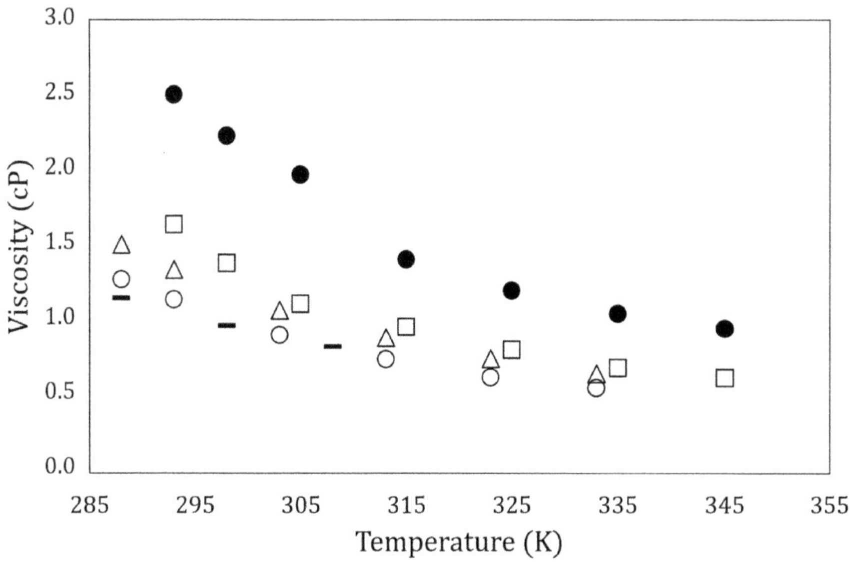

Figure 12.7 Effect of temperature on the viscosity of different nanofluids: ○ TiO_2–water (1.0 vol%),[58] △ TiO_2-water (4.0 vol%),[58] □ Al_2O_3-water (1.3 vol%),[59] ● Al_2O_3-water (2.8 vol%),[59] and – Al_2O_3-water (1.0 vol%).[46]

condenser, a power supply, and a data acquisition system (DAQ). The test section can basically be a chamber made from glass or another translucent material to facilitate bubble visualization, which contains the boiling heating surface and a temperature sensor to measure the liquid temperature. The heating element can vary in shape and size, and it can be a wire, a ribbon, a horizontal cylinder, or a copper block, *etc.* Figures 12.8 and 12.9 show examples of the two most used experimental set-ups that use a wire heater (Figure 12.9) and copper block (Figure 12.10) as heating surfaces.

As shown in Figure 12.8, the heating wire is horizontally immersed in a pool of nanofluid, attached to lead wires. There are many options of materials that can be used, but wires with the highest melting points are preferable, with Ni–Cr alloy being the most applied.[3] Small step changes in voltage are provided to the wire from the power supply and measurements of current and voltage are recorded by a data acquisition system. The temperature of the heating surface is obtained from the relationship between the resistance of wire (determined from the current and voltage) and temperature, calculated from the data sheet of the supplier or *via* a calibration curve. The resistance of the lead wires should be accounted for, and the resistance of the heating wire is the total resistance less the resistance of the lead wires. The boiling curve is a plot of the heat flux (q'') *versus* the wall superheat ($T_s - T_{sat}$), where T_s is the wire heater temperature and T_{sat} is the saturation temperature of the nanofluid, measured by a thermocouple immersed in solution. The heat flux is calculated as follows:

$$q'' = \frac{Vi}{\pi dL} \tag{12.10}$$

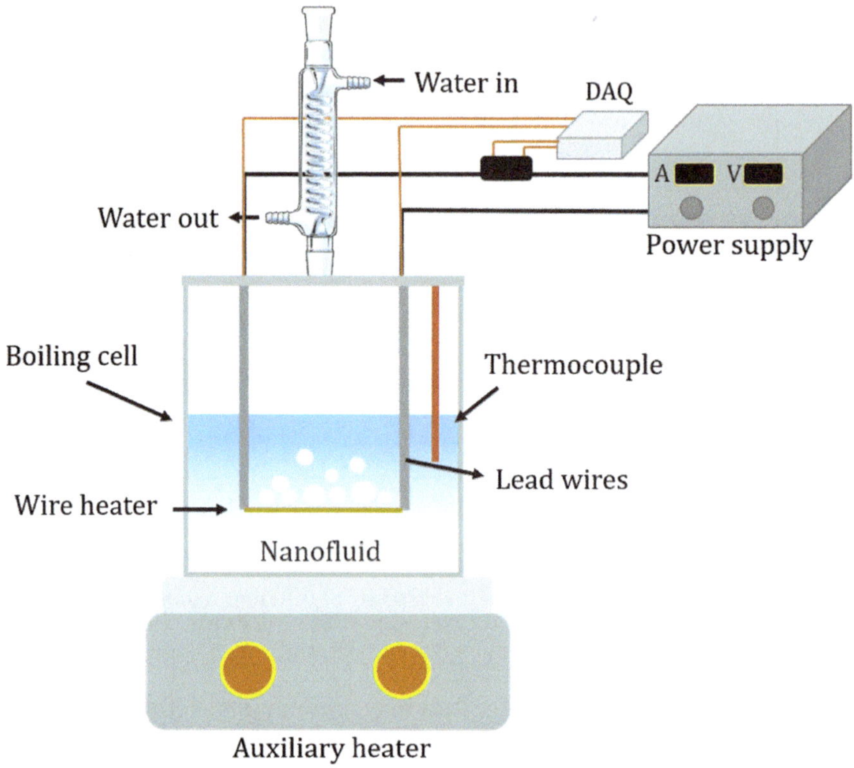

Figure 12.8 Experimental set-up using a heater wire as a boiling heating surface.

where V is the voltage, i is the current, d is the wire diameter and L is the wire length.

The configuration shown in Figure 12.9 is slightly different. The heating surface is a copper block, placed at the bottom of the boiling cell, which contains a set of heaters inserted into it. The heaters are connected to a power supply that provides the voltage step change and current, similar to as mentioned above. Thermocouples are distributed in several positions along the copper block to determine the temperature of the heating surface. Assuming one-dimensional and steady state heat conduction, the heat flux can be calculated using the Fourier law as follows:[20]

$$q = kA\left(\frac{T_1 - T_2}{\Delta x}\right) \tag{12.11}$$

$$q'' = \frac{q}{A'} \tag{12.12}$$

where T_1 and T_2 are temperatures measured by two thermocouples placed along the copper block, Δx is the distance between them, A is the cross-section area of the copper block and A' is the boiling surface area.

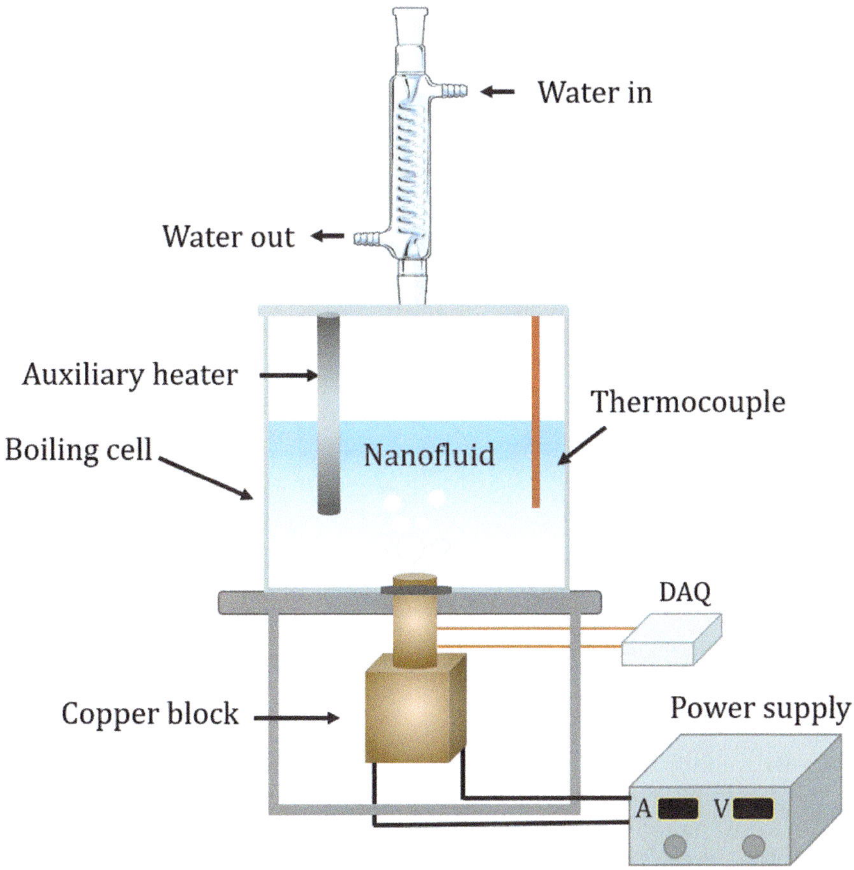

Figure 12.9 Experimental set-up using a copper block as a boiling heating surface.

The wall superheat can be determined using:[20]

$$\Delta T_e = T_s - T_{Sat} = \left(T_0 - \frac{q''d}{k}\right) - T_{Sat} \tag{12.13}$$

where T_0 is the temperature measured by the thermocouple nearest to the boiling surface and d is the distance between the thermocouple and the surface.

12.5.2 Effect of the Roughness of the Heater Surface and Nanoparticle Concentration

Surface modification methods are applied with the main goals of increasing surface area and/or surface porosity. Extending the surface area obviously leads to increased boiling heat transfer, while enhancing porosity increases the number of active nucleation sites that lead to the formation and departure of more bubbles. Surface roughness is not only changed by

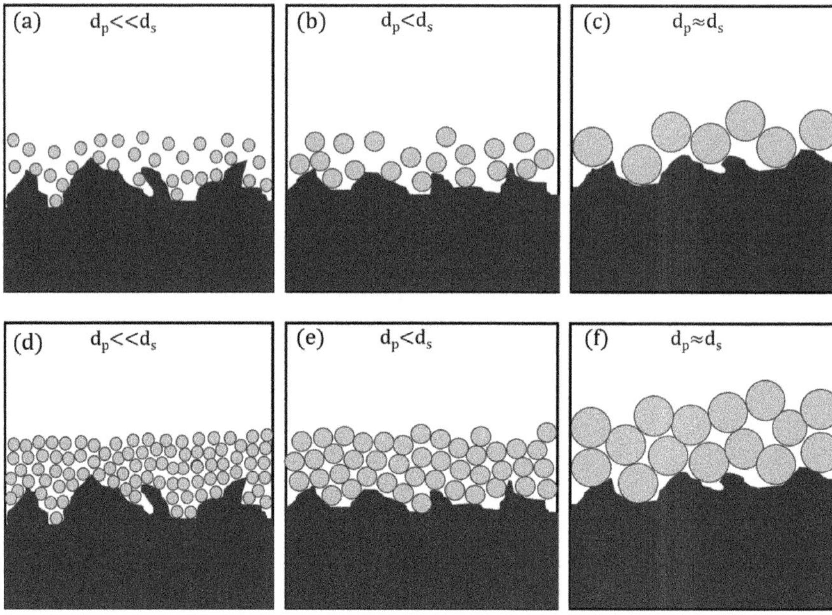

Figure 12.10 Effect of the relationship between particle diameter (d_p) and the diameter of the surface cavities (d_s) on the boiling characteristics of nanofluids.

intentional modification of surface topography or coating techniques. In fact, the most important mechanism responsible for improving boiling heat transfer using nanofluids is the deposition of nanoparticles on a heating surface. The layer formed by nanoparticles at the heating surface has a complex impact on boiling behavior, influencing the number of active nucleation sites, bubble diameter and frequency of departure. Hence, HTC and CHF may be highly dependent on nanoparticle size and concentration.

The roughness of a surface can be calculated by the deviations in the direction of the normal vector of a real surface from its ideal form. Usually, the parameters used for its definition are the average deviation of the workpiece from the centerline, Ra and the root mean square average of the profile deviation Rq, given by:

$$\mathrm{Ra} = \frac{1}{L_\mathrm{p}} \int_0^L |Z(x)| \mathrm{d}(x) \tag{12.14}$$

$$\mathrm{Rq} = \sqrt{\frac{1}{L_\mathrm{p}} \int_0^L Z(x)^2 \mathrm{d}(x)} \tag{12.15}$$

where $Z(x)$ represents the coordinates of the roughness profile and L_p is the maximum longitudinal distance from one peak to the valley on the distribution.

Rostamian and Etesami[61] evaluated the variation in the roughness on a circular flat plate during the boiling of SiO_2–water nanofluids. They tested three different concentrations of nanoparticles (0.005, 0.007 and 0.01 vol%) and noticed that as the boiling proceeded the roughness of the heating surface increased. Analysis of the HTC showed that it increased up to an optimum NP concentration of 0.007 vol% and then deteriorated. This enhancement at low concentrations was attributed to the enhancement in the active nucleation sites due to the deposition of nanoparticles, which make the nucleation sites smaller in size (multiplying their number). As the concentration increased, the layer of nanoparticles sedimented on the surface was enlarged and the thermal resistance was higher, thus decreasing the HTC. CHF increased for all concentrations, ascribed to the enhancement in the wettability of the surface after nanoparticle deposition, which delayed the appearance of dry spots and overheating of the boiling surface.[61]

Dareh *et al.*[42] investigated the effect of surface roughness on the pool boiling of alumina nanofluids using polished and machined copper surfaces. For the polished surface, the addition of nanoparticles decreased boiling HTC in comparison to the pure base fluid (water), and the HTC became worse for higher nanoparticle concentrations. For the machined surface, the HTC enhanced for all the nanofluids tested, and the higher the nanoparticle concentration, the higher the HTC. These outcomes can be explained by the fact that the roughness of the machined surface was greater than the size of the nanoparticles, and their settling increased the number of nucleation sites, while the polished surface featured cavities with similar dimensions to those of the nanoparticles, resulting in the clogging of the nucleation sites by nanoparticle deposition impairing bubble formation. CHF enhanced for all the nanofluids investigated in comparison with pure water, and it was always greater for the polished surface than the machined surface.[42]

Das *et al.*[62] also studied the pool boiling of Al_2O_3–water (0.1–4 vol%) nanofluids. Their experiments were carried out using a cartridge heater with a surface roughness varying from $Ra = 0.387 \mu m$ (smoother heater) to $Ra = 1.15$ μm (roughened heater). Their findings were different to those of Dareh *et al.*,[42] as although the HTC deteriorated for the smoother surface similar to in the study of Dareh *et al.*,[42] for the rough surface the HTC was lower than for the pure water results. In addition, boiling performance dropped with an increase in nanoparticle concentration. The rough surface presented better HTC than the smooth surface. The authors suggested that the reason behind the reduction in HTC was the change in the surface characteristics promoted by nanoparticle sedimentation. This was attributed to trapped nanoparticles, which had a diameter that was one to two orders of magnitude lower than the disturbed boiling heat transfer of the surface roughness tested.[62]

These two works mentioned above showed that the impact of nanoparticle concentration on HTC is directly related to the ratio between the dimensions of nanoparticles and the roughness of the surfaces. Therefore, the different conclusions of the authors in fact are not contradictory. The concentration of nanoparticles together with nanoparticle size play an essential role in understanding boiling behaviour. Indeed, the ratio between the size of the nanoparticles and the cavities of the heating surface dictates whether HTC improves or deteriorates, which can be better comprehended by examining Figure 12.10. At low concentrations (Figure 12.10(a)–(c)), if nanoparticles are much smaller than the cavities (Figure 12.10(a)), their deposition enhances the density of nucleation sites and heat transfer improves. As the diameter of nanoparticles increases (Figure 12.10(b)), the density of active nucleation sites reduces up to the point that the nanoparticles have the same order of magnitude as the cavities on the heating surface (Figure 12.10(c)) promoting their clogging and impairing of bubble formation. However, at high concentrations (Figure 12.10(d)–(f)), more nanoparticles might settle, forming a thicker layer that closes many nucleating sites and increases thermal resistance, leading to lower HTC, which may explain the results reported by Das *et al.*[62]

The work of Wen *et al.*[30] is in agreement with the aforementioned phenomena. They investigated the effect of surface morphology on the boiling of alumina nanofluids at low concentrations of 0.001, 0.01 and 0.1 vol%. Analyzing the surfaces after boiling they noticed that the smooth surface increased its roughness by over two-fold, while the rough surface remained practically the same. The roughness of the surface at the end of boiling was reflected in the HTC, which enhanced for the smooth surface but did not exhibit significant change for the rough surface, showing that the modification of the surface may play an important role in boiling efficiency. The authors concluded that the divergency of the HTC results in the literature is due to the different surface modification that occurs during the boiling process, which is dependent on the relationship between the nanoparticles dispersed in the nanofluids and the morphology of the surface, besides the thermophysical properties of the nanoparticles and the heating surface.[30]

The effect of concentration of nanoparticles on HTC is shown in Figure 12.11, which presents the results of Al_2O_3–water-based[36] and TiO_2–water-based[63] nanofluids for a copper heating surface. Ahmed and Hamed[36] investigated the effect of concentration of alumina–water-based nanofluids (0.01, 0.1 and 0.5 vol%) on the on pool boiling *via* observation of the deposition and roughness of the surface. The HTC increased for an Al_2O_3 concentration of 0.01 vol%, similar to the HTC of water at a concentration of 0.1 vol%, and deteriorated at 0.5 vol%. The enhancement in HTC could be attributed to the effect of thermal conductivity of the nanoparticles and their settling on the copper block surface. In addition, the drop in the HTC may be due to an insulation effect caused by the deposition of nanoparticles, which prevented liquid from making contact with the surface and deactivated some nucleation sites. Interestingly, the authors reported that the uniformity of the coating of the nanoparticles on the heating surface was dependent on the concentration of the nanofluid.[36] Suriyawong and Wongwises[63] observed a similar pattern of HTC with nanofluid concentration. For a very low

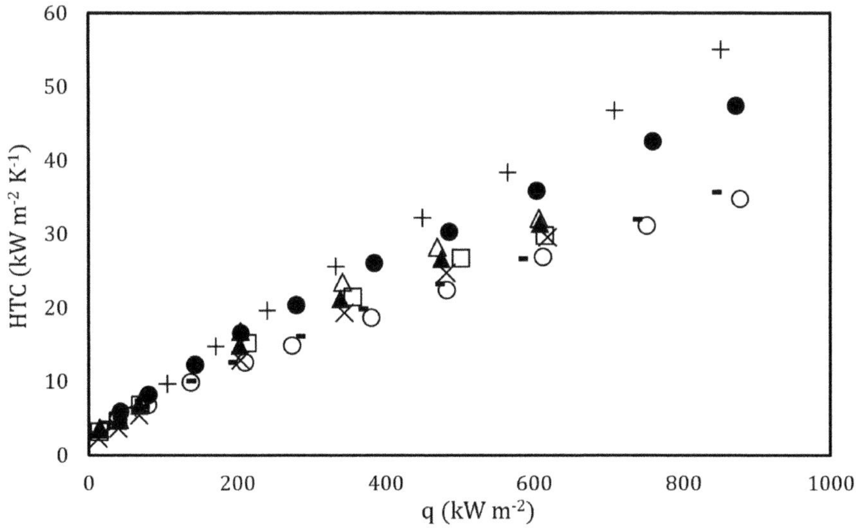

Figure 12.11 HTC of nanofluids: ● water,[63] +TiO_2–water (0.0001 vol%),[63] – TiO_2–water (0.005 vol%),[63] ○ TiO_2–water (0.01 vol%),[63] ▲ water,[36] △ Al_2O_3–water (0.01 vol%),[36] □ Al_2O_3–water (0.1 vol%),[36] and × Al_2O_3–water (0.5 vol%).[36]

concentration of TiO_2 of 0.0001 vol%, the HTC was enhanced by around 15% compared with the base fluid, but with higher concentrations the HTC observed was worse than in water. The general conclusion from these results is that nanofluids improve boiling performance at very low concentrations, while for higher concentrations they seem to deteriorate the heat transfer and the advantages of higher thermal conductivity are overshadowed by their impairing of liquid contact and deactivation of nucleation sites.

Moreover, all the concentrations of carbon black (CB) investigated by Sezer *et al.*[64] improved HTC in comparison with pure water, as presented in Figure 12.12. Heat transfer performance was assessed using a copper block as a boiling heater. Nanofluids composed from deionized water and carbon black (CB) at concentrations of 0.001, 0.005 and 0.01 wt% were tested. HTC and CHF enhanced with increasing nanoparticle concentration by around 190% and 68%, respectively, for 0.01 wt% of CB. Better HTCs were ascribed to the enlarged surface area and roughness promoted by nanoparticle coating, together with a greater active nucleation site density. Indeed, scanning electron microscope (SEM) images showed that a layer of many mesopores was formed upon the deposition of nanoparticles on the heating surface. The size of the cavities in this layer was smaller in nanofluids boiling with higher concentrations, generating smaller bubbles with higher departure frequency. This mesoporous layer significantly increased the roughness of the surface, up to approximately 9-fold. CHF enhancement was attributed to the high thermal conductivity of the nanoparticles deposited on the heating surface, which improved heat transfer and delayed the formation of dry spots.[64]

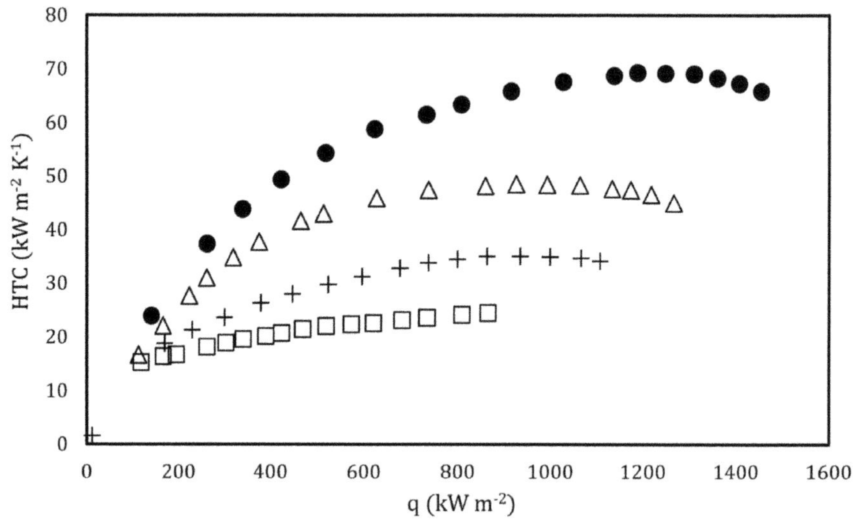

Figure 12.12 HTC of nanofluids: □ water,[64] + CB–water (0.001 vol%),[64] △ CB–water (0.005 vol%),[64] and ● CB–water (0.01 vol%).[64]

In a study by Shahmoradi *et al.*,[25] boiling experiments were conducted using water-based nanofluids containing Al_2O_3 (0.001–0.1 vol%). Except for the concentration of 0.001 vol%, HTC deteriorated with nanoparticle loading. The roughness of the surface was much lower than the nanoparticle diameter, which blocked some cavities on the heating surface and increased thermal resistance. In contrast, the CHF increased with increasing nanofluid concentration. The layer formed by nanoparticles being deposited enhanced wettability and in turn delayed the CHF appearance.[25]

The effect of surface geometry on the boiling of CuO–water nanofluids was assessed using different modified heating surfaces with parallel fins, varying the spacing between the fins, their width and height.[34] As expected, it was found that that the surface with the smallest space (highest number of fins) exhibited the highest HTC due to the highest surface area being available for heat transfer and improved frequency of bubble detachment. More interestingly, the authors evaluated the fouling thermal resistance with time on both plain and finned surfaces using 0.1 wt% CuO–water nanofluids. They reported that at the beginning of the boiling process the deposition of nanoparticles on the heating surface increased HTC *via* the creation of some new active nucleation sites and there was no fouling thermal resistance, but as time went by, the fouling thermal resistance increased and the HTC dropped, up to a point where an equilibrium state was established. Both surfaces showed the same pattern, but in the presence of fins the fouling thermal resistance was lower and so was the reduction in the HTC.[34]

12.5.3 Effect of Adding Surfactant to Nanofluids

The impact of surfactant addition to nanofluids for boiling heat transfer applications has been intensively studied.[15,22,24,33] The use of surfactants enhances the dispersion and stability of nanoparticles in solution, which may have an effect on boiling efficiency, moreover, it alters the thermophysical properties of nanofluids, such as viscosity and surface tension. The addition of surfactants reduces the surface tension, which directly influences the bubble dynamics. The departure of bubbles formed on the heating surface only occurs when the buoyancy and surface tension are equalized. If the surface tension is low, the frequency of bubbles leaving the surface increases and the heat transfer improves.[33]

The effect of surfactant on the boiling heat transfer of CuO and ZnO water-based nanofluids was investigated using sodium dodecyl sulfate (SDS).[33] Low concentrations of nanoparticles were used (0.01 and 0.02 wt%) combined with low concentrations of surfactant (0.01–0.04 wt%). The results were compared with nanofluids without surfactant and pure water. A negative impact of nanofluids on boiling performance was observed without surfactant, related to the smoother surface at the end of the process. Mixing SDS with the nanofluids significantly reduced their surface tension, which led to an increase in HTC. The diameter of the bubbles was smaller after surfactant addition and the frequency of bubble detachment increased. The maximum HTC was verified at a SDS concentration of 0.02 wt%; above this concentration a layer of bubbles was formed, insulating the heating surface and deteriorating the HTC.[33]

Peng *et al.*[22] investigated the effects of three different types of surfactant (SDS, cetyl-trimethyl ammonium bromide (CTAB) and sorbitan monooleate (Span-80)) on the nucleate pool boiling of refrigerant-based nanofluids (Cu-R113). Experimental assays were performed on a horizontal surface, using nanofluids with a concentration in the range of 0.1–1 wt% and a surfactant concentration of 200–5000 ppm (by weight). Overall, the presence of surfactant improved boiling heat transfer, but at high concentrations it had the opposite effect. A maximum enhancement of HTC was achieved for concentrations of 2000, 500 and 1000 ppm for SDS, CTAB and Span-80, respectively, improving the HTC by 29–67%. The lower HTC values at high surfactant doses can be explained by the increase in viscosity reducing the micro-convection that occurs near to the heating surface. The best surfactant was in the order of SDS > CTAB > Spam-80. Without surfactant, the increase in nanoparticle concentration increased the HTC, but in the presence of surfactant (keeping its concentration constant) the increase in nanoparticle concentration decreased the HTC.

Etedali *et al.*[65] also evaluated the influence of several surfactants on the pool boiling of SiO_2–water nanofluids on a copper surface. Polysorbate 20 (PS20), CTAB and sodium lauryl sulfate (SLS) were used at a concentration ratio of 1/20 with the nanoparticle mass in the nanofluids. Improved HTCs were obtained for water–surfactant compared to pure water and SiO_2–water with or without surfactant. The addition of surfactants has been reported to reduce surface tension, favouring the departure of bubbles from the heating surface, hence improving heat transfer. In nanofluids containing CTAB and

SLS, by increasing the nanoparticle concentration the HTC decreased. For nanofluids with PS20, the increase in SiO_2 concentration from 0.01–0.1 vol% improved boiling heat transfer, but a further increase (up to 1 vol%) resulted in a significant reduction in the HTC. The ability of the surfactants tested to improve the HTC was in the order of SLS > CTAB > PS20, which are anionic, cationic and non-ionic, respectively.[65]

The impact of a dispersant agent on the boiling of CuO–water nanofluid performance was assessed using 0.1 wt% of SDS, sodium dodecyl benzene sulfonate (SDBS) or Triton X-100.[24] Experiments were conducted using nanofluids with concentrations in the range of 0.1–0.4 wt%. It was revealed that the HTC deteriorated for nanofluids without surfactant in comparison with base fluid, however, when surfactants were added the HTC was better than water. Upon the addition of surfactant, a higher number of active nucleation sites and the formation of smaller bubbles were clearly observed.[24]

In relation to the CHF, as for HTC, the CHF increases with surfactant addition up to a maximum concentration.[15] Murshed *et al.*[15] investigated the effect of sodium dodecyl benzene-sulfonate (NaDBS) on the boiling performance of water–carbon nanotubes (CNTs). From a surfactant concentration of 1 : 20 (NaDBS : CNTs) to 1 : 5 (NaDBS : CNTs) the CHF improved by almost 500% compared with the base fluid, but above this concentration it drastically dropped to less than 1/3 of its values. The enhancement in the CHF was attributed to the deposition of nanoparticles on the heating surface. The burnout heat flux (BHF) was also determined using nanofluids with a CNT concentration of 0.1 vol% and varying the surfactant concentration. At any concentration, the surface tension of the nanofluid always reduced in the presence of surfactant. The authors obtained a non-linear correlation that showed the strong relationship between the BHF point and the surface tension, with BHF being higher for lower surface tension values. In addition, similar to the CHF, the BHF showed maximum enhancement (265%) at an intermediate surfactant concentration of 1 : 5 (NaDBS : CNTs).[15]

12.5.4 Effect of Contact Angle and Surface Wettability

Boiling heat transfer is primarily influenced by bubble dynamics, which are related to bubble growth, or diameter, and frequency of departure. In the case of nanofluids, when the bubbles leave the surface, the nanoparticles are accumulated on the heating surface, which means that the main mechanism of nanoparticle deposition is the evaporation of the bubbles and not gravitational force. At a high concentration of nanoparticle sedimentation on a solid surface, the surface becomes smoother, and consequently the contact angle of the bubbles and heating surface decreases, leading to a reduction in the number of active nucleation sites. This relationship between contact angle (θ) and nucleation site density (n'') was proposed by Wang and Dhir:[66]

$$n'' \propto N_c(1 - \cos\theta)(T_S - T_{Sat})^6 \tag{12.16}$$

where N_c is the number of microcavities per unit of surface area.

Furthermore, at a low contact angle, the adhesion energy, defined as the horizontal component of the surface tension acting against the bubble growth, increases, resulting in a longer time between bubble formation and detachment, or lower frequency of departure. The reduction in the number of active sites added to the lower bubble departure frequency negatively affects the boiling HTC. Moreover, the HTC can improve due to the suction of liquid caused by bubble departure, referred to as the capillary wicking effect.[2,67]

In contrast, the CHF is inversely dependent on the contact angle, therefore it increases when the contact angle reduces. Nanoparticle deposition on the heating surface creates a porous layer that traps liquid and impairs bubble generation and the formation of a vapor layer on the surface. Hence, the onset of nucleate boiling is delayed, and the CHF is enhanced.[2] It is important to mention that the enhancement in the CHF has been reported to be a consequence of surface modification *via* the deposition of nanoparticles, which improves wettability, rather than the thermophysical properties of the nanofluids themselves. For that reason, when the surface remains unchanged the CHF does not alter.[2] A study by Coursey and Kim corroborated the abovementioned phenomena.[68] The authors showed that the addition of nanoparticles was responsible for CHF improvement (improved wettability) only when the surface was fouled. Moreover, they conducted boiling experiments using pure base fluids but with a surface with similar wettability and observed similar CHF enhancement.[68]

The wettability effect can be associated with the relationship between the contact angle, surface tension (σ_{LV}), adhesion tension ($\sigma_{SV} - \sigma_{SL}$), and roughness using the modified Young's equation:[2,69]

$$\cos\theta = \frac{\sigma_{SV} - \sigma_{SL}}{\sigma_{LV}} r \tag{12.17}$$

where r is the roughness factor, defined by the ratio of the effective contact area by the smooth contact area having the same geometric shape and sizes. Wettability enhances when the adhesion tension increases, therefore decreasing the contact angle.[2,69] Figure 12.13 illustrates the effects of contact angle and adhesion tension on bubble formation.

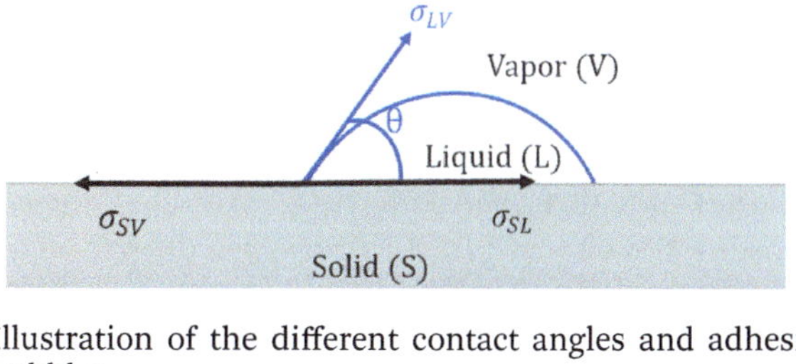

Figure 12.13 Illustration of the different contact angles and adhesion tension on bubbles.

Kim *et al.*[70] performed boiling experiments using different water-based nanofluids composed of Al_2O_3 or ZrO_2 or SiO_2, at concentrations of 0.001–0.01 vol%. They assessed the wettability changes by measuring the contact angles of sessile droplets. They observed a nanoparticle layer on the heating surface after nanofluid boiling, which resulted in better wettability (or lower contact angle). They concluded that this enhancement was a result of the combination of the augmentation of the adhesion tension together with an increase in surface roughness. With an improvement in wettability, the dry patch on the surface was rewetted and CHF was delayed. Concerning the HTC, a decrease in contact angle led to a decrease in the number of active sites, which reduced the bubble generation and deteriorated the HTC.[70]

In work by Sarafraz *et al.*[24] the contact angles of nanofluids boiling with different concentrations were examined. The authors reported that by increasing nanoparticle loading, the contact angles of the bubbles decreased, ascribed to more nanoparticles being deposited on the heating surface. The surface became smoother upon the coating of nanoparticles and the wettability was increased.[24]

Rostamian and Etesami[61] evaluated the bubble behaviour of SiO_2–water nanofluids throughout the boiling process. They showed that the contact angle of the bubbles decreased with time, which was consistent with an increase in the nucleation sites that occurred due to the motion of the bubbles as the heat flux increased.[61] Jeong *et al.*[71] observed a significant decrease in the contact angle for alumina nanofluids compared with those of water. This was attributed to the improvement in the hydrophilicity of the surface promoted by the nanoparticles and surfactant settling, enhancing the wettability of the boiling surface, which was interpreted as being the main mechanism of CHF enhancement. Liu *et al.* also related the increased CHF they obtained with the lower contact angle of nanofluids.[72]

12.6 Flow Boiling Experiments with Nanofluids

Although the industrial applications of nanofluids are mainly performed in flow boiling configurations, the number of studies applying nanofluids on convective boiling setups is much smaller than the ones on pool boiling. Flow boiling of nanofluids can be applied for systems with elevated energy density, being able to significantly increase boiling heat transfer performance and maintain a more uniform temperature distribution. Regarding published studies, more attention has been on internal flow boiling, therefore this will be the focus of the following discussion.

As in pool boiling experiments, the main goals in flow boiling are to enhance HTC and delay the occurrence of the boiling crisis. Nevertheless, the flow boiling of nanofluids faces the following main challenges: clogging of the deposition layers, flow instability and pressure drop.[5]

Patra *et al.*[73] conducted flow boiling experiments using Al_2O_3–water and SiO_2–water (concentration of 0.001 and 0.01 vol%) in a vertical quartz tube. HTC was enhanced for nanofluids with low concentrations (0.001 vol%), but was worse than the base fluid at higher nanofluid concentration (0.01 vol%). This was interpreted to be due to the increased thermal resistance caused by the sedimentation of nanoparticles on the heating wall. In addition, the dynamics of bubbles were different for nanofluids and pure water. At high heat fluxes, the coalescence of bubbles was more noticeable for water flow boiling than for nanofluids. Hence, water boiling exhibited a small number of larger bubbles and these bubbles grew to slug flow and then formed dry patches on the heating surface, conducting to the CHF. Moreover, in nanofluid boiling the bubbles remained much smaller, promoting better heat transfer and delaying the formation of dry spots. Moreover, compared to pure water, the detachment time of the bubbles was lower for all of the nanofluids studied, therefore so was the bubble detachment diameter.[73]

Water-based nanofluids with 0.2 wt% of Al_2O_3 were studied in a single microchannel.[32] The flow pattern of pure water was very unstable, with large oscillation amplitudes and random bubble coalescence. The authors reported that this flow instability was mitigated by the nanofluids, with no deposition on the surface observed. The behaviour of nanofluid flow was approximately stable, ascribed to the miniature bubbles generated, formed due to the lower surface tension of the nanofluids. Moreover, the pressure drop was considerably reduced, and HTC improved, as a consequence of the higher evaporation of the thin film liquid. The occurrence of dry patches was prevented by the structural disjoining pressure.[32]

Lee *et al.*[23] conducted an experiment to investigate the vertical upward flow boiling performance of nanofluids composed of graphene oxide and water at 0.01 vol%. The CHF of their nanofluids was improved by approximately 100%. The authors tested different mass fluxes and reported that CHF enhanced with mass flux, but enhancement was not linear due to different depositions of nanoparticles observed. Such deposition of nanoparticles on the inner surface of the tube occurred during nucleate boiling and increased the wettability of the heated surface, which was considered to be responsible for CHF enhancement. The contact angles were measured and, in fact, the contact angle after boiling of the nanofluids was smaller than after the use of water.[23]

Wang *et al.*[74] investigated the subcooled flow boiling of Al_2O_3–water and AlN–water nanofluids at concentrations of 0.1 and 0.5 vol%, with the objective of evaluating the effect of different parameters on CHF. Overall, the results showed an increase in CHF for all nanofluids in comparison with pure water, achieving a maximum enhancement of 18%. The greater CHF was explained by the deposition of nanoparticles on the heating wall, which created a nanolayer that was responsible for increasing the thickness and adhesion of the film liquid layer (better wettability). CHF increased by increasing the mass flux, heating diameter and outlet pressure and decreasing the heating length. In contrast, the effect of nanoparticle type and concentration on CHF was considered negligible. Nanoparticle type had no influence on the enhancement of CHF as they had equivalent mean diameters and their deposition features was verified to be similar. The cavities of the heating surface were much larger than the diameters of the nanoparticles, which deposited on the surface, filled the cavities, and enhanced the wettability. At higher concentrations, the deposition pattern remained the same, since the cavities were already closed, and higher concentrations led to no changes in the values of CHF.[74]

Experiments of subcooled flow boiling using nanofluids of CuO and water (0.5, 1.0 and 1.5 vol%) were performed using several mass fluxes.[19] Two boiling regimes were observed, forced convection and nucleate boiling. HTC deteriorated in all nanofluids studied in comparison to water, with the higher the concentration of CuO the worse the HTC. These outcomes were elucidated by the verification of the fouling resistance, which linearly decreased with time due to the precipitation of nanoparticles on the heating surface.[19]

In general, investigations on nanofluid flow boiling are still scarce, being more directed toward internal flow boiling using metal oxide nanoparticles.[5] Increasing the concentration of nanoparticles generally has a negative effect on HTC due to increased thermal resistance. CHF is almost always enhanced using nanofluids due to the improved wettability as a result of nanoparticle deposition. However, due to their fouling effect, it is more likely to result in clogging of the channels.

12.7 Conclusions

In this chapter, boiling heat transfer using nanofluids and the main parameters that influence their performance were presented and discussed. Regarding boiling efficiency, it can be concluded that HTC and CFH are dependent on the heating surface features and/or properties of the nanofluids. In this sense, the possibility and ease of altering the properties of nanofluids represent an appealing advantage of using nanofluids for boiling heat transfer. Nanoparticle deposition on the heating surface has been reported by nearly all the investigations in this field, for both pool and flow boiling. The type and concentration of nanoparticles impact on the characteristics of the deposition layer formed and its thickness. Globally, the main conclusions are as follows:

(i) HTC is highly influenced by nanoparticle concentration and the ratio between the nanoparticle diameter and the diameter of surface cavities. Usually, at high concentrations, HTC deteriorates due to high thermal resistance promoted by nanoparticle deposition. Moreover, at low concentrations, HTC may be enhanced due to the effect of thermal conductivity being more pronounced than the extra thermal resistance and the presence of nanoparticles may increase (higher HTC) or decrease (lower HTC) the number of active nucleation sites depending on the size of nanoparticles and roughness of the surface.

(ii) Most of the studies that have used nanofluids have shown CHF improvement. These findings may be explained by the changes in the topography of the heating surface due to nanoparticle deposition rather than by the nanofluids themselves. It has been reported that CHF is dependent on the nanoparticle layer formed on the heating surface, which in turn alters its wettability, roughness, and porosity.

(iii) Deposition of nanoparticles can reduce the contact angle of bubbles and the heating surface, resulting in higher adhesion tension leading to longer time of bubbles growth and departure and a reduced number of active nucleation sites. The lower frequency of generation and detachment of bubbles reduces fluid motion and deteriorates the boiling HTC. In contrast, lower contact angles improve the wettability of the surface, thus enhancing CHF.

(iv) Addition of surfactant reduces surface tension, increasing the frequency of bubbles departure and reducing their diameter. This increased fluid motion leads to an enhancement in the number of active nucleation sites and may improve HTC. High concentrations of surfactants have the opposite effect, due to an increase in viscosity that reduces micro-convection or *via* the formation of an insulation layer of bubbles on the heater surface.

(v) In flow boiling, nanoparticle deposition has been verified in most of the investigations conducted in the field. Nanofluids considerably affect the dynamics of bubble formation, where it has been shown that the bubbles formed are frequently smaller and the frequency of departure is higher. Consequently, nanofluid flow boiling may be more stable, and the occurrence of CHF is delayed. Pressure drop, which is a major concern in flow boiling, can be significantly reduced along the flow using nanofluids in comparison with the base fluid.

Finally, the application of nanofluids in boiling heat transfer was proven to be a promising alternative for the improvement in the performance of thermal systems. There are clearly many benefits in using nanofluids, especially for CHF enhancement. However, their cost and large-scale production still limit their application. Besides this, more systematic investigations should be carried out for a longer boiling period to evaluate the effect of nanoparticle deposition over time.

Acknowledgements

This work has been supported by Fundação para a Ciência e a Tecnologia (FCT), through IDMEC, under LAETA, project UIDB/50022/2020 and through project PTDC/NAN-MAT/29989/2017.

References

1. M. S. Kamel, F. Lezsovits, A. M. Hussein, O. Mahian and S. Wongwises, *Int. Commun. Heat Mass Transfer*, 2018, **98**, 59–66.
2. D. Ciloglu and A. Bolukbasi, *Appl. Therm. Eng.*, 2015, **84**, 45–63.

3. J. Barber, D. Brutin and T. Lounes, *Nanoscale Res. Lett.*, 2011, **6**, 280.
4. A. Bejan, *Convection Heat Transfer*, John Wiley & Sons, Inc., New Jersey, 4th edn, 2013.
5. J. Chen, S. Ahmad, J. Cai, H. Liu, K. T. Lau and J. Zhao, *Energy*, 2021, **215**, 119114.
6. S. U. S. Choi, *Am. Soc. Mech. Eng., Fluids Eng. Div., FED*, 1995, **231**, 99–105.
7. Y. M. Yang and J. R. Maa, *Int. J. Heat Mass Transfer*, 1984, **27**, 145–147.
8. F. P. Incropera, D. P. Dewitt, T. L. Bergman and A. S. Lavine, *Fundamentals of Heat and Mass Transfer*, John Wiley & Sons, Inc., USA, 6th edn, 2007.
9. S. Nukiyama, *Int. J. Heat Mass Transfer*, 1966, **9**, 1419–1433.
10. A. E. Bergles and W. M. Rohsenow, *J. Heat Transfer*, 1964, **86**, 365–372.
11. S. van Stralen and R. Cole, *Boiling phenomena*, McGraw-Hill/Hemisphere, New York, 1979.
12. W. M. Rohsenow, *Trans. ASME*, 1952, **74**, 969.
13. I. L. Pioro, *Int. J. Heat Mass Transfer*, 1999, **42**, 2003–2013.
14. N. Zuber, *Trans. ASME*, 1958, **80**, 711–720.
15. S. M. S. Murshed, C. A. Nieto De Castro, M. J. V. Loureno, M. L. M. Lopes and F. J. V. Santos, *Renewable Sustainable Energy Rev.*, 2011, **15**, 2342–2354.
16. J. H. Lienhard, *A Heat Transfer Textbook*, Prentice Hall, Englewood Cliffs, NJ, 2nd edn, 1987.
17. P. J. Berenson, *J. Heat Transfer*, 1961, **83**, 351–358.
18. A. L. Bromley, *Chem. Eng. Prog.*, 1950, **46**, 221–227.
19. M. M. Sarafraz and F. Hormozi, *Exp. Therm. Fluid Sci.*, 2014, **52**, 205–214.
20. M. R. Raveshi, A. Keshavarz, M. S. Mojarrad and S. Amiri, *Exp. Therm. Fluid Sci.*, 2013, **44**, 805–814.
21. A. Amiri, M. Shanbedi, H. Amiri, S. Z. Heris, S. N. Kazi, B. T. Chew and H. Eshghi, *Appl. Therm. Eng.*, 2014, **71**, 450–459.
22. H. Peng, G. Ding and H. Hu, *Exp. Therm. Fluid Sci.*, 2011, **35**, 960–970.
23. S. W. Lee, K. M. Kim and I. C. Bang, *Int. J. Heat Mass Transfer*, 2013, **65**, 348–356.
24. M. M. Sarafraz and F. Hormozi, *Int. J. Therm. Sci.*, 2015, **90**, 224–237.
25. Z. Shahmoradi, N. Etesami and M. N. Esfahany, *Int. Commun. Heat Mass Transfer*, 2013, **47**, 113–120.
26. M. Kole and T. K. Dey, *Appl. Therm. Eng.*, 2012, **37**, 112–119.
27. S. L. Song, J. H. Lee and S. H. Chang, *Exp. Therm. Fluid Sci.*, 2014, **52**, 12–18.
28. S. Z. Heris, *Int. Commun. Heat Mass Transfer*, 2011, **38**, 1470–1473.
29. T. Il Kim, W. J. Chang and S. H. Chang, *Int. J. Heat Mass Transfer*, 2011, **54**, 2021–2025.
30. D. Wen, M. Corr, X. Hu and G. Lin, *Int. J. Therm. Sci.*, 2011, **50**, 480–485.

31. X. Tang, Y. H. Zhao and Y. H Diao, *Exp. Therm. Fluid Sci.*, 2014, **52**, 88–96.
32. L. Xu and J. Xu, *Int. J. Heat Mass Transfer*, 2012, **55**, 5673–5686.
33. S. N. Shoghl and M. Bahrami, *Int. Commun. Heat Mass Transfer*, 2013, **45**, 122–129.
34. Z. Li, M. M. Sarafraz, A. Mazinani, T. Hayat, H. Alsulami and M. Goodarzi, *Int. J. Heat Mass Transfer*, 2020, **156**, 119780.
35. T. Okawa, M. Takamura and T. Kamiya, *Int. J. Heat Mass Transfer*, 2012, **55**, 2719–2725.
36. O. Ahmed and M. S. Hamed, *Int. J. Heat Mass Transfer*, 2012, **55**, 3423–3436.
37. H. M. Ali, M. M. Generous, F. Ahmad and M. Irfan, *Appl. Therm. Eng.*, 2017, **113**, 1146–1151.
38. T. Lee, J. H. Lee and Y. H. Jeong, *Int. J. Heat Mass Transfer*, 2013, **56**, 101–106.
39. A. Norouzipour, A. Abdollahi and M. Afrand, *Powder Technol.*, 2019, **345**, 728–738.
40. A. R. Yagnem and S. Venkatachalapathy, *Thermochim. Acta*, 2019, **672**, 93–100.
41. M. S. Kamel and F. Lezsovits, *Int. Commun. Heat Mass Transfer*, 2020, **114**, 104587.
42. F. R. Dareh, M. Haghshenasfard, M. Nasr Esfahany and H. Salimi Jazi, *Heat Mass Transfer*, 2018, **54**, 1653–1668.
43. S. M. S. Murshed, K. C. Leong and C. Yang, *Int. J. Therm. Sci.*, 2005, **44**, 367–373.
44. S. M. S. Murshed, K. C. Leong and C. Yang, *NanoSingapore 2006 IEEE Conference on Emerging Technologies – Nanoelectronics – Proceedings*, 2006, pp. 155–158.
45. J. A. Eastman, U. S. Choi, S. Li, L. J. Thompson and S. Lee, *MRS Online Proc. Libr.*, 1997, **457**, 3–11.
46. W. Duangthongsuk and S. Wongwises, *Exp. Therm. Fluid Sci.*, 2009, **33**, 706–714.
47. C. Selvam, D. M. Lal and S. Harish, *Thermochim. Acta*, 2016, **642**, 32–38.
48. L. S. Sundar, E. V. Ramana, M. K. Singh and A. C. M. Sousa, *Int. Commun. Heat Mass Transfer*, 2014, **56**, 86–95.
49. P. Garg, J. L. Alvarado, C. Marsh, T. A. Carlson, D. A. Kessler and K. Annamalai, *Int. J. Heat Mass Transfer*, 2009, **52**, 5090–5101.
50. B. Bakthavatchalam, K. Habib, R. Saidur, B. B. Saha and K. Irshad, *J. Mol. Liq.*, 2020, **305**, 112787.
51. H. Zhu, C. Zhang, S. Liu, Y. Tang and Y. Yin, *Appl. Phys. Lett.*, 2006, **89**, 023123.
52. M. H. Esfe, A. Karimipour, W. M. Yan, M. Akbari, M. R. Safaei and M. Dahari, *Int. J. Heat Mass Transfer*, 2015, **88**, 728–734.
53. S. M. S. Murshed, K. C. Leong and C. Yang, *Int. J. Therm. Sci.*, 2008, **47**, 560–568.
54. N. Jamshidi, M. Farhadi, D. D. Ganji and K. Sedighi, *Int. J. Eng., Trans. B*, 2012, **25**, 201–209.

55. K. Bashirnezhad, S. Bazri, M. R. Safaei, M. Goodarzi, M. Dahari, O. Mahian, A. S. Dalkiliça and S. Wongwises, *Int. Commun. Heat Mass Transfer*, 2016, **73**, 114–123.
56. H. Li, L. Wang, Y. He, Y. Hu, J. Zhu and B. Jiang, *Appl. Therm. Eng.*, 2015, **88**, 363–368.
57. P. K. Namburu, D. P. Kulkarni, A. Dandekar and D. K. Das, *Micro Nano Lett.*, 2007, **2**, 67–71.
58. A. M. Abdullah, A. R. Chowdhury, Y. Yang, H. Vasquez, H. J. Moore, J. G. Parsons, K. Lozano, J. J. Gutierrez, K. S. Martirosyan and M. J. Uddin, *J. Mol. Liq.*, 2020, **297**, 111982.
59. B. C. Pak and Y. I. Cho, *Exp. Heat Transfer*, 2007, **11**, 151–170.
60. E. Fabre and S. M. S. Murshed, *Renewable Sustainable Energy Rev.*, 2021, **151**, 111593.
61. F. Rostamian and N. Etesami, *Int. Commun. Heat Mass Transfer*, 2018, **95**, 98–105.
62. S. K. Das, N. Putra and W. Roetzel, *Int. J. Heat Mass Transfer*, 2003, **46**, 851–862.
63. A. Suriyawong and S. Wongwises, *Exp. Therm. Fluid Sci.*, 2010, **34**, 992–999.
64. N. Sezer, S. A. Khan and M. Koç, *Int. J. Heat Mass Transfer*, 2019, **137**, 599–608.
65. S. Etedali, M. Afrand and A. Abdollahi, *Int. J. Therm. Sci.*, 2019, **145**, 105977.
66. C. H. Wang and V. K. Dhir, *J. Heat Transfer*, 1993, **115**, 659–669.
67. A. Mourgues, V. Hourtané, T. Muller and M. Caron-Charles, *Int. J. Heat Mass Transfer*, 2013, **57**, 595–607.
68. J. S. Coursey and J. Kim, *Int. J. Heat Fluid Flow*, 2008, **29**, 1577–1585.
69. R. N. Wenzel, *J. Phys. Chem.*, 1949, **53**, 1466–1467.
70. S. J. Kim, I. C. Bang, J. Buongiorno and L. W. Hu, *Appl. Phys. Lett.*, 2006, **89**, 153107.
71. Y. H. Jeong, W. J. Chang and S. H. Chang, *Int. J. Heat Mass Transfer*, 2008, **51**, 3025–3031.
72. Z. H. Liu and L. Liao, *Int. J. Heat Mass Transfer*, 2008, **51**, 2593–2602.
73. N. Patra, P. Ghosh, R. S. Singh and A. Nayak, *Int. J. Heat Mass Transfer*, 2019, **135**, 331–344.
74. Y. Wang, K. Deng, J. Wu, G. Su and S. Qiu, *Int. J. Heat Mass Transfer*, 2018, **122**, 212–221.

Section IV: Applications and Impact of Nanofluids

CHAPTER 13

Thermal Management Applications of Nanofluids

R. R. RIEHL*[a] AND S. M. S. MURSHED[b]

[a] GamaTech Thermal Solutions, R. Córdoba 34, 12235-390, São José dos Campos, SP, Brazil; [b] IDMEC, Department of Mechanical Engineering, Instituto Superior Técnico, University of Lisbon, 1049-001 Lisbon, Portugal
*Email: roger.riehl@gtts.com.br

13.1 Introduction

The application of nanofluids is a relatively recent area of investigation that has shown promising results to improve the performance of thermal control device performance. Nanofluids, nanoparticles dispersed in regular working fluids (called a base fluid), exhibit improved thermal conductivity when applied in thermal systems. For example, by adding 5% by mass of nanoparticles to the working fluid, the liquid thermal conductivity can be increased by up to 20%.[1] Early work in this area has already presented important contributions using nanofluids, composed of water (as the base fluid) and copper oxide (CuO) nanoparticles with sizes of around 25 nm.[2–4] These investigations then proposed possible mechanisms of the enhanced thermal conductivity of nanofluids, with one of the important mechanisms being the formation of a solid-like nanolayer over the surface of nanoparticles, which acts as a connection between the solid nanoparticles and the base fluid.[5] The investigations performed in the early stages of nanofluid development highlighted the potential in using them in several thermal control applications with great improvement in the heat transfer coefficient, especially when liquid single-phase thermal control was used. Such improvement was obtained using

Soft Matter Series No. 16
Fundamentals and Transport Properties of Nanofluids
Edited by S. M. Sohel Murshed

Published by the Royal Society of Chemistry, www.rsc.org

nanofluids circulating in a loop driven by a regular pump, which present different thermal behavior compared to systems where the nanofluid needs to go through an evaporation/condensation cycle.[6]

Besides the characterization of the properties of nanofluids, their applications in numerous thermal management applications have also been studied over the past few decades. An important development of nanofluids applied to passive thermal control devices, such as heat pipes (HPs) and pulsating heat pipes (PHPs), revealed the importance and capabilities of using this working fluid to enhance the overall heat dissipation and transport in that type of device.[7–11] This special type of application has been shown to be very promising, as direct use of nanofluids without any major change in the design of the thermal device is required, allowing direct comparison between the usage of a nanofluid and a regular working fluid.[1,12] However, due to issues related to the stability of nanofluids over time, their application in a more general manner has raised concerns related to their potential use in a wide range of commercial equipment and devices.

Besides the fact that the combination of a base fluid and solid nanoparticles must present chemical compatibility, the resulting nanofluid must also present stability in its structure to be able to perform as expected over a long time period. For that, minimal sedimentation of its solid material that usually causes a lack in performance needs to be achieved.[13] Also, minimal generation of non-condensable gases (NCGs) due to decomposition of the base fluid in the presence of the solid material needs to be observed.[14]

The long-term stability of nanofluids is one of the main challenges for their application in thermal management systems. Although studies have been reported on the influence of the stability of nanofluids on their thermal conductivity,[15–20] very few studies have been conducted on the issue of stability for heating and cooling applications such as in convection heat transfer performance in various systems. Among a handful of studies, the stability effects of nanofluid on the thermal performance of HPs were critically reviewed by Cacua *et al.*[21] It was demonstrated that the deposition of nanoparticles on an evaporator surface and wick or groove structures was directly responsible for the thermal performance of HPs, with stability playing a major role in their deposition. Ensuring long-term stability is critical for not only thermal performance, but also for the continuous operation and longevity of the systems.

The primary application of nanofluids has been in thermal control devices that are widely used in numerous industries, from automotive to aerospace. From one area to another, the degree of reliability required increases exponentially, as in industrial applications maintenance is frequent, whereas in aerospace the device needs to be certified to continuously operate for years. Therefore, there is a very high level of reliability required for aerospace applications that a thermal control device operating with a nanofluid needs to meet, which involves the processes and technologies applied, along with the certification that long-term operation can be met.[22] Since aerospace applications set the highest quality levels for any given thermal control device, other designed devices for industrial and domestic use can be obtained for the same purpose with lower quality requirements. Nanofluids have also been considered by the automotive industry to assist in vehicle engine thermal management, as modern designs have stricter tolerance, a higher compression rate, *etc.*, to achieve high power. In this case, the cooling system needs to exhibit higher performances compared to previous vehicle engine generations, which indicates that the use of nanofluids is highly recommended.[23–25]

Given the importance of this subject in future thermal management applications, the objective of this work is to detail the fundamentals that need to be met to better design thermal control systems that utilize nanofluids. Analysis related to the use of nanofluids is also presented, focusing on the evaluation of the improvement obtained compared to the base fluid.

13.2 Evaluation of the Thermophysical Properties of Nanofluids

With the addition of solid nanoparticles in a base fluid to form a nanofluid, the thermophysical properties require proper adjustment to represent a gain and/or loss of any characteristics due to such an addition. Such an adjustment requires taking into account the proportion of solid nanoparticles added, which is usually considered as a percentage by mass or volume of the base fluid itself. All common thermophysical properties, such as the thermal conductivity, viscosity, density, and specific heat of nanofluids, are mainly measured using commercial devices. The thermophysical property that is widely discussed among researchers is thermal conductivity, since the addition of solid nanoparticles can lead to a substantial improvement in this property, depending on the materials used and their chemical compatibility. However, most of the reported thermal conductivity data of nanofluids are not very accurate, thus the results of this property are very scattered and inconsistent.[26] Therefore, measurements and the reported results on thermophysical properties of nanofluids are not discussed. However, predictions of these properties are commonly performed using existing models or rules and some popular and selective models, which are briefly presented.

Usually, the Maxwell model is applied,[27] as this is one of the most straightforward uses for a nanofluid with reasonable accuracy for the obtained results, where eqn (13.1) presents the effective thermal conductivity of a homogeneous solution:

$$k_\mathrm{n} = \frac{k_\mathrm{p} + 2k_\mathrm{l} + 2\left(k_\mathrm{p} - k_\mathrm{l}\right)\phi}{k_\mathrm{p} + 2k_\mathrm{l} - \left(k_\mathrm{p} - k_\mathrm{l}\right)\phi} k_\mathrm{l}. \tag{13.1}$$

It should be noted that the Maxwell model is only applied to suspensions of spherical particles and is good for diluted suspensions. However, as other shapes of nanoparticles, such as a tubes, needles, and flakes, are also commonly used in nanofluids, it is important to use a suitable model.

Another classical model, referred to as the Hamilton and Crosser model,[26] can be used for any shape of particles. The Hamilton and Crosser model is expressed as in eqn (13.2):

$$k_n/k_l = \left[\frac{k_p + (n-1)k_l - (n-1)\phi(k_l - k_p)}{k_p + (n-1)k_l + \phi(k_l - k_p)}\right] \tag{13.2}$$

where $n = 3/\psi$ is the shape factor and ψ is the particle sphericity.

Since not only the thermal conductivity is affected when solid nanoparticles are added to a base fluid, other important properties that need to be properly evaluated are density and viscosity. In this case, proper consideration and evaluation of the solid particles in a liquid must be taken, according to what is presented by the two-phase theory,[28] represented as:

$$\frac{1}{\rho_n} = \left(\frac{\phi}{\rho_p} + \frac{1-\phi}{\rho_l}\right), \tag{13.3}$$

and the dynamic viscosity of nanofluids can then be calculated using a power law based on Brinkman's model:[1]

$$\mu_n = \mu_l \frac{1}{(1-\phi)^{2.5}} \tag{13.4}$$

Eqn (13.4) has been selected as it presents a better correlation when comparing the experimental data with the predicted results. Eqn (13.3) and (13.4) are important to calculate the overall pressure drop that a system presents, which directly result in the selection of the pump (in the case of a mechanical system) or capillary evaporator (in the case of a passive system). When considering a mechanical system, the size of the pump directly impacts the power consumption and how robust the design must be to overcome the overall pressure drop. In the case of a capillary evaporator used in passive systems like HPs, due to limitations in the capillary pumping pressure generation that these devices present, failure to properly evaluate the effects of the transport properties when using nanofluids can lead to severe loss in performance, which can be a serious problem when considering aerospace applications.

With the use of a nanofluid, both density and dynamic viscosity present higher values when compared to the properties of the base fluid alone. Although this can be a minor issue in a system using a mechanical pump, in a capillary-driven system it can signify a substantial increase in the overall pressure drop that directly impacts the temperature distribution throughout the loop. In this case, with higher temperatures, a thermal device can show a lack of capability in maintaining the required levels and the use of a nanofluid can present a disadvantage rather than an advantage, which must be properly evaluated upon defining the set parameters.

Since the addition of solid nanoparticles in the base fluid has a direct impact on the overall thermal performance of a given device, several models have been proposed[29] that claim to be able to predict the expected results to be obtained. However, most of the time those models yield different results even for the same solid nanoparticle material, sizes, concentration, and also for the same base fluid. Other thermophysical properties must also comply with adequate verification related to their application, as one must have in mind that nanofluids are intended for a sustainable improvement in the performance of thermal control devices.

13.3 Nanofluids Stability

Although there has been no rigorous study and conclusive findings on the impact of the stability of nanofluids in terms of their performance in thermal management systems, particularly in heat pipes,[21] such devices have been presented to be a promising direct application for nanofluids. However, the stability status of nanofluids has been found to greatly influence thermophysical properties and their steady-state cooling operation. Thus, it is necessary to understand and accurately assess the stability of nanofluids. Stability is routed to the synthesis (also termed as preparation) of nanofluids, which is commonly performed *via* two routes: one-step and two-step methods. Simply, the one-step method is where nanoparticles are directly synthesized in base fluids while the two-step method is where pre-synthesized/purchased nanopowders are dispersed in base fluids. Although the two-step method is used to prepare nanofluids most of the time, the prepared nanofluids exhibit poor stability. Therefore, it is important to perform some additional procedures and operations on the prepared nanofluids (before their property characterization and applications) to enhance their stability. Most of the reported studies have described the performing of operations such as sonication, the addition of surfactants, or nanoparticle surface modification to some extent. Details on the importance of stability and their rigorous assessments are well documented in the literature[24–30] and are not discussed further here.

13.4 Chemical Compatibility in a Nanofluid

As a very important subject that is often forgotten or not even considered by many researchers, NCG generation is related to the decomposition of the working fluid in the presence of other materials, such as the presence of solid nanoparticles in a base fluid. The decomposition rate can be properly verified according to the combination of the nanoparticle material and the base fluid, related to the temperature to which the nanofluid is subjected. Therefore, the decomposition rate can be verified applying the Arrhenius model, which can be calculated as follows:[14]

$$K = Ae^{\frac{-E_{act}}{RT}} \tag{13.5}$$

The importance of this process must not be disregarded during the evaluation of a nanofluid, which has been defined during extensive investigations performed with HPs. Several investigations have been conducted, especially in the 1970s and 1980s, resulting in important contributions to better understand how a given material and working fluid degrade over time regarding the chemical reactions occurring during operation.[14]

Studies have been reported using known pairs of materials and base fluids, such as Al_2O_3 and water.[29] Researchers have constantly used such combinations to gather data, which can be hiding potential issues as NCGs are generated and certainly lead to misinterpretation of the obtained results. Very often, due to the use of incompatible pairs, a thermal device presents different results from one day to another, or, at its limit, it ceases its operation due to the accumulation of NCGs.[14] Also, to make nanofluids more stable, surfactants are added to them to prevent particle sedimentation, which has been reported to increase the quality of a given nanofluid. However, upon adding another substance to the nanofluid, there is an increase in the uncertainty regarding chemical compatibility, which turns the nanofluid application into a difficult task to properly analyze the obtained results without the bias of presenting NCGs that certainly cause misleading interpretation or even wrong conclusions.

There are several sources of incompatible materials and working fluid that can be easily accessed. Table 13.1 presents a summarized list of solid materials and their compatibility with some working fluids, which can be related to several other combinations of materials and working fluids.[31]

Interactions between the working fluid and solid material must be properly evaluated to avoid and minimize the effects of decomposition that result in NCG generation, which can be extremely harmful to the thermal device if not properly chosen.

Based on the model presented in eqn (13.5), the degradation rate of two well-known compatible pairs widely used in HPs is shown in Figure 13.1.

Table 13.1 Compatibility between solid materials and working fluids (compiled by GamaTech).

Working fluid	Compatible material	Incompatible material
Water	Stainless steel, copper, silica, nickel, titanium	Aluminum, inconel
Pentane	302 stainless steel, aluminum	304 and 306 stainless steel
Ammonia	Aluminum, stainless steel, cold-rolled steel, iron, nickel	Copper
Methanol	Stainless steel, iron, copper, brass, silica, nickel	Aluminum
Acetone	Aluminum, stainless steel, copper, brass, silica	
Freon-11	Carbon steel, stainless steel	Aluminum
Freon-21	Iron	Aluminum
Freon-123	Aluminum	
Heptane	Aluminum	
Dowtherm	Stainless steel, copper, silica	

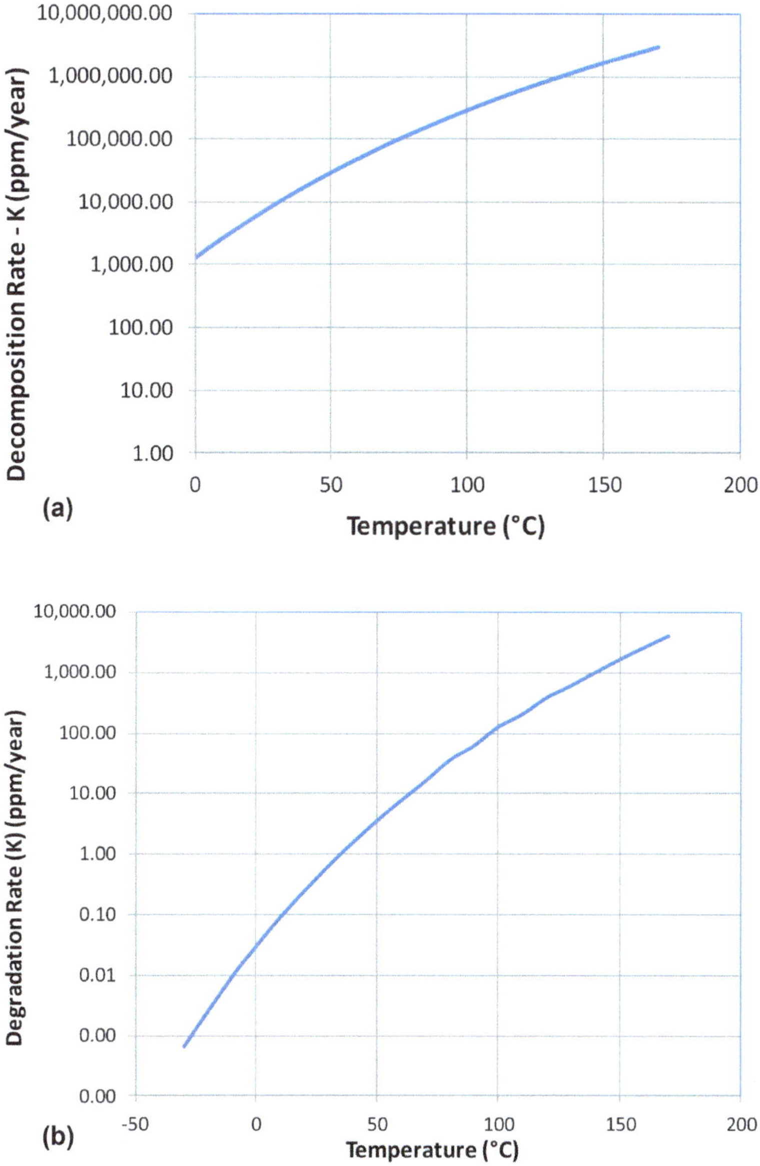

Figure 13.1 The decomposition rate of compatible combinations: (a) copper–water and (b) aluminum–ammonia nanofluids (Image by GamaTech).

It is important to mention that even when utilizing compatible combinations of working fluid (or base fluid) and solid materials, NCG generation occurs, based on the presented calculations. Given the lifetime expectancy of a thermal device, they have to go through the aging process to generate the amount of NCG based on the degradation rate, which is then removed from the device before the final sealing. Decomposition rate verification, as shown in Figure 13.1, is a normal practice during HP manufacturing that can potentially be useful when evaluating the nanofluid chemical compatibility for a given application.

13.5 Nanofluids Applications in Thermal Control Devices

From the available publications, there have been several investigations towards the use of nanofluids in thermal devices, where the enhancement of the thermal properties of this working fluid allows important improvements to be made in the performance of the devices. Table 13.2 summarizes some of the thermal applications currently being considered for nanofluids, as well as the gains observed during the operation of the related devices. It is important to mention that heat pipes have gained much attention from other thermal devices, as the application of the nanofluid is a direct measure that can be taken without the need for major changes in their design.

In work reporting the effects of nanofluids in loop heat pipes (LHPs),[32] experimental and simulated results were used to compare the behavior of the thermal control device during operation compared with base fluid only.

In general, the LHP operates by absorbing heat from the source (such as an electronic component), which is then transferred to the saturated

Table 13.2 Examples of nanofluid applications.

Researcher	Potential application	Main results
Riehl[32]	Oscillating heat pipes, loop heat pipes and liquid cooling loop	A significant decrease in the heat source temperature was observed using a CuO–water nanofluid compared with water. The heat source temperatures presented a decrease of ≤20% during operation with the nanofluid.
Wong and de Leon[33]	Nuclear energy	The use of a nanofluid increases the in-vessel retention capabilities of nuclear reactors by as much as 40%.
	Automotive – radiators	The use of highly thermally conductive nanofluids in radiators led to a reduction in the frontal area of the radiator by ≤10%.
Taylor *et al.*[34]	Solar collector systems	Efficiency improvements of ≤10% improvements were theoretically/experimentally reported.
Kumaresan and Venkatachalapathy[35]	Heat pipes and thermosyphons	Optimal performance was obtained at a tilt angle of around 60° for wick heat pipes and in the vertical position for thermosyphon heat pipes.
Kavusi and Toghraie[36]	Heat pipes	Using a nanofluid instead of water (base fluid) led to increased thermal efficiency and reduced heat at the wall of the heat pipe.

working fluid present in its internal volume. Administrating a small amount of heat is enough to evaporate the saturated working fluid present in the porous wick. Since the vapor can only flow through the vapor grooves machined in the wick, it flows towards the vapor line that connects the evaporator to the condenser. In the condenser (heat sink), the working fluid is condensed back to the liquid phase, flowing towards the evaporator by the liquid line. Coupled to the evaporator is the compensation chamber, which is responsible for self-regulating the amount of working fluid present in the loop, depending on the amount of heat applied to the evaporator. It is also responsible for establishing the operating temperature of the LHP at which the entire loop works. The geometric characteristics of the LHP are presented in Table 13.3, which was built using 316L stainless steel. Figure 13.2 presents the schematics of an LHP.

Eqn (13.1)–(13.3) were used to calculate the NiO–water nanofluid properties to verify their effects on the thermal behavior of the LHP, to better assess the influence of the addition of solid nanoparticles to the base fluid and how this would impact the overall performance of the LHP.

Table 13.3 Geometric characteristics of the LHP (elaborated by GamaTech).

Capillary evaporator	Liquid line	Vapor line
Heat removal area: 3.0 cm^2	OD/ID (mm): 3.0/1.5	OD/ID (mm): 3.0/1.5
Number of grooves: 10	Length: 300 mm	Length: 200 mm
Nickel wick structure	Compensation chamber	Condenser
Mean pore size: 3 μm	Total volume: 6 cm^3	OD/ID (mm): 3.0/1.5
Porosity: 75%		Length: 500 mm

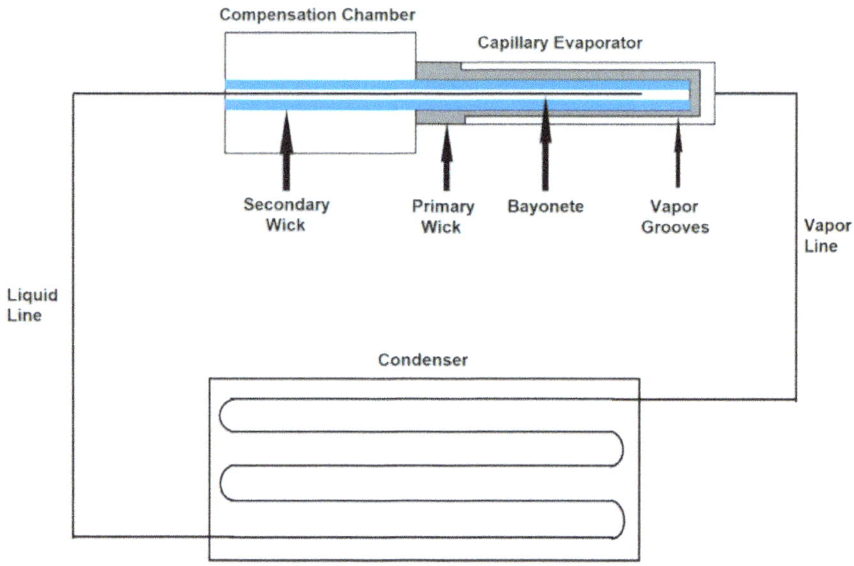

Figure 13.2 Schematics of an LHP (Image by GamaTech).

The objective was to investigate the direct effect on the temperature of the heat source when using the nanofluid because the addition of solid nanoparticles also increases the density and viscosity of the fluid, directly impacting the overall pressure drop of the loop.

Figure 13.3 presents a comparison between the simulation and experimental results for the LHP operating with deionized water and a NiO–water nanofluid, with an addition of 3.5% of NiO nanoparticles (99.5% purity) with a mean diameter of 40 nm. The nanofluid was obtained by mixing the base fluid and the nanoparticles in an ultrasonic bath for 24 h.[32]

The simulated and experimental results present acceptable agreement, showing that the heat source temperature (evaporator) of the LHP is lower when only using the base fluid (water). When using the NiO–water nanofluid, the evaporator temperature increases due to an increase in the overall pressure drop observed throughout the LHP. Since the increase in the pressure drop has a direct impact on the performance of the LHP, as the capillary pressure generated in the evaporator is limited, it finds more resistance to drive the working fluid, thus impacting its temperature. In this specific case, when using a capillary evaporator where the working fluid is evaporated in a porous wick, the use of a nanofluid can result in an undesirable impact on the overall performance of the device. This is particularly true in thermal control devices that utilize wick porous structures to generate capillary pumping pressure to drive the working fluid through the loop. Since the flow through the porous structure results in elevated pressure drops, it is enhanced with higher viscosity working fluids, such as nanofluids.

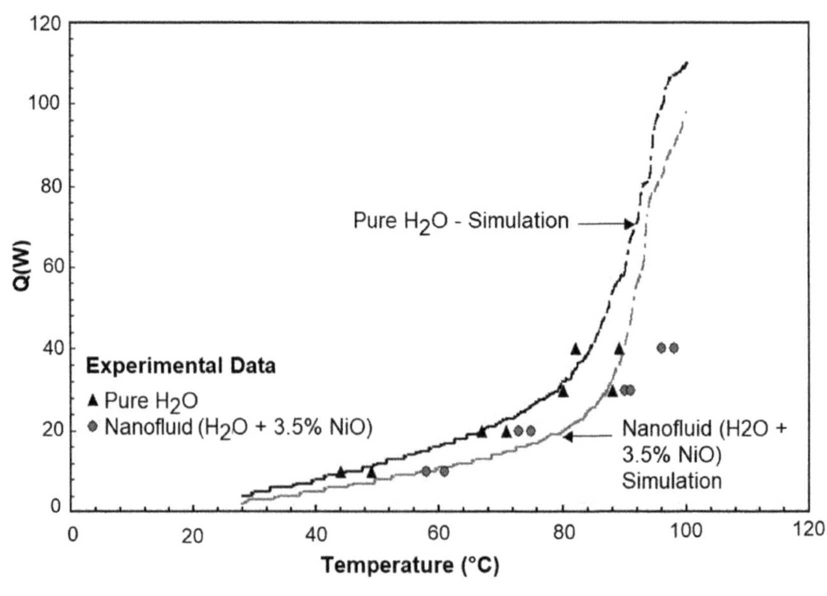

Figure 13.3 Comparison between the experimental and simulated results of LHP operation (Image by GamaTech).

However, in applications where the thermal control device does not present a porous wick structure, the nanofluid is an important factor in the enhancement of its performance. In this case, pulsating/oscillating heat pipes (PHPs or OHPs) are devices that benefit from the use of nanofluids, as they are built with meandering smooth tubes bent with several curves that form parallel channels. They operate through liquid plugs and vapor slugs generated in the saturated working fluid that transfer the heat from the source to the sink and can operate in any orientation despite the effects of gravity.[6,12,32,37] Research efforts have been made on the use of nanofluids in PHPs/OHPs, reporting important results,[37] as they have already been used at the product level in thermal management devices for the cooling of electronics. The PHP/OHP can operate under different temperature conditions depending on the selected working fluid and can operate under conditions adverse to the gravity orientation. The selection of the working fluid must therefore follow the procedures as those for regular heat pipes, verifying the chemical compatibility with the housing material and the expected performance. The selection of the correct working fluid can result in better thermal control capability by the PHP/OHP and the HP, which must be chosen by the designers or engineers.

13.6 Present Research using Nanofluids in PHPs/OHPs

Due to their capability in performing under adverse conditions, along with their reliability during the heat transport from the source to the sink, PHPs/OHPs have been successfully integrated into several projects[6,12,37] over the last few years, presenting different configurations, as shown by Figure 13.4. For the presented configurations, the thermal control of printed circuit boards (PCBs) was accomplished using PHPs/OHPs, built using oxygen-free copper tubing with water, for a working fluid inventory of 50% of their internal volume. Due to the nature of the application of PHPs/OHPs, their condenser was connected to a liquid cooling system that operated as a heat sink, which was able to dissipate the heat load to the ambient environment in a remote session.

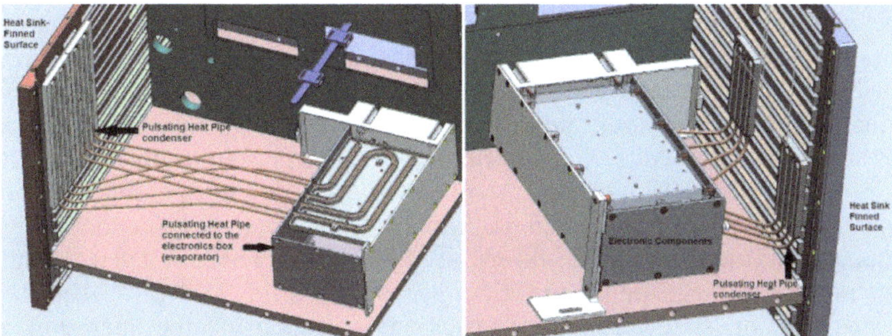

Figure 13.4 Integration of PHPs for electronics cooling (Image by GamaTech).

These special thermal control devices present a very complex operation, even though their construction is really simple, and are low in cost when compared to regular HPs. They can be built without major technological needs, and can operate with high heat loads regardless of gravity orientation. However, the same procedures applied to the manufacture of an HP must also be applied to a PHP/OHP, to bring them to a level of reliability that is needed for a complex thermal management system.

Figure 13.5 presents an image and layout of the dimensions of a copper PHP/OHP applied to the thermal control of electronics that were tested for comparison purposes using pure water and a CuO–water nanofluid with the addition of 3.5 wt% of solid nanoparticles. The CuO nanoparticles used exhibited a purity of 95% and a particle size of 29 nm. For this thermal control device verification, the evaporator session was connected to a heat source, the adiabatic session was thermally insulated and the condenser session was open to a forced-air source at a velocity of 2 $m\,s^{-1}$.

Tests were conducted using a skin heater to simulate the heat source, and thermocouples were used to read the temperatures to evaluate the operation of the PHP/OHP. The evaporator and adiabatic sessions were fully insulated and the condenser was open to the environment, where air blown at 3.0 $m\,s^{-1}$ was responsible for the heat dissipation.

The PHP/OHP was charged with 80% of its internal volume with working fluid and tested horizontally. Figure 13.6 presents the results of the PHP/OHP tested with pure water and Figure 13.7 presents the results of the CuO–water nanofluid. It can be seen that the PHP/OHP started presenting intensive pulsations when the heat load was 40 W, which was consistent with the design of the device that predicted this as the power at which they would occur. Since this device operates with intense slug/plug formation, its design must account for heat loads at which such flow pattern transitions take place to present adequate operation. Due to the characteristics of this thermal control device, before the intensive pulsations begin, the device operates in a film evaporation pattern, where greater temperature differences between the evaporator and condenser are observed.

When operating with the CuO–water nanofluid, as presented by Figure 13.7, the intense pulsations begin at much lower heat loads. Using the nanofluid, the slug/plug formation begins much earlier, resulting in smaller temperature differences between the evaporator and condenser from the beginning. The nanoparticles are believed to act as nucleation sites that enhance vapor formation, thus enhancing the pulsation frequencies, even at lower power levels. Thus, the nanofluid appears to play an important role in improving PHP/OHP operation. However, they limit the operation of this type of device due to their highest heat load, as a more viscous nanofluid imposes additional pressure drops, as discussed previously. Even though this might be a negative factor for the use of nanofluids in PHPs/OHPs, this is not a major issue compared to those observed during the operation of an LHP.

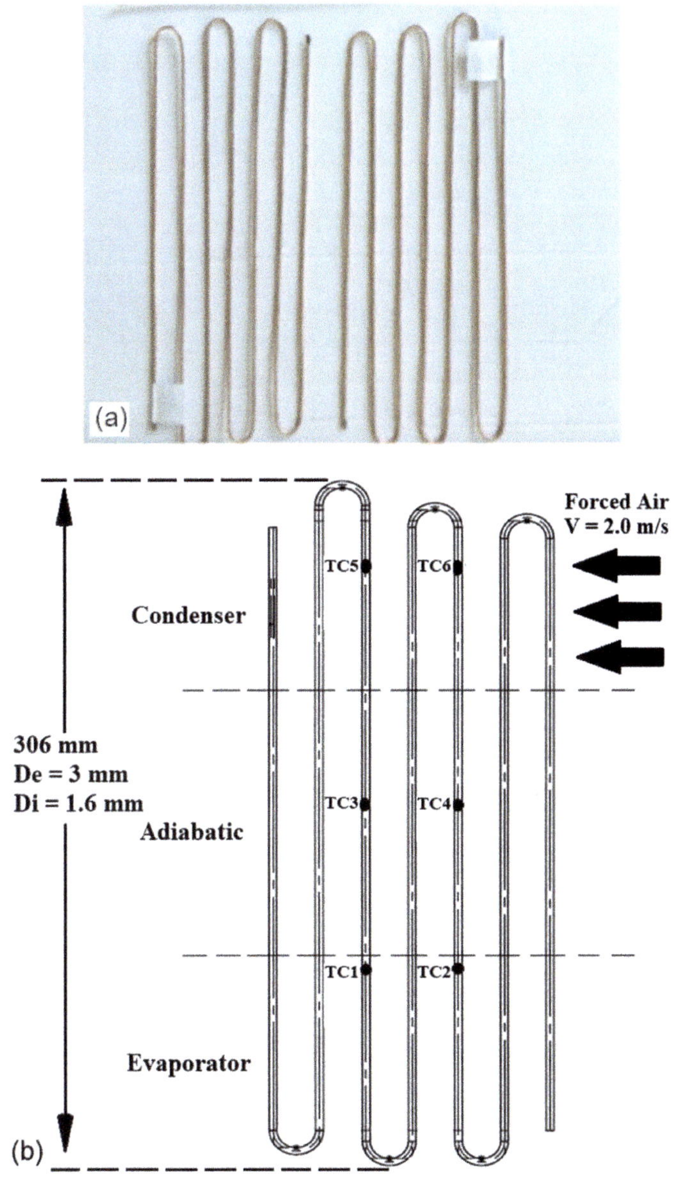

Figure 13.5 Copper PHP/OHP applied to the thermal control of electronics (Image by GamaTech). (a) Real PHP/OHP and (b) dimensions and location of instruments.

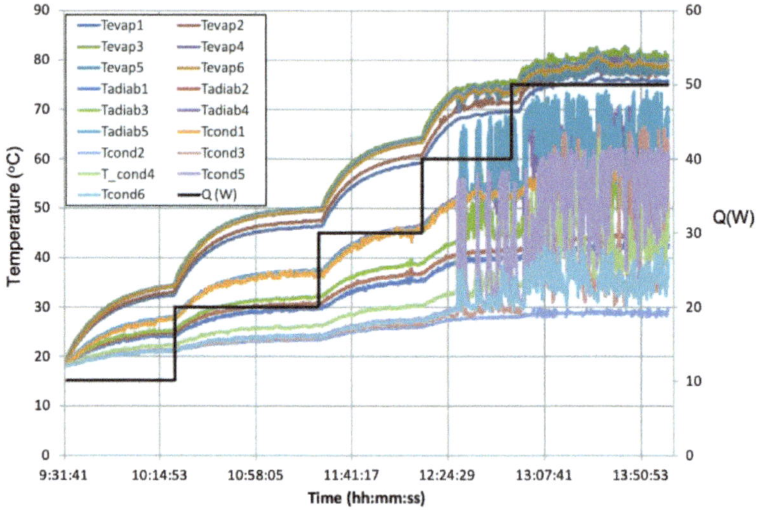

Figure 13.6 PHP/OHP operating with pure water (Image by GamaTech).

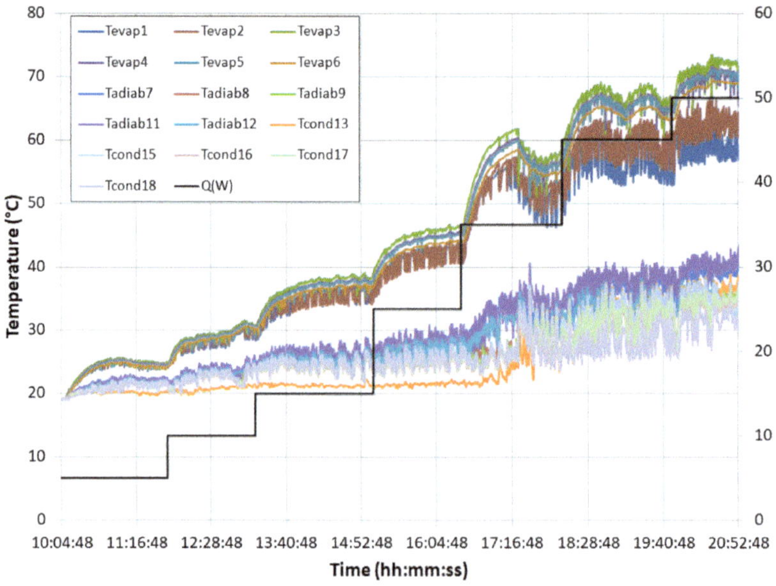

Figure 13.7 PHP/OHP operating with a CuO–water nanofluid (Image by GamaTech).

13.7 Conclusions

Over the years, nanofluids have demonstrated their capability in the enhancement of the performances of thermal devices or systems. Particularly for direct application in heat pipes, the fast development of thermal control devices employing nanofluids has emerged, showing their potential in several applications with substantial augmentation in the thermal capabilities of systems. However, proper consideration must be taken into account when evaluating the following aspects of the applications of nanofluids in thermal management:

a. transport properties must be carefully considered when using capillary-driven devices, such as heat pipes, loop heat pipes, and pulsating heat pipes;
b. the thermal design of these devices must take into account the properties of the nanofluids, as they are influenced by the solid nanoparticles;
c. chemical compatibility between the solid nanoparticles and base fluid, as well as the other materials (housing, porous wick) of the device, must be verified to avoid the generation of non-condensable gases that can potentially lead to device failure;
d. the use of high-purity solid nanoparticles along with the base fluid, is the key to higher performance when using nanofluids.

Continuous development of this technology can certainly result in devices with higher heat transport capabilities, which can find new applications. Nanofluids have been demonstrated to play important roles in the present and future designs of thermal management systems.

Nomenclature List

ϕ	Mass fraction of solid nanoparticles
ρ_n	Nanofluid density
ρ_l	Liquid density
ρ_p	Particle density
μ_n	Nanofluid dynamic viscosity
μ_l	Liquid dynamic viscosity
ψ	Particle sphericity
A	Empirical relationship between the temperature and rate coefficient (s^{-1})
D_e	External diameter
D_i	Internal diameter
E_{act}	Activation energy of the reaction, based on the combination of the working fluid and the materials involved ($J\,mol^{-1}$)
K	Decomposition rate (eqn (4))
k_n	Nanofluid effective thermal conductivity
k_l	Liquid thermal conductivity
k_p	Particle thermal conductivity
Q	Heat load
R	Universal gas constant

T	Absolute temperature
TC	Thermocouple
T_{evap}	Evaporator temperature
T_{adiab}	Adiabatic temperature
T_{cond}	Condenser temperature

Acknowledgements

This work has been partially supported by Fundação para a Ciência e a Tecnologia (FCT), through IDMEC, under LAETA, project UIDB/50022/2020 and through project PTDC/NAN-MAT/29989/2017.

References

1. Y. Xuan and W. Roetzel, *Int. J. Heat Mass Transfer*, 2000, **43**, 3701.
2. R. Chein and G. Huang, *Appl. Therm. Eng.*, 2005, **25**, 3104.
3. D. Wen and Y. Ding, *J. Nanopart. Res.*, 2005, **7**, 265.
4. Y. Xuan and Q. Li, *Int. J. Heat Fluid Flow*, 2000, **21**, 158.
5. W. Yu and S. U. S. Choi, *J. Nanopart. Res.*, 2003, **5**, 167.
6. M. H. Buschmann, A. Huminic, S. Mancin, R. R. Riehl and G. Huminic, *J. Nanofluids*, 2019, **8**(2), 253.
7. Y. Ji, H. Ma, F. Su and G. Wang, *Exp. Therm. Fluid Sci.*, 2011, **35**(4), 724.
8. J. Qu and H. Wu, *Int. J. Therm. Sci.*, 2011, **50**(10), 1954.
9. J. Qu, H. Wu and P. Cheng, *Int. Commun. Heat Mass Transfer*, 2009, **37**(2), 111.
10. T. Parametthanuwat, S. Rittidech, A. Pattiya, Y. Ding and S. Witharana, *Nanoscale Res. Lett.*, 2011, **6**(1), 315.
11. N. K. Gupta, A. K. Tiwari and S. K. Ghosh, *Exp. Therm. Fluid Sci.*, 2018, **90**, 84.
12. R. R. Riehl, *J. Nanofluids*, 2019, **8**, 30.
13. W. Yu and H. Xie, *J. Nanomater.*, 2011, **2012**, 1687.
14. R. R. Riehl and S. M. S. Murshed, *Heat Transfer Eng.*, 2020, **1**, 1.
15. S. M. S. Murshed, C. A. Nieto de Castro, M. J. V. Lourenço, M. L. M. Lopes and F. J. V. Santos, *Renewable Sustainable Energy Rev.*, 2011, **15**, 2342.
16. L. Fedele, L. Colla, S. Bobbo, S. Barison and F. Agresti, *Nanoscale Res. Lett.*, 2011, **6**, 300.
17. S. M. S. Murshed, C. A. Nieto de Castro and M. J. V. Lourenço, *J. Nanofluids*, 2012, **1**, 175.
18. Babita, S. K. Sharma and G. S. Mital, *Exp. Therm. Fluid Sci.*, 2016, **79**, 202.
19. C. A. Nieto de Castro, S. I. C. Vieira, M. J. V. Lourenço and S. M. S. Murshed, *J. Nanofluids*, 2017, **6**, 804.
20. K. Cacua, S. M. S. Murshed, E. Pabón and R. Buitrago, *J. Therm. Anal. Calorim.*, 2020, **140**, 109.
21. K. Cacua, R. Buitrago-Sierra, B. Herrera, E. Pabón and S. M. S. Murshed, *J. Therm. Anal. Calorim.*, 2019, **136**, 1597.

22. S. Oldenburg, A. Siekkine, T. Darlington and R. Baldwin, SAE Technical Paper 01-3128, 2007.
23. M. Bai, Z. Xu and Z. J. Ly, SAE Technical Paper 01-1821, 2008.
24. W. Yu and H. Xie, *J. Nanomaterials*, 2012, **2012**, 435873.
25. R. Saidur, K. Y. Leong and H. A. Mohammad, *Energy Rev.*, 2010, **1**, 1646.
26. S. M. S. Murshed, K. C. Leong and C. Yang, *Appl. Therm. Eng.*, 2008, **28**, 2109.
27. J. Koo and C. Kleinstreuer, *J. Nanopart. Res.*, 2004, **6**, 577.
28. V. P. Carey, *Liquid-Vapor Phase-Change Phenomena*, Taylor and Francis, New York, NY USA, 2nd edn, 2008.
29. E. W. Marcelino, R. R. Riehl and D. de Oliveira, *presented in part at ITherm – The Intersociety Conference on Thermal and Thermomechanical Phenomena in Electronic Systems*, 2016.
30. S. M. S. Murshed and C. A. Nieto de Castro, *Nanofluids: Synthesis, Properties, and Applications*, Nova Science Publishers, Inc, New York, USA, 2014.
31. G. Peterson, *An Introduction to Heat Pipe Modeling, Testing and Applications*, John Wiley and Sons, New York, USA, 1994.
32. R. R. Riehl, *presented in part at Heat Powered Cycles Conference*, Newcastle, UK, September 2006.
33. K. V. Wong and O. de Leon, *Adv. Mech. Eng.*, 2010, **1**, 1.
34. R. Taylor, S. Coulombe, T. Otanicar, *et al.*, *presented in part at ASME 2012 3rd Micro/Nanoscale Heat & Mass Transfer International Conference*, Atlanta, GA, USA, 2012.
35. G. Kumaresan and S. Venkatachalapathy, *Frontiers in Heat Pipes*, 2012, **3**, 1.
36. H. Kavusi and D. Toghraie, *Adv. Powder Technol.*, 2017, **28**, 3074.
37. R. R. Riehl, *presented in part at IX Minsk International Seminar Heat Pipes, Heat Pumps, Refrigerators, Power Sources*, Minsk, Belarus, September 2015.

CHAPTER 14

Applications of Nanofluids in Solar Thermal Systems

K. KHANAFER,[a] M. E. H. ASSAD[b] AND K. VAFAI*[c]

[a] Mechanical Engineering Department, University of Michigan, Flint MI 48502, USA; [b] Mechanical Engineering Department, University of Sharjah, UAE; [c] Mechanical Engineering Department, University of California, Riverside, CA 92521
*Email: vafai@engr.ucr.edu

14.1 Introduction

Traditional cooling fluids, such as water and oil, have been used in various applications over the past few decades. Due to their low thermal characteristics and latest advancement in nanotechnology, Choi *et al.*[1] was first to propose the concept of nanofluids by dispersing nanoparticles ranging in diameters from 1 nm to 100 nm. Since then, nanoparticles have been utilized significantly by many researchers in various areas. Such areas include conversion, energy storage, energy saving, and drug-loading nanoparticles. Several studies and reviews on increasing the thermal efficiency of thermal systems using nanofluids have been reported in the literature.[2–18] Shafahi *et al.*[6] investigated numerically the thermal performance of a cylindrical heat pipe using various nanoparticles, namely Al_2O_3, CuO, and TiO_2. The findings of their study showed that the thermal performance of the heat pipe increased with the addition of the nanoparticles, related to a reduction in thermal resistance. Furthermore, the authors documented that the size of the particles was found to have a profound effect on the temperature gradient along the heat pipe. Jang and Choi[9] investigated theoretically the effect

Soft Matter Series No. 16
Fundamentals and Transport Properties of Nanofluids
Edited by S. M. Sohel Murshed

Published by the Royal Society of Chemistry, www.rsc.org

of Brownian motion in augmenting the thermal conductivity of nanofluids. The authors claimed that they found a significant difference between solid/solid composites and solid/liquid suspensions in the size-dependent effective thermal conductivity of nanofluids. Wang *et al.*[11] developed a method to determine the effective thermal conductivity of a nanofluid using the fractal theory and the effective medium approximation. The size of the nanoparticles and the surface adsorption were considered in their study. The developed model was found to predict well the thermal conductivity of the measured values of metallic oxide nanoparticles.

Nanofluids have been shown to improve heat transfer compared to standard cooling fluids in several studies.[4] However, conflicting findings on the efficiency of nanofluids have been documented. Khanafer and Vafai[4] illustrated different thermal conductivity improvements for various volume fractions and diameters of nanoparticles. Table 14.1 shows the effective thermal conductivity augmentation for metallic and non-metallic nanoparticles based on experimental studies cited in the literature. The authors highlighted inconsistencies in published experimental data regarding the thermal characteristics of nanofluids. According to the authors, the thermal characteristics of nanofluids may be determined using classical equations for low volume fractions of nanoparticles at room temperature. Furthermore, based on experimental data given in the literature for different factors such as temperature, diameter size, and base fluid characteristics, temperature-dependent correlations for viscosity and effective thermal conductivity were constructed in that study.

Table 14.2 shows an overview of the relationships reported by Khanafer and Vafai.[4] Okonkwo *et al.*[18] provided a detailed analysis of nanofluid synthesis and its uses in various thermal systems, including solar cells, heat exchangers, and electronics cooling devices. The authors suggested that the effect of introducing nanoparticles on the specific heat capacity of nanofluids be investigated further. The authors also advocated for more research into nanofluid uses at temperatures above 100 °C.

Recently, Li *et al.*[16] published a review on the uses of nanofluids. The authors provided a comprehensive discussion of several thermal conductivity models for nanofluids. They also explored the effects of various factors on the thermal conductivity of nanofluids. The authors also highlighted a wide range of nanofluid uses, including solar energy, medicine, and aerospace. Khanafer and Vafai[15] studied nanofluid applications in several types of solar devices. Their evaluation covered the most current developments in nanofluids in solar collectors, solar stills, and photovoltaic (PV) systems. The authors demonstrated that by using nanoparticles, the efficiency of a PV system might be significantly increased, lowering production costs. The authors also discussed the safety of using nanoparticles in solar systems. Sajid and Ali[19] published a review paper on nanofluid applications in different heat exchangers, including radiators, plates, circular heat exchangers, shell and tube heat exchangers, and heat sinks. The authors compiled a list of recent research on the Nusselt number, thermal conductivity, friction

Table 14.1 Thermal conductivity results of nanofluids in terms of various parameters.

Ref.	Base fluid	Particle	d_p, nm	ϕ_p, %	Thermal conductivity enhancement
18	Water	Al_2O_3	38.4	4	9% (21 °C), 16% (36 °C), 24% (51 °C)
18	Water	CuO	28.6	4	14% (21 °C), 26% (36 °C), 36% (51 °C)
19	Water	Al_2O_3	131	4	24% (51 °C)
20	Water	Al_2O_3	13	4.3	32.4% (31.8 °C)
20	Water	Al_2O_3	13	4.3	29.6% (46.8 °C)
20	Water	Al_2O_3	13	4.3	26.2% (66.8 °C)
20	Water	SiO_2	12	2.3	1.1% (31.8 °C)
20	Water	SiO_2	12	2.3	1% (46.8 °C)
21	Water	CuO	23.6	3.4	12%
21	EG[a]	CuO	23.6	4	23%
21	Water	Al_2O_3	38.4	4.3	11%
21	EG[a]	Al_2O_3	38.4	5	19%
22	Water	CuO	23	9.7	34%
22	EG[a]	CuO	23	14.8	54%
22	Water	Al_2O_3	28	5.5	16%
22	EG[a]	Al_2O_3	28	5	24.5%
23	Water	Al_2O_3	11	1	14.8% (70 °C)
23	Water	Al_2O_3	47	1	10.2% (70 °C)
23	Water	Al_2O_3	150	1	4.8% (60 °C)
23	Water	Al_2O_3	47	4	28.8% (70 °C)
23	Water	Al_2O_3	47	1	3% (21 °C)
24	Water	CuO	29	6	36% (28.9 °C)
24	Water	CuO	29	6	50% (31.3 °C)
24	Water	Al_2O_3	36	6	28.2%
24	Water	Al_2O_3	47	6	26.1%
25	Water	Al_2O_3	20	5	15%
26	Water	Al_2O_3	11	5	8%
26	Water	Al_2O_3	20	5	7%
26	Water	Al_2O_3	40	5	10%
27	Water	Cu	100	7.5	78%
28	Water	Au	10–20	0.03	21%
28	Water	Ag	60–80	0.001	17%

[a]EG: ethylene glycol.

factor, and pressure drop that looked at the impacts of nanoparticle volume fraction, size, and shape. The authors also provided and evaluated many correlations employed in various research for experimental validation. Sezer *et al.*[20] provided an overview of nanofluid stability, thermal characteristics, and classification. Several factors impacting the effective thermal conductivity of nanofluids were noted and explored. In their review, the authors stated that the stability of nanoparticles posed a significant problem for the heat transfer enhancement of nanofluids and, as a result, their applications. Salari *et al.*[21] reviewed the use of nanofluids in solar thermal systems. Their work examined and analysed the energy, exergy, economic, and environmental features of these systems. In that study, the impact of nanofluid-based PV-thermal (PVT) systems on the environment (*e.g.*, CO_2 emissions) was examined. According to various studies referenced in their paper,

Table 14.2 Correlations for Al_2O_3–water nanofluid thermal characteristics. Reproduced from ref. 4 with permission from Elsevier, Copyright 2011.

Physical properties	Room temperature	Temperature dependence
Density	$\rho_{eff} = (1-\phi_p)\rho_f + \phi_p\rho_p$	Al_2O_3–water $\rho_{eff} = 1001.064 + 2738.6191\phi_p - 0.2095T$ $0 \le \phi_p \le 0.04,\ 5 \le T(^\circ C) \le 40$
Specific heat	$c_{eff} = \frac{(1-\phi_p)\rho_f c_f + \phi_p\rho_p c_p}{\rho_{eff}}$	N/A
Thermal expansion coefficient	$\beta_{eff} = \frac{(1-\phi_p)(\rho\beta)_f + \phi_p(\rho\beta)_p}{\rho_{eff}}$ $\beta_{eff} = (1-\phi_p)\beta_f + \phi_p\beta_p$	Al_2O_3–water $\beta_{eff} = \left(-0.479\phi_p + 9.3158\times10^{-3}T - \frac{4.7211}{T^2}\right)\times10^{-3}$ $0 \le \phi_p \le 0.04,\quad 10\,^\circ C \le T \le 40\,^\circ C$
Viscosity	N/A	Al_2O_3–water $\mu_{eff} = -0.4491 + \frac{28.837}{T} + 0.574\phi_p - 0.1634\phi_p^2$ $+ 23.053\frac{\phi_p^2}{T^2} + 0.0132\phi_p^3 - 2354.735\frac{\phi_p}{T^3}$ $+ 23..498\frac{\phi_p^2}{d_p^2} - 3.0185\frac{\phi_p^3}{d_p^2}$ $1\,\% \le \phi_p \le 9\,\%,\ 20 \le T(^\circ C) \le 70,\ 13\text{ nm} \le d_p \le 131\text{ nm}$
Thermal conductivity	Al_2O_3–water and CuO–water $\frac{k_{eff}}{k_f} = 1.0 + 1.0112\phi_p + 2.4375\phi_p\left(\frac{47}{d_p(\text{nm})}\right) - 0.0248\phi_p\left(\frac{k_p}{0.613}\right)$	Al_2O_3–water $\frac{k_{eff}}{k_f} = 0.9843 + 0.398\phi_p^{0.7383}\left(\frac{1}{d_p(\text{nm})}\right)^{0.2246}\left(\frac{\mu_{eff}(T)}{\mu_f(T)}\right)^{0.0235}$ $- 3.9517\frac{\phi_p}{T} + 34.034\frac{\phi_p^2}{T^3} + 32.509\frac{\phi_p}{T^2}$ $0 \le \phi_p \le 10\,\%,\ 11\text{ nm} \le d \le 150\text{ nm},\ 20\,^\circ C \le T \le 70\,^\circ C$

nanofluids were shown to boost the efficiency of PVT devices from both energy and exergy assessments.

14.2 Nanofluids and their Use in Energy Systems

The use of nanofluids in solar collectors such as flat plate collectors, solar stills, PVT systems, and parabolic trough solar collectors (PTSC) is discussed in this section. Solar collectors have low energy efficiency in general. Many attempts have been made in the last few decades to increase the thermal efficiency of the solar collector by adding nanoparticles to the base fluid. A review of the literature reveals a large information on nanofluids as working fluids in solar collectors.

14.2.1 Flat Plate Solar Collector (FPSC)

The effectiveness of traditional flat plate collectors is heavily influenced by incident solar radiation and climatological variables.[22] Furthermore, conventional heat transfer fluids used in solar collectors have been well established in the literature to exhibit low absorptive properties and low thermal conductivity. A schematic representation of the flat plate collector is shown in Figure 14.1

Nanoparticles are used to mimic the thermal and radiative properties of standard heat transfer fluids.[22–25] In an experimental investigation, Ahmadlouydarab *et al.*[26] looked at the influence of TiO_2–water nanofluids on the thermal efficiency of a flat plate collector. In their study, they looked at several volume fractions of TiO_2 nanoparticles, ranging from 0.1% to 5%, with the effectiveness of the solar collector raised by 45% and 17% for TiO_2 volume fractions of 5% and 2.5%, respectively. Tyagi *et al.*[24] theoretically investigated the feasibility of using an Al_2O_3–water nanofluid in a direct absorption solar collector (DASC) system. According to the numerical data published in that study, the Al_2O_3–water nanofluid boosted the absorption of incoming radiation by more than 9 times when compared to pure water. Furthermore, the efficiency of the DASC system was 10% higher than that of

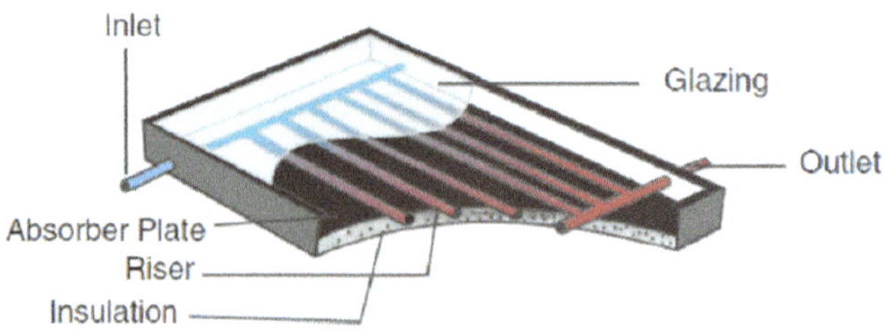

Figure 14.1 Flat plate collector schematic design.[18] Reproduced from ref. 18, https://doi.org/10.1007/s10973-020-09760-2, under the terms of the CC BY 4.0 license, http://creativecommons.org/licenses/by/4.0/.

a flat plate collector. Tong *et al.*[24] used Al_2O_3 and CuO nanoparticles to examine the thermal properties of a flat plate solar collector. Using Al_2O_3 and CuO nanofluids, the thermal efficiency of the flat plate was increased by 21% and 16%, respectively, according to the results (Figure 14.2). Furthermore, the addition of a 1% volume fraction of Al_2O_3 and a 0.5% volume fraction of CuO enhanced exergy efficiency by 56.9% and 49.6%, respectively.

In experimental research, Mondragon *et al.*[28] studied the performance of a flat plate collector using an Al_2O_3–water nanofluid. The researchers demonstrated that using a 1% volume fraction of nanoparticles boosted the heat transfer coefficient by 2.3% in theory. The thermal efficiency of the solar collector, however, reduced from 47% when using water to 41.5% when using an Al_2O_3–water nanofluid, as shown in Figure 14.3. The decline in solar collector thermal performance was attributable to the deposition of nanoparticles on the absorber tube due to the low velocity of the nanofluid. Heat transmission barriers were thought to be created by the deposited nanoparticle layers.

In flat plate and U-tube solar collectors, Kang *et al.*[29] tested the thermal performance of an Al_2O_3–water nanofluid compared to water, with the nanoparticles enhancing the thermal performance of both solar collectors.

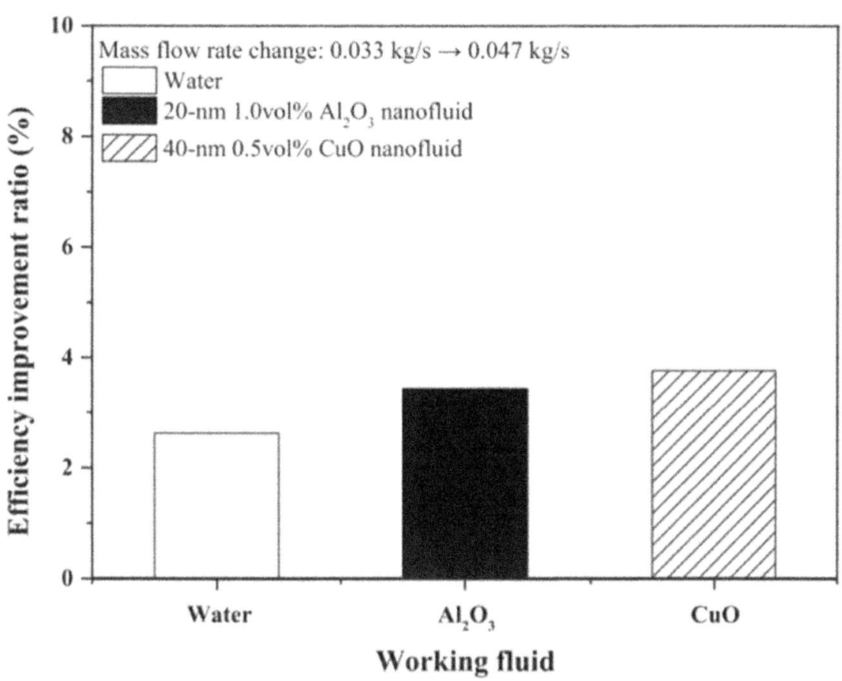

Figure 14.2 Efficiency enhancement ratio for different diameters of nanoparticles at various volume fractions.[27] Reproduced from ref. 27 with permission from Elsevier, Copyright 2019.

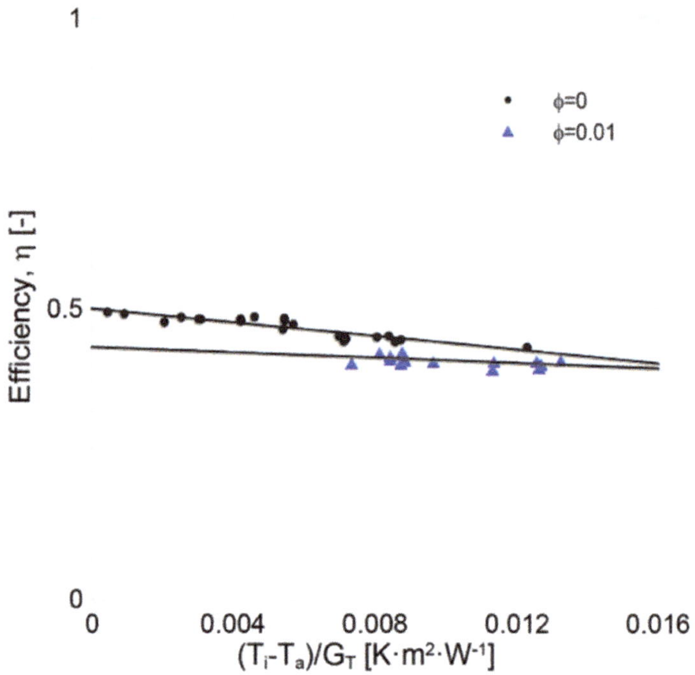

Figure 14.3 Comparison between the solar collector efficiency of water and an Al_2O_3–water nanofluid.[28] Reproduced from ref. 28, https://doi.org/10.1371/journal.pone.0212260, under the terms of the CC BY 4.0 license, https://creativecommons.org/licenses/by/4.0/.

They also discovered that the maximum thermal efficiency of the flat plate and U-tube solar collectors were 74.9% and 72.4%, respectively, when using a 1% volume fraction of nanoparticles with a 20 nm diameter. The thermal efficiency of a flat solar collector was reduced as the particle diameter increased, which was an interesting finding from their experiments (Figure 14.4).

14.2.2 Solar Stills

The solar still is the most specific technology for extracting fresh water using solar energy. The amount of absorbed energy and the operating temperature have a significant impact on the performance of a solar still. Adding nanoparticles to the base fluid increases the thermal efficiency of solar stills, according to several research studies recently published in the literature. A review on the use of nanofluids in solar stills to improve heat and mass characteristics was published by Bait and Si-Ameur.[30] The authors proposed that more research be done on the impacts of nanoparticle size and quantity and the stability and structure of nanoparticles. The effect of employing an Al_2O_3–water nanofluid on the productivity of a single slope solar still was

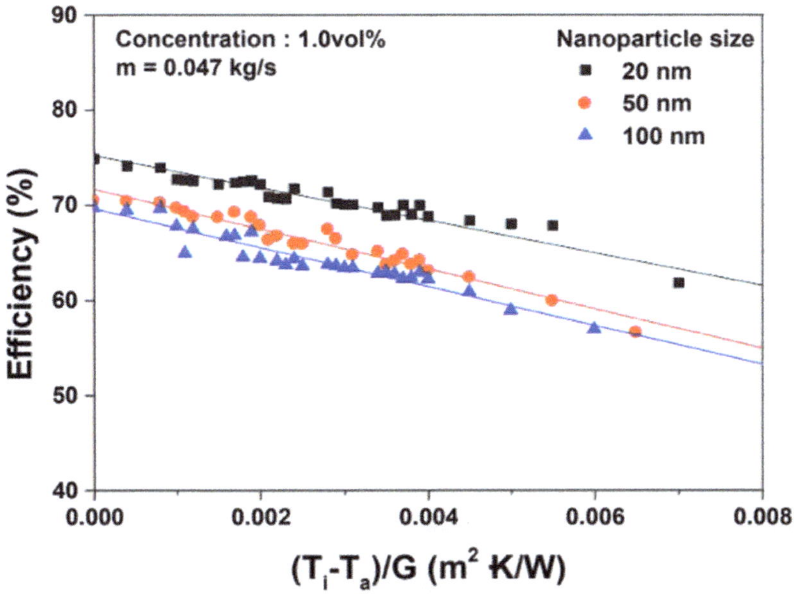

Figure 14.4 The effect of nanoparticle size on flat plat solar collector thermal performance.[29] Reproduced from ref. 29, https://doi.org/10.3390/en10111911, under the terms of the CC BY 4.0 license, https://creativecommons.org/licenses/by/4.0/.

studied quantitatively by Rashidi *et al.*[31] They used an entropy generation study to analyse the solar still system using the 2nd rule of thermodynamics. According to the authors' findings, increasing the volume percentage of nanoparticles from 0% to 5% increased productivity by 25% (Figure 14.5). Furthermore, when the volume percentage of nanoparticles was increased from 0% to 5%, the viscosity and thermal entropy production increased by approximately 95% and 25%, respectively.

Iqbal *et al.*[32] published a review of the possible uses of nanofluids in improving the productivity and thermal performance of solar stills. The authors in that review documented that Al_2O_3 nanoparticles represented the most used nanoparticles in solar stills due to their availability and low cost. Moreover, the authors reviewed and discussed the challenges associated with integrated systems.

14.2.3 Solar Parabolic Trough Collector (PTCs)

Due to their uses in many energy sectors, such as power generation, hydrogen production, solar cooling, and heat production, solar PTCs are widely utilized in industry.[33–35] In the literature, many strategies have been explored to increase the thermal efficiency of solar PTCs. The most often used approaches are employing nanofluids and turbulators.[36] However,

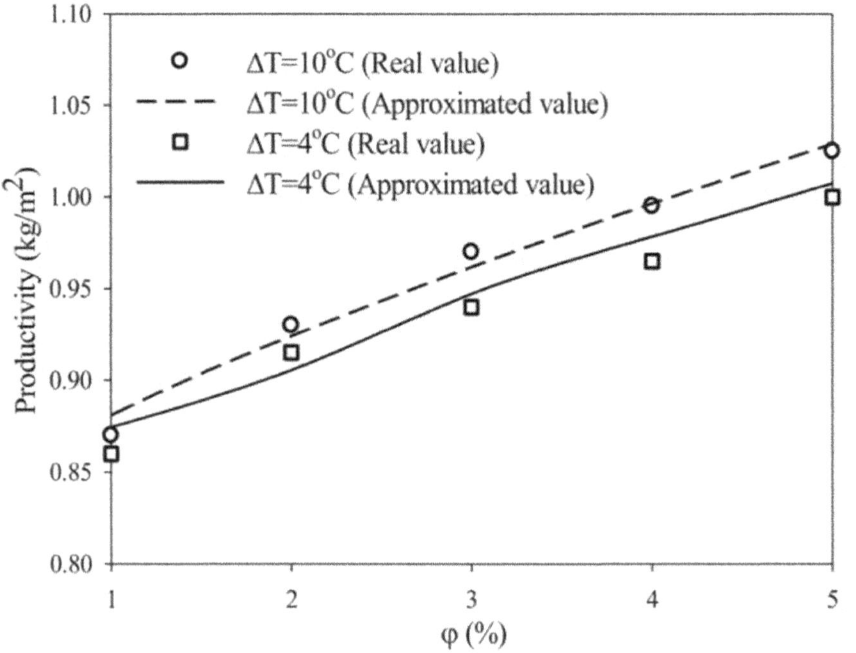

Figure 14.5 The impact of changing the nanoparticle volume percentage on the production of a solar still.[31] Reproduced from ref. 31 with permission from Elsevier, Copyright 2018.

turbulator processes show a significant pressure decrease down the tube, making them less desirable. The effect of utilizing nanofluids on the thermal performance of PTCs has been studied computationally and experimentally in several research studies. Subramani *et al.*[37] investigated the impact of employing Al_2O_3/deionised (DI)–H_2O nanofluids on the thermal performance of a PTC in an experimental investigation. In that investigation, two volume fractions of nanoparticles (0.05% and 0.5%) were employed. The thermal efficiency of the solar collector was enhanced by up to 56% when utilizing a volume fraction of 0.5% and a mass flow rate of 0.05 kg s^{-1}, according to the experimental data reported. Mwesigye *et al.*[38] used a computational model to investigate the impact of utilizing a synthetic oil–Al_2O_3 nanofluid on the thermal performance of a PTC. In that investigation, different volume fractions of nanoparticles were employed (0–8%). The research results showed that using nanofluids, the thermal efficiency of the PTC was increased by 7.6%.

Bellos and Tzivanidis[39] conducted a comparison study to see how different nanoparticles distributed in Syltherm 800 oil affected the thermal performance of a PTC. In the investigation, several flow rates and intake temperatures were evaluated, with Cu shown to be the most effective nanoparticle compared to other nanoparticles (Fe_2O_3, TiO_2, Al_2O_3, and SiO_2). The thermal efficiency gain was 0.31%, 0.54%, and 0.74% for Cu nanoparticles volume fractions of 2%, 4%, and 6%, respectively, at an intake temperature of 600 K and a volume flow rate of 150 L min^{-1}. Ghasemi and Ranjbar[40] investigated the effect of several nanofluids on the thermal performance of a solar PTC using a computational model employing CuO–water and Al_2O_3–water nanofluids. The average Nusselt number increased as the volume percentage of nanoparticles increased in both nanofluids, according to the data. Furthermore, the researchers discovered that using Al_2O_3–water and CuO–water nanofluids at a volume percentage of 3% increased the heat transfer coefficient by up to 28% and 35%, respectively.

14.2.4 Photovoltaic Thermal (PVT) Systems

PVT systems, sometimes referred to as hybrid solar collectors, are commonly used to convert solar energy into thermal and electrical power. In the previous decade, several studies have been published on the effect of employing nanofluids as cooling fluids to improve the performance of hybrid PVT systems. Hissouf *et al.*[41] used energy balance equations to do a computational simulation to see how the performance of a hybrid system would be affected by utilizing Cu–water and Al_2O_3–water nanofluids. At a Cu volume fraction of 2%, the results showed that a Cu–water nanofluid increased both electrical and thermal efficiency by 1.9% and 4.1%, respectively. However, compared to pure water, an Al_2O_3–water nanofluid showed improvements of 2.7% and 1.2%, respectively. Radwan *et al.*[42] used microchannel heat transfer and nanofluids to simulate the performance of a low concentrated PVT system. For various volume fractions of nanoparticles, both Al_2O_3–water and SiC–water nanofluids were examined, with both nanofluids showing a considerable drop in cell temperature. Furthermore, it was found that the SiC-based nanofluid reduced cell temperature more effectively than the Al_2O_3-based nanofluid.

14.3 Energy and Exergy Analysis

As can be noticed from the above literature review, the working fluid of a flat plate solar collector plays a crucial role in improving its thermal performance; moreover the thermodynamic analysis, including energy and exergy analysis, is a mandatory tool to determine which parameters could result in enhancing the performance of flat plate solar collectors. In this regard, a summary of the literature on the use of nanofluids, including their nanoparticle types, base fluid and concentration, in flat plate solar collectors, is given in Table 14.3, where the energy and exergy efficiencies are based on the mathematical formulations given in the energy and exergy analysis section below.

Due to the energy absorbed by the absorber plate, the fluid in the flat plate solar collector heats up, with the heat gain of the fluid defined as:

$$\dot{Q} = \dot{m}c_p(T_o - T_i) \tag{14.1}$$

Table 14.3 Summary of nanofluidic–FPSC thermal performance.

Ref.	Material	Base fluid	Vol%	Energy efficiency	Exergy efficiency	Remarks
46	CeO_2	H_2O	0.0667%	10.74	N/A	Improvement compared to water
47	Al_2O_3	H_2O	3%	74.39	7.7	Exact efficiency values
48	WO_3	H_2O	0.0667%	13.48	N/A	Improvement compared to water
49	Graphene	H_2O	0.2%	16.97	21.46	Improvement compared to water
49	SiO_2	H_2O	0.3%	4.08	5.74	Improvement compared to water
49	TiO_2	H_2O	0.35%	5.09	6.96	Improvement compared to water
49	CuO	H_2O	0.18%	12.64	16.67	Improvement compared to water
49	MWCNT	H_2O	0.2%	23.47	29.32	Improvement compared to water
49	Al_2O_3	H_2O	0.2%	8.28	10.86	Improvement compared to water
50	MWCNT	Distilled H_2O	0.1%	76.92	N/A	Exact efficiency value
51	Al_2O_3	H_2O	0.3%	83.5	20.3	Exact efficiency values
52	SWCNT	H_2O	—	95.12	26.25	Exact efficiency values
27	CuO	H_2O	0.5%	73.9	49.6	Energy efficiency values are exact. Exergy efficiency value is improved compared to water
38	Al_2O_3	H_2O	1%	77.5	59.6	Energy efficiency values are exact. Exergy efficiency value is improved compared to water

Table 14.3 (*Continued*)

Ref.	Material	Base fluid	Vol%	Energy efficiency	Exergy efficiency	Remarks
53	Al_2O_3	H_2O	0.1%	70.7–73.7	15.3–15.8	Exact efficiency values
54	Fe_3O_4	H_2O	2%	83.97	N/A	Exact efficiency value
55	TiO_2	H_2O	2%	48.67	N/A	Exact efficiency value
56	Al_2O_3	H_2O	0.1423%	N/A	10.68	Exact efficiency value
57	Fe	H_2O/ propylene glycol	0.5%	59.5	N/A	Exact efficiency value
47	MWCNT	Distilled H_2O	0.1%	70.67	23.35	Thermosiphon flat plate solar collector. Exact efficiency values
43	Al_2O_3	Distilled H_2O	1%	16.63	N/A	Thermosiphon flat plate solar collector. Exact efficiency value
58	MWCNT	H_2O	0.05%	59	11.5	Maximum efficiency values
59	Al_2O_3	H_2O	0.1%	83.17	18.73	Maximum efficiency values
60	TiO_2	H_2O	0.1%	22	N/A	Improvement compared to water
61	CeO_2	H_2O	0.01%	21.5	N/A	Improvement compared to water
62	CuO_2	H_2O	0.1%	55.1	N/A	Improvement compared to water
63	Al_2O_3	H_2O	0.1%	2.16	N/A	Improvement compared to water
64	Graphene nanoplatelets	H_2O	0.1 wt%	13.3	N/A	Improvement compared to water
65	Graphene nanoplatelets	H_2O	0.025–0.1 wt%	18.2	N/A	Improvement compared to water
66	CuO	H_2O	5%	17.98	N/A	Improvement compared to water

Table 14.3 (*Continued*)

Ref.	Material	Base fluid	Vol%	Energy efficiency	Exergy efficiency	Remarks
66	Al_2O_3	H_2O	5%	14.51	N/A	Improvement compared to water
67	CeO_2	H_2O	0.05%	28.07	N/A	Improvement compared to water
68	CuO	H_2O	0.1%	50	N/A	Improvement compared to water
68	Al_2O_3	H_2O	0.1%	16	N/A	Improvement compared to water
69	Al_2O_3	H_2O	N/A	4.47	N/A	Improvement compared to water
69	SiO_2	H_2O	N/A	4.65	N/A	Improvement compared to water
69	CuO	H_2O	N/A	5.22	N/A	Improvement compared to water
70	CuO/MWCNTs	H_2O	0.75–1 wt%	—	21	Improvement compared to water
70	MgO/ MWCNTs	H_2O	0.7–1 wt%	—	20	Improvement compared to water
71	Al_2O_3/TiO_2	H_2O	0.1 wt%	26	N/A	Improvement compared to water
72	CuO	H_2O	0.05%	6.35	N/A	Improvement compared to water
73	Al_2O_3	Distilled H_2O	3%	7	N/A	Improvement compared to water
74	SiO_2	EG/H_2O	0.1%	4–8	N/A	Improvement compared to water
75	SiO_2	Distilled H_2O	0.2 and 0.4 vol%	23.5	N/A	Improvement compared to water
76	Graphene nanoplatelets	Deionized H_2O	0.01 and 0.02 wt%	18.87	N/A	Improvement compared to water
77	Al_2O_3	Double distilled H_2O	0.1–0.3%	18	N/A	Improvement compared to water
78	Al_2O_3	H_2O	0.1 wt%	23.6	N/A	Improvement compared to water

Table 14.3 *(Continued)*

Ref.	Material	Base fluid	Vol%	Energy efficiency	Exergy efficiency	Remarks
79	Al_2O_3	H_2O	0.1 and 0.3%	22	N/A	Improvement compared to water
80	SiO_2	Deionized H_2O	0.2–0.6%	6–8	N/A	Improvement compared to water
29	Al_2O_3	H_2O	1%	14.8	N/A	Improvement compared to water
81	Al_2O_3/CuO	EG/H_2O	0.5–2 wt%	45	N/A	Improvement compared to water
82	Al_2O_3	H_2O	0.2 and 0.4 wt%	17	N/A	Improvement compared to water
82	CuO	H_2O	0.2 and 0.4 wt%	13.3	N/A	Improvement compared to water
82	ZrO_2	H_2O	0.2 and 0.4 wt%	9.2	N/A	Improvement compared to water

where $\dot{m}$ is the fluid mass flow rate, c_p is the specific heat of the fluid, T_i is the fluid inlet temperature and T_o is the fluid outlet temperature.

Eqn (14.1) can be written using the equation derived by Hottel–Whillier as:

$$\dot{Q} = F_R A_p [I\tau\alpha - U_L(T_i - T_a)] \quad (14.2)$$

where A_p is the absorber plate surface area, F_R is the heat removal factor, I is the solar irradiance, T_a is the ambient temperature, U_L is the overall heat loss coefficient, and $\tau\alpha$ is the transmittance–absorptance coefficient.

F_R is defined as:

$$F_R = \frac{\dot{m}c_p}{U_L A_p}\left\{1 - \exp\left[-\frac{F U_L A_p}{\dot{m}c_p}\right]\right\} \quad (14.3)$$

where F is the collector efficiency factor. The solar collector thermal efficiency is defined as:

$$\eta = \frac{\dot{Q}}{I A_p} \quad (14.4)$$

Exergy analysis is a useful method for improving the performance of a system. When a system component or process is brought reversibly into thermal, mechanical, and chemical equilibrium with its surroundings, it gives a measure of the maximum possible work that can be taken from the component or process. The general formula for exergy balance is as follows:

$$E_{\text{in}} = E_{\text{out}} + E_{\text{s}} + E_{\text{d}} \tag{14.5}$$

where E_{in} is the inlet exergy rate, E_{out} is the outlet exergy rate, E_{s} is the stored exergy rate, and E_{d} is the destruction exergy rate. Under steady-state conditions, the stored exergy rate is equal to zero. Hence, eqn (14.5) can be rewritten as:

$$E_{\text{in}} = E_{\text{out}} + E_{\text{d}} \tag{14.6}$$

The fluid flow and radiation of the sun contribute to the input exergy rate. The rate of inflow exergy owing to fluid flow is given by:[43]

$$E_{\text{in,f}} = \dot{m}c_{\text{p}}\left[T_{\text{i}} - T_{\text{a}} - T_{\text{a}}\ln\left(\frac{T_{\text{i}}}{T_{\text{a}}}\right)\right] \tag{14.7}$$

The inlet exergy rate due to solar radiation is obtained as:[44]

$$E_{\text{in,rad}} = IA_{\text{p}}\left[1 - \frac{T_{\text{a}}}{T_{\text{S}}}\right] \tag{14.8}$$

where T_{S} is the temperature of the sun, which is around 4350 K.[45] The exergy rate due to the outlet fluid flow is expressed as:

$$E_{\text{out,f}} = \dot{m}c_{\text{p}}\left[T_{\text{o}} - T_{\text{a}} - T_{\text{a}}\ln\left(\frac{T_{\text{o}}}{T_{\text{a}}}\right)\right] \tag{14.9}$$

The exergy destruction rate is influenced by heat loss from the absorber plate, the temperature difference between the Sun and the absorber plate, the solar radiation loss from the collection surface to the absorber plate, and the temperature difference between the fluid and the absorber plate. These rates of exergy destruction for the above mentioned exergy losses can be expressed as:

$$E_{\text{Q}} = U_{\text{L}}A_{\text{p}}\left[T_{\text{p}} - T_{\text{a}}\right]\left[1 - \frac{T_{\text{a}}}{T_{\text{p}}}\right] \tag{14.10}$$

$$E_{\Delta T1} = (\tau\alpha)_{\text{eff}} IA_{\text{p}}T_{\text{a}}\left(\frac{T_{\text{S}} - T_{\text{p}}}{T_{\text{p}}T_{\text{S}}}\right) \tag{14.11}$$

where $(\tau\alpha)_{\text{eff}}$ is the effective transmittance–absorptance coefficient.

$$E_{\text{r}} = IA_{\text{p}}\left[1 - (\tau\alpha)_{\text{eff}}\right]\left(\frac{T_{\text{S}} - T_{\text{a}}}{T_{\text{S}}}\right) \tag{14.12}$$

and

$$E_{\Delta T2} = \dot{m}c_{\text{p}}T_{\text{a}}\left[\ln\left(\frac{T_{\text{o}}}{T_{\text{i}}}\right) - \frac{T_{\text{o}} - T_{\text{i}}}{T_{\text{p}}}\right] \tag{14.13}$$

The exergy efficiency of the solar collector can thus be defined as:

$$\eta_{ex} = \frac{E_{out,f} - E_{in,f}}{E_{in,rad}} = \frac{\dot{m}c_p\left[T_o - T_i + T_a \ln\left(\frac{T_i}{T_o}\right)\right]}{IA_p\left[1 - \frac{T_a}{T_S}\right]} \tag{14.14}$$

14.4 Concluding Remarks and Future Work

In the literature, nanofluid applications have garnered much attention due to their extraordinary thermal properties. Nanofluids may enhance solar thermal systems such as solar collectors, solar stills, thermal energy storage systems, and PVT systems. In solar thermal systems, nanofluids pose a number of challenges, including stability, dependability, safety, and nanoparticle cost reduction. According to various studies published in the literature, the use of nanofluids improves the thermal and electrical efficiency of solar systems. Nanofluids in PV systems also result in lower cell temperatures. However, many experiments on increasing the thermal properties of nanofluids have been conducted at low temperatures. More studies on nanofluids should thus be conducted at higher temperatures (>100 °C) to test the stability and effect of high temperatures on the thermal properties of nanofluids. Moreover, more experimental studies are also required to analyse the performance of hybrid nanofluids at different temperatures, the size and shape of nanoparticles, and the stability of the hybrid nanofluids. Additional studies are required to be conducted to improve the specific heat capacity of nanofluids. Existing studies show that the addition of nanoparticles leads to a decrease in the effective specific heat compared with the base fluid. The cost of preparing nanofluids is relatively high and therefore more investigations are essential to develop cost-effective approaches to prepare nanofluids.

References

1. S. U. S. Choi, *J. Heat Transfer*, 2009, **131**, 1–9.
2. K. Khanafer, K. Vafai and M. Lightstone, *Int. J. Heat Mass Transfer*, 2003, **46**, 3639–3653.
3. A. R. A. Khaled and K. Vafai, *Int. J. Heat Mass Transfer*, 2005, **48**, 2172–2185.
4. K. Khanafer and K. Vafai, *Int. J. Heat Mass Transfer*, 2011, **54**, 4410–4428.
5. M. Shafahi, V. Bianco, K. Vafai and O. Manca, *Int. J. Heat Mass Transfer*, 2010, **53**, 1438–1445.
6. M. Shafahi, V. Bianco, K. Vafai and O. Manca, *Int. J. Heat Mass Transfer*, 2010, **53**, 376–383.
7. J. A. Eastman, S. U. S. Choi, S. Li, L. J. Thompson and S. Lee, Fall meeting of the Materials Research Society (MRS), Boston, USA, 1996.

8. J. A. Eastman, S. U. S. Choi, S. Li, W. Yu and L. J. Thompson, *Appl. Phys. Lett.*, 2001, **78**, 718–720.
9. S. P. Jang and S. U. S. Choi, *Appl. Phys. Lett.*, 2004, **84**, 4316–4318.
10. S. Lee and S. U. S. Choi, International Mechanical Engineering Congress and Exhibition, Atlanta, USA, 1996.
11. B. X. Wang, L. P. Zhou and X. F. Peng, *Int. J. Heat Mass Transfer*, 2003, **46**, 2665–2672.
12. S. K. Das, N. Putra and W. Roetzel, *Int. J. Heat Mass Transfer*, 2003, **46**, 851–862.
13. S. K. Das, N. Putra and W. Roetzel, *Int. J. Multiphase Flow*, 2003, **29**, 1237–1247.
14. S. K. Das, N. Putra, P. Thiesen and W. Roetzel, *J. Heat Transfer*, 2003, **125**, 567–574.
15. K. Khanafer and K. Vafai, *Renewable Energy*, 2018, **123**, 398–406.
16. J. Li, X. Zhang, B. Xu and M. Yuan, *Int. Commun. Heat Mass Transfer*, 2021, **127**, 105543.
17. A. Ibn Ali and B. Salam, *SN Appl. Sci.*, 2020, **2**, 1636.
18. E. Okonkwo, I. Wole-Osho, I. Almanassra, Y. Abdullatif and T. Al-Ansari, *J. Therm. Anal. Calorim.*, 2021, **145**, 2817–2872.
19. M. U. Sajid and H. M. Ali, *Renewable Sustainable Energy Rev.*, 2019, **103**, 556–592.
20. N. Sezer, M. Atieh and M. Koç, *Powder Technol.*, 2019, **344**, 404–431.
21. A. Salaria, A. Taheri, A. Farzanehnia, M. Passandideh-fard and M. Sardarabadi, *J. Cleaner Prod.*, 2021, **282**, 124318.
22. Z. Su, S. Gu and K. Vafai, *Int. Commun. Heat Mass Transfer*, 2017, **87**, 169–174.
23. R. A. Taylor, P. E. Phelan, T. P. Otanicar, C. A. Walker and M. Nguyen, *J. Renewable Sustainable Energy*, 2011, **3**, 023104.
24. H. Tyagi, P. Phelan and R. Prasher, *J. Sol. Energy Eng.*, 2009, **131**, 041004.
25. T. P. Otanicar, P. E. Phelan, R. S. Prasher, G. Rosengarten and R. A. Taylor, *J. Renewable Sustainable Energy*, 2010, **2**, 033102.
26. M. Ahmadlouydarab, M. Ebadolahzadeh and H. M. Ali, *Phys. A*, 2020, **540**, 123109.
27. Y. Tong, H. Lee, W. Kang and H. Cho, *Appl. Therm. Eng.*, 2019, **159**, 113959.
28. R. Mondragon, D. Sanchez, R. Cabello, R. Llopis and E. Julia, *PLoS One*, 2019, **14**, 0212260.
29. W. Kang, Y. Shin and H. Cho, *Energies*, 2017, **10**, 1911–1926.
30. O. Bait and M. Si-Ameur, *Sol. Energy*, 2018, **170**, 694–722.
31. S. Rashidi, S. Akar, M. Boland and E. Rahmat, *Renewable Energy*, 2018, **115**, 400–410.
32. A. Iqbal, M. Mahmoud, E. Sayed, K. Elsaid, M. Abdelkareem, H. Alawadhi and A. G. Olabi, *J. Environ. Manage.*, 2021, **277**, 111415.
33. Q. Wang, H. Yang, S. Zhong, Y. Huang, M. Hu and J. Cao, *et al.*, *Sol. Energy*, 2020, **196**, 663–672.

34. K. Shahverdi, R. Loni, B. Ghobadian, M. J. Monem, S. Gohari and S. Marofi, *et al.*, *Energy Convers. Manage.*, 2019, **187**, 205–220.
35. G. Quinones, C. Felbol, C. Valenzuela, J. M. Cardemil and R. A. Escobar, *Sol Energy*, 2020, **197**, 292–310.
36. W. Fuqiang, C. Ziming, T. Jianyu, Y. Yuan, S. Yong and L. Linhua, *Renewable Sustainable Energy Rev.*, 2017, **79**, 1314–1328.
37. J. Subramani, P. K. Nagarajan, S. Wongwises, S. A. El-Agouz and R. Sathyamurthya, *Environ. Prog. Sustainable Energy*, 2018, **37**, 1149–1159.
38. A. Mwesigye, Z. Huan and J. P. Meyer, *Appl Energy*, 2015, **156**, 398–412.
39. E. Bellos and C. Tzivanidis, *J. Therm. Anal. Calorim.*, 2019, **135**, 597–608.
40. S. E. Ghasemi and A. A. Ranjbar, *J. Mol. Liq.*, 2016, **222**, 159–166.
41. M. Hissouf, M. Feddaoui, M. Najima and A. Charef, *Sol. Energy*, 2020, **199**, 115–127.
42. A. Radwan, M. Ahmed and S. Ookawara, *Energy Convers. Manage.*, 2016, **119**, 289–303.
43. T. J. Kotas, *The exergy method of thermal plant analysis*, Elsevier, 2013.
44. E. Torres-Reyes, J. Cervantes-de Gortari, B. Ibarra-Salazar and M. Picon-Nunez, *Exergy, Int. J.*, 2001, **1**, 46–52.
45. S. Farahat, F. Sarhaddi and H. Ajam, *Renewable Energy*, 2009, **34**, 1169–1174.
46. M. Sharafeldin and G. Grof, *J. Cleaner Prod.*, 2018, **185**, 347–356.
47. A. M. Genc, M. A. Ezan and A. Turgut, *Appl. Therm. Eng.*, 2018, **130**, 395–407.
48. M. A. Sharafeldin, G. Gróf and O. Mahian, *Energy*, 2017, **141**, 2436–2444.
49. S. K. Verma, A. K. Tiwari and D. S. Chauhan, *Energy Convers. Manage.*, 2017, **134**, 103–115.
50. M. Eltaweel and A. A. Abdel-Rehim, *Case Stud. Therm. Eng.*, 2019, **14**, 100416.
51. Z. Said, R. Saidur, M. Sabiha, A. Hepbasli and N. Rahim, *J. Cleaner Prod.*, 2016, **112**, 3915–3926.
52. Z. Said, *et al.*, *J. Cleaner Prod.*, 2015, **92**, 343–353.
53. Z. Said, R. Saidur and N. Rahim, *J. Cleaner Prod.*, 2016, **133**, 518–530.
54. M. Ameri and M. S. Eshaghi, *Therm. Sci. Eng. Prog.*, 2018, **8**, 223–235.
55. F. Kiliç, T. Menlik and A. Sözen, *Sol. Energy*, 2018, **164**, 101–108.
56. E. Shojaeizadeh, F. Veysi and A. Kamandi, *Energy Build.*, 2015, **101**, 12–23.
57. A. L. Owolabi, H. H. Al-Kayiem and A. T. Baheta, *Int. J. Energy Res.*, 2017, **41**, 650–657.
58. M. N. A. Negm, A. A. Abdel-Rehim and A. A. Attia, *ASME Int. Mech. Eng. Congr. Expo.*, 2016, **50626**, V008T10A097.
59. S. Arora, G. Fekadu and S. Subudhi, *J. Solar Energy Eng.*, 2019, **141**, 041008.
60. A. Sacithra and A. Manivannan, *Heat and Mass Transfer.*, 2019, **55**, 1783–1799.
61. P. M. J. Stalin, T. Arjunan, M. Matheswaran and N. Sadanandam, *J. Therm. Anal. Calorim.*, 2019, **135**, 29–44.

62. M. Mirzaei, *Environ. Prog. Sustainable Energy*, 2019, **38**, 260–267.
63. E. C. Okonkwo, I. Wole-Osho, D. Kavaz, M. Abid and T. Al-Ansari, *Sustainable Energy Technol. Assess.*, 2020, **37**, 100636.
64. O. A. Alawi, H. M. Kamar, A. Mallah, S. Kazi and N. A. C. Sidik, *Sol. Energy*, 2019, **191**, 360–370.
65. N. Akram, *et al.*, *J. Therm. Anal. Calorim.*, 2019, **138**, 609–621.
66. H. Asgharian, E. Baniasadi and C. O. Colpan, *Int. J. Exergy.*, 2019, **30**, 63–85.
67. P. M. J. Stalin, T. Arjunan, M. Matheswaran, H. Dolli and N. Sadanandam, *J. Therm. Anal. Calorim.*, 2020, **139**, 3219–3233.
68. Z. O. Hajabdollahi, M. Mirzaei and K. C. Kim, *Heat Transfer Res.*, 2019, **50**, 945–965.
69. Z. Hajabdollahi, H. Hajabdollahi and K. C. Kim, *J. Therm. Anal. Calorim.*, 2020, **140**, 991–1002.
70. Z. Said, M. A. Abdelkareem, H. Rezk, A. M. Nassef and H. Z. Atwany, *Powder Technol.*, 2020, **364**, 795–809.
71. E. Farajzadeh, S. Movahed and R. Hosseini, *Renewable Energy*, 2018, **118**, 122–130.
72. J. J. Michael and S. Iniyan, *Energy Convers. Manage.*, 2015, **95**, 60–169.
73. P. Visconti, P. Primiceri, P. Costantini, G. Colangelo and G. Cavalera, *Int. J. Smart Sens. Intell. Syst.*, 2016, **9**, 1220–1242.
74. S. S. Meibodi, A. Kianifar, H. Niazmand, O. Mahian and S. Wongwises, *Int. Commun. Heat Mass Transfer*, 2015, **65**, 71–75.
75. M. Faizal, R. Saidur, S. Mekhilef, A. Hepbasli and I. Mahbubul, *Clean Technol. Environ. Policy*, 2015, **17**, 1457–1473.
76. A. Ahmadi, D. D. Ganji and F. Jafarkazemi, *Energy Convers. Manage.*, 2016, **126**, 1–11.
77. A. Hawwash, A. K. A. Rahman, S. Nada and S. Ookawara, *Appl. Therm. Eng.*, 2018, **130**, 363–374.
78. M. Mirzaei, S. M. S. Hosseini and A. M. M. Kashkooli, *Appl. Therm. Eng.*, 2018, **134**, 68–77.
79. L. S. Sundar, M. K. Singh, V. Punnaiah and A. C. Sousa, *Renewable Energy*, 2018, **119**, 820–833.
80. H. J. Jouybari, S. Saedodin, A. Zamzamian, M. E. Nimvari and S. Wongwises, *Renewable Energy*, 2017, **114**, 1407–1418.
81. M. S. Tahat and A. C. Benim, *Trans Tech Publ*, 2017, **374**, 148–156.
82. Y. Devarajan and D. B. Munuswamy, *Energy Fuels*, 2016, **30**, 9908–9913.

CHAPTER 15

Socio-economic and Environmental Impacts of Nanofluids

E. B. ELCIOGLU,[a] A. TURGUT*[b] AND S. M. S. MURSHED[c]

[a] Department of Mechanical Engineering, Eskisehir Technical University, 26555 Eskisehir, Turkey; [b] Department of Mechanical Engineering, Dokuz Eylul University, 35390 Izmir, Turkey; [c] IDMEC, Instituto Superior Técnico, Universidade de Lisboa, 1049-001 Lisboa, Portugal
*Email: alpaslan.turgut@deu.edu.tr

15.1 Introduction

The definition of nanotechnology may be given based on the perspective of Richard P. Feynman, who foresaw the potential of small-scale science. He asked "*[...] what would happen if we could arrange atoms the way we want [...]*" in his revolutionary paper "There is Plenty of Room at the Bottom",[1] wherein he touched on the possibilities that could be brought about by the atom-by-atom manipulation of materials. Nanotechnology is a diverse and multi-disciplinary field that covers study and research on phenomena, materials, systems, and operations on the nanometre scale. Nanoparticles are ultra-fine materials at a size range on the order of nanometres ($\sim 10^{-9}$ m), and in general, defined as particles of 1–100 nm in size, although the need to treat sub-micrometre-sized particles (particularly smaller than the lower limit of the wavelength of visible light) is emphasized.[2] The term nanofluids refers to the mixing of nanoparticles and common industrial fluids (*i.e.*, base fluid),

Soft Matter Series No. 16
Fundamentals and Transport Properties of Nanofluids
Edited by S. M. Sohel Murshed

Published by the Royal Society of Chemistry, www.rsc.org

to obtain properties much better suited to applications than those of base fluids alone. The term nanofluids was coined by Choi and Eastman,[3] who stressed that nanofluids were *"the best hope for enhancement of heat transfer"* due to their higher thermal conductivities compared to those of common industrial liquids. The main focus of this chapter is the use of nanofluids in heat transfer applications, and possible environmental, health-related, and socio-economic outcomes that may potentially be brought about by the use of nanofluids. With this motivation, this chapter aims to draw a realistic picture showing both the promising and challenging points in the path to nanofluid industrialization.

15.1.1 Nanofluids: Overview of Current Status

Nanofluids exhibit promising features in many fields, including energy transfer. Efforts have been made to understand preparation,[4–7] various properties such as thermal conductivity,[8–23] viscosity,[14,18,22,24–27] specific heat,[28,29] surface tension,[30,31] and contact angles,[32] as well as the stability[33,34] of nanofluids. In addition to properties,[5,7,35–39] the heat transfer performance,[5,21,35,37,40–54] pressure drop,[52,55] pumping power,[37] and challenges[6,15,56–58] of nanofluids have been studied. Mechanisms that are responsible for the improvement in conduction heat transfer, particularly thermal conductivity, have been identified (for example, see the literature[49,59–61]). Nanofluids have found potential use in industrially-relevant areas (*e.g.*, see the review by Okonkwo *et al.*[42]) and systems including heat exchangers,[62,63] photovoltaic/thermal systems,[62,64–66] solar energy harvesting,[62,64,65,67,68] electronics cooling,[57] spray cooling,[69] refrigeration systems,[42] minimum quantity lubrication machining,[70] and the optimization of thermal units processes.[71] Besides heat transfer applications, nanofluids have been under research for biomedical applications.[72]

The current status of nanofluids research and development is summarized in Figure 15.1, which shows that the advantages, challenges, and development-usage cycle issues are intertwined.

15.1.2 On the Way to Industrialization

Understanding the behaviour of nanofluids in any application requires recognizing microscopic forces that act on a single nanoparticle, between interacting nanoparticles, and on the nanofluid as a whole.[73] Among the complex phenomena that affect nanofluid properties, colloidal instability is a major and quantifiable phenomenon. It originates from the colliding of nanoparticles due to attractive (can be amplified at high temperatures)[3] and repulsive forces. Tackling this issue thus requires measures from chemistry, materials science, and thermal engineering perspectives. The agglomeration–thermal conductivity dependence as well as temporal variations in dispersed concentration are discussed in the literature.[44,74] Nanoparticle agglomeration has been regarded mostly as undesired. However, it can also be considered advantageous, for example by Wen *et al.*,[44] who pointed out that there was an

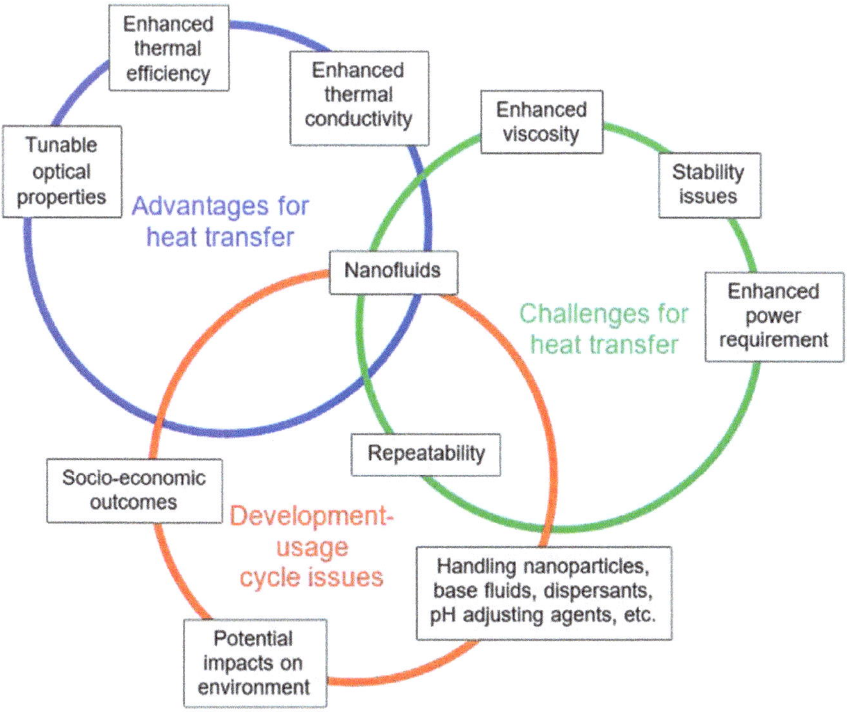

Figure 15.1 Advantages, challenges, and development-usage cycle issues of nanofluids.

optimum degree of aggregation maximizing the enhancement in thermal conductivity. Dependence of thermal conductivity on the level of nanoparticle aggregation makes the thermal performance heavily dependent on colloidal stability and highlights the importance of stability issues.

15.1.3 Challenges: Preparation to Performance

The major issue with nanofluids is sustaining the obtained improvements over time. Temporal stability of nanofluids is mainly responsible for the deterioration of performance and for posing serious challenges for their sustainable applications. Cuce *et al.*[65] highlighted some challenges in nanofluids applications, such as high production costs, nanoparticle agglomeration and stability, increased pump power requirements, corrosion on thermal components for long-term usage, and health concerns. Guo[49] pointed out that the issues to be solved to enable the wider usage of nanofluids include long-term stability, nanofluid preparation, uncertainties, need for systematic heat transfer enhancement studies, understanding nanofluid heat transfer on the atomic and nanoscale levels, and so on. Among other challenges, a lack of a common protocol in the preparation and analysis of nanofluids, as well as a lack of scale-up production also need to be overcome.[15] Sharaf *et al.*[75] reported that for solar nanofluids, storage time, working temperature, and thermal cycling are among the issues that still need to be resolved.

It is important to point out efforts making such materials development and transition accessible to researchers and future users, so that all parties can contribute to the know-how and industrialization of nanofluids. Inconsistencies and scattering in nanofluid data emphasize the importance of comprehensive (large-scale) assessments using sufficiently large datasets. In this regard, the database presented by Mondejar *et al.*[76] should be mentioned, which covers more than 8000 thermodynamic and transport property data of different nanofluids. It is critically important that along with comprehensive investigations regarding nanofluids technology (*e.g.*, theoretical assessments) environmental, health, and socio-economic impact investigations should be performed simultaneously.

15.2 Socio-economic Impacts

Technological developments aim at improving quality of life, providing benefits to society, and delivering solutions to existing problems as making ways for new areas of employment and possible economic developments at the same time. Nanotechnology is a highly dynamic field with high pervasiveness and innovations for a high number of applications.[77] Brignon[78] studied socio-economic analyses as a tool for evaluating nanotechnology impact and provided the definition of socio-economic analysis as a variant for regulatory impact analysis aimed at "balancing the benefits and costs of action or lack of action". Palmberg *et al.*[79] discussed the socio-economic impacts of nanotechnology in terms of market forecasts, companies, products and applications, and jobs. Foladori and Invernizzi[80] stressed that nanotechnology has been advancing at a time where the social differences in the world are alarming. They highlighted that nanotechnology revolution might give rise to a widening of the gap between the poor and rich in the world (as for other technologies). These studies report the nanotechnology–socio-economic impact relationship and may also pave the way for such evaluations on nanofluids. To the best of the authors' knowledge, a socio-economic impact analysis of nanofluids, which is an emerging area in nanotechnology, has not yet been performed, while some works (for example ref. 81) have indicated that improved devices would have a socio-economic impact on developing countries due to the increased lifetime of devices, by saving money and decreasing losses. We highlight the necessity of such evaluations and herein examine the latest status of nanofluids from environmental, socio-economic, availability and accessibility points of view.

15.2.1 Need for Conscious Decisions on the Environment

Energy is indispensable for life. However, meeting the energy needs of large populations requires efficient and environmentally friendly decisions.

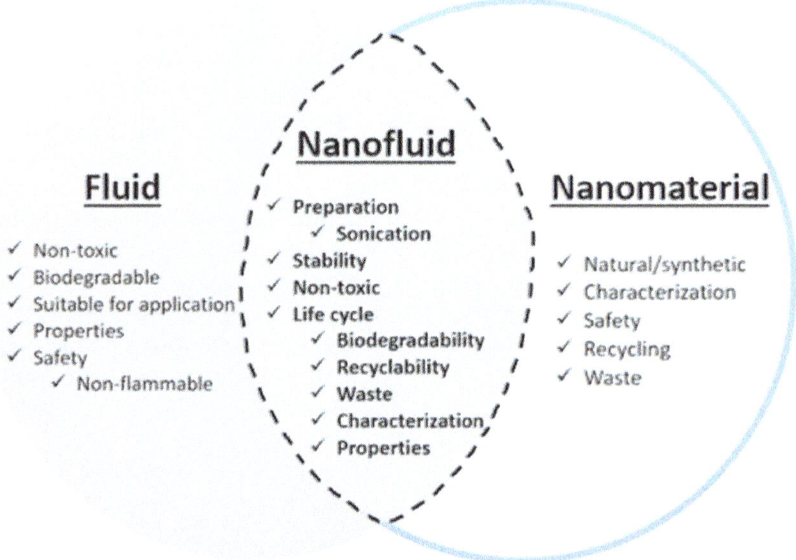

Figure 15.2 Aspects of nanofluids use and preparation by adhering to safety, health and environmental regulations. Adapted from ref. 82, https://doi.org/10.3390/nano11081871, under the terms of the CC BY 4.0 license, https://creativecommons.org/licenses/by/4.0/.

Nanofluids with their improved characteristics are considered important in a variety of energy applications due to their potential high thermal efficiencies. Systematic research is still needed to assess impacts other than heat transfer performance, to gain insight on the handling, storage, usage, preparation, waste management, recycling and reuse of nanofluids. A study[82] recently highlighted current regulations on nanomaterials and their applicability to nanofluids and suggested some points, as illustrated in Figure 15.2, to be considered during the preparation of nanofluids, in accordance with safety, health and environmental regulations. However, more extensive and systematic protocols on the handling, use and disposal of nanoparticles and nanofluids from these serious issues related to health and the environment need to be addressed and adopted as soon as possible.

15.2.2 Availability and Accessibility of Nanofluids: Economic Impacts

Once development of sustainable and repeatable routes for new materials are confirmed (*via* their technology readiness level (TRL)), their industrialization may begin. Such processes should exhibit high efficiency, sufficient lifespan, eco-friendliness, and also lower costs compared to their existing counterparts.[83] Some efforts have also been made to investigate nanofluids from an economic point of view. For example, Faizal *et al.*[84] stated that SiO_2

nanofluids provide greater yearly cost savings and a shorter payback period compared with a water-based flat-plate solar collector system. Nanofluid cost calculation methods and definitions have been proposed[85–90] along with a comparison of system costs in the presence and absence of nanofluids. Such evaluations have been performed for solar[84,91–104] and other energy applications.[89,105–109] For other studies on evaluating nanofluids and their corresponding system costs, readers are referred to the literature.[88,110–120] Alirezaie *et al.*[121] evaluated water-based Ag, MgO, multi-walled carbon nanotube (MWCNTs), and double-walled CNT nanofluids based on their efficiency-price index (EPI) and reported that MgO–water nanofluids exhibited the highest EPI among the other materials considered. Esfe *et al.*[122,123] and Alirezaie *et al.*[124] performed economic evaluations using the price-performance factor (PPF) defined proportional to the ratio of thermal conductivity ratio to the price of nanofluids. Xuan *et al.*[125] defined separate PPFs for laminar and turbulent flow conditions.

In general, nanofluid performance has been regarded advantageous with reasonable payback periods, while there were several issues that require careful consideration. Due to the use of strong chemicals, undesired outcomes, including corrosion and/or erosion of the host equipment,[83,126,127] may be faced with nanofluids, which should be considered during the economic analysis of some nanofluid-based heat transfer systems. Wciślik[85] pointed out that nanofluids with surfactants for high temperature systems require the use of ultrasonic cleaners and mechanical stirrers, increasing the associated operational costs. Salari *et al.*[128] mentioned that nanofluid-based photovoltaic thermal systems are not economic due to the high costs of nanoparticles, but they expect that with time and increased demand, nanoparticle costs might decrease, making such systems more economic. Based on the mentioned studies, it can be concluded that in-depth economic analyses should accompany nanofluid thermal performance investigations, and should be updated as devices evolve and materials opportunities become more diverse.

15.3 Human and Environmental Impacts

15.3.1 Need for Environmental Impact Examination

To fully assess the potential of nanofluids on the path to industrialization, the impact of the use of nanofluids at the academic, research, industry and environmental level should be evaluated. The use of nanofluids should be optimized in a way that the impacts of nanofluid components, nanofluid preparation methods and their usage are known, quantifiable, and made environmentally friendly in the short and long term. Figure 15.3 shows a nanofluids environmental impact assessment map, showing points that must be systematically assessed. This map should be updated, as the long-term usage of components, devices and performing the efforts may change as nanofluid technology advances, and new items discovered along the path

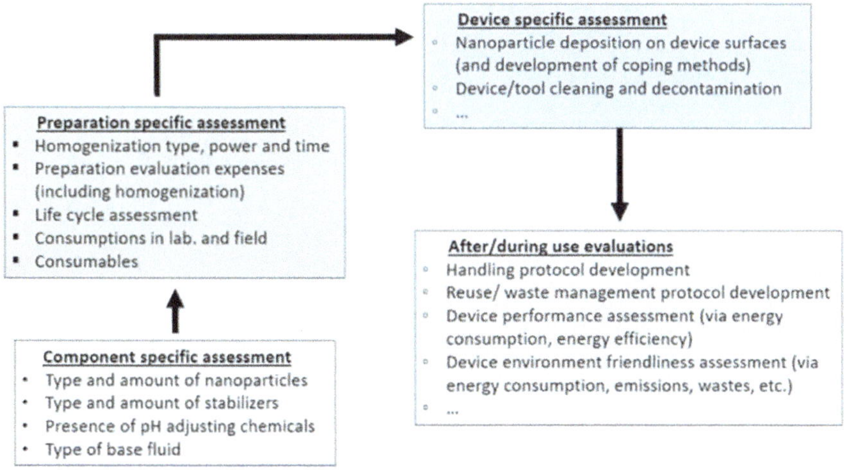

Figure 15.3 Environmental impact assessment map of nanofluids.

may be added. The environmental impact of nanofluids includes the combination of environmental impacts of its components[129] and other stabilizing additives – if present.

Although the use of water (the most commonly used base fluid,[76] which is non-toxic, non-flammable and readily available) and some other base fluids has matured, potential issues brought about by handling, using, storing, and being exposed to them, nanoparticles, and surfactants, *etc.*, should be considered. Zainon and Azmi[130] compared ethylene glycol (EG), propylene glycol (PG), and renewable bio-glycol (BG) and stressed that PG was safer than EG, but both are petroleum-based and non-renewable, while BG is non-toxic, biodegradable and domestically produced.

It is also important to highlight the conditions of use (*e.g.*, high and low temperatures). Elsaid *et al.*[129] suggested that it was convenient to focus on the effects originating from the nanoparticle type (and resulting chemical, toxicological and environmental characteristics, among others), nanoparticle concentration, and nanofluid preparation methods. They noted that nanofluid preparation methods, nanoparticles sourced from natural resources, and nanoparticle–base fluid interactions are among the factors that have environmental impacts. Bretado-de los Rios *et al.*[62] proposed that for nanofluid assessment in energy systems, nanoparticle prices, nanofluid preparation, and pump power requirements should be taken into account. They also suggested that the use of nanofluids should be justified considering economic (*e.g.*, payback period), environmental, and technical aspects.

Green nanotechnology refers to efforts incorporating nature and minimizing the risks caused by the use of nanomaterials on humans and the environment while encouraging substitution of existing conventional products with environmentally friendly nanoscale products.[130] Zainon and Azmi[130] reported eco-friendly nanofluids produced using plants, fruits and

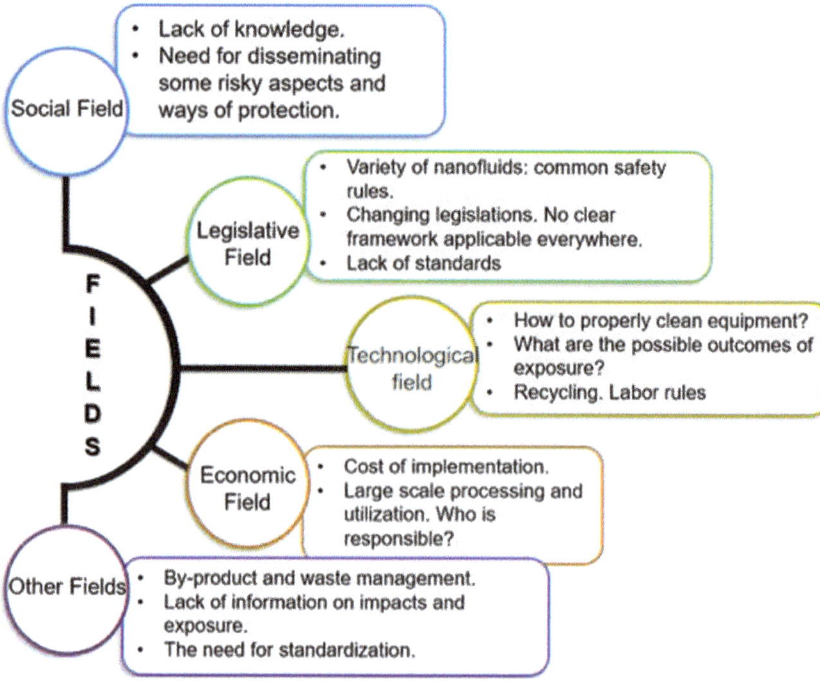

Figure 15.4 Summary of a brainstorming session on the political, social, technology, ecology, and legal issues associated with nanofluids.[131]

waste materials that exhibit promising thermophysical properties compared to other nanofluids. As the environmentally friendly green nanofluids[83] field matures, they may become promising heat transfer fluids. Hernandez *et al.*[131] presented a report on nanofluid health, safety and environmental impacts with respect to negative aspects on political, social, technology, ecology, and legal issues. Figure 15.4 illustrates the contributions of the participants to a workshop at which these issues were discussed.

15.3.2 Potential Health Risks

The implementation of nanoparticles in a number of nanotechnology-related practices is considered advantageous and beneficial to society. However, there have been some concerns about the health and environmental impacts of nanoparticles, especially with regards to toxicity,[132] since every compound when in too high a concentration may have some adverse, even toxic effects.[133] Nanofluids must also be examined with regards to health and usage cycle issues. Due to the use of nanoparticles in a wide range of research and applications, humans are exposed to nanoparticles, which may cause short-term and/or long-term effects, including toxicity.[134] Some nanoparticles that exhibit flammable and pyrophoric properties are risky[83] to handle during their processes. For example, Said *et al.*[135] reviewed nanofluid-cooled thermal photovoltaic systems and reported that the toxicity of nanoparticles used should be considered and that the establishment of protocols, standards, and regulations were needed for the use of nanofluids as coolants for their future use in thermal photovoltaic systems.

15.3.3 Environment and Sustainability

The synthesis and research of nanoparticles is becoming more common, resulting in increased human–nanoparticle and environment–nanoparticle interactions.[134] Effects on the environment and sustainability can be evaluated by means of the reduction of CO_2[100,103] and/or other emissions[95,101,136] by systems using nanofluids. For example, Faizal *et al.*[84] analysed a SiO_2 nanofluid flat-plate solar collector system and reported that embodied energy emissions (CO_2, SO_x, and NO_x) with a SiO_2 nanofluid were lower than those using water alone. Taborda *et al.*[137] studied the transport of heavy crude oil through pipelines and studied the use of SiO_2-based nanofluids to reduce the viscosity of crude oil, and hence reduce the consumption of naphtha, which would reduce environmental and operational risks. The use of nanofluids is meaningful for increasing the performance and efficiency of systems with reduced emissions as mentioned by some of the works in the literature. For example, Salari *et al.*[128] reported that the effects of nanoparticle production should also be accounted for, as nanoparticles might contaminate the environment.

15.4 Conclusions

Innovations have always been transportive, as scientific and technological breakthroughs have high potential to enhance the quality of lives by approaching problems faced by societies as well as improving industrial systems using efficient and environmentally friendly ways. Innovations must be respectful to, and conscious of, their potential and impact on health and the environment. Nanofluids exhibit high potential since they are nanoparticle-enhanced versions of commercial heat transfer fluids, and they exhibit outstanding and tuneable properties. This allows for the development of new systems and applications once sustainable preparation and handling routes are developed and precise tuneability is possible.

The current state of nanofluids research is at a point where despite numerous efforts having been made resulting in an enormous amount of data, serious challenges remain to be tackled. As for all technologies under development, the socio-economic impacts must be studied. This is to be an ongoing effort, as when new parties start using nanofluids, their socio-economic impact will be widened. Nanofluids, as mixtures of nanoparticles, base fluids, and sometimes other chemicals, must be handled, prepared, used, stored safely and be environmentally friendly. So far, most of the system-specific research predicts that nanofluids are more advantageous (with higher thermal efficiencies), economical (with reasonable payback periods), and respectful to environment (with lower emissions) compared to their base fluid counterparts, although some research has concluded the opposite. Moreover, nanofluids should be approached from a cost and benefit analysis point of view, to decide whether their use is suitable or not for a certain application. Such analysis should include nanofluid (or their components) costs, whether or not they are renewable, potential health consequences upon exposure, emissions resulting from nanofluid-based systems, the recycling of nanofluids, and practical challenges, such as the necessity of increased pump power requirements, among others.

Acknowledgements

This work was supported by the Fundação para a Ciência e Tecnologia (FCT), Portugal through project grant number: PTDC/NAN-MAT/29989/2017.

References

1. R. Feynmann, *Eng. Sci.*, 1960, **23**, 22.
2. *Nanoparticle Technology Handbook*, ed. M. Hosokawa, K. Nogi, M. Naito and T. Yokoyama, Elsevier, Oxford, UK, 2007.
3. S. U. S. Choi and J. A. Eastman, presented in *ASME International Mechanical Engineering Congress & Exposition*, San Fransisco, CA, 1995.
4. M. Jama, T. Singh, S. M. Gamaleldin, M. Koc, A. Samara, R. J. Isaifan and M. A. Atieh, *J. Nanomater.*, 2016, **2016**, 6717624.
5. K. S. Suganthi and K. S. Rajan, *Renewable Sustainable Energy Rev.*, 2017, **76**, 226.
6. N. A. C. Sidik, H. A. Mohammed, O. A. Alawi and S. Samion, *Int. Commun. Heat Mass Transfer*, 2014, **54**, 115.
7. D. K. Devendiran and V. A. Amirtham, *Renewable Sustainable Energy Rev.*, 2016, **60**, 21.
8. P. Keblinski, R. Prasher and J. Eapen, *J. Nanoparticle Res.*, 2008, **10**, 1089.
9. G. Paul, M. Chopkar, I. Manna and P. K. Das, *Renewable Sustainable Energy Rev.*, 2010, **14**, 1913.
10. A. Hilo, A. R. A. Talib, S. R. Nfawa, M. T. H. Sultan, M. F. A. Hamid and M. I. N. Bheekhun, *J. Adv. Res. Fluid Mech. Therm. Sci.*, 2019, **55**, 74.
11. X. Huaqing, Y. Wei, L. Yang and C. Lifei, *Nanoscale Res. Lett.*, 2011, **6**, 124.
12. J. Buongiorno, D. C. Venerus and N. Prabhat, *et al.*, *J. Appl. Phys.*, 2009, **106**, 094312.
13. S. Özerinç, S. Kakaç and A. G. Yazıcıoğlu, *Microfluid. Nanofluidics*, 2010, **8**, 145.
14. A. Turgut, Ş. Sağlanmak and S. Doğanay, *J. Fac. Eng. Archit. Gazi Univ.*, 2016, **31**, 95.
15. I. Gonçalves, R. Souza, G. Coutinho, J. Miranda, A. Moita, J. E. Pereira, A. Moreira and R. Lima, *Appl. Sci.*, 2021, **11**, 2525.

16. A. Kotia, S. Borkakoti, P. Deval and S. K. Ghosh, *Heat Mass Transfer*, 2017, **53**, 2199.
17. R. Prasher, P. Bhattacharya and P. E. Phelan, *Phys. Rev. Lett.*, 2005, **94**, 3.
18. L. Yang, J. Xu, K. Du and X. Zhang, *Powder Technol.*, 2017, **317**, 348.
19. S. Simpson, A. Schelfhout, C. Golden and S. Vafaei, *Appl. Sci.*, 2018, **9**, 87.
20. M. H. Ahmadi, A. Mirlohi, M. A. Nazari and R. Ghasempour, *J. Mol. Liq.*, 2018, **265**, 181.
21. W. Yu, D. M. France, J. L. Routbort and S. U. S. Choi, *Heat Transfer Eng.*, 2008, **29**, 432.
22. K. Apmann, R. Fulmer, A. Soto and S. Vafaei, *Materials*, 2021, **14**, 1291.
23. M. Pavía, K. Alajami, P. Estellé, A. Desforges and B. Vigolo, *Adv. Colloid Interface Sci.*, 2021, **294**, 102452.
24. I. M. Mahbubul, R. Saidur and M. A. Amalina, *Int. J. Heat Mass Transfer*, 2012, **55**, 874.
25. H. Khodadadi, S. Aghakhani, H. Majd, R. Kalbasi and S. Wongwises, *Int. J. Heat Mass Transfer*, 2018, **127**, 997.
26. E. B. Elçioğlu, A. Güvenç Yazıcıoğlu and S. Kakaç, *J. Therm. Sci. Technol.*, 2014, **34**, 137.
27. L. S. Sundar, K. V. Sharma, M. T. Naik and M. K. Singh, *Renewable Sustainable Energy Rev.*, 2013, **25**, 670.
28. H. Riazi, T. Murphy, G. B. Webber, R. Atkin, S. S. M. Mostafavi and R. A. Taylor, *Int. J. Therm. Sci.*, 2016, **107**, 25.
29. I. M. Shahrul, I. M. Mahbubul, S. S. Khaleduzzaman, R. Saidur and M. F. M. Sabri, *Renewable Sustainable Energy Rev.*, 2014, **38**, 88.
30. P. Estellé, D. Cabaleiro, G. Żyła, L. Lugo and S. M. S. Murshed, *Renewable Sustainable Energy Rev.*, 2018, **94**, 931.
31. S. Tanvir and L. Qiao, *Nanoscale Res. Lett.*, 2012, **7**, 226.
32. M. Hernaiz, V. Alonso and P. Estellé, *et al.*, *J. Colloid Interface Sci.*, 2019, **547**, 393.
33. A. Ghadimi, R. Saidur and H. S. C. Metselaar, *Int. J. Heat Mass Transfer*, 2011, **54**, 4051.
34. E. B. Elçioğlu, E. Şimşek and T. Okutucu-Özyurt, in *Microscale and Nanoscale Heat Transfer: Analysis, Design, and Application*, ed. M. Rebay, S. Kakaç and R. M. Cotta, CRC Press, Taylor & Francis Group, FL, 2016, 6, pp. 155–181.
35. K. Khanafer and K. Vafai, *Int. J. Heat Mass Transfer*, 2011, **54**, 4410.
36. P. C. Mukesh Kumar and B. Rajappa, *Mater. Today Proc.*, 2020, **21**, 415.
37. R. S. Vajjha and D. K. Das, *Int. J. Heat Mass Transfer*, 2012, **55**, 4063.
38. J. M. Munyalo and X. Zhang, *J. Mol. Liq.*, 2018, **265**, 77.
39. M. Gupta, V. Singh, R. Kumar and Z. Said, *Renewable Sustainable Energy Rev.*, 2017, **74**, 638.
40. S.-H. Lee, H. J. Kim and S. P. Jang, *J. Mech. Sci. Tecnnology*, 2017, **31**, 975.
41. L. Yang and K. Du, *J. Therm. Anal. Calorim.*, 2020, **140**, 2033.

42. E. C. Okonkwo, I. Wole-Osho, I. W. Almanassra, Y. M. Abdullatif and T. Al-Ansari, *An updated review of nanofluids in various heat transfer devices*, Springer International Publishing, 2020.
43. M. S. Kamel and F. Lezsovits, *Therm. Sci.*, 2019, **23**, 109.
44. D. Wen, G. Lin, S. Vafaei and K. Zhang, *Particuology*, 2009, **7**, 141.
45. A. Sergis and Y. Hardalupas, *Nanoscale Res. Lett.*, 2011, **6**, 1.
46. M. Raja, R. Vijayan, P. Dineshkumar and M. Venkatesan, *Renewable Sustainable Energy Rev.*, 2016, **64**, 163.
47. A. Rahimi, A. D. Saee, A. Kasaeipoor and E. H. Malekshah, *Int. J. Numer. Methods Heat Fluid Flow*, 2019, **29**, 834.
48. W. Yu, D. M. France, E. V. Timofeeva, D. Singh and J. L. Routbort, *Int. J. Heat Mass Transf.*, 2012, **55**, 5380.
49. Z. Guo, *J. Enhanc. Heat Transfer*, 2020, **27**, 1.
50. B. Bakthavatchalam, K. Habib, R. Saidur, B. B. Saha and K. Irshad, *J. Mol. Liq.*, 2020, **305**, 112787.
51. G. S. Sokhal, *Mater. Today Proc.*, 2020, **37**, 3652.
52. G. S. Sokhal, D. Gangacharyulu and V. K. Bulasara, *Chem. Eng. Commun.*, 2018, **205**, 257.
53. S. Kakaç and A. Pramuanjaroenkij, *Int. J. Heat Mass Transfer*, 2009, **52**, 3187.
54. R. Agromayor, D. Cabaleiro, A. A. Pardinas, J. P. Vallejo, J. Fernandez-Seara and L. Lugo, *Materials*, 2016, **9**, 455.
55. M. Awais, N. Ullah, J. Ahmad, F. Sikandar, M. M. Ehsan, S. Salehin and A. A. Bhuiyan, *Int. J. Thermofluids*, 2021, **9**, 100065.
56. R. Saidur, K. Y. Leong and H. A. Mohammed, *Renewable Sustainable Energy Rev.*, 2011, **15**, 1646.
57. K. R. Aglawe, R. K. Yadav and S. B. Thool, *Mater. Today Proc.*, 2020, **43**, 366.
58. D. H. Sukarno, *J. Phys. Conf. Ser.*, 2017, **795**, 012020.
59. J. W. Gao, R. T. Zheg, H. Ohtani, D. S. Zho and G. Chen, *Nano Lett.*, 2009, **9**, 4128.
60. P. Kelinski, S. R. Phillpot, S. U. S. Choi and J. A. Eastman, *Int. J. Heat Mass Int. J. Thermofluids*, 2002, **45**, 855.
61. C. Kleinstreuer and Y. Feng, *Nanoscale Res. Lett.*, 2011, **6**, 229.
62. M. S. Bretado-de los Rios, C. I. Rivera-Solorio and K. D. P. Nigam, *Renewable Sustainable Energy Rev.*, 2021, **142**, 110855.
63. S. K. Gupta, S. Gupta, T. Gupta, A. Raghav and A. Singh, *Mater. Today Proc.*, 2021, **44**, 229.
64. M. Ghalandari, A. Maleki, A. Haghighi, M. S. Shadloo, M. A. Nazari and I. Tlili, *J. Mol. Liq.*, 2020, **313**, 113476.
65. E. Cuce, P. M. Cuce, T. Guclu and A. B. Besir, *J. Therm. Sci.*, 2020, **29**, 513.
66. M. Chen, X. Chen and D. Wu, in *Advances in Microfluidics and Nanofluids*, ed. S. Murshed, IntechOpen, London, UK, 2021, vol. 7, pp. 1–22.
67. O. A. Alawi, H. M. Kamar, A. R. Mallah, H. A. Mohammed, S. N. Kazi, N. A. C. Sidik and G. Najafi, *J. Clean. Prod.*, 2021, **291**, 125725.

68. M. U. Sajid and Y. Bicer, *Sol. Energy*, 2020, **207**, 974.
69. M. Sanches, G. Marseglia, A. P. C. Ribeiro, A. L. N. Moreira and A. S. Moita, *Symmetry*, 2021, **13**, 788.
70. S. Chinchanikar, S. S. Kore and P. Hujare, *J. Manuf. Process.*, 2021, **68**, 56.
71. S. Salari and S. M. Jafari, *Trends Food Sci. Technol.*, 2020, **97**, 100.
72. M. Sheikhpour, M. Arabi, A. Kasaeian, A. R. Rabei and Z. Taherian, *Nanotechnol., Sci. Appl.*, 2020, **13**, 47.
73. F. Yu, Y. Chen, X. Liang, J. Xu, C. Lee, Q. Liang, P. Tao and T. Deng, *Prog. Nat. Sci. Mater. Int.*, 2017, **27**, 531.
74. I. M. Mahbubul, E. B. Elcioglu, M. A. Amalina and R. Saidur, *Powder Technol.*, 2019, **345**, 668.
75. O. Z. Sharaf, R. A. Taylor and E. Abu-Nada, *Phys. Rep.*, 2020, **867**, 1.
76. M. E. Mondejar, M. Regidor, J. Krafczyk, C. Ihmels, B. Schmid, G. M. Kontogeorgis and F. Haglind, *J. Mol. Liq.*, 2021, **333**, 115140.
77. F. Kreuchauff and N. Teichert, *KIT Work. Pap. Ser. Econ.*, 2014, **53**, 1–23.
78. J. M. Brignon, *J. Phys. Conf. Ser.*, 2011, **304**, 012069.
79. C. Palmberg, H. Dernis and C. Miguet, *OECD Science, Technology and Industry Working Papers, 2009/07*, 2009, 112.
80. G. Foladori and N. Invernizzi, *Sci. Technol. Stud.*, 2005, **18**, 67.
81. M. Farhan, M. S. Hameed, H. M. Suleman and M. Anwar, *Arab. J. Sci. Eng.*, 2019, **44**, 5733.
82. M. J. Lourenço, J. Alexandre, C. Huisman, X. Paredes and C. Nieto de Castro, *Nanomaterials*, 2021, **11**, 1871.
83. L. H. Kumar, S. N. Kazi, H. H. Masjuki and M. N. M. Zubir, *Chem. Eng. J.*, 2022, **429**, 132321.
84. M. Faizal, R. Saidur, S. Mekhilef, A. Hepbasli and I. M. Mahbubul, *Clean Technol. Environ. Policy*, 2015, **17**, 1457.
85. S. Wciślik, *Chem. Pap.*, 2017, **71**, 2395.
86. A. Zeiny, H. Jin, G. Lin, P. Song and D. Wen, *Renew. Energy*, 2018, **122**, 443.
87. D. Das, P. Kalita and O. Roy, *Renewable Sustainable Energy Rev.*, 2018, **84**, 111.
88. S. Mukherjee, P. C. Mishra and P. Chaudhuri, *J. Mol. Liq.*, 2020, **299**, 112200.
89. S. M. Seyyedi, A. S. Dogonchi, M. Hashemi-Tilehnoee, M. Waqas and D. D. Ganji, *Appl. Therm. Eng.*, 2020, **168**, 114789.
90. S. R. Shamshirgaran, H. H. Al-Kayiem, K. V. Sharma and M. Ghasemi, *Sustainability*, 2020, **12**, 9119.
91. O. Mahian, A. Kianifar, S. A. Kalogirou, I. Pop and S. Wongwises, *Int. J. Heat Mass Transfer*, 2013, **57**, 582.
92. R. Sathyamurthy, A. E. Kabeel, E.-S. El-Agouz, D. Rufus, H. Panchal, T. Arunkumar, A. M. Manokar and D. G. P. Winston, *Int. J. Energy Res.*, 2019, **43**, 3295.
93. F. A. Essa, A. H. Elsheikh, A. A. Algazzar, R. Sathyamurthy, M. K. A. Ali, M. A. Elaziz and K. H. Salman, *Process Saf. Environ. Prot.*, 2020, **136**, 259.

94. Z. Hajabdollahi, H. Hajabdollahi and K. C. Kim, *J. Therm. Anal. Calorim.*, 2020, **140**, 991.
95. P. M. J. Stalin, T. V. Arjunan, M. M. Matheswaran, H. Dolli and N. Sadanandam, *J. Therm. Anal. Calorim.*, 2020, **139**, 3219.
96. L. S. Sundar, A. H. Misganaw, M. K. Singh, A. M. B. Pereira and A. C. M. Sousa, *Renew. Energy Focus*, 2020, **35**, 10.
97. T. P. Otanicar and J. S. Golden, *Environ. Sci. Technol.*, 2009, **43**, 6082.
98. M. Faizal, R. Saidur, S. Mekhilef and M. A. Alim, *Energy Convers. Manag.*, 2013, **76**, 162.
99. A. F. Owolabi, H. H. Al-Kayiem and A. T. Baheta, *Int. J. Energy Res.*, 2016, **41**, 650.
100. L. Sahota and G. N. Tiwari, *Energy Convers. Manag.*, 2017, **148**, 413.
101. A. Abadeh, O. Rejeb, M. Sardarabadi, C. Menezo, M. Passandideh-Fard and A. Jemni, *Energy*, 2018, **159**, 1234.
102. A. Alashkar and M. Gadalla, *Int. J. Energy Res.*, 2018, **42**, 2875.
103. M. I. Hussain and J.-T. Kim, *Int. J. Green Energy*, 2018, **15**, 596.
104. S. Eslami, A. Gholami, A. Bakhtiari, M. Zandi and Y. Noorollahi, *Energy Convers. Manag.*, 2019, **200**, 112107.
105. H. Hajabdollahi and Z. Hajabdollahi, presented in Proceedings of the Institution of Mechanical Engineers, *Part E: Journal of Process Mechanical Engineering*, 2016.
106. M. Yousefi, D. Hooshyar, J. H. Kim and H. Lim, *Int. Commun. Heat Mass Transfer*, 2017, **82**, 63.
107. V. Kumar and J. Sarkar, *J. Therm. Anal. Calorim.*, 2020, **141**, 1477.
108. S. Mukherjee, P. C. Mishra and P. Chaudhuri, *JOM*, 2020, **72**, 3958.
109. P. P. Prajapati and V. K. Patel, *Therm. Sci. Eng. Prog.*, 2020, **17**, 100381.
110. N. H. Abu-Hamdeh and K. H. Almitani, *Sol. Energy*, 2016, **134**, 202.
111. S. W. Sharshir, A. W. Kandeal, M. Ismail, G. B. Abdelaziz, A. E. Kabeel and N. Yang, *Appl. Therm. Eng.*, 2019, **160**, 113997.
112. M. Ma, Y. Zhai, P. Yao, Y. Li and H. Wang, *Powder Technol.*, 2020, **373**, 702.
113. A. Alashkar and M. Gadalla, presented in Proceedings of the ASME 2017 *International Mechanical Engineering Congress and Exposition IMECE2017*, Tampa, Florida, USA, 2017.
114. W. Kang, Y. Shin and H. Cho, *Energies*, 2017, **10**, 1911.
115. M. O. Lari and A. Z. Sahin, *Energy Convers. Manag.*, 2017, **149**, 467.
116. A. H. A. Al-Waeli, H. A. Kazem, K. Sopian and M. T. Chaichan, *Int. J. Sustain. Energy*, 2018, **37**, 558.
117. G. Diglio, C. Roselli, M. Sasso and U. J. Channabasappa, *Geothermics*, 2018, **72**, 112.
118. S. K. Singh and J. Sarkar, *Int. Commun. Heat Mass Transfer*, 2018, **98**, 41.
119. A. H. A. Al-Waeli, H. A. Kazem, M. T. Chaichan and K. Sopian, *Therm. Sci. Eng. Prog.*, 2019, **11**, 213.
120. M. A. Ehyaei, A. Ahmadi, M. E. H. Assad, A. A. Hachicha and Z. Said, *Sol. Energy*, 2019, **187**, 175.
121. A. Alirezaie, M. H. Hajmohammad, A. Alipour and M. Salari, *Energy*, 2018, **157**, 979.

122. M. H. Esfe, S. Esfandeh, M. Afrand, M. Rejvani and S. H. Rostamian, *Appl. Therm. Eng.*, 2018, **133**, 452.
123. M. H. Esfe, S. Esfandeh and M. Rejvani, *J. Therm. Anal. Calorim.*, 2018, **131**, 1437.
124. A. Alirezaie, M. H. Hajmohammad, M. R. H. Ahangar and M. H. Esfe, *Appl. Therm. Eng.*, 2018, **128**, 373.
125. Z. Xuan, Y. Zhai, M. Ma, Y. Li and H. Wang, *J. Mol. Liq.*, 2021, **323**, 114889.
126. H. Jiang, W. Wang, D. Chu, W. Lu, D. Cheng, Y. Huang and Y. Wu, *J. Nucl. Mater.*, 2017, **494**, 361.
127. S. Fotowat, S. Askar, M. Ismail and A. Fartaj, *Sustain. Energy Technol. Assess.*, 2017, **24**, 39.
128. A. Salari, A. Taheri, A. Farzanehnia, M. Passandideh-fard and M. Sardarabadi, *J. Clean. Prod.*, 2021, **282**, 124318.
129. K. Elsaid, A. G. Olabi, T. Wilberforce, M. A. Abdelkareem and E. T. Sayed, *Sci. Total Environ*, 2021, **763**, 144202.
130. S. N. M. Zainon and W. H. Azmi, *Micromachines*, 2021, **12**, 176.
131. L. Hernandez Lopez, L. Menendez Monzonis, L. B. Vicente, J. Kaur and M. Buschmann, *Report about nanofluid's health, safety and environmental impact*, Nanouptake COST Action CA15119, 2019, Castello de la Plana, Spain.
132. W. G. Kreyling, M. Semmler-Behnke and W. Möller, *J. Nanoparticle Res.*, 2006, **8**, 543.
133. B. Nowack, *Nano Today*, 2009, **4**, 11.
134. L. Yang, W. Ji, M. Mao and J.-N. Huang, *J. Clean. Prod.*, 2020, **257**, 120408.
135. Z. Said, S. Arora and E. Bellos, *Renewable Sustainable Energy Rev.*, 2018, **94**, 302.
136. H. Kaya and K. Arslan, *Heat Mass Transfer*, 2019, **55**, 581.
137. E. A. Taborda, V. Alvarado and F. B. Cortés, *Energy Convers. Manag.*, 2017, **148**, 30.

Subject Index